Virus and Virus-like Diseases of Bulb and Flower Crops

Edited by G. Loebenstein, R.H. Lawson and A.A. Brunt

Virus and Virus-like Diseases of Bulb and Flower Crops

Edited by

GAD LOEBENSTEIN
Department of Virology, Agricultural Research Organization, The Volcani Center, Bet Dagan, Israel

ROGER H. LAWSON
Floral and Nursery Plants Research Unit, USDA/ARS, Beltsville, MD, USA

ALAN A. BRUNT
Microbiology and Crop Protection Department, Horticulture Research International, Littlehampton, Sussex, UK

JOHN WILEY & SONS
Chichester • New York • Brisbane • Toronto • Singapore

A Co-publication with

BALABAN
Rehovot • Philadelphia • Santa Maria Imbaro

Published by John Wiley & Sons
Baffins Lane, Chichester
West Sussex PO19 1UD, United Kingdom

Other Wiley Editorial Offices

John Wiley & Sons, Inc., 605 Third Avenue,
New York NY 10158-0012, USA

Jacaranda Wiley Ltd., G.P.O. Box 859, Brisbane
Queensland 4001, Australia

John Wiley & Sons (Canada) Ltd., 22 Worcester Road,
Roxdale, Ontario M9W IL1, Canada

John Wiley & Sons (SEA) Pte Ltd., 37 Jalam Pemumpin 05-04,
Block B, Union Industrial Building, Singapore 2057

Co-publishers

Balaban Science Publishers
PO Box 2039, Rehovot, Israel

Library of Congress Cataloging-in-Publication Data

Virus and virus-like diseases of bulb and flower crops/edited by Gad Loebenstein, R.H. Lawson, and A.A. Brunt
p. cm.
Includes bibliographical references (p.) and index.
ISBN 0-471-95293-1 (alk. paper)
1. Bulbs–Diseases and pests. 2. Plants. Ornamental–Diseases and pests. 3. Flowers–Diseases and pests. 4. Virus diseases of plants. 5. Plant diseases. I. Loebenstein, G. (Gad) II. Lawson, R.H. (Roger) III. Brunt, A.A. (Alan Alac)
SB608.885V57 1995
635.9'28–dc20 95-18716
CIP

A catalogue record for this book is available from the British Library

Printed by Abudalo Press, Jerusalem

Contributors

J. Albouy
Station de Pathologie Végétale, INRA, Versailles, France

K. Bech
Danish Institute of Plant and Soil Science, Department of Plant Pathology and Pest Management, Lyngby, Denmark

M.G. Bellardi
Istituto di Patologia Vegetale, Bologna, Italy

A. Bertaccini
Istituto di Patologia Vegetale, Bologna, Italy

L. Bos
DLO Research Institute for Plant Protection (IPO-DLO), PO Box 9060, 6700 GW Wageningen, The Netherlands

I. Bouwen
DLO Research Institute for Plant Protection (IPO-DLO), PO Box 9060, 6700 GW Wageningen, The Netherlands

A.A. Brunt
Microbiology and Crop Protection Department, Horticulture Research International, Littlehampton, Sussex, UK

J.T. Burger
Vegetable and Ornamental Plant Institute, P/Bag X293, Pretoria 0001, South Africa

J. Cohen
Department of Virology, Agricultural Research Organization, The Volcani Center, Bet Dagan 50250, Israel

A.F.L.M. Derks
Bulb Research Centre, PO Box 85, 2160 AB Lisse, The Netherlands

A. Franck
Department of Virology, Agricultural Research Organization, The Volcani Center, Bet Dagan 50250, Israel

R. Gafny
Department of Virology, Agricultural Research Organization, The Volcani Center, Bet Dagan 50250, Israel

A. Gera
Department of Virology, Agricultural Research Organization, The Volcani Center, Bet Dagan 50250, Israel

J. Hammond
Florist and Nursery Plants Research Unit, USDA–ARS, Beltsville, MD 20705-2350, USA

R. Hammond
Molecular and Plant Pathology Laboratory, Bldg. 011A, ARS-BARC-W, Beltsville, MD 20705–2350, USA

R.D. Hartman
Twyford Plant Laboratories, 11850 Twitty Road, Sebring, FL 33870, USA

H.-T. Hsu
US Department of Agriculture, Agricultural Research Service, US National Arboretum, Floral and Nursery Plants Research Unit, Beltsville, MD 20705, USA

K. Husted
Danish Institute of Plant and Soil Science, Department of Plant Pathology and Pest Management, Lyngby, Denmark

M. Klein
Department of Entomology, Agricultural Research Organization, The Volcani Center, Bet Dagan 50250, Israel

R.H. Lawson
US Department of Agriculture, Agricultural Research Service, US National Arboretum, Floral and Nursery Plants Research Unit, Beltsville, MD 20705, USA

D.-E. Lesemann
Federal Biological Research Centre for Agriculture and Forestry, Institute of Biochemistry and Plant Virology, Braunschweig, Germany

V. Lisa
Istituto di Fitovirologia applicata CNR, Strada delle Cacce 73, Torino, Italy

G. Loebenstein
Department of Virology, Agricultural Research Organization, The Volcani Center, Bet Dagan 50250, Israel

D.Z. Maat
DLO Research Institute for Plant Protection, PO Box 9060, 6700 GW Wageningen, The Netherlands

A.J. Malter
Market Research Department, Ministry of Agriculture, Tel Aviv, Israel

W.P. Mowat
Formerly of Scottish Crop Research Institute, Invergowrie, Dundee DD2 5DA, UK

B. Raccah
Department of Virology, Agricultural Research Organization, The Volcani Center, Bet Dagan 50250, Israel

S. Spiegel
Department of Virology, Agricultural Research Organization, The Volcani Center, Bet Dagan 50250, Israel

A. Stein
Department of Virology, Agricultural Research Organization, The Volcani Center, Bet Dagan 50250, Israel

E. Teverovsky
Department of Virology, Agricultural Research Organization, The Volcani Center, Bet Dagan 50250, Israel

C.I.M. Van der Vlugt
Bulb Research Centre, PO Box 85, 2160 AB Lisse, The Netherlands

A. van Zaayen
Inspection Service for Floriculture and Arboriculture (NAKB), Johan de Wittlaan 12, 2517 JR Den Haag, The Netherlands

M.B. Von Wechmar
Department of Microbiology, University of Cape Town, P/Bag, Rondebosch 7700, South Africa

F.W. Zettler
Plant Pathology Department, 1453 Fifield Hall, University of Florida, Gainesville, FL 32611-0680, USA

Preface

Ornamental horticulture has experienced a marked growth in the past 40 years and has developed into one of the major economic branches of modern agriculture. Development of this industry is based on the supply of high quality flowers and plants, and control of viral and other diseases is a prerequisite in success.

This book was conceived in 1992, when the three editors met in Prague during the 8th International Symposium on Virus Diseases of Ornamental Plants. Our first objective was to define the intended audience. The book has been written for researchers, teachers, students, general practitioners, and clinicians as well as growers.

The book has several chapters that present general topics of virus classification, transmission, purification, detection, diagnosis, control, etc. The objective of these chapters is to present the general audience with an overview of each major topic. We have included these chapters as an adjunct to more specific crop-oriented information since the basic concepts in virology have relevance to the specific diseases described in the body of the work.

We have attempted to include chapters describing the major flower bulb, cut flower, flowering potted plant, and some of the most important bedding plants. The list of authors includes internationally known virologists. These individuals were invited to contribute based on their recognized knowledge and experience with specific crops. Each chapter, therefore, has its own personal touch, though the editors tried to give some uniformity to the crop-oriented sections.

We have attempted to include the most up-to-date information. We recognize, however, that rapid developments in methodology, particularly in virus detection and disease diagnosis, will likely change quickly. Nevertheless, we believe that the principles that are presented here will be applicable when new methodology is applied to control virus diseases in bulb and flower crops.

We hope that this book will contribute to further development and improvement of the ornamental industry.

February 1995 *G. Loebenstein, R.H. Lawson, A.A. Brunt*

Contents

Pot and Bedding Plants

1

The Economic Importance of Ornamentals

A.J. Malter*
Market Research Department, Ministry of Agriculture, Tel Aviv, Israel

Ornamental horticulture has experienced tremendous growth in the past 40 years and has developed into one of the major economic branches of modern agriculture. It has become a global industry with commercial production in almost every country, and international trade has expanded rapidly. Innovations in product development and in production, handling, and transport technology have enabled the year-round supply of high-quality flowers and plants on a mass scale. The huge success of the industry to date has further stimulated demand but has also raised expectations in the trade and among consumers regarding product quality and availability. Recent commercial success has also encouraged the large-scale entry of new producers, for whom floriculture exports may have great economic potential that is unmatched by any other agricultural crop.

---

*On leave at the Department of Marketing, Graduate School of Business, University of Wisconsin-Madison.

Size of the Industry

The economic importance of ornamental horticulture can be shown in a number of ways. In terms of the absolute size of the industry, the Rabobank in the Netherlands estimated that world-wide sales of ornamental horticultural products in 1990 were more than $50 billion. Within this total, the most important category was cut flowers with retail sales of $24.7 billion, followed by $14.3 billion in sales of pot plants (Haak et al., 1992), $7.6 billion of tree nursery products (wholesale value) (Hack et al., 1992), $900 million of flower bulbs (wholesale value) (de Kleijn and Heybroek, 1992), and $1.6 billion of other propagation material (young plants $1.1 billion, flower seeds $330 million, and tissue culture plantlets $160 million) (de Kleijn et al., 1992).

Due to the perception of ornamental horticultural products as non-essential goods for luxury consumption and their relatively complicated and capital-intensive production and handling requirements, most of the world's commercial production, consumption, and trade occurs in and among wealthy northern hemisphere countries. In terms of geographic distribution, the most important region for the consumption of ornamental horticultural products is Western Europe, with more than 350 million consumers and retail cut flower sales estimated at $12.5 billion in 1990. The two other major regions of consumption in terms of market value are the US, with 260 million consumers and cut flower sales of over $6 billion, and Japan, with 124 million consumers and cut flower sales of about $6 billion (van Doesburg, 1992).

Data from the US show that cut flower sales, though frequently used for cross-country comparisons, comprise only a small part of the total sales of ornamental horticultural products. Retail expenditures on all ornamental horticultural products in the US in 1993 were estimated at $37.6 billion, of which "floriculture" products were $13.5 billion and "environmental horticulture" products were $24.1 billion. Cut flowers were the largest single category of floriculture sales ($6.4 billion), followed by bedding plants ($3.4 billion), flowering pot plants ($2.6 billion), and foliage pot plants ($1.1 billion). The leading category of environmental horticultural products was evergreens ($8.4 billion), followed by shade trees and flowering trees ($3.9 billion), turfgrass ($3.6 billion), fruit and nut plants ($1.7 billion), and flower bulbs ($500 million) (Johnson, 1993).

Per Capita Consumption

Large differences exist in the per capita consumption of ornamental horticultural products between regions and among countries in the same region, reflecting national differences in both levels of economic development and consumption habits. In 1991, per capita consumption of cut flowers ranged from $155 in Switzerland to $120 in Norway, $93 in Japan, $81 in Germany, $73 in Holland, $47 in the US, $38 in the United Kingdom, and only $27 in Spain (Flower Council of Holland, 1993). Since per capita expenditure levels are a function of both the price level of cut flowers and the number of stems purchased, a country like Holland, where every person purchases an average of 150 stems of flowers per year (nearly three times more than in any other country) but at relatively low retail prices (Haak et al., 1992), ranks only in the middle of the per capita expenditure table.

Leading Exporters

The leading exporters are either industrialized countries or developing countries in which the ornamental horticultural sector represents a pocket of high-level economic development capable of competing in the international market. In 1990, the largest exporter of cut flowers was Holland, with 59% of world exports, followed by Colombia (10%), Italy (6%), Israel (4%), Spain (2%) and Kenya (1%). Holland was also the leading exporter of pot plants in 1990, with 48% of world exports, followed by Denmark (16%), Belgium (10%), Germany (5%), and Italy (4%) (Flower Council of Holland, 1993). The largest exporters, such as Holland and Colombia in the case of cut flowers, depend heavily on a huge nearby import market to absorb the bulk of their exports: Germany receives nearly 50% of Holland's exports and the US receives almost 80% of Colombia's exports (Asocolflores, 1993; Vakblad voor de Bloemisterij, 1994).

The massive development of ornamental horticulture in Holland and its emergence as the dominant leader in many categories of international trade is a relatively recent phenomenon. In 1960, the leading exporter of cut flowers was Italy, which at that time exported nearly twice the volume and 30% more in value than did Holland. However, during the 1960s, cut flower exports from Holland grew six-fold, while exports from Italy increased by only 25%. As a result, by 1970 Dutch exports were nearly three times larger in volume and value than Italian exports. By 1992, the volume of Dutch cut flower exports had grown by another eleven times and reached 381,000 tons, or nearly 30 times the volume exported by Italy, which had remained at about the same level as in 1970. Elsewhere in Europe, cut flower exports from Spain grew during the 1970s and 1980s to equal those from Italy in terms of volume but reached only one-third the value of Italian exports (more than 90% of Spanish exports were lower-value spray carnations). It is important to note, however, that export figures do not reveal the true size and economic importance of ornamental horticulture in either Italy or Spain (or in France, Germany, or many other countries) because unlike Holland, Colombia, Israel, or Africa, production in these countries is directed mainly towards their large and rapidly growing domestic markets. By the early 1990s, Spain even developed into a net importer of ornamental horticultural products (AIPH, 1993).

Leading Producers

Colombia and Ecuador

During the 1970s and 1980s, significant cut flower export industries developed in countries such as Colombia and Israel, and ornamental horticulture production continued to develop rapidly and change structurally in large northern hemisphere countries such as the US and Japan. In the developing country of Colombia, agriculture is of major economic importance and accounted for 17% of the gross domestic product in 1990 (World Bank, 1992). Colombian floriculture, which uses nearly three times as many workers per hectare as the more capital-intensive Dutch floriculture (Hamrick, 1994), generates over 75,000 direct production jobs and about 50,000 jobs in various supporting industries. Though there were only about 450 cut flower growers cultivating 4,300 hectares in 1992 (three times more of each than in 1980), their exports of $300 million (Asocolflores, 1993) ranked third behind Colombia's two largest traditional agricultural exports: coffee ($1.285 billion) and bananas ($400 million) (FAO, 1993). Colombian floriculture is viewed as a great success which many developing countries, especially in Latin America and East Africa, are trying to duplicate.

The development of Colombian floriculture has been a major contributor to the explosive growth in US cut flower imports since the early 1970s and the parallel decline in US production of many types of cut flowers. By the early 1990s, Colombia supplied over 70% of US cut flower imports and was especially dominant in the major commodity crops. For example, US imports of pompon (spray) chrysanthemums grew from 2 million bunches in 1971 to 94 million bunches in 1992, of which 88% came from Colombia. Colombia is also the major supplier of US imports of standard carnations (97%), spray carnations (88%), and roses (69%). Attempts have been made to diversify into other crops such as bulbflowers, but so far Colombia has not developed into a significant supplier. Since the mid-1980s many Colombians and others have been investing in floriculture production in Ecuador, which by 1992 became the leading source of US gypsophilia imports (44%, compared to 39% from Colombia) and the second largest supplier of US rose imports (15%) (F-S Market News, 1993; Smith, 1993).

United States

Despite losing a significant market share to imported cut flowers in the past two decades, US floriculture remains a large, diverse, and dynamic industry. In 1992, the top 36 producing states (of 50) had over 10,000 commercial floriculture growers with a production area of 20,000 hectares and wholesale revenue of nearly $3 billion. Production was more heavily concentrated in a few major states such as California ($666 million, including 58% of US cut flower production), Florida ($526 million, including 56% of US foliage plant production and 76% of US cut foliage production), New York ($230 million), Texas ($150 million), Ohio ($128 million), and Michigan ($128 million). The production of bedding plants and potted flowering plants is more evenly distributed around the country with significant concentrations near major population centers in northern states such as New York, Michigan, Ohio, and Pennsylvania. In response to the huge growth in cut flower imports in recent years, US floriculture production has largely shifted away from cut flowers and potted foliage plants (which together accounted for 57% of total US production value in 1981) and toward bedding plants (now the leading production sector) and potted flowering plants (which together accounted for 65% of total US production value in 1992). The major exception to this development has been fresh cut roses: despite a massive increase in imported roses since the 1970s, US production actually grew by 25% from 428 million stems in 1971 to 534 million stems in 1992 (USDA, 1993, Smith, 1993).

Japan

Japan is also a large and growing floricultural producer, though a relatively small participant in international trade. In 1990, more than 81,000 growers with an area of over 16,600 hectares produced over 5.3 billion stems of cut flowers. These totals represent increases since the mid-1980s of 10% in the number of growers, 30% in the acreage, and 18% in the quantity of production (JETRO, 1992). Floriculture attracted new growers with higher profitability than traditional crops such as rice and vegetables and government incentives to diversify away from rice production (Egberts, 1992). However, floriculture is still the smallest sector in Japanese horticulture in terms of production value, reaching only 60% of thc value of fruit production, 20% of vegetable production, and only 4.5% of the total agricultural income (Ohkawa, 1993).

Though Japanese cut flower imports have increased rapidly, nearly tripling between 1986–1991, they accounted for only 7.5% of the Japanese market in 1991. Japanese imports consist primarily of one or two specific products from a few main suppliers who each fill a special product or seasonal niche in the

Japanese market. Orchids from Thailand and Singapore, chrysanthemums (the most important product in Japan) from Taiwan, and cut greens (especially ferns and bear grass) from the US accounted for two-thirds of the volume of Japanese cut flower imports in 1990. Other suppliers finding a niche in the Japanese market in recent years include Holland (tulips, lilies, and other bulbflowers), Australia (anigozanthos and waxflowers), New Zealand (zantedeschias and cymbidium orchids), Costa Rica (lower-cost ferns), and Colombia (lower-cost carnations). The Japanese market pays relatively high prices for high-quality cut flowers because most flowers are used for gifts (either business or personal), and the Japanese flower-arranging style emphasizes the beauty of a single bloom or a small number of selected flowers. However, imports must satisfy stringent phytosanitary standards and pass through a complex distribution system in order to succeed in Japan (JETRO, 1992).

Israel

Israel was a pioneer in developing "off-season" produce exports to Europe and North America in the winter months. Cut flower exports from Israel grew rapidly from less than 200 million stems and $20 million per year in the early 1970s to more than 600 million stems and $70 million per year by the late 1970s. In the early 1980s the floriculture sector experienced a downturn and underwent a period of consolidation, after which exports began to grow again, though at a slower pace than previously. By the early 1990s, Israeli exports reached 1 billion stems and $150 million per year (Malter, 1992).

Though the total volume of Israel's cut flower exports has hardly increased since the mid-1980s, there have been major changes in the product mix which have greatly strengthened Israel's competitive position. After beginning in the early 1970s with a base of mostly roses and gladiolas, Israeli floriculture developed major new crops such as spray carnations and gypsophilia. Roses remain the leading crop today in terms of both volume and revenue. In contrast, the importance of gladiolas has declined sharply, and spray carnations and gypsophilia peaked in the mid-1980s and have been decreasing ever since. However, these products have been replaced by a long list of "new crops", led by waxflower, standard carnations, limonium, aster, solidaster, solidago, phlox, gerberas, and many others. Some of these crops were either developed by Israel into commercial products or adopted by Israel for production in the European "off-season" (i.e., winter or early spring). These new crops (i.e., everything other than roses, spray carnations, gypsophilia, and gladiolas) have increased in importance from less than

10% of Israel's export quantity in the mid-1970s to nearly 50% in the early 1990s (Malter, 1992).

Though Israel's exports of major crops such as roses and carnations are now surpassed by up-and-coming producers (some with Israeli assistance) in Latin America, East Africa, and other Mediterranean countries, Israel is still a leader in floriculture innovation. Unlike these other producers who mostly copy the relatively simple production of the major crops, Israel invests heavily in breeding new crops, developing new production technologies, and providing extension advisory services to growers. Israel is also one of the few countries outside of Europe, North America, and Japan that is developing large-scale exports of a wide range of ornamental products besides cut flowers, including potted flowering plants, flower bulbs and corms, vegetative cuttings, young plants, and other propagation material.

The Netherlands

The absolute and relative economic importance of floriculture is perhaps greater in Holland than in any other country. Over 70,000 workers are employed in all facets of the floricultural industry, from input suppliers and growers through auction houses, wholesalers, exporters, and retailers. In a country of only 15 million inhabitants, flowers and plants are sold in over 10,000 retail sales outlets. In 1992, over 5,000 Dutch companies exported nearly $4 billion in cut flowers, pot plants, flower bulbs, and nursery stock (compared to Dutch imports of less than $275 million) (Flower Council of Holland, 1993; PVS, 1993). The production value of Dutch ornamental horticulture ($4 billion) exceeded that of fresh fruits and vegetables ($2.8 billion) and of field crops ($1.7 billion), but was slightly less than the value of dairy products ($4.2 billion) and far less than the value of other livestock products ($6.7 billion) (Ministry of Agriculture, 1992). In terms of overall economic importance, agriculture accounts for only 4% of Holland's gross domestic product, compared to industry (31%), and services (65%) (World Bank, 1992).

Significant changes have occurred in the production and sales patterns of ornamental horticulture since the 1950s, many of which can be seen in the casc of Holland. Though the production value of all branches of Dutch horticulture increased substantially over the past few decades, the share of flower bulbs in the overall production value of ornamental horticulture decreased from 54% in 1955 to only 13% in 1992, while the combined share of cut flowers and pot plants grew from 37% to 76%, thus becoming the dominant sector. The share of nursery stock remained remarkably stable during this period and was 10–11% in most years. Within the cut flowers and pot plants sector, the share of cut flowers in the total sales turnover of the Dutch auctions declined from 90% in 1970 to 70% in 1993 while the share of pot plants increased from 10% to 30%. The sales of pot plants have been growing at the fastest average rate (albeit from a smaller initial base), followed by cut flowers and flower bulbs (which began as the largest sector) (PVS, 1992, 1993; VBN, 1993).

All sectors of Dutch ornamental horticulture enjoyed a remarkable pace of continuous rapid growth from the 1950s through the 1990s. For the 25-year period from 1958 through 1983, the nominal sales turnover of flowers and plants in the Dutch auctions grew at a phenomenal average rate of 15.3% per year! However, since the 1968 to 1973 period, when the average annual turnover growth reached a peak of 19.2 %, sales turnover has been increasing at a steadily decreasing rate and averaged only 5.4% during the 1988 to 1993 period. There was even one year of negative growth when, for the first time after four decades of uninterrupted expansion, sales turnover decreased by 1.5% in 1992. In 1993, sales recovered and increased by 4.2% (PVS, 1992, 1993; VBN, 1993), but it would appear that double-digit increases in annual sales are mainly a thing of the past.

There have also been major changes in the commercial assortment of ornamental horticultural products from the 1950s to the 1990s as a result of breeding new varieties, technological advances, and changes in consumer preferences. Regarding cut flowers in Europe, a comparison of sales data from the Dutch auctions in 1958 and 1993 in Table 1 shows that carnations dropped from first place to third (despite the development since the 1960s of new spray varieties), while roses climbed from second place to first. However, since Dutch carnation production decreased substantially during the 1980s and since significant quantities of imported carnations bypass the Dutch auctions, the gap between roses and carnations in the overall European market may not be as large as it appears from auction sales figures (PVS, 1992; VBN, 1993).

Spray chrysanthemums ("pompon" varieties) rose from fifth place in 1958 to second place in 1993, while bulb flowers such as tulips and freesias slipped from third and fourth places, respectively, to fourth and six places during the same period. Indeed, bulb flowers, which comprised half of the "Top 10" products in 1958, suffered the greatest decline as crops such as narcissus, iris, and gladiola all dropped out of the Top 10 by the late 1980s. Other crops that also dropped off the list of leaders were sering (lilacs) and large-flowered chrysanthemums. These were replaced in the 1993 rankings by products that developed rapidly in the past 25 years, such as lilies (fifth place), gerbera (seventh place), cymbidium orchids (eighth place), gypsophilia (ninth place) and alstroemeria (tenth place) (PVS, 1992; VBN, 1993).

Table 1.1. Top 10 cut flowers in Dutch auctions, 1958–1993, by turnover (sales turnover in million Dfl., quantity sold in million stems, price in Dfl. cents per stem)

Rank	Flower type	Turnover	Quantity	Average price
1958				
1	Carnation	15.3	134.6	11.4
2	Rose	11.1	89.6	12.4
3	Tulip	8.9	117.6	7.5
4	Freesia	6.2	79.4	7.9
5	Spray chrys.	5.1	46.6	11.0
6	Sering	2.8	9.1	31.1
7	Large chrys.	2.7	15.8	17.1
8	Narcissus	2.0	55.7	3.6
9	Iris	1.2	19.3	6.3
10	Gladiola	0.9	27.2	3.3
1993				
1	Rose	902.4	2343.7	39.0
2	Spray chrys.	580.8	1262.3	46.0
3	Carnation	275.8	971.7	28.0
4	Tulip	265.0	872.4	30.0
5	Lily	236.8	329.0	72.0
6	Freesia	145.2	567.5	26.0
7	Gerbera	133.4	364.0	37.0
8	Cymbidium	102.5	59.9	171.0
9	Gypsophilia	82.0	173.3	47.0
10	Alstroemeria	78.7	222.0	35.0

Sources: PVS, 1992; VBN, 1993.

Pricc and Quality

The data for each crop in Table 1.1 represent an average price over a whole year for all quality classes. There are, of course, considerable price variations depending on the season and the quality of the specific product. Now that many products can be supplied on a year-round basis, quality has become the crucial factor that differentiates one product and one producer from another. However, the "quality" of fresh produce is a very complicated concept and one which often cannot be precisely defined. Quality is a very subjective concept, with different product characteristics assuming varying importance at each stage in the supply chain from growers to consumers. Moreover, every individual consumer has his/her own personal preferences, sense of esthetics, and ideas about what constitutes "good quality".

Nevertheless, in order to facilitate communication and price-formation between sellers and buyers, the flower trade has developed some objective measures for quantifiable product characteristics which may best reflect the overall "quality" of each product. Naturally, one would expect a positive correlation between quality and price; i.e., that higher quality would receive a higher price. Dutch auction sales data for cut flowers (Bloemenveiling Holland, 1994) show that this is generally the case, with certain qualifications. The auctions have chosen one key measurable physical characteristic for each major crop to indicate to buyers the general quality level of the produce offered for sale, which is then modified for each sales lot by additional remarks of the auction's quality inspectors. For example, roses and carnations are classified by stem length, chrysanthemums by weight, lilies by the number of flower buds on each stem, and gerberas by flower diameter. On a yearly basis, roses generally draw higher prices for increasingly longer lengths, though this relationship is accentuated by the fact that, due to seasonal variations in light and temperature conditions, Dutch growers produce relatively greater quantities of long-stemmed roses in the winter (when market prices are higher) and a larger proportion of short-stemmed roses (of the same varieties) in the summer (when prices are lower).

In contrast to roses, the average annual price of spray chrysanthemums rises steadily as stem weight increases from 30 grams to 60 grams, but then decreases with each incremental increase in weight beyond 60 grams. This apparent paradox results from the fact that 60 grams is the most common stem weight produced by Dutch growers in the mid-winter months when prices are at their peak, while a larger quantity of stems heavier than 60 grams is produced in the summer months when prices are much lower. Examining sales for each month separately shows that prices do actually continue to increase as stem weight increases beyond 60 grams. However, prices rise at a declining rate until they stabilize and then begin to decrease as stem weight exceeds 80 grams. The European trade has come to view 60–70 grams as the optimal weight for one spray chrysanthemum stem. Heavier stems are considered to be "too woody" and do not take up water well, causing shorter vase-life and thus resulting in lower prices.

A third type of quality-price relationship can be seen in products such as standard carnations and Asiatic hybrid lilies. In the case of standard carnations, prices are virtually identical (about 20 Dfl. cents in 1993) for stems of 40, 45, and

50 cm, which apparently are all too short to constitute desirable quality for this product. When stem length increases further to 55 cm, however, the average price increases by about 50% to 30 Dfl. cents. Prices continue to increase at a decreasing rate until stem length reaches 65 cm (which apparently is the upper bound of desirable quality for this product), at which point the price (40 Dfl. cents) is double that of 50 cm stems (20 cents). There is virtually no further increase in price for an additional increase in stem length from 65–70 cm. In the case of Asiatic hybrid lilies, prices increase proportionately from 30 Dfl. cents to 60 Dfl. cents as the number of flower buds per stem increases from 2 to 4. Prices increase at a much slower rate with each additional bud beyond four per stem, reaching only about 70 Dfl. cents for a stem with seven buds and even decreasing as the bud count increases from 7 to 8.

With many crops it is well-known that virus-tested plants produce flowers of higher quality than plants infected by viruses. Thus, virus-tested roses (see Chapter 38) and carnations have longer stems, and virus-tested lilies give more flower buds per stem than infected plants. In order to survive in increasingly competitive markets and to maximize profits, producers will have to start with healthy plant material and consider the quality preferences of the trade and of consumers.

Production Trends in Other Northern Countries

Since 1960, the major producers of ornamental horticultural products in the northern hemisphere have generally been decreasing their outdoor production area while increasing their covered production area, as occurred for example in Italy. At the same time, ornamental horticultural producers in countries such as Germany and the US have decreased their production of cut flowers (which are now mainly imported) and developed strong industries based on producing nursery stock, flowering pot plants, and bedding plants for their domestic markets and neighboring countries (Johnson, 1990; AIPH, 1993; USDA, 1993). European and North American producers of these products enjoy comparative advantages that protect them from import competition due to the large physical size of some of these products, the extreme sensitivity of some products to long-distance handling, and rigid phytosanitary regulations. However, in some instances, some of these countries' producers of relatively simple and freely traded products — such as Florida growers of cut greens (especially of leatherleaf ferns) — have remained internationally competitive and even rank among the leading exporters in their product category.

Conclusions

Ornamental horticulture has become a major branch of modern agriculture and is of large and growing importance in many countries around the world. Producers in every country seek to exploit their comparative advantages and to find their niche in domestic and export markets. Though competition has become intense and market growth has slowed, the demand for ornamental horticultural products has remained remarkably buoyant even during periods of economic recession. Consumption in most countries remains underdeveloped, and even on major floral occasions such as St. Valentine's Day and Mother's Day, only a tiny percent of consumers buy flowers and plants, indicating much room for future sales growth. Meanwhile, breeding and new technologies are causing continuous changes in the product assortment and are providing endless commercial opportunities for daring entrepreneurs who are willing to invest and take chances. For floricultural producers in high-cost northern hemisphere countries, technical innovations adopted by distant competitors still cannot completely offset their huge advantage of being able to provide fresher products to increasingly quality-conscience markets, which is of critical importance for such highly sensitive and perishable products as fresh cut flowers and plants. The ability of growers to produce high-quality products from virus-tested plant material will also play a critical role in determining their economic viability in such competitive conditions.

Selected Literature

AIPH (International Association of Horticultural Producers). 1993. Yearbook of the International Horticultural Statistics, Institut für Gartenbauokonomie der Universität Hannover.

Asocolflores (Association of Colombian Flower Exporters). 1993. Floriculture in Colombia, Bogota.

Bloemenveiling Holland. 1994. Aanvoer- en prijsinformatie (for various crops), Product and Market Information Department, Naaldwijk, Holland.

Egberts, E. 1992. Japanese floriculture combines tradition and change, FloraCulture International, November/December, pp. 14–17.

FAO (Food and Agriculture Organization of the United Nations). 1993. Trade Yearbook 1992. Volume 46, FAO Statistics Series No. 115, Rome.

Flower Council of Holland. 1993. Facts and figures about the Dutch horticultural industry 1992, Leiden, the Netherlands.

F-S (Federal-State) Market News. 1993. Ornamental crops national market trends, Agricultural Marketing Service, USDA, San Fransisco, various issues.

Haak, M., Tap, H. and Heybroek, A.M. 1992. A view of international competitiveness in the floristry industry, Rabobank Nederland, International Agribusiness Research, Utrecht, June 1992.

Hack, M.D., van de Zwaan, A.G. and Heybroek, A.M. 1992. International competitiveness in the arboricultural sector, Rabobank Nederland, International Agribusiness Research, Utrecht.

Hamrick, D. 1994. For Colombia, the business of flowers is change, FloraCulture International. 4(1):24–29.

JETRO (Japan External Trade Resources Organization). 1992. The Japanese market for cut flowers, Tradescope, April, pp. 7–18.

Johnson, D.C. 1990. Floriculture and environmental horticultural products: A production and marketing statistical review, 1960–88, Statistical Bull. Number 817, US Dept. Agric., Econ. Res. Serv., Washington, DC.

Johnson, D.C. 1993. Financial performance trends and economic outlook for the US greenhouse, turfgrass and nursery industries, Agricultural Outlook '94: Conference Proceedings, US Dept. Agric., Washington, DC.

de Kleijn, E., Boers, G.J. and Heybroek, A.M. 1992. International competitiveness in the propagating material sector, Rabobank Nederland, International Agribusiness Research, Utrecht.

de Kleijn, E. and Heybroek, A.M. 1992. A view of international competitiveness in the flower bulb industry, Rabobank Nederland, International Agribusiness Research, Utrecht.

Malter, A.J. 1992. Onat hayetsu shel prachim 1991/92, Leket dochot mispar 5, Department of Market Research, Ministry of Agriculture, Tel Aviv (in Hebrew).

Ministry of Agriculture. 1992. Facts and Figures 1992: Highlights of Dutch Agriculture, Nature Management and Fisheries, Ministry of Agriculture, Nature Management and Fisheries of the Netherlands, The Hague.

Ohkawa, K. 1993. Advances in Japanese ornamental horticulture, Chronica Hortic., 33(1):3–4.

PVS (Commodity Board for Ornamental Horticultural Products). 1993. Tuinbouw statistiek 1992: teelt, handel, industrie, The Hague.

PVS (Commodity Board for Ornamental Horticultural Products). 1992. Jaarverslag, The Hague.

Smith, J. 1993. The changing floriculture industry: A statistical overview by the Society of American Florists statistics committee, Society of American Florists.

USDA (United States Department of Agriculture). 1993. Floriculture Crops, National Agricultural Statistics Service, Agricultural Statistics Board, Washington, DC.

Vakblad voor de Bloemisterij. 1994. Duitsland opnieuw belangrijkste groeimarkt, Vakblad voor de Bloemisterij, 49(4):15.

Van Doesburg, J. 1992. Lecture for the Flower Board of Israel, VBN, lecture in Tel Aviv, March 25, 1992.

VBN (Association of Dutch Flower Auctions). 1993. Statistiekboek, Leiden, the Netherlands.

World Bank. 1992. World Development Indicators, World Development Report 1992: Development and the Environment, Oxford University Press, New York.

2

Viruses of Ornamentals: Historical Perspectives

L. Bos

DLO Research Institute for Plant Protection (IPO-DLO), Box 9060, 6700 GW Wageningen, The Netherlands

Introduction

Since time immemorial, diseases have been haunting man. They have led to suffering of man himself and caused damage to his domesticated animals and crops needed for subsistance and for lending charm to life (such as ornamentals). Although long ascribed to supernatural forces, diseases were gradually related with nasty, "venomous" or "slimy" agents, often called "*viruses*", that spread in populations and were thus infectious. Towards the end of the nineteenth century, Pasteur and Koch learned to associate infection with micro-organisms, and Koch's postulates (ca 1880) enormously enhanced the isolation and characterization of disease-inciting micro-organisms or "germs". An ever increasing variety of them could be observed with the light microscope in diseased tissues, grown in pure culture on artificial media in the laboratory, then studied in detail, and thereafter reintroduced into healthy organisms to reproduce the disease under study for proof of their causal involvement (Bos, 1981).

However, soon after 1880 many diseases were found not to yield to the then modern microbiological approaches. Classical investigations on the highly contagious mosaic disease of tobacco (Fig. 2.1) were performed by Mayer (1882, 1886) and Beijerinck (1898) in the Netherlands and by Ivanovsky (1892) in Russia, proving the involvement of another type of pathogenic agent. It was contagious, but (1) was invisible even with the light microscope, (2) could pass through a porcelain filter (while bacteria could not), and retain infectivity, but (3) could not be cultivated for study on artificial media. Mayer (1882) suggested that the infectant had an enzyme-like nature. Beijerinck (1898), calling it a *contagium vivum fluidum*, was the first to realize its basic difference from micro-organisms, whereas Ivanovsky (1903) kept insisting on its microbial nature.

Late in the last century, the new category of disease agents was called *filterable viruses*, but the adjective "filterable" gradually fell into disuse early this century. With rapidly increasing information on fungi and bacteria, the term *viruses* remained for those agents that were invisible and could not be cultivated on artifical media but were infectious, as by sap inoculation, insect transmission, and — in case of failure of all such endeavors — by grafting.

Fig. 2.1. Mosaic patterns in leaves of tobacco caused by the highly infectious sap-transmissible tobacco mosaic virus.

Further Development of the Virus Concept

After much long-lasting speculation about the nature of viruses, a breakthrough occurred in 1936 when the first virus (tobacco mosaic virus, TMV) was isolated by Stanley from diseased plants as a pure chemical. One year later Bawden and Pirie determined that TMV is nothing but a nucleoprotein. The next crucial event in the history of virology was the discovery by Kausche and coworkers in 1939 of corpuscular particles of TMV using the newly developed electron microscope. These developments led to *further precision of the virus concept in terms of molecular biology*. Viruses can be defined as small packages of host-alien genetic information encoded in a strand of nucleic acid, wrapped up in a rigid protein coat (Fig. 3.1, Chapter 3).

Further studies of ultrathin sections of plants with witches' broom and aster yellows by electron microscopy, begun by Doi et al. (1967) in Japan, revealed a new type of infectious agents. Such diseases had first been studied in the US, for example in peach with "peach yellows" by Erwin Smith (1888), and in a very wide range of ornamentals and other plant species with "aster yellows" by Kunkel (1926) (Fig. 8.11; see color plate). Though first ascribed to virus infections because of their graft- and leafhopper-transmissibility, increasing evidence associated them with minute pleiomorphic microorganisms (*mycoplasmas*, Chapter 7), which are also filterable, and more or less behave like viruses in their relationships with plants and the insects that transmit them. That is why they are often called *virus-like pathogens*. In contrast to viruses, mycoplasmas contain much more genetic information and do have a physiology of their own, and they differ from bacteria in that they have no rigid cell wall and no organized and bounded nucleus.

The smaller parasitic agents are, the more they depend upon their host. At the lowest extreme of parasitism are the *viroids*, first detected in 1967, consisting only of nucleic acid in such small amounts that they depend entirely on host enzymes for their multiplication (Chapter 6). Viroids have increasingly been found to cause several diseases of plants including ornamentals, such as chrysanthemum stunt.

In recent years new classes of pathogens are being found that range between true viruses and viroids. The *satellite viruses*, such as of tobacco necrosis virus, have spherical particles and are composed of coat protein and nucleic acid that only codes for one protein, hence they need a helper virus for multiplication. *Satellite RNAs*, such as of cucumber mosaic virus, are somewhat similar to satellite viruses, but they do not code for a coat protein of their own and are therefore co-encapsidated with the nucleic acid of the helper virus in its protein coat. Both satellite viruses and satellite RNAs can modify the symptoms of the helper virus.

Within the range of parasitic agents, viruses, particularly the viroids, can no longer be considered as (micro)organisms. That is why many plant virologists find it difficult to accept a *species concept* for viruses, as adopted by the International Committee for the Taxonomy of Viruses (ICTV). Viruses live an entirely "borrowed life" and really are "at the threshold of life".

Historical Records of Plant Virus Disease

In retrospect, the oldest record of a plant virus disorder may be the brief poem by a Japanese empress in the year 752 on the yellowing of the leaves of *Eupatorium* "looking like the effect of frosting during summer" (Inouye and Osaki, 1980), possibly the vein yellowing caused by the whitefly-transmitted tobacco leafcurl geminivirus (Fig. 8.8; see color plate), which affects many plant species, including ornamentals. The oldest disease now known to be caused by a mycoplasma may be the green flowering of tree peony, *Paeonia suffruticosa*, first recorded by Ou-yong Xiu (1031, quoted by Wang and Maramorosch, 1988). Such abnormal plants were widely acclaimed for ornamental purposes during the Song dynasty (960–1227) in China. However, the first well-documented example of a plant virus disease is that of color-breaking in tulip flowers, often found in Dutch still-lifes during the early part of the 17th century. The phenomenon was first described in 1576 by Carolus Clusius (Charles de l'Ecluse), soon after the introduction of the species in the Netherlands as a new ornamental from the Middle East; patterns of color-breaking were so attractive that the value of single bulbs producing a "broken" tulip flower soared to fortunes (Fig. 2.2; see color plate). Around 1637 the gambling trade suddenly collapsed, perhaps because some tulip growers had learned how to transmit the abnormality to healthy plants by bulb grafting, as apparent from a document of 1643. Interestingly, Clusius already noticed that the flower "breaking" may lead to the loss of even-colored varieties and that the color beautification is "*an endeavour of the tulips to please their master's eyes before bidding him a last farewell prior to evanescence*". It was not until McKay (1926) first detected sap- and aphid-transmission of the abnormality, that it is really known as an important, naturally spreading threat to the cultivation of tulips.

Another historical example is leaf variegation in woody ornamentals where the phenomenon, often ascribed to and easily confused with genetic aberration, is

Fig. 2.3. Mosaic in *Abutilon striatum* var *spurium*, since 1869 known to be graft transmissible and now ascribed to a whitefly-transmitted geminivirus.

commercially exploited to date. Attractively variegated plants of *Abutilon striatum* var. *spurium*, often known as var. *thompsonii* (Fig. 2.3), were imported into Europe from South America in 1868 as a vegetatively propagated indoor ornamental. The following year the variegation was discovered to be graft-transmissible and thus infectious. Graft transmissibility of a similar abnormality of laburnum observed by Masters (1877) in Britain led to his report with the remarkable title "Action of scion on stock". The report on similar variegations of several ornamentals by Baur (1904) in Germany is a phytopathological classic.

Virus Identification and Molecular Biology

Plant virology has rapidly expanded since World War II thanks to molecular biology and to a wide range of tools that have become available to isolate an increasing number of viruses from plants, to study and characterize them in the laboratory, to visualize them *in vitro* and in diseased tissues, and to classify and name them.

Virus *identification* (that is characterization and recognition) was first by the symptoms they specifically cause on plants, by their host ranges, their vectors (agents that transmit them) and their relationships with these vectors, and by the persistence of their infectivity in expressed sap. However, these *biological properties* that were studied by variable host or test plant reactions were soon deemed unreliable. Molecular biology later allowed the determination of *intrinsic* (that is, physico-chemical) *virus properties* that can be assessed in the laboratory. Molecular biology has also helped to define more precisely what viruses really are.

Technology and methodology for routine and large-scale detection of viruses have advanced immensely during the last few decades. Particle morphology and size can be assessed with the electron microscope. In modern virology, the study of the content and composition of viral proteins and nucleic-acids and of their amino-acid and nucleotide sequences has become the order of the day. For several viruses the complete genetic code has already been revealed. Highly sensitive serological methods exploiting particular amino-acid configurations (serological determinants or epitopes) on the coat-protein particle are particularly fit for large-scale application and automatization.

An unanswered question, however, is how far serological specificity represents biological identity. Only a small part of the total genetic information is expressed in coat-protein composition and structure, and only part of that in serological specificity. That is why there is increasing interest in methods to determine directly the genetic information encoded in the viral nucleic acid. For rapid diagnosis, however, full nucleic-acid sequence determination is impracticable. For this purpose, small bits of information can be used to trap matching pieces of viral nucleic acid in a sample collected from a field-grown plant with a laboratory-made, hopefully representative probe of replica (complementary) nucleic acid (cDNA).

As for serology, the question re-emerges whether these nucleic-acid-based techniques sufficiently cover the complex of biological characteristics of the virus, that is, whether the molecular-biological characteristics used can be considered *representative parameters* of biological identity. Biological identity of viruses — that is, the symptoms they cause, the ranges of plant species and cultivars they infect, and the ways they are spread — is what counts in agricultural practice. Growers are, and must be, particularly interested in biological and ecological behavior of viruses, in their strains, and in how viruses vary (change) with time. These characteristics are often not covered by the molecular-biological parameters used for virus detection and diagnosis, or they are so specific that only single characteristics are recognized while the others are ignored.

Viruses No More Beyond the Grasp of Man?

The number of viruses found associated with diseases of ornamentals and other crops has increased tremendously, as has information on their *ecology*, that is on their sources of infection, the ways in which they are spread, and on how they interact with crops. When now dealing with viruses of plants, we know reasonably well what we are talking about and how to counteract their involvement in causing disease and economic damage. The viruses are no more "beyond the microscope" and no more "beyond the grasp of man". However, viruses are part of nature, which is dynamic in essence. Nature remains more vigilant and versatile than man, and will continuously haunt us with new viral problems in our crops. These arise as a consequence of (1) virus mutation, (2) the coming to the fore of viruses or virus strains that previously were unimportant or escaped attention as a consequence of the creation of new ecological niches through continuous change in agro-ecosystems, and (3) the increasing spread of viruses as a result of international trade and traffic of plant propagation materials (Bos, 1992). Despite tremendous progress, we must keep on the alert for viruses. This requires input by molecular biologists to provide and improve laboratory technology, as well as by ecological virologists, who are aware of virus behavior in the field, and in their dynamic and ever-changing relationships with crops, vectors, and wild vegetation. Nature and man's relationships with nature and with viruses will keep changing. That is why the history of virology is open-ended.

Selected Literature

Baur, E. 1904. On the etiolgy of infectious variegation. (English translation from German.) Phytopathological Classics, American Phytopathological Society 7, 1942:55–62.

Bos, L. 1981. Hundred years of Koch's Postulates and the history of etiology in plant virus research. Neth. J. Plant Pathol. 87:91–110.

Bos, L. 1992. New plant virus problems in developing countries: a corollary of agricultural modernization. Adv. Virus Res. 41:349–407.

Inouye, T. and Osaki, T. 1980. The first record in the literature on the possible plant virus disease that appeared in "Manyōshū", a Japanese classic anthology, as far back as the time of the 8th century. (Japanese, with English summary). Ann. Phytopathol. Soc. Jpn. 46:49–50.

Kunkel, L.O. 1926. Studies on aster yellows. Am. J. Bot. 13:646–705 (also in: Contrib. Boyce Thompson Inst. 1:181–240).

McKay, M.B. and Warner M.F., 1933. Historical sketch of tulip mosaic or breaking. The oldest known plant virus disease. National Horticultural Magazine 12:179–216.

Masters, M.T. 1877. Action of scion on stock. Gardeners' Chronicle 7:730.
Wang, M.Q. and Maramorosch, K. 1988. Earliest historical record of a tree mycoplasma disease: Beneficial effect of mycoplasma-like organisms on peonies. In: Mycoplasma Diseases of Crops. Basic and Applied Aspects, K. Maramorosch and S.P. Raychaudhuri (eds.), Springer Verlag, New York, pp. 349–355.

3

What is a Virus?

G. Loebenstein
Department of Virology, Agricultural Research Organization,
The Volcani Center, Bet Dagan 50250, Israel

Viruses are infectious agents with submicroscopic particles, the sizes of which are in the nanometer range (1 nm $=10^{-6}$ mm). They contain one type of nucleic acid (RNA or DNA) encapsidated in a protein coat, replicate only inside living cells, and may induce diseases. Viruses therefore are intracellular obligatory parasites, which are relatively simple in structure; in principle they can be described as a package of genetic material, encapsidated in a protective shell.

Further characteristics of viruses which distinguish them from other obligatory parasites such as mycoplasmas, rickettsias, bacteria and fungi, are:

1. Viruses multiply not by binary fission, but by separate synthesis of the individual nucleic acid and protein components which are later assembled into particles.

Contribution No. 1375-E, 1994 series from the Agricultural Research Organization, The Volcani Center, Bet Dagan, Israel.

2. Viruses, unlike other pathogens which have both RNA and DNA, contain a single type of nucleic acid (either RNA or DNA).

3. The proteins associated with virus replication are synthesized on the ribosomes of host cells.

4. For their replication, viruses require the energy-synthesizing machinery of the host cell.

The sizes of virus particles range between 15–2000 nm, whereas the size of bacteria ranges from 1000–3000 nm, and the diameter of a tobacco mesophyll cell is approximately 8×10^6 nm (0.08 mm). The relative sizes of viruses, viroids, mycoplasma and viral RNA are indicated in Fig. 3.1 and are compared with a bacterial cell of *Escherichia coli* represented by the outer frame.

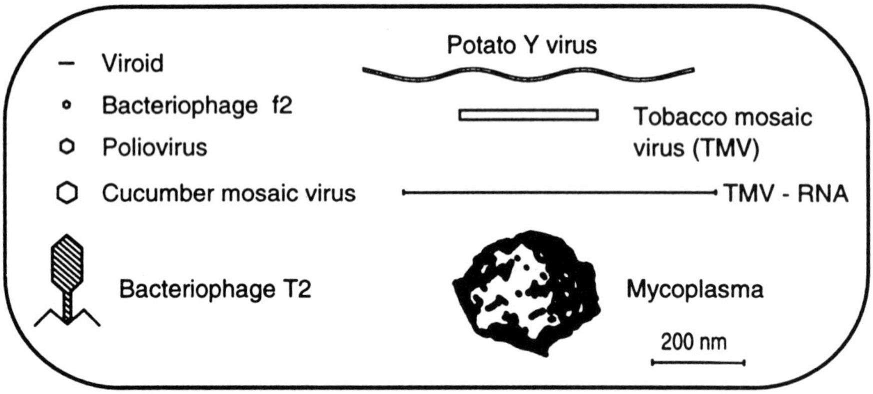

Fig. 3.1. Sizes and shapes of virus particles in comparison with a bacterial cell of *Escherichia coli* (surrounding frame).

Selected Literature

Francki, R.I.B., Milne, R.G. and Hatta, T. 1985. Atlas of Plant Viruses. CRC Press, Boca Raton, FL.

Matthews, R.E.F. 1991. Plant Virology, 3rd ed. Academic Press, New York.

4

Plant Virus Names

Alan A. Brunt
Horticulture Research International, Littlehampton, Sussex, UK

Introduction

During the first half of this century, many systems had been proposed for the nomenclature and classification of plant viruses. Two of these used a numerical nomenclature (e.g., Johnson, 1927; Smith, 1937), and six others a binomial system (Palm, 1922; Holmes, 1939; Valleau, 1940; Thornberry, 1941; Fawcett, 1942). Although little supported by practicing virologists, three more binomial systems of nomenclature and classification were proposed between 1966 and 1970 (see Francki, 1981). Better known viruses had thus been given several different epithets and, up to 1970, at least 11 names had been proposed for tobacco mosaic virus (Francki, 1981). These early numerical and binomial systems, however, were inevitably based on the biological properties of the viruses alone, an observation which prompted Bawden (1939), a notable early critic, to state that "any lasting system of classification should be based on the intrinsic properties of the virus rather than those of host plants". As these proposed systems of nomenclature and taxonomy were unsound, none was widely accepted and all were subsequently discarded.

During the latter half of the century, there has been an increasingly rapid accretion of information on the physico-chemical properties of viruses and their component proteins and nucleic acids. Such data have been shown to be reliable taxonomic criteria and so are now used as the basis of virus classification and as a valuable adjunct to virus nomenclature.

The International Committee on Taxonomy of Viruses (ICTV) has had international responsibility for virus taxonomy since 1951, and during the past 43 years has published five reports of its activities (see Francki et al., 1991). With the guidance of ICTV, a stable taxonomic system has been developed during the past four decades or so; currently, viruses (now designated species) are placed into groups (or genera) which, if they have obvious affinities, are grouped into families. ICTV now recognizes 47 genera of plant viruses, 25 of which are grouped into families (Mayo and Martelli, 1993).

Virus Names

The names of individual plant viruses are now often derived from the common or, occasionally, the generic name of the host species in which the virus was first found, or is of the greatest importance, together with a description of the characteristic symptom(s) of infection. For convenience, virus names are often abbreviated; it has been proposed that such sigla should be standardized for definite species within a genus (Hull et al., 1991).

Many plant viruses induce leaf chlorosis of various types in infected plants. Abutilon mosaic (AbMV), dahlia mosaic (DMV), cucumber mosaic (CMV), alfalfa mosaic (AlfMV), and many other viruses induce mosaic patterns of chlorotic and normal green areas and this is so indicated in their names; the characteristic symptomatology of other viruses is similarly indicated in their names, such as turnip yellow mosaic virus (TYMV), which causes the production of extensive and conspicuous chlorotic areas, and beet western yellows virus (BWYV), which characteristically causes complete or nearly complete leaf chlorosis. The chlorosis induced by some viruses such as *Pittosporum* vein-banding or maize streak, as described in their names, is mainly adjacent to leaf veins; the veinal chlorosis induced by viruses such as narcissus yellow stripe virus (NYSV) and beet necrotic yellow vein virus (BNYVV) is often especially prominent, and this is appropriately indicated in their names. Carnation mottle virus (CarMV) is typical of viruses which cause leaf mottling of infected host species.

Other viruses such as tomato ringspot (TomRSV), tobacco ringspot (TRSV), and hydrangea ringspot (HRV) induce leaf chlorosis which is usually in the form of chlorotic rings; as indicated in its name, the rings induced by viruses such as hibiscus chlorotic ringspot (HCRV) are especially conspicuous. The ringspots caused by some viruses such as tomato black ring (TBRV), carnation etched ring (CarERV), and *Prunus* necrotic ringspot (PNRSV) are black, semi-necrotic, or necrotic, respectively. As evident from their names, the chlorosis induced by some viruses such as *Kalanchoë* top-spotting (KTSV) is mainly restricted to spots on upper, immature leaves. Other viruses, such as *Impatiens* necrotic spot (INSV), induce necrotic spots in leaves.

Some viruses cause both leaf chlorosis and irregular pigmentation ("breaking") of flowers. Well-known examples in which symptoms in flowers are indicated in the name of the virus include tulip breaking (TBV), Rembrandt tulip breaking (RTBV), tulip band-breaking (TBBV), and tulip chlorotic blotch (TCBV). Other viruses, however, such as turnip mosaic (TuMV) and bean yellow mosaic (BYMV) viruses also cause flower breaking in a range of infected host species.

Tobacco necrosis virus (TNV) is so named because in tobacco and other infected host species it characteristically causes necrotic leaf spots and stem necrosis, and tomato spotted wilt virus (TSWV) is an appropriate name for a virus which in tomato and other hosts induces leaf chlorosis, necrotic spotting, and wilting of plants. Some viruses, notably broad bean wilt virus (BBWV), cause wilting of plants, and others such as tobacco streak virus (TSV) cause systemic necrotic streaking of stems in tobacco and other host species.

Tomato aspermy virus (TAV), a virus which causes severe flower distortion in chrysanthemum, is so named because infected tomato plants often produce seedless fruits.

Wound tumor virus (WTV) is an appropriate name for the virus which causes the formation of tumors at sites of wounding, usually the hypocotyl, of infected clovers. The name of cocoa swollen shoot virus (CSSV) originates from the characteristic stem and root swellings (usually in addition to leaf chlorosis) induced in infected cocoa plants by many strains of the virus. Pea enation mosaic virus (PEMV), as indicated in its name, induces the formation of leaf enations in some infected host species.

Some viruses, such as plum pox virus (PPV), papaya ringspot virus (PRSV), and melon necrotic spot virus (MNSV), cause mottling of the fruits of infected hosts, and this is so indicated in their names.

Although many viruses induce symptoms in their infected hosts, others such as carnation latent (CarLV) and narcissus latent (NLV) usually induce no obvious symptoms; such viruses, however, can cause significant losses of yield and quality of infected crops. Narcissus late season yellows virus (NLSYV) is so named because infected plants remain symptomless for much of the growing season but develop conspicuous leaf chlorosis as they approach maturity.

Current Practice

Of the 47 genera of plant viruses now recognized by ICTV (Mayo and Martelli, 1993), 22 contain species that naturally infect ornamental plant species. Some such better known taxa include the cucumovirus, alfamovirus, potyvirus, tombusvirus, carmovirus, nepovirus, fabavirus, tospovirus, potexvirus, and ilarvirus genera (see Chapter 5). For those unfamiliar with specific viruses but with a general knowledge of plant virus taxonomy, it is very helpful for the name of the genus to be included in the virus name. It is thus now becoming increasingly common for this to be done, and reference is now often made, for example, to cucumber mosaic cucumovirus, poinsettia mosaic tymovirus, tulip breaking potyvirus, tomato spotted wilt tospovirus, tobacco streak ilarvirus, arabis mosaic nepovirus and narcissus mosaic potexvirus. This improvement in virus nomenclature has much support, and is thus likely to continue.

Selected Literature

Bawden, F.C. 1939. Plant Viruses and Virus Diseases, Chronica Botanica, Leiden.

Fawcett, H.S. 1942. Virus nomenclature. Chronica Botanica 7:7–8.

Francki, R.I.B. 1981. Plant virus taxonomy. In: Handbook of Plant Virus Infections; Comparative Diagnosis, E. Kurstak (ed.), Elsevier, North Holland Biomedical Press, Amsterdam, pp. 3–16.

Holmes, F.O. 1939. Handbook of Phytopathogenic Viruses. Burgess, Minneapolis, MN.

Hull, R., R.G. Milne and M.H.V. van Regenmortel. 1991. A list of proposed standard acronyms for plant viruses and viroids. Arch. Virol. 120:151–164.

Johnson, J. 1927. The classification of plant viruses. Res. Bull. Wisconsin Exp. Stat. 76:1–16.

Mayo, M.A. and Martelli, G.P. 1993. New families and genera of plant viruses. Arch. Virol. 133:496–498.

Palm, B.T. 1922. Bulletin of the Deli Proefstation, Medan (Sumatra) 15:17–20.

Smith, K.M. 1937. A Textbook of Plant Virus Diseases. Churchill, London.

Thornberry, H.H. 1941. A proposed system of virus nomenclature and classification. Phytopathology 31:23.

Valleau, W.D. 1940. Classification and nomenclature of tobacco viruses. Phytopathology 30: 820–830.

5

Major Genera of Plant Viruses

Alan A. Brunt
Horticulture Research International, Littlehampton, Sussex, UK

Introduction

The International Committee on Taxonomy of Viruses (ICTV) currently recognizes 47 genera (previously designated groups) of plant viruses (Mayo and Martelli, 1993). Of these, 25 are grouped into 10 families (Table 5.1). The remaining 22 (Table 5.2) are not yet assigned to higher taxa, but some with similar particle morphology and genome properties are informally placed in groups which might eventually attain the status of families (Martelli, 1992). There are, of course, many other viruses (now designated species) with unknown affinities which have yet to be assigned to genera.

Of the 24 genera containing one or more species that occur naturally in ornamental plants, 12 are assigned to families (Table 5.1) and 12 are not yet so assigned (Table 5.2). The major properties of each of the 24 are briefly discussed, but further information on the biology and physico-chemical properties and taxonomy of the viruses is reviewed in Kurstak (1981) and Francki et al. (1991).

Table 5.1. Families and genera of plant viruses containing virus species that infect ornamental plants

Genome type		Family	Genera	Virus species infecting ornamental plants
1. Monopartite	(+) ssRNA	*Tombusviridae*	*Tombusvirus*	+
			Carmovirus	+
		Sequiviridae	*Sequivirus*	–[a]
			Waikavirus	–
		Potyviridae	*Potyvirus*	+
			Bymovirus[b]	–
			Rymovirus	–
	(–) ssRNA	*Rhabdoviridae*	*Cytorhabdovirus*	–
			Nucleorhabdovirus	+
2. Bipartite	(+) ssRNA	*Comoviridae*	*Comovirus*	–
			Fabavirus	+
			Nepovirus	+
	(+) dsRNA	*Partitiviridae*	*Alphacryptovirus*	+
			Betacryptovirus	–
	(+) dsDNA	*Geminiviridae*	Subgroup 1	–
			Subgroup 2	–
			Subgroup 3	+
3. Tripartite	(+) ssRNA	*Bromoviridae*	*Bromovirus*	–
			Cucumovirus	+
			Ilarvirus	+
			Alfamovirus	+
	(+/–) ssRNA	*Bunyaviridae*	*Tospovirus*	+
4. Multipartite	(+) dsRNA	*Reoviridae*	*Phytoreovirus*	–
			Fijivirus	–
			Oryzavirus	–

[a]Genus contains no important viruses that naturally infect ornamental plant species.
[b]Bymoviruses have bipartite genomes.

Table 5.2. Plant virus genera not yet assigned to families

Genome type		Particle morphology	Genus	Virus species infecting ornamental plants
1. Monopartite	(+) ssRNA	Isometric	*Luteovirus*	–[a]
			Sobemovirus	–
			Tymovirus	+
			Necrovirus	+
			Machlomovirus	–
			Marafivirus	–
		Anisometric	*Tobamovirus*	+
			Potexvirus	+
			Carlavirus	+
			Closterovirus	+
			Capillovirus	+
			Trichovirus	–
		No particles	*Umbravirus*	–
	(+) dsDNA	Isometric	*Caulimovirus*	+
		Bacilliform	*Badnavirus*	+
2. Bipartite	(+) ssRNS	Isometric	*Enamovirus*	+
			Dianthovirus	+
			Idaeovirus	–
		Rod-shaped	*Tobravirus*	+
			Furovirus[b]	–
3. Tripartite	(+) ssRNA	Rod-shaped	*Hordeivirus*	–
4. Quadripartite	(+) ssRNA	Filamentous	*Tenuivirus*	–

[a]Genus contains no important viruses that naturally infect ornamental plant species.
[b]Beet necrotic yellow virus, an unassigned species, has four or five genomic RNAs.

Plant Virus Genera Assigned to Families

Genera with monopartite ssRNA genomes

Tombusviridae

The family designation is derived from *tombus*, the name of the first of its two genera to be described.

1. *Carmovirus* genus — The generic designation is an acronym derived from the name of the type species, *car*nation *mo*ttle *virus* (CarMV). Viruses infecting ornamental plant species include: carnation mottle virus (CarMV), hibiscus chlorotic ringspot virus (HCRV), pelargonium flower break virus (PFBV), saguaro cactus virus (SCV), and narcissus tip necrosis virus (possible species).*

Species have isometric particles c. 33 nm in diameter (Fig. 5.1) which have a sedimentation coefficient of c. 122 S, a buoyant density in CsCl of 1.35 g cm^{-3} and an M_r of 8.2×10^6. The capsid contains 180 protein subunits, each of 38 kDa, and encapsidates the genomic ssRNA of c. 4 kb and two subgenomic RNAs of 1.7 and 1.5 kb. The genomic RNA of some species has been fully sequenced and infectious transcripts produced; it encodes a protein of 27 kDa and two potential read-through polyproteins of 86 and 98 kDa. The 1.7 and 1.5 kb subgenomic RNAs produce proteins, respectively, of 7 kDa and 38 kDa, the latter being coat protein; both subgenomic RNAs and genomic RNA are needed for systemic infection of plants. Three dsRNAs (4.0, 1.7, and 1.5 kbp) corresponding to the three virus-specific ssRNAs are found in infected plants.

The species occur worldwide in a range of plant genera. Particles occur *in vivo* in nuclei and in cytoplasm associated with cytoplasmic membranes. Others are soil-borne, but probably abiotically. Species are experimentally transmissible by mechanical inoculation of sap.

Type virus: carnation mottle virus (CarMV)

Diagnostic species: *Chenopodium amaranticolor*, *C. quinoa*, local chlorotic lesions

Propagation species: *Dianthus caryophyllus* (carnation); *Chenopodium quinoa*

Purification: See Hollings and Stone (1970)

Serology: strongly immunogenic

---

*To avoid future confusion, possible members (now described as unassigned species by ICTV) are not designated by sigla (Hull et al., 1991).

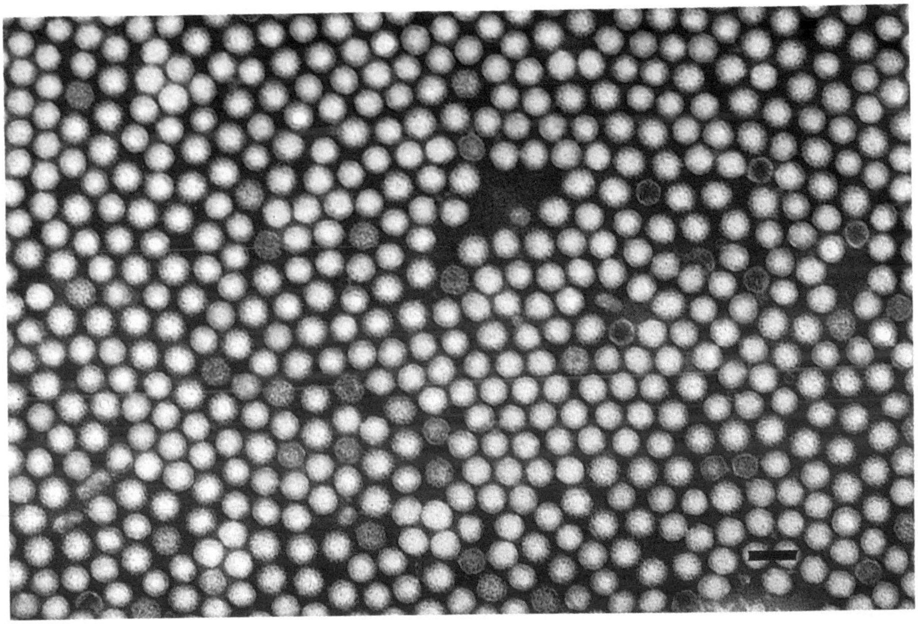

Fig. 5.1. Particles of carnation mottle carmovirus; bar scale=50 nm.

Properties: A_{260}/A_{280} = 1.50–1.60
Specific absorbance at 260 nm=5.0

2. *Tombusvirus* genus – The generic designation is an acronym derived from the name of the type species, *tom*ato *bus*hy stunt *virus* (TBSV). Other species also infecting ornamental plants include: carnation Italian ringspot virus (CarIRV), cymbidium ringspot virus (CyRSV), pelargonium leaf curl virus (PLCV), and petunia asteroid mosaic virus (PAMV).

Species have isometric particles c. 30 nm in diameter which are morphologically similar to those of carmoviruses, have a sedimentation coefficient of 131–140 S, a buoyant density in CsCl of 1.35 g cm^{-3} and M_r of 9×10^6. The particles contain 17% nucleic acid and 83% protein. The capsid contains 180 subunits, each of c. 43 kDa, and encapsidates a single molecule of genomic ssRNA (4.7 kb), two subgenomic ssRNAs (2.1 and 0.9 kb), defective interfering RNA (0.4-0.7 kb) and, sometimes, a satellite RNA (0.62 kb). The 3′ tails of genomic, interfering, and satellite RNAs are not polyadenylated, and their 5′ ends probably have no cap structure or covalently bound genome-linked protein

(VPg). The capsid proteins and nucleic acids of species have extensive amino acid and nucleotide homology. The genomic RNA has four open reading frames (ORFs): ORF 1 encodes a protein of 33 kDa and, by continuous read-through with ORF 2, a protein of 92 kDa (the replicase); ORF 3 is translated via the larger subgenomic RNA (2.1 kb) to produce a polypeptide of 41 kDa (coat protein), and ORF 4 is translated via the smaller subgenomic RNA (0.9 kb) to produce a polypeptide of 22 kDa (function unknown).

Infected plants contain three ds RNAs (4.7, 2.1, and 0.9 kbp) corresponding to genomic and subgenomic RNAs. The species occur worldwide in a range of plant genera. All induce the formation of compact membranous inclusions ("multivesicular bodies") in cytoplasm during early stages of infection, and some cause peripheral vesiculation of chloroplasts; amorphous inclusions in cytoplasm consist of excess coat protein. Virus particles occur in cytoplasm, which sometimes protrudes into cytoplasm, nuclei, and mitochondria.

One species (cucumber necrosis tombusvirus) is transmitted by zoospores of the chytrid fungus *Olpidium radicale*. Others are soil-borne, probably abiotically, and others are reported to be seed-borne. All are experimentally transmissible by mechanical inoculation of sap.

Type virus: tomato bushy stunt virus (TBSV)

Diagnostic species: *Gomphrena globosa*, reddish lesions; *Chenopodium amaranticolor*, whitish local necrotic dots with chlorotic haloes; *Nicotiana glutinosa*, necrotic local lesions

Propagation species: *Datura stramonium*, *Nicotiana clevelandii*

Purification: See Gallitelli et al. (1985)

Serology: strongly immunogenic

Properties: A_{260}/A_{280} = 1.64 (corrected for light scattering)
Specific absorbance at 260 nm=4.5

Potyviridae

The family name is derived from *potyvirus*, its best-known genus.

1. *Potyvirus* genus — The generic designation is an acronym derived from the name of the type species, *pot*ato *Y virus* (PVY). The genus contains almost 200 viruses; better-known species naturally infecting ornamental plant species include: alstroemeria mosaic virus (AlMV), bean yellow mosaic virus (BYMV), bidens mottle virus (BiMoV), carnation vein mottle virus (CVMV), clover yellow vein virus (ClYVV), commelina mosaic virus (ComMV), dasheen mosaic virus

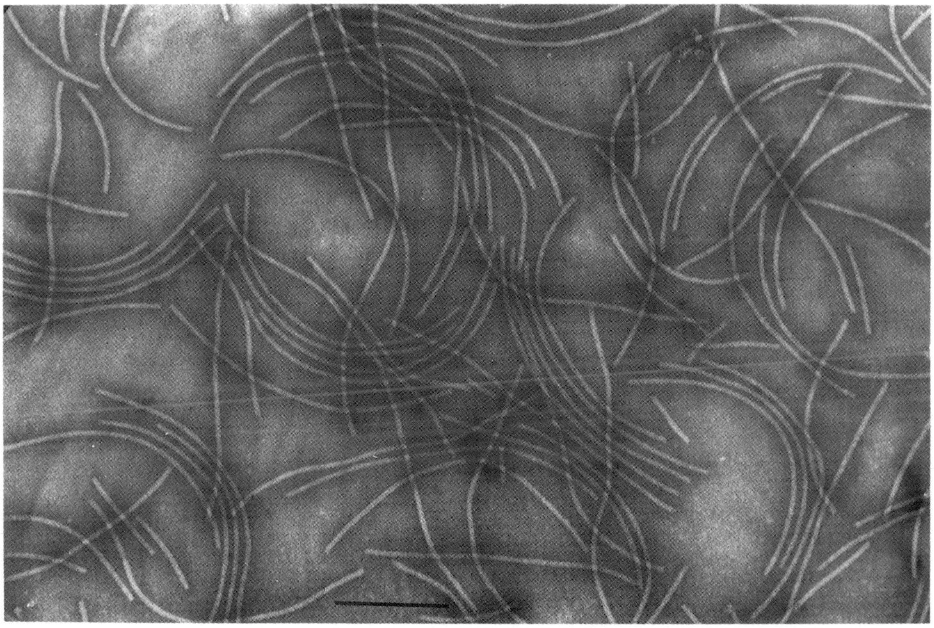

Fig. 5.2. Particles of pepper veinal mottle potyvirus; bar scale=250 nm. Micrograph courtesy of Rothamsted Experimental Station, Harpenden, UK.

(DsMV), dendrobium mosaic virus (DeMV), helenium Y virus (HVY), hippeastrum mosaic virus (HiMV), iris mild mosaic virus (IMMV), iris severe mosaic virus (ISMV), narcissus yellow stripe virus (NYSV), ornithogalum mosaic virus (OrMV), tulip chlorotic blotch virus (TCBV), tulip breaking virus (TBV), turnip mosaic virus (TuMV), and wisteria vein mosaic virus (WVMV).

Species have flexuous filamentous particles measuring 11–13 × 680–900 nm (Fig. 5.2) which sediment as a single component with a sedimentation coefficient of 150–160 S and have a buoyant density in CsCl of 1.31 g cm^{-3} and an M_r of 70×10^6. Capsid protein subunits, each of 32–36 kDa, are helically arranged with a helical pitch of 3.4 nm. The coat protein of the type species contains 267 amino acids. Each particle contains c. 5% by weight of linear positive-sense ssRNA of 8.5–10 kb which is polyadenylated at its 3′ terminus and has a VPg of c. 24 kDa at the 5′ terminus. The genomic RNA replicates in cytoplasm with the formation of a polyprotein which is subsequently cleaved to produce eight proteins; 34 kDa (movement protein or protease), 54 kDa (helper component), 42 kDa (unknown function), c. 70 kDa (cylindrical inclusion protein), 6 kDa (unknown function), 32–36 kDa (coat protein), 49 kDa (nuclear inclusion protein A, proteinase), and 58 kDa (nuclear inclusion protein B, polymerase).

Individual viruses usually have narrow natural host ranges, but a few, such as BYMV and TuMV, infect numerous plant species. Infected plants invariably contain cylindrical or conical cytoplasmic inclusions ("pinwheels") consisting of a single protein of c. 70 kDa; some viruses also induce the formation of crystalline nuclear inclusions consisting of two proteins (49 and 58 kDa) in near equimolar concentrations and/or amorphous inclusions which are probably mainly aggregated "helper component" protein (HC-Pro) of c. 54 kDa. The viruses are transmitted by aphids in the non-persistent manner.

Type virus: potato Y virus (PVY)
Diagnostic species: *Nicotiana glutinosa*, mottling; *Chenopodium amaranticolor*, chlorotic lesions
Propagation species: *N. glutinosa, N. tabacum* cv. Samsun NN
Purification: See Moghal and Francki (1976)
Serology: strongly immunogenic
Properties: A_{260}/A_{280} = 1.21–1.22 (corrected for light scattering)
Specific absorbance at 260 nm=2.3

Rhabdoviridae family

The name is derived from the Greek *rhabdos* (rod) which describes the morphology of the virus particles.

1. *Nucleorhabdovirus* genus — The genus is so described because its members replicate in the nuclei of their hosts. The type species is potato yellow dwarf virus (PYDV); another member infecting ornamental species is eggplant mottled dwarf virus (of which Pittosporum yellow vein virus is a synonym). The family also includes as unassigned species many viruses which occur in ornamental plant species but which have yet to be characterized, including: *Callistephus chinensis* chlorosis virus, carnation bacilliform virus, chrysanthemum vein chlorosis virus, *Euonymus* fasciation virus, gerbera symptomless virus, *Gomphrena* virus, *Iris germanica* leaf stripe virus, lupin yellow vein virus, orchid fleck virus, pelargonium vein clearing virus, saintpaulia leaf necrosis virus, *Dendrobium* leaf streak virus, and *Phalaenopsis* chlorotic spot virus.

The species have bacilliform particles mostly measuring 70–95×200–350 nm (Fig. 5.3) which sediment as a single component of c. 1000 S; partially degraded

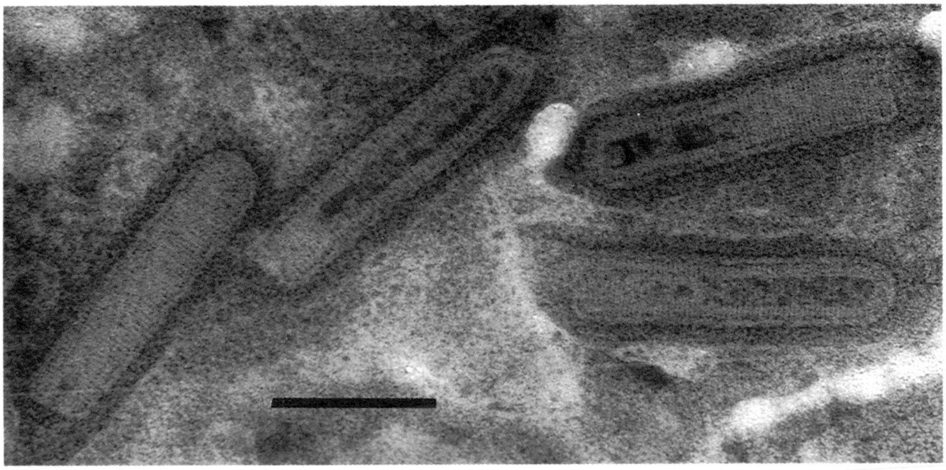

Fig. 5.3. Particles of broccoli necrotic yellows rhabdovirus; bar scale = 100 nm.

particles are bullet-shaped. The particles contain 1–2% RNA, 65–75% protein, 15-25% lipid, and 3% protein. A lipid envelope containing matrix or M protein (22–25 kDa) with protruding spikes (peplomers) of up to 12 nm consisting of trimers of the glycoprotein or G protein (71–92 kDa) encloses each nucleocapsid. The latter is 30–70 nm in diameter and has helical symmetry with a pitch of 4.5–5.0 nm. It consists of negative-sense ssRNA (11–15 kb) together with large or L (135 kDa), non-structural or NS (40 kDa), and nucleocapsid or N (55–64 kDa) proteins. The nucleocapsid is infectious and has transcriptase activity. The negative-sense genome codes for at least 5 ORFs.

Nucleorhabdoviruses occur worldwide in a wide range of hosts, but each virus has a very restricted host range. Most are transmitted by leafhoppers, plant hoppers, aphids, mites, or lace bugs. Those that have been so investigated have been shown to replicate both their arthropod vector and host plant. In both, particles of nucleorhabdoviruses occur in large aggregates in perinuclear spaces.

Type virus: potato yellow dwarf virus (PYDV)

Diagnostic species: *Nicotiana rustica*, yellow primary lesions, later systemic mottling and yellowing

Propagation species: *N. rustica*

Purification: See Black (1970)

Serology: moderately immunogenic

Genera with bipartite ssRNA, dsRNA, or dsDNA genomes

Comoviridae family

The family designation is derived from the name of *comovirus*, its better known genus.

1. *Fabavirus* genus — The generic designation is derived from the Latin *faba* (bean), as in *Vicia faba*, a major host of the virus. The type species is broad bean wilt virus 1 (BBWV1). Another virus infecting ornamental species is broad bean wilt virus 2 (BBWV2).

The type species has isometric particles c. 30 nm in diameter (Fig. 5.4) which sediment as three components (designated T, M, and B) with sedimentation coefficients, respectively, of 55 S, 95 S, and 120 S; have a buoyant density in CsCl of 1.28 (T), 1.42 (M), and 1.48 (B) g cm^{-3}; and M_r ($\times 10^6$) of 3.4 (T), 5 (M), and 6.1 (B). The capsid contains two polypeptides (40 and 24 kDa). T particles contain no nucleic acid, but M and B particles each contain a single molecule of linear positive-sense ssRNA; that in M particles is 4.5 kb (RNA 1) and that in B particles 6.3 kb (RNA 1). Both RNAs are required for infectivity.

Fabaviruses have very extensive natural host ranges. The cytoplasm of infected cells usually contains numerous particles which often occur in crystalline or paracrystalline assays. They are transmitted in the non-persistent manner by aphids.

Type virus: broad bean wilt virus (BBWV)

Diagnostic species: *Chenopodium amaranticolor*, *C. quinoa*, chlorotic lesions, systemic epinasty with leaf distortion; *Vicia faba*, systemic vein clearing

Propagation species: *C. quinoa*, *Vicia faba*

Purification: Doel (1975)

Serology: Strongly immunogenic

Properties: $A_{260}/A_{280} = 1.65$

Specific absorbance at 260 nm=4.5

2. *Nepovirus* genus — The generic designation is an acronym derived from an early description of the viruses as *ne*matode-transmitted *po*lyhedral *viruses*. The type species is tobacco ringspot virus (TobRSV). Other well-known viruses infecting ornamental plant species include: arabis mosaic virus (ArMV), cherry leaf roll virus (CLRV), hibiscus latent ringspot virus (HLRSV), raspberry ringspot virus (RRSV), tomato black ring virus (TBRV), tomato ringspot virus (TomRSV), and strawberry latent ringspot virus (possible species).

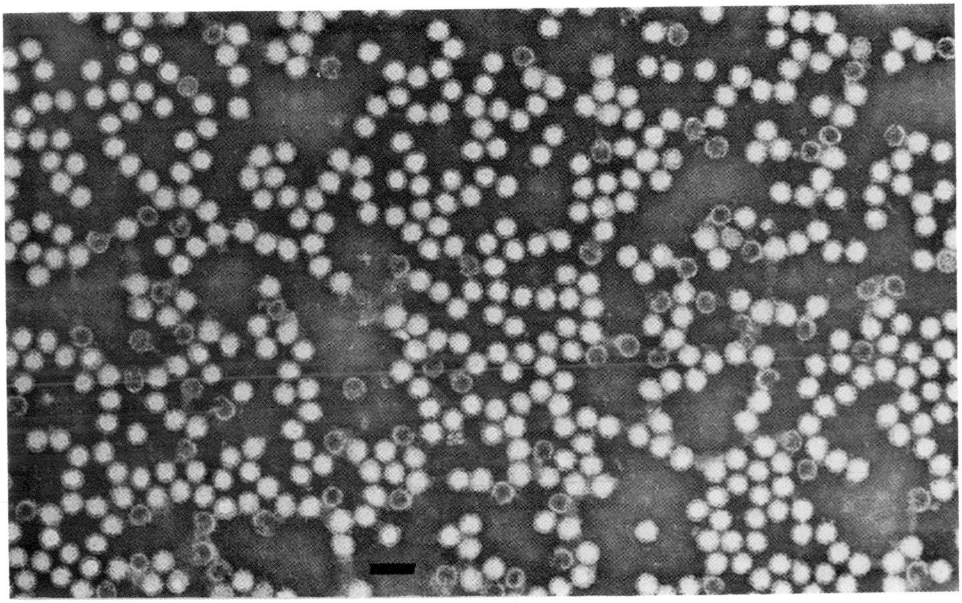

Fig. 5.4. Particles of broad bean wilt fabavirus; bar scale=50 nm.

Species have isometric particles, sometimes hexagonal in appearance, c. 28 nm in diameter (Fig. 5.5) which sediment as three components with sedimentation coefficients of 49–56 S ("top" component; T), 86–128 S ("middle" component; M) and 115–134 S ("bottom" component; B); have buoyant densities in CsCl of 1.28 (T), c. 1.5 (M), and 1.52 (B) g cm^{-3}; and an M_r ($\times 10^6$) of 3.3 (T), 4.6–5.8 (M), and 6 (B).

The capsid contains 60 subunits, each of a single polypeptide of 55–60 kDa. T particles contain no nucleic acid, but M and B particles contain 27–40% and 42–46% by weight of linear positive-sense ssRNA of, respectively, 3.9–7.2 kb and 7.2–8.4 kb. The 5′ terminus of each has a covalently linked VPg (3–6 kDa) and a polyadenylated 3′ terminus. Some virus isolates also contain one or two satellite RNAs (up to 1.4 or 0.46 kb) which, when present, can greatly affect symptom severity in infected plants. Both genomic RNAs are necessary for infection. Each RNA is translated into a polyprotein that is subsequently cleaved into a polymerase, a proteinase, VPg and an NTB-binding protein and the smaller the coat protein and, possibly, a movement protein.

The viruses have wide natural host ranges, including many ornamental species. Virus particles occur in cytoplasm, tubules, and in plasmodesmata.

Infected cells also contain cytoplasmic vesiculated bodies which are possibly sites of virus replication and/or assembly. The viruses are each transmitted by one or more species of longidorid nematodes (especially *Xiphinema* and *Longidorus* species); nematodes, especially if non-feeding, can remain viruliferous for several weeks but cannot transmit after molting. The viruses are also seed-borne in many hosts.

The viruses have wider taxonomic affinities with the *Picornaviridae* and *Potyviridae*.

Type virus: tobacco ringspot virus (TobRSV)

Diagnostic species: *Chenopodium amaranticolor*, *C. quinoa*, necrotic or chlorotic lesions; *N. glutinosa, N. tabacum* necrotic lesions, systemic ringspots

Propagation species: *Cucumis sativus*, *Nicotiana tabacum*

Purification: Rezaian et al. (1973)

Serology: Strongly immunogenic

Properties: A_{260}/A_{280} of separated components 0.91 (T), 1.37 (M), and 1.88 (B)
Specific absorbance at 260 nm=7.0

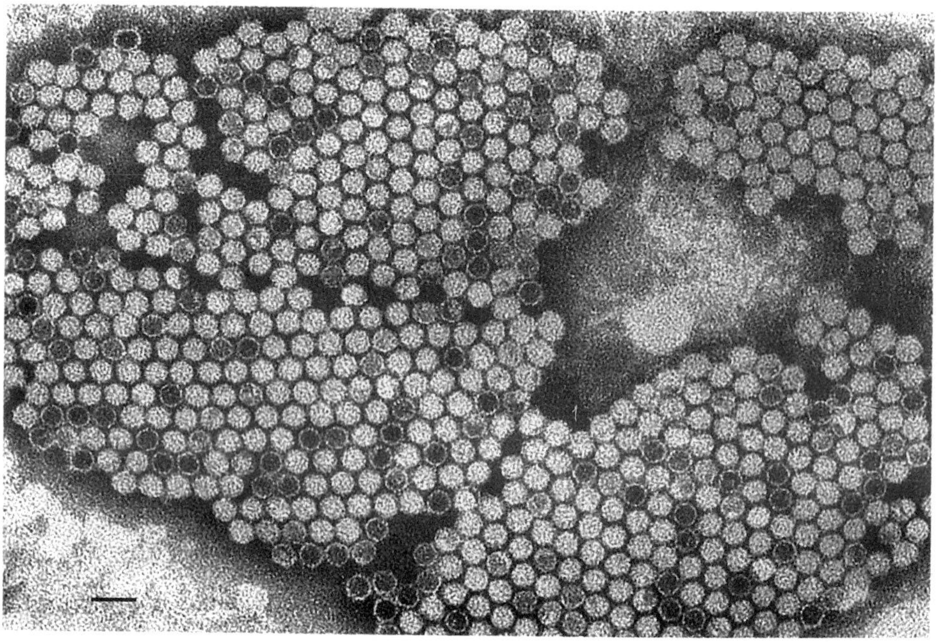

Fig. 5.5. Particles of arabis mosaic nepovirus; bar scale=50 nm.

Geminiviridae family

The family name is derived from the Latin *gemini* (twins) which describes the characteristic occurrence of particles in pairs.

Subgroup III genus — The type species of the genus is bean golden mosaic virus (BGMV). Other viruses infecting ornamental species include: abutilon mosaic virus (AbMV), euphorbia mosaic virus (EuMV), honeysuckle yellow vein mosaic virus (HYVMV), jatropha mosaic virus (JMV), and tobacco leaf curl virus (TLCV).

The species have geminate, incomplete icosahedral particles c. 18×30 nm (Fig. 5.6) which sediment as a single component with a sedimentation coefficient of 70 S. The capsid (polypeptide size of 27–30 kDa) of each particle encapsidates one molecule of ssDNA (2.5–2.8 kb); both encapsidated and complementary strands of the genome encode genes. The larger of the ssDNAs encodes four genes (one on the viral and three on the complementary strand) and the smaller for two genes (one on each strand). The larger ssDNA encodes the coat protein and replication enzymes and the smaller the movement protein. Genomic ssDNA replicates via dsDNA and particles accumulate in large numbers in nuclei.

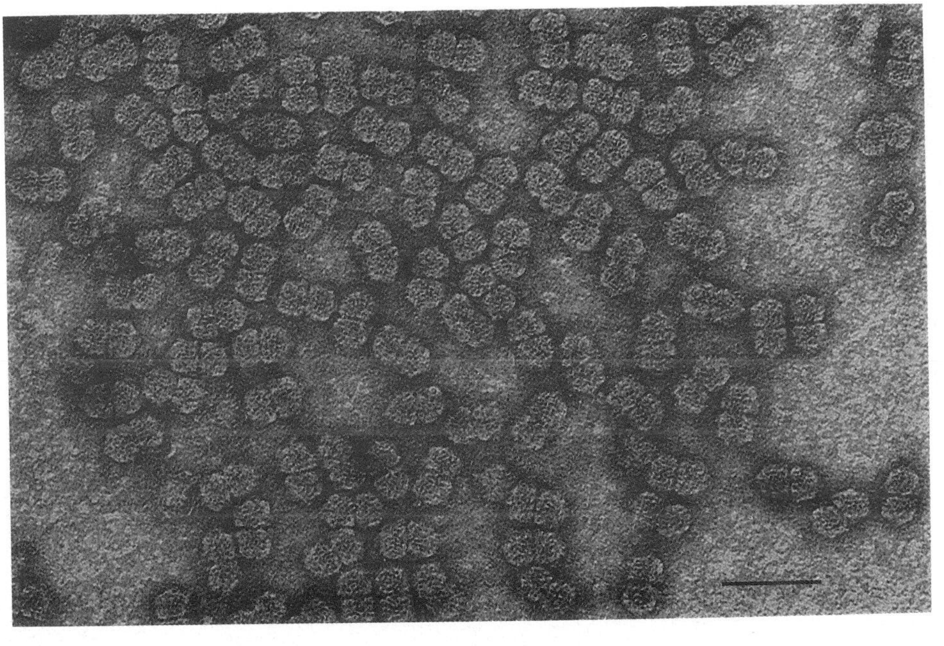

Fig. 5.6. Particles of maize streak bigeminivirus (subgroup 3 genus); bar scale=50 nm. Micrograph courtesy of Rothamsted Experimental Station, Harpenden, UK.

Individual viruses have narrow host ranges and infect only dicotyledonous species. The viruses are transmitted in the persistent manner by the whitefly *Bemisia tabaci*.

Type virus: bean golden mosaic virus (BGMV)

Diagnostic species: *Phaseolus vulgaris* cv. Top Crop, *P. lunatus*, systemic vein chlorosis and golden mosaic; *P. vulgaris* Red Mexican 35, local lesions

Propagation species: *Phaseolus vulgaris* cv. Top Crop

Purification: Goodman et al. (1977)

Serology: immunogenic

Properties: A_{260}/A_{280} = 1.4

Specific absorbance at 260 nm=6.3

Partitiviridae family

The family name is derived from the Latin *partitus* (divided) which describes the bipartite genomes of members.

Alphacryptovirus genus — The generic designation of the genus is derived from its previous description as subgroup 1 of the cryptovirus group, so named because its constituent viruses cause no obvious diseases and can thus be described as cryptic. The type species is white clover cryptic virus 1 (WCCV1). Possible members infecting ornamental species are carnation cryptic 2, garland chrysanthemum cryptic, and poinsettia cryptic viruses.

The species have isometric virus particles 30 nm in diameter (Fig. 5.7) which have a buoyant density in CsCl of 1.392 g cm^{-3} and consist of a capsid (subunits of c. 55 kDa) encapsidating two molecules of linear dsRNA of 1.7 and 2.0 kbp. The larger genomic dsRNA of the type species encodes the virion-associated polymerase and the smaller the coat protein.

Individual viruses have narrow host ranges and usually occur in very low concentration in cytoplasm and/or nuclei of hosts which remain symptomless. The viruses are transmitted by seed or pollen but not by grafting.

Type virus: white clover cryptic virus 1 (WCCV1)

Diagnostic species: None, occurs smptomlessly

Propagation species: *Trifolium repens*

Purification: Boccardo et al. (1985)

Serology: Strongly immunogenic

Properties: A_{260}/A_{280} = 1.31

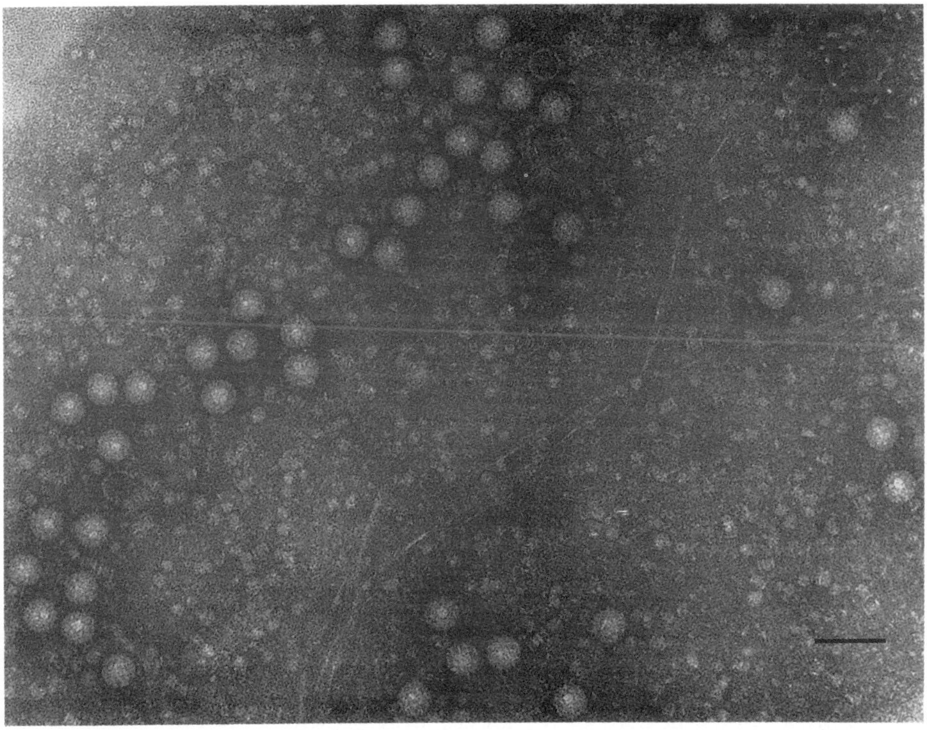

Fig. 5.7. Particles of white clover 1 alphacryptovirus; bar scale=50 nm. Micrograph courtesy Rothamsted Experimental Station, Harpenden, UK.

Genera with tripartite ssRNA genomes

Bromoviridae family

The name is derived from *bro*me *mo*saic virus, the type species of one (*bromovirus*) of the four genera included in the family.

1. *Alfamovirus* genus — The generic designation is an acronym of the type and only member of the genus, *al*falfa *mo*saic *virus* (AMV).

The species has virus particles which are mostly bacilliform (Fig. 5.8) with a diameter of 18 nm but lengths of four main sizes ranging from 30 to 60 nm (designated components Ta, Tb, M, and B). Tb, M, and B particles are necessary for infection, but Ta (and other minor spheroidal components) are not.

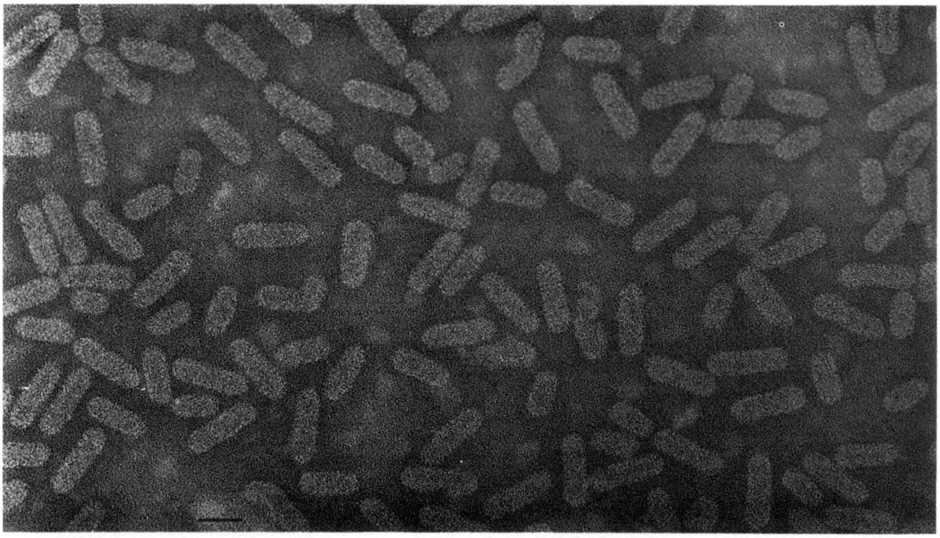

Fig. 5.8 Particles of alfalfa mosaic alfamovirus; bar scale=50 nm. Micrograph courtesy Rothamsted Experimental Station, Harpenden, UK.

The particles have sedimentation coefficients of 66 (Ta), 73 (Tb), 82 (M), and 94S (B); a mean buoyant density for all components of 1.28 g cm^{-3} and an M_r ($\times 10^6$) of 3.8 (Ta), 4.3 (Tb), 5.2 (M), and 6.9 (B). The A 260/280 absorption ratio of all components is 1.75, and their extinction coefficients are 4.7 (Ta), 4.8 (Tb and M), and 5.1 (B). All particle types are precipitated below pH 6. The capsid polypeptide of most strains contains c. 220 amino acids with a total weight of 24.3 kDa.

The virus contains four ssRNAs designated RNA 1 (3.64 kb), RNA 2 (2.60 kb), RNA 3 (2.04 kb), and RNA 4 (0.88 kb). RNAs 1-4 are separately encapsidated in, respectively, components B, M, Tb, and Ta. All the RNAs are capped at the 5′ terminus but are not 3′ polyadenylated. RNA 1 and RNA 2 each encode single polypeptides of 110–126 kDa and 90–95 kDa, respectively, both of which function in viral RNA synthesis. RNA 3 encodes a movement protein of c. 30 kDa and RNA 4 the capsid protein of 24.3 kDa.

The virus occurs worldwide in a very wide range of plant species, causing no conspicuous symptoms in many, but severe diseases in some. Cytoplasm of infected plants sometimes contains particles aggregated into whorls or rafts. The virus is transmitted in the non-persistent manner by c. 20 aphid species and through seed in some species.

Type virus: alfalfa mosaic virus (AMV)
Diagnostic species: *Chenopodium amaranticolor*, *C. quinoa*, chlorotic lesions, systemic chlorotic and necrotic flecks; *Nicotiana tabacum*, necrotic or chlorotic lesions, systemic mottling (some strains)
Propagation species: *Nicotiana glutinosa*
Purification: van Vloten-Doting and Jaspers (1972)
Serology: poorly immunogenic
Properties: A_{260}/A_{280} = 1.7–1.8 (unfractionated preparation)
Specific absorbance at 260 nm=4.7 (Ta), 4.8 (M+Tb), 5.1 (B)

2. *Ilarvirus* genus — The generic name is derived from an earlier description of the better-known species as *i*sometric *la*bile *r*ingspot *viruses*. The type species is tobacco streak virus (TSV). Other viruses infecting ornamental plant species include: apple mosaic virus (ApMV), prunus necrotic ringspot virus (PNRSV), lilac ring mottle virus (LRMV), and hydrangea mosaic virus (HdMV).

Species have several types of quasi-isometric or bacilliform particles 26–35 nm in diameter (Fig. 5.9) which are readily disrupted by neutral chloride salts and sodium dodecyl sulphate. They sediment as several components with sedimentation coefficients of 80–120 S, but all have a buoyant density in CsCl of 1.36 g cm^{-3}. The capsid has 180 structural subunits, each of 24 kDa. Particles contain three genomic and one subgenomic ssRNA molecules. RNA 1 (2.94 kb) and RNA 2 (2.77 kb) are encapsidated in separate particles, and RNA 3 (2.04 kb) and the subgenomic RNA 4 (0.85 kb) are contained together in the same particle. The 5′ termini of all are capped, but their 3′ termini are not polyadenylated. RNA 1 and RNA 2 each encode single polypeptides of 110–126 kDa and 90–95 kDa, respectively, both of which are involved in viral RNA synthesis. RNA 3 encodes a polypeptide (movement protein) of 30–35 kDa, and its subgenomic RNA 4 encodes the capsid protein. The genomic RNAs replicate via full length negative-sense RNAs in cytoplasm in association with membranes.

Some ilarviruses occur predominantly in woody plants, but the type species also infects a wide range of herbaceous species. The viruses are commonly seed-borne and some also transmitted via pollen; pollen transmission can be facilitated by thrips.

Type virus: tobacco streak virus (TSV)
Diagnostic species: *Beta patellaris*, necrotic lesion; *Vigna unguiculata*, chlorotic or necrotic lesions; *Nicotiana tabacum* cv. Turkish, upper leaves become toothed
Propagation species: *Nicotiana tabacum*, *Cucumis sativus*

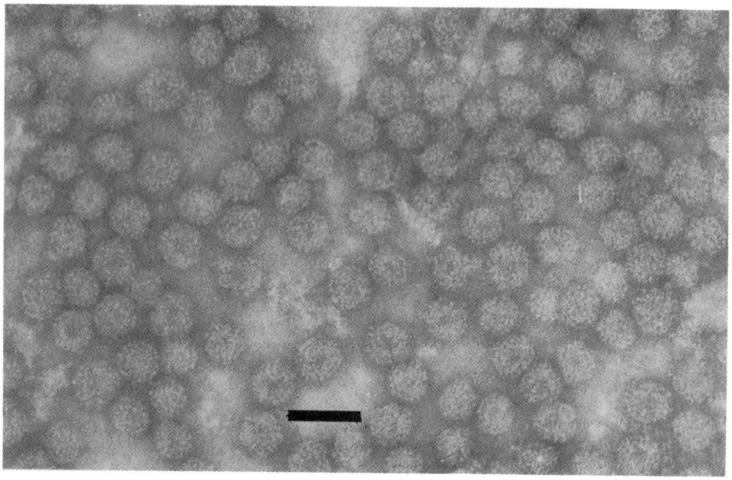

Fig. 5.9. Particles of tobacco streak ilarvirus; bar scale=50 nm.

Purification: Fulton (1967)
Serology: Strongly immunogenic
Properties: A_{260}/A_{280} = 1.56 (unfractionated preparation)
Specific absorbance at 260 nm=5.1 (unfractionated preparation)

3. *Cucumovirus* genus — The generic name is an acronym derived from the name of its type species, *cucu*mber *mo*saic *virus* (CMV). Another virus infecting ornamental plant species is tomato aspermy virus (TAV).

The species have isometric particles mostly c. 30 nm in diameter (Fig. 5.10). They sediment as a single component with a sedimentation coefficient of 100 S, have a buoyant density in CsCl of 1.37 g cm^{-3} and an M_r of 6×10^6, and are readily disrupted by neutral chloride salts and anionic detergents. The capsid protein has 180 structural subunits, each of 26 kDa. The particles contain three genomic and one subgenomic linear, positive-sense ssRNA molecules. Like those of other genera in the family, RNA 1 (3.41 kb) and RNA 2 (3.06 kb) are contained in separate particles and RNA 3 (2.19 kb) and its subgenomic RNA 4 (1.03 kb) together in a third type of particle. The 5′ termini of all RNAs are capped, but the 3′ termini are not polyadenylated. RNA 1 encodes a polypeptide of 111 kDA and RNA another of 94 kDa; both are involved in viral RNA synthesis. RNA 3 encodes a movement protein of 30 kDa and the subgenomic RNA 4, the coat protein. Virus particles of some isolates also contain satellite RNAs of 0.33-0.39 kb which, when present, can exacerbate or ameliorate disease

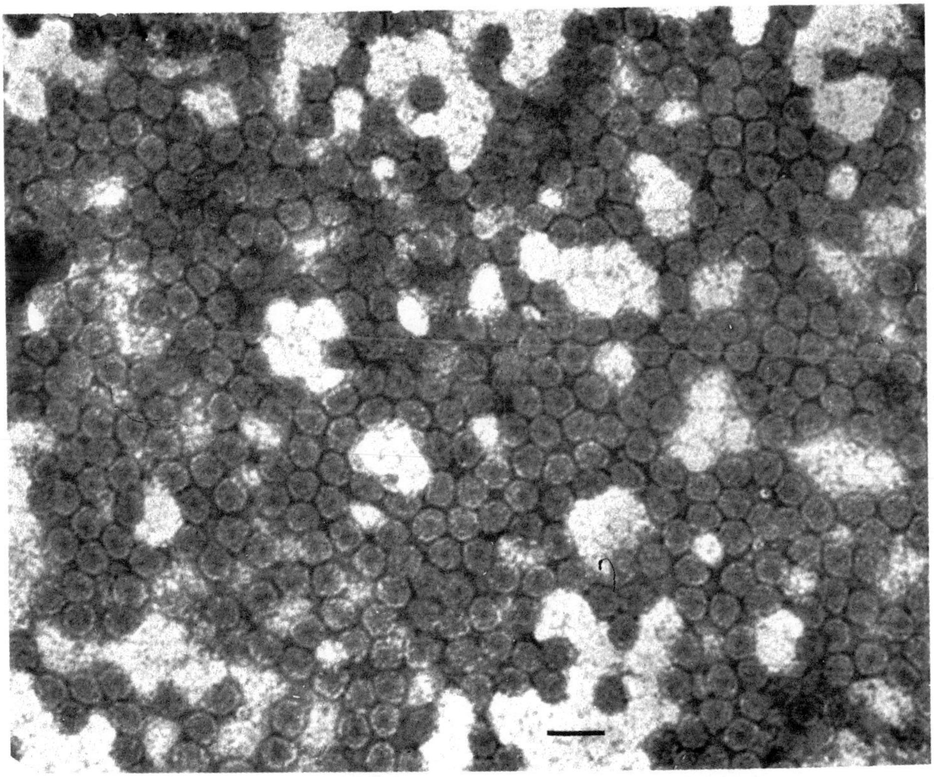

Fig. 5.10. Particles of cucumber mosaic cucumovirus; bar scale=50 nm.

symptoms. Virus particles assemble in cytoplasm and occur rarely in nuclei and vacuoles.

Cucumber mosaic virus has a very extensive natural host range (probably more than 1,000 species), but that of tomato aspermy virus is somewhat more restricted. Both viruses occur worldwide and are transmissible in the non-persisent manner by aphids and through seed of some plant species.

Type virus: cucumber mosaic virus (CMV)
Diagnostic species: *Chenopodium amaranticolor*, *C. quinoa* chlorotic lesions; *Cucumis sativus*, systemic mosaic; *Vigna unguiculata*, necrotic lesions
Propagation species: *Nicotiana glutinosa*, *N. tabacum*
Purification: Scott (1963); Lot et al. (1972)
Serology: poorly immunogenic aldehyde fixation improves immunogenicity.
Properties: A_{260}/A_{280} = 1.3 (unfractionated)
Specific absorbance at 260 nm=5

Bunyavaridae family

The family name is derived from *Bunya*mwera, Uganda (where the vertebrate-infecting type species occurs) and *virus*.

1. *Tospovirus* genus — The generic designation is an acronym of the type member *to*mato *spo*tted wilt *virus* (TSWV). Another member infecting ornamental species is impatiens necrotic spot virus (INSV).

The species have quasi-isometric or pleomorphic particles 80–120 nm in diameter (Fig. 5.11) which have a sedimentation coefficient of 350–500 S, a buoyant density in CsCl of 1.20 g cm^{-3}, and an M_r of c. 350×10^6. Particles contain c. 75% protein, 2% carbohydrate, 1–2% ss-RNA, and 20–30% lipid. They are enveloped in a bilayered lipid membrane c. 5 nm thick, and they contain four structural polypeptides; two glycoproteins designated G1 (78 kDa) and G2 (58 kDa) are external, another (designated N) of c. 30 kDa is the nucleocapsid protein, and the fourth (200 kDa) is the large (L) transcriptase protein. Particles contain three ss-RNA molecules (designated Large, Medium and Small) of c. 8.90 (L), 4.82 (M), and 2.92 (S) kb, none of which is poly-

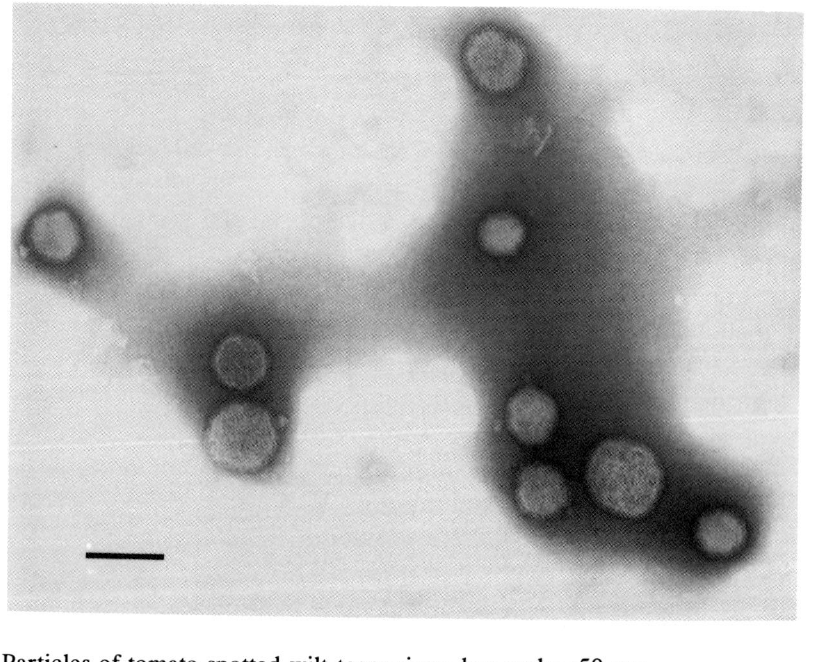

Fig. 5.11. Particles of tomato spotted wilt tospovirus; bar scale=50 nm.

adenylated. RNAs S and M are ambisense and encode non-structural proteins in the positive-sense sequences. The virus-complementary RNA of the M molecule encodes the two glycoproteins (G1 and G2), and that of the mRNA of the S molecule encodes the nucleocapsid (N) protein. Proteins encoded in RNA-S are translated from two subgenomic RNAs which, unlike genomic RNAs, have a cap structure at their 5′ terminus. RNA-S has a stable hairpin loop structure in the intergenic region at which both subgenomic RNAs terminate.

The type virus has a very extensive natural host range. In infected cells, virus particles frequently occur in cisternae of the endoplasmic reticulum and nucleocapsid material often accumulates in dense masses in cytoplasm. Species within the genus are transmitted by up to 13 species of thrips which can only acquire virus at the nymph stage but then retain virus for life. The Western Flower Thrips (*Frankliniella occidentalis*) is a very efficient vector.

Type virus: tomato spotted wilt virus (TSWV)

Diagnostic species: *Petunia hybrida*, necrotic lesions; *Nicotiana glutinosa*, necrotic lesions, systemic necrotic patterns

Propagation species: *Nicotiana glutinosa*, *Tropaeolum majus*

Purification: van den Hurk et al. (1977)

Serology: Immunogenic; serological variants

Plant Virus Genera Not Yet Assigned to Families

Genera with monopartite ssRNA or dsDNA genomes

1. *Tymovirus* genus — The name of the genus is derived from the name of its type species, *t*urnip *y*ellow *mo*saic *virus* (TYMV).

Although this genus contains several agriculturally important viruses, only one unassigned species (poinsettia mosaic virus) occurs in ornamental species. Species have isometric particles c. 29 nm in diameter (Fig. 5.12) which sediment as two components (“Top” and “Bottom”) with sedimentation coefficients of 54 (T) or 115 S (B), and have buoyant densities in CsCl of 1.29 (T) or 1.42 (B) g cm^{-3} and an M_r of 3.6×10^6 (T) or 5.6×10^6 (B). The capsid consists of 180 protein subunits, each of 20 kDa. “T” component particles contain no nucleic acid, but those of the “B” component each contain 35% by weight of a single molecule of linear positive-sense ssRNA of 6.3 kb with a 5′ cap structure and a tRNA-like structure at its 3′ terminus. Some particles also contain a subgenomic ssRNA of 0.70 kb (corresponding to the 3′ genomic RNA) which encodes

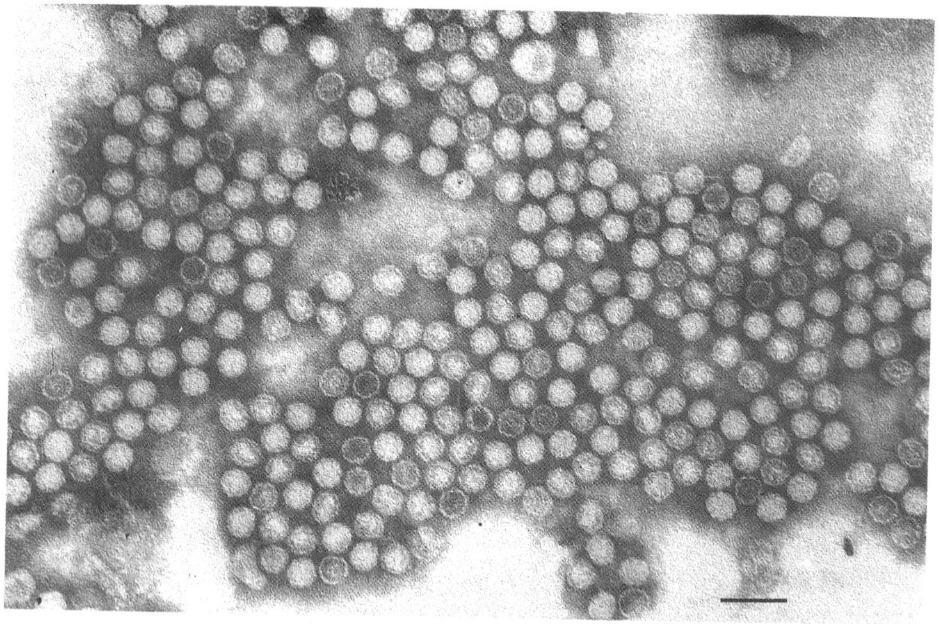

Fig. 5.12. Particles of turnip yellow mosaic tymovirus; bar scale=50 nm.

the capsid protein. Both genomic and subgenomic RNAs contain 32–41% cytidylic acid. Genomic RNA contains 3 ORFs which encode proteins of 150 and 195 kDa.

Individual viruses have restricted natural host ranges; poinsettia mosaic virus, considered to be a possible member of the genus, is known to occur only in poinsettias. Most species induce clumping of chloroplasts; the peripheries of the latter have double-membraned vesicles which contain a polymerase and are probably the sites of viral replication. Capsid protein is produced in cytoplasm and assembles with genomic RNA at the orifices of the vesicles.

Type virus: turnip yellow mosaic virus (TYMV)

Diagnostic species: *Brassica campestris* spp. *pekinensis*, bright yellow, yellow-green mosaic, or mild mosaic

Propagation species: *Brassica campestris* spp. *pekinensis*

Purification: Matthews (1960)

Serology: Strongly immunogenic

Properties: A_{260}/A_{280} = 0.81 (T), 1.51 (B)
Specific absorbance at 260 nm=0.96 (T), 9.6 (B)

2. *Necrovirus* genus — The generic designation is derived from tobacco *necrosis virus* (TNV), the type species. The genus also contains two unassigned or possible species (carnation yellow stripe virus and lisianthus necrosis virus).

Species have isometric particles c. 28 nm in diameter (Fig. 5.13) which have a sedimentation coefficient of 180 S, a buoyant density in CsCl of 1.40 g cm^{-3}, and M_r of 7.6×10^6. The capsid consists of 180 protein subunits, each of 23–33 kDa, which encapsidates a single molecule of genomic ssRNA of 3.66–3.71 kb. Of three virus-specific dsRNAs occurring in infected plants, one 3.76 kbp is probably the replicative form of genomic RNA and the other two (1.50 and 1.35 kbp) replicative forms of subgenomic RNAs.

The type member has a wide natural host range, but in many susceptible species, the virus is restricted to roots. The virus is transmitted naturally externally on zoospores of the chytrid fungus *Olpidium brassicae* and experimentally by mechanical inoculation. Virus particles are often found *in vivo* as crystal-like aggregates.

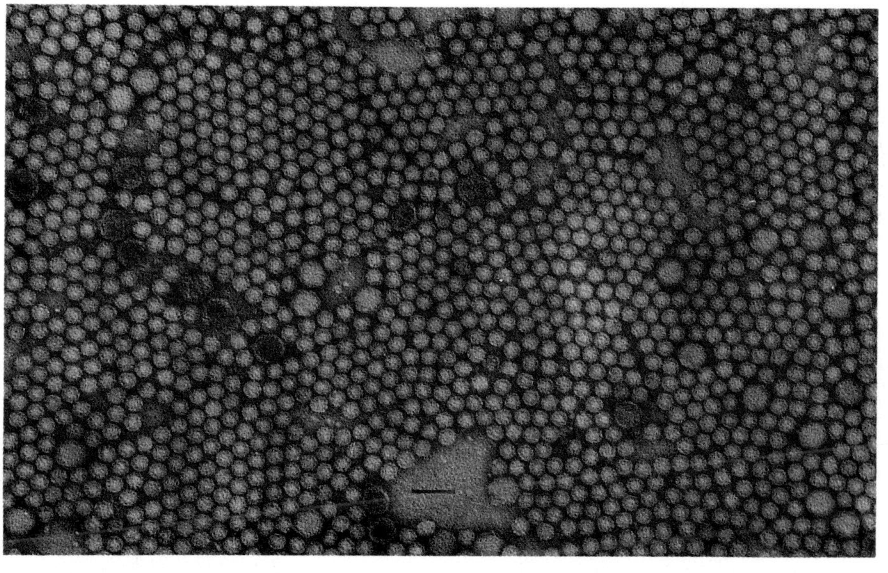

Fig. 5.13. Particles of tobacco necrosis necrovirus (larger) in mixture with tobacco necrosis satellite virus (smaller); bar scale=50 nm. Micrograph courtesy Rothamsted Experimental Station, Harpenden, UK.

Type virus: tobacco necrosis virus (TNV)
Diagnostic species: *Chenopodium amaranticolor*, *C. quinoa*, *Cucumis sativus*, *Nicotiana tabacum* — local lesions
Propagation species: *Nicotiana tabacum*, *Phaseolus vulgaris*, *Vigna unguiculata*
Purification: Kassanis (1970)
Serology: Strongly immunogenic
Properties: $A_{260}/A_{280} = 1.5$
Specific absorbance at 260 nm=5–5.5

3. *Tobamovirus* genus — The name of the genus is an acronym derived from the name of its type species, *to*bacco *mo*saic *virus* (TMV). Other tobamoviruses infecting ornamental plant species include: frangipani mosaic virus (FrMV), odontoglossum ringspot virus (ORSV), Sammons Opuntia virus (SOV), and tomato mosaic virus (ToMV).

Species included in this genus have rod-shaped particles mostly measuring c. 18×300 nm (Fig. 5.14) which sediment as a single component with a sedimentation coefficient of c. 195 S and have a buoyant density in CsCl of 1.33 g cm^{-3} and an M_r of 40×10^6. The capsid contains c. 2,000 protein subunits, each of M_r 17.5, which are helically arranged with a pitch of 2.3 nm. Each particle contains a single molecule of positive-sense ssRNA of 6.395 kb which at its 5′ and 3′ termini, respectively, has a cap structure with an affinity for histidine and a tRNA-like amino acid acceptor; particles of some species also contain a subgenomic RNA which encodes the viral replicase. The genomic RNA replicates in cytoplasm via a replicative intermediate or replicative form; three proteins of 180, 126, and 30 (the transport protein) kDa are produced. Capsid protein is synthesised via a subgenomic RNA, the nucleotide sequence of which is homologous with the 3′ terminus of genomic RNA; in some species this monocistronic mRNA is encapsidated with genomic RNA.

Each species has a narrow or moderately wide natural host range. The viruses characteristically induce in infected plants cytoplasmic inclusions, the main components of which are the 126 kDa protein and virus particles which often occur in cystalline arrays; the 30 kDa transport protein is mainly localized in plasmodesmata. The viruses are characteristically very stable *in vitro* and are transmitted readily by mechanical inoculation; some are also seed-borne.

Homologies in non-structural proteins induced by the type species suggest that the genus has wider affinities with the *Bromoviridae*, *Tombusviridae*, tobravirus genus, and sindbis virus.

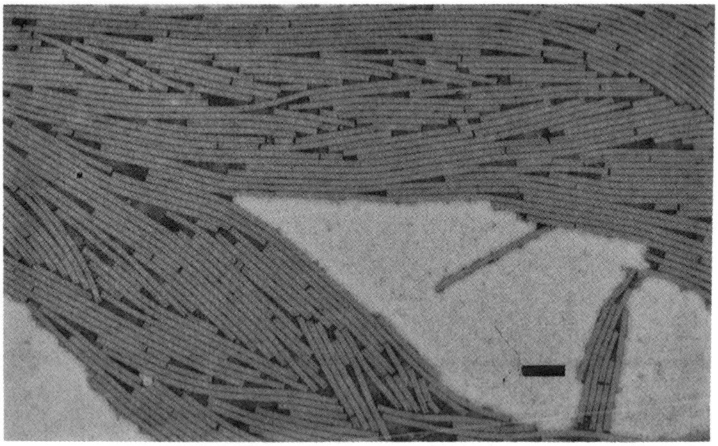

Fig. 5.14. Particles of tobacco mosaic tobamovirus; bar scale=100 nm.

Type virus: tobacco mosaic virus (TMV)
Diagnostic species: *Nicotiana glutinosa*, necrotic local lesions; *N. tabacum*, systemic mosaic
Propagation species: *Nicotiana tabacum* cv. Samsun
Purification: Gooding and Hebert (1967)
Serology: Strongly immunogenic
Properties: A_{260}/A_{280} = 1.19–1.21
Specific absorbance at 260 nm=2.8-3.2

4. *Potexvirus* genus — The name of the genus is an acronym derived from the name of the type species, *po*tato *X virus* (PVX).

Better known viruses in the genus that infect ornamental species include: cactus X virus (CVX), commelina X virus (CoVX), cymbidium mosaic virus (CyMV), hydrangea ringspot virus (HRSV), lily X virus (LVX), narcissus mosaic virus (NMV), nerine X virus (NVX), tulip X virus (TVX), viola mottle virus (VMV), and white clover mosaic virus (WCMV).

Species have stable flexuous filamentous particles measuring 13×470–580 nm (Fig. 5.15) which sediment as a single component with a sedimentation coefficient of 115-130 S, have a buoyant density in CsCl of 1.31 g cm^{-3} and M_r of c. 3.5×10^6. The capsid protein subunits, each of 18–23 kDa, are helically arranged with a helical pitch of 3.4 nm. Each particle of the type species contains c. 6% by weight of positive-sense ss-RNA of 6.44 kb which has a cap structure

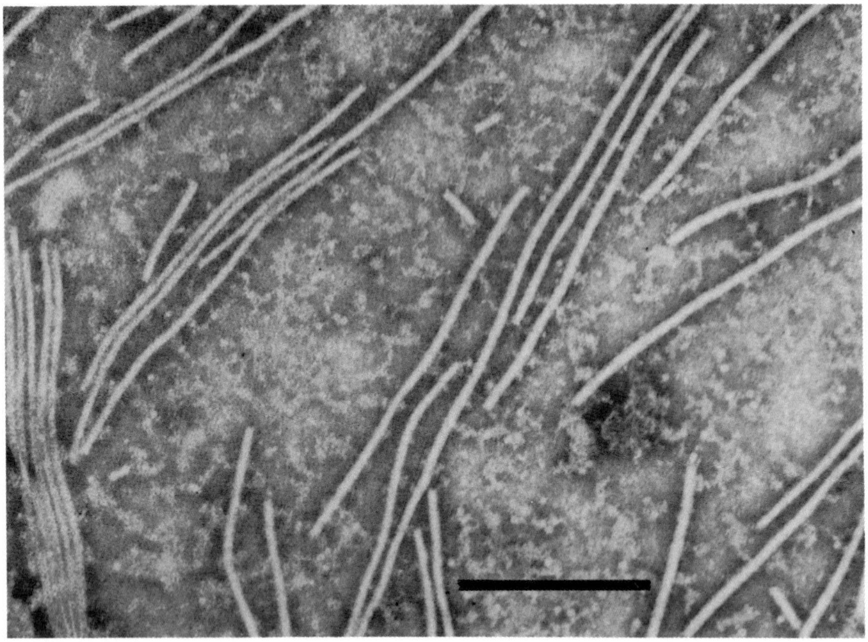

Fig. 5.15. Particles of potato X potexvirus; bar scale=250 nm.

and a polyadenylated tail at its 5′ and 3′ termini, respectively. The genomic RNA contains five ORFs; coat protein is encoded in a subgenomic RNA.

Most viruses have narrow natural host ranges; although they have no known specific vectors, they are usually very infectious and can be transmitted from infected to healthy plants by contact. Infected plants contain cytoplasmic inclusions, the major component of which is virus particles.

Type virus: potato X virus (PVX)
Diagnostic species: *Gomphrena globosa*, white necrotic lesions with red halo
Propagation species: *Nicotiana glutinosa*
Purification: Francki and McLean (1968)
Serology: Strongly immunogenic
Properties: A_{260}/A_{280} = 1.2.
Specific absorbance at 260 nm=2.97

5. *Carlavirus* genus — The name of the genus is an acronym derived from the name of the type species, *car*nation *la*tent *virus* (CarLV).

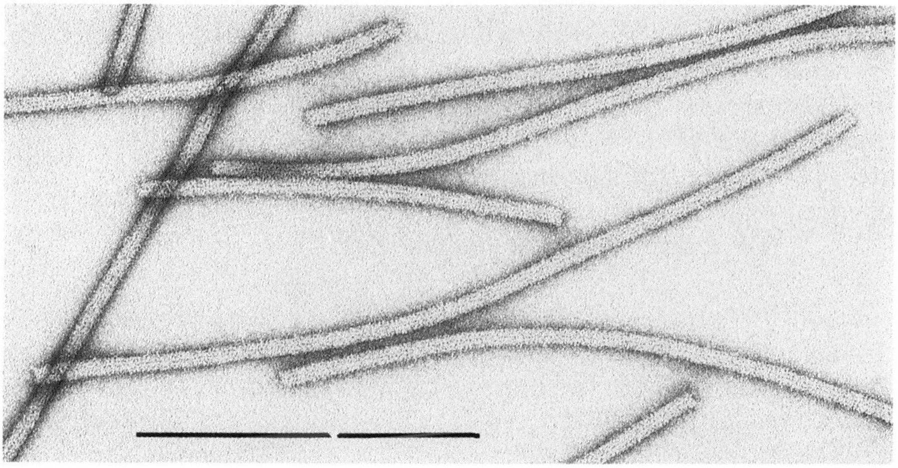

Fig. 5.16. Particles of carnation latent carlavirus; bar scale=250 nm. Micrograph courtesy of R.G. Milne, Turin, Italy.

Well-known viruses in the genus that infect ornamental species include: cactus 2 virus (CV2), chrysanthemum B virus (CVB), helenium S virus (HVS), honeysuckle latent virus (HuLV), hydrangea latent virus (HdLV), kalanchoë latent virus (KLV), lilac mottle virus (LiMV), lily symptomless virus (LSV), nerine latent virus (NeLV), and passiflora latent virus (PLV).

Species have slightly flexuous particles measuring 12×610–700 nm (Fig. 5.16) which sediment as a single component with a sedimentation coefficient of 147–176 S and have a buoyant density in CsCl of 1.3 g cm^{-3} and an M_r of 60×10^6; some species also have shorter particles which possibly encapsidate subgenomic RNAs. The capsid protein subunits, each of 31–36 kDa, are helically arranged with a helical pitch of 3.4 nm. Each particle contains 5.6% by weight of positive sense ssRNA of 7.4–7.7 kb which is polyadenylated at its 3′ terminus but lacks a VPg or cap structure at its 5′ terminus. Some species also have two subgenomic RNAs (2.1–3.3 kb and 1.3–1.6 kb) which are possibly encapsidated in shorter particles. Genomic RNA contains six large ORFs which code for polypeptides of 223 (probably the viral replicase), 25 (the transport protein), 34 (coat protein), and 11 kDa (possibly facilitates aphid transmissibility).

Individual viruses have restricted natural host ranges, but some infect a wide range of experimental hosts. The viruses are transmitted by aphids in the non-persistent manner; three viruses naturally occurring in leguminous species are also seed-borne. All are mechanically transmissible.

Type virus: carnation latent virus (CarLV)
Diagnostic species: *Chenopodium amaranticolor*, chlorotic local lesions, systemic mottle
Propagation species: *Dianthus barbatus*, *Chenopodium amaranticolor*
Purification: Wetter and Paul (1961)
Serology: Strongly immunogenic
Properties: A_{260}/A_{280} = 1.37
Specific absorbance at 260 nm=2.1

6. *Capillovirus* genus — The generic name is derived from the latin description of virus particles (capillus=a hair) of members. The type species is apple stem grooving virus (ASGV). Possible species infecting ornamental plant species are lilac chlorotic leaf spot (LCLV) and nandina stem pitting (NSPV) viruses.

Species have filamentous particles measuring c. 12×640 nm which sediment as a single component with a sedimentation coefficient of 100 S. The capsid protein (27 kDa) is helically arranged and that of each particle encapsidates one molecule of positive-sense ssRNA of 6.5 kb, the RNA is polyadenylated at its 3′ terminus.

The species have restricted natural host ranges. Their vectors are unknown, but some are seed-borne.

Type virus: apple stem grooving virus (ASGV)
Diagnostic species: *Malus sylvestris* cv. Virginia Crab, elongated grooves; *Nicotiana glutinosa*, systemic yellow mosaic; *Phaseolus vulgaris*, purple brown spots
Propagation species: *Chenopodium quinoa*, *Nicotiana glutinosa*
Purification: see Lister (1970)
Serology: moderately immunogenic
Properties: A_{260}/A_{280} = 1.18

7. *Closterovirus* genus — The generic name is derived from the Greek *Kloster* (spindle or thread) which describes the very long thread-like particles of species. The type member is beet yellows virus (BYV). Other viruses infecting ornamental plant species are carnation necrotic fleck virus (CNFV) and, possibly, dendrobium vein necrosis virus (DVNV).

Species have very long flexuous filamentous particles measuring c. 12×1200–2000 nm (Fig. 5.17) which sediment as a single component with a sedimentation coefficient of 96–140 S and have a buoyant density in CsCl of c. 1.32 g cm^{-3} or in Cs_2SO_4 of c. 1.25 g cm^{-3} (but particles of most species are degraded by CsCl).

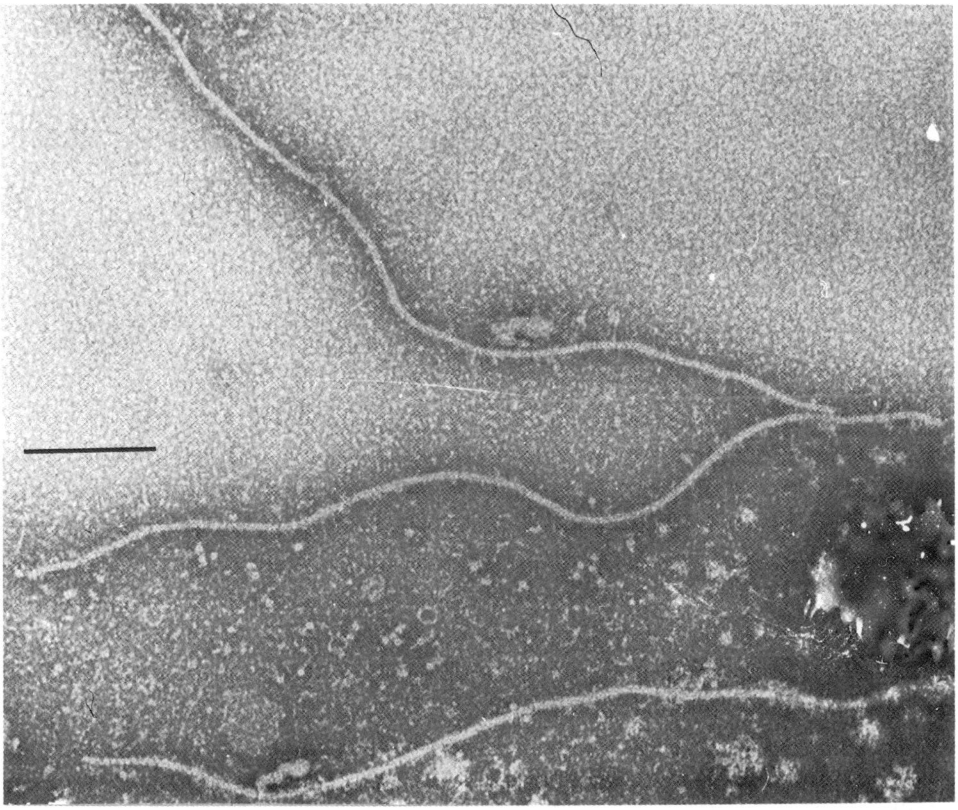

Fig. 5.17. Particles of carnation necrotic fleck closterovirus; bar scale=250 nm.

Capsid subunits each consist of a single polypeptide of 22–43 kDa and are helically arranged with a helical pitch of 3.4–3.8 nm. Particles each contain a single molecule of linear, positive sense ssRNA; that of the type species is c. 14.5 kb, but the size of other species is commensurate to particle length and ranges from 7.5–20 kb. The 5′ terminus is probably capped, but the 3′ end is not polyadenylated.

Genomic RNA of the type species contains at least eight ORFs. Replication, the strategy of which has yet to be determined, probably occurs in cytoplasm, possibly in association with membranous vesicles and vesiculated mitochondria. Nevertheless, some ORFs are probably expressed via subgenomic RNAs as indicated by the occurrence of dsRNAs in infected plants.

Individual virus species have restricted host ranges. They are transmitted in the semi-persistent manner by aphids (although some unassigned species are transmitted by whiteflies or mealybugs). The virus particles characteristically occur as aggregates in phloem cells or as clusters with membranous vesicles to form conspicuous fibrous masses.

Type virus: beet yellows virus (BYV)

Diagnostic species: *Beta vulgaris*, vein clearing; *Chenopodium foliosum*, stunting, distortion and vein-clearing

Propagation species: *Beta vulgaris*, *Montia perfoliata*

Purification: Horne et al. (1959)

Serology: Moderately immunogenic

Properties: A_{260}/A_{280} = 1.73

Specific absorbance at 260=2.2

8. *Caulimovirus* genus — The generic designation is an acronym of the type species, *cauli*flower *mo*saic *virus* (CaMV). Other viruses infecting ornamental species include carnation etched ring (CERV), dahlia mosaic (DaMV), and mirabilis mosaic (MMV) viruses and the possible species aquilegia necrotic mosaic and petunia vein-clearing viruses.

The species have isometric particles c. 50 nm in diameter (Fig. 5.18) which sediment as a single component with a sedimentation coefficient of 200–250 S, have a buoyant density in CsCl of 1.37 g cm^{-3} and an M_r of 20×10^6. The capsid subunits, each of 57 kDa, are phosphorylated and degrade rapidly *in vitro* mainly to polypeptides of 42 and 37 kDa.

Particles each contain one molecule of open circular dsDNA of 8.02–8.06 kbp which has one specific-site discontinuity in the transcribed strand (a) and two in the non-transcribed strand (b). The b strand is non-coding; the a strand has 6–8 ORFs and is transcribed in the nucleus to produce two major transcripts (19 and 35 S). The former, originating from ORF VI, is translated to a protein of 62 kDa which accumulates in characteristic intracellular inclusions; the 35 S transcript is possibly the mRNA of several other ORFs and is probably a template for replication of viral replication by reverse transcription. ORF V probably codes for the replicase, ORF IV for the coat protein ORF II for the protein conferring aphid transmissibility, and ORF 1 for the movement protein.

Each of the species has a narrow host range; infected plants contain spherical or ovoid cytoplasmic inclusions containing aggregates of the 62 kDa protein and accumulations of virus particles. The viruses are transmitted in the semi-persistent manner by aphids and experimentally by mechanical inoculation of sap.

Type virus: cauliflower mosaic virus (CaMV)

Diagnostic species: *Brassica oleracea*, systemic mosaic, vein yellowing

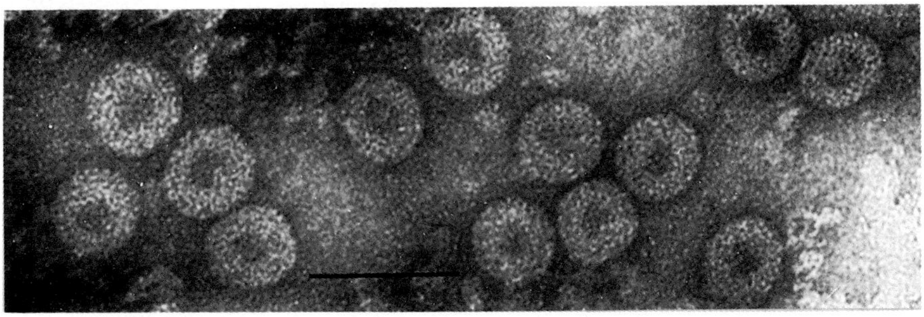

Fig. 5.18. Particles of cauliflower mosaic caulimovirus; bar scale=50 nm.

Propagation species: *Brassica campestris*
Purification: Gardner and Shepherd (1980)
Serology: Strongly immunogenic
Properties: A_{260}/A_{280} = 1.4
Specific absorbance at 260 nm=7

9. *Badnavirus* genus — The generic name is derived from an earlier description of *ba*re *DNA viruses*, but also indicates that they induce severe diseases (bad) in infected plants.

The type species is commelina yellow mottle virus (CoYMV). Other viruses infecting ornamental species include: canna yellow mottle virus (CaYMV), kalanchoë top-spotting virus (KTSV), schefflera ringspot virus (SRSV), aucuba ringspot virus (possible member), and yucca bacilliform virus (possible member).

The species have unenveloped bacilliform particles measuring c. 30×130 nm (Fig. 5.19) which sediment as a single component with a sedimentation coefficient of c. 220 S and have a buoyant density in CsCl of 1.37 g cm^{-3}. The capsid contains two polypeptides of 40 and 35 kDa. Each particle contains one molecule of open circular dsDNA with a single discontinuity in each strand; the genome of the type member has 7.49 kbp. The strategy of replication has yet to be determined but, like that of caulimoviruses, probably involves reverse transcription.

Individual species have very restricted natural and experimental host ranges which are often limited to a few species within a genus. Virus particles occur in the cytoplasm of all types of cell. Most viruses are transmitted in the semi-persistent manner by mealybugs and disseminated in vegetatively-propagated

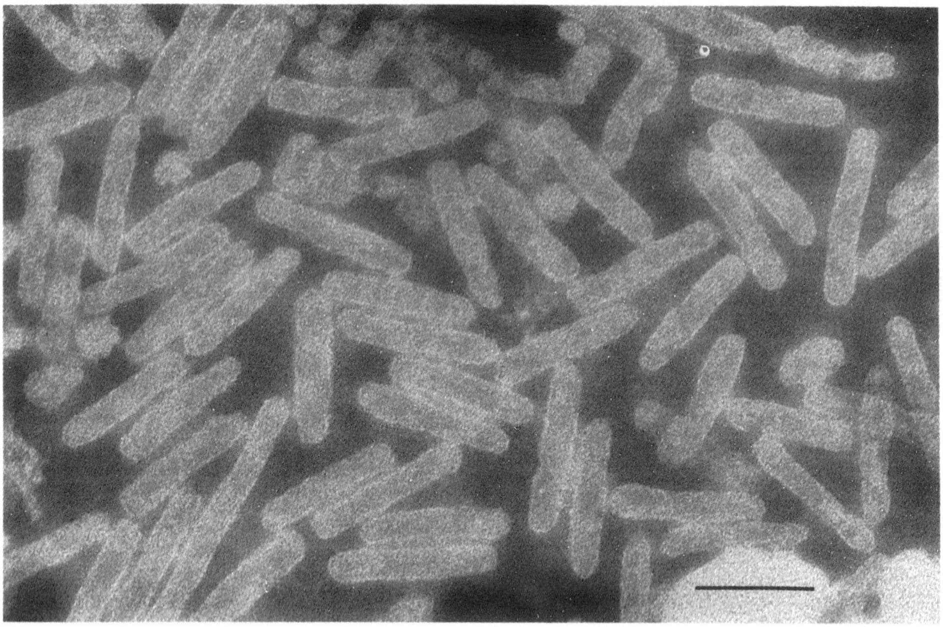

Fig. 5.19. Particles of cocoa swollen shoot badnavirus; bar scale = 100 nm.

plants. One (KTSV) is reported to be seed and/or pollen transmitted. The viruses occur mainly in tropical plant species.

Type virus: commelina yellow mottle virus (CoYMV)
Diagnostic species: *Commelina diffusa*, yellow mottling of leaves
Propagation species: *Commelina diffusa*
Purification: Lockhart and Khaless (1988)
Serology: immunogenic
Properties: $A_{260}/A_{280}=1.35$ (unaffected)

Virus genera with isometric particles and bipartite ssRNA genomes

1. *Dianthovirus* genus — The generic designation is derived from the generic name of *Dianthus caryophyllus*, the major host of the type species, carnation ringspot virus (CRSV). CRSV is the only one that infects ornamental plants.

Species have isometric particles 31–34 nm in diameter (Fig. 5.20) which have a sedimentation coefficient of 133 S, a buoyant density in CsCl of 1.37 g cm^{-3}, and an M_r of 7.1×10^6. The capsid, with subunits of 40 kDa, encapsidates two molecules of positive-sense genomic ssRNAs of c. 4.23 and 1.41 kb; the larger

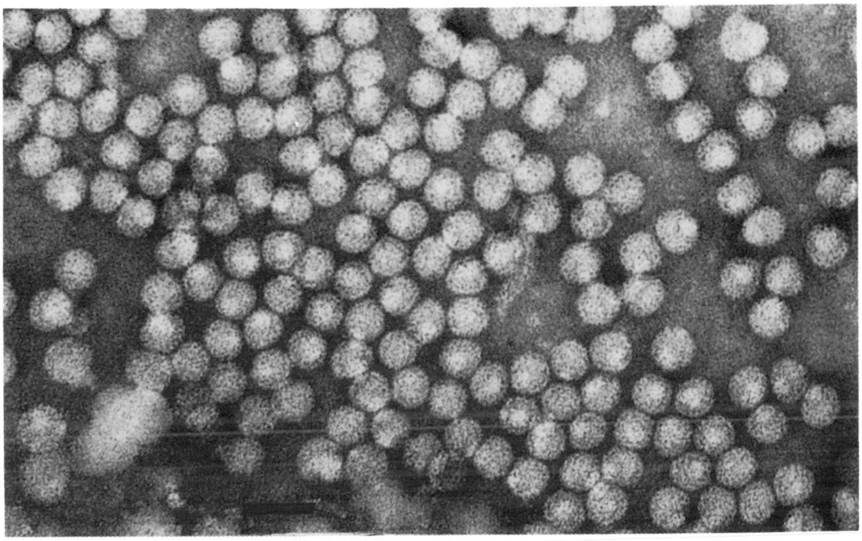

Fig. 5.20. Particles of carnation ringspot dianthovirus; bar scale=50 nm.

RNA encodes the coat protein (40 kDa) and polypeptides of 27 and 57 kDa and the smaller a polypeptide of 35 kDa. Both RNAs have a 5′ cap structure, but neither is 3′ polyadenylated or contains a VPg.

Infected plants contain three dsRNAs; two correspond to the genomic ssRNAs and one to a subgenomic RNA derived from the larger genomic RNA.

The viruses have moderately wide host ranges. Particles occur, both scattered and in aggregates, in cytoplasm of infected plants. One species (red clover necrotic mottle virus) is transmissible experimentally by zoospores of *Olpidium radicale*; others are transmissible experimentally by mechanical inoculation of sap.

Type virus: carnation ringspot virus (CRSV)

Diagnostic species: *Dianthus barbatus*, flecks and ringspots; *Gomphrena globosa*, necrotic rings followed by systemic flecking and mottling

Propagation species: *Nicotiana clevelandii*, *Phaseolus vulgaris*, *Vigna unguiculata*

Purification: Tremaine et al. (1976)

Serology: very immunogenic

Properties: A_{260}/A_{280} = 1.67 (corrected for light scattering)
Specific absorbance at 260 nm=6.46

2. *Enamovirus* genus — The generic designation is an acronym derived from the name of the type species, pea *ena*tion *mo*saic *virus* (PEMV). PEMV, the only species of the genus, has isometric particles c. 28 nm in diameter (Fig. 5.21) which sediment as two components ("Bottom" and "Top") with sedimentation coefficients of 112 (B) and c. 99 S (T) and an M_r of 5.7×10^6 (both B and T); the B component has a buoyant density in CsCl of 1.42 g cm^{-3}, but the T component is disrupted in caesium salts. The capsid consists of a major polypeptide of 22 kDa and a minor of 28 kDa which is associated with aphid transmissibility. Particles contain two molecules of genomic ssRNAs of 3.66 and 4.79 kb; those of some strains also contain a satellite RNA of 0.8 kb. Both genomic RNAs have a VPg (17.5 kDa) at their 5′ termini, but neither is 3′ polyadenylated. Virus particles occur in the nuclei of infected plants.

The larger genomic RNA encodes proteins of 88, 30, and 147 kDa, and the smaller the capsid protein.

The virus naturally infects numerous species in the Papilionaceae but few in other families. It is transmitted in the persistent manner by aphids.

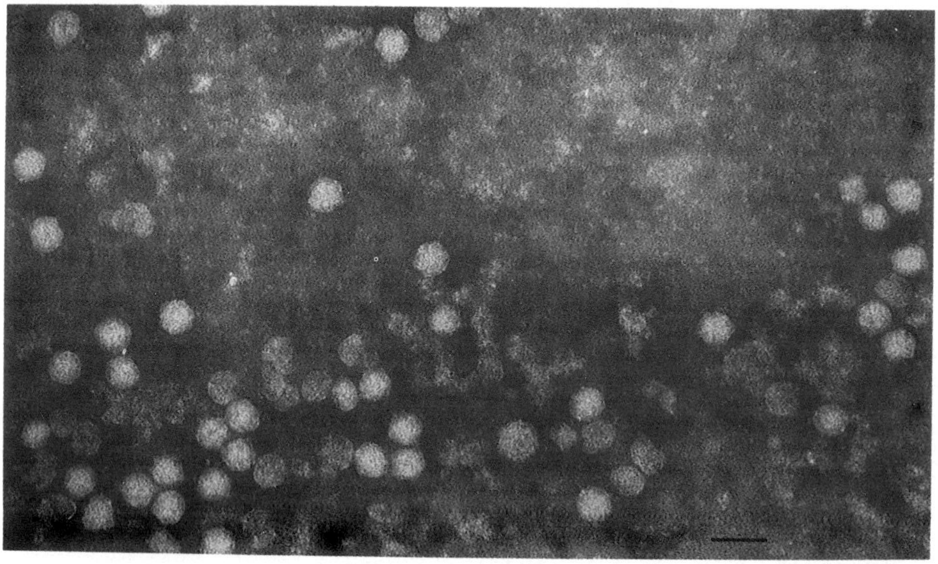

Fig. 5.21. Particles of pea enation mosaic enamovirus; bar scale=50 nm. Micrograph courtesy of Rothamsted Experimental Station, Harpenden, UK.

Type virus: pea enation mosaic virus (PEMV)
Diagnostic species: *Pisum sativum*, *Vicia faba*, slight vein-clearing followed by chlorotic flecks. Chronically infected plants develop enations
Propagation species: *Pisum sativum*
Purification: Mahmood and Peters (1973)
Serology: moderately immunogenic
Properties: $A_{260}/A_{280} = 1.63$
Specific absorbance at 260 nm = 7.2

3. *Tobravirus* genus — The generic designation is an acronym of the name of the type species, *tob*acco *ra*ttle *virus* (TRV).

The type member is the only species of the genus to infect ornamental crops. The virus has rod-shaped particles (Fig. 5.22) c. 22 nm in diameter and of two predominant lengths 180–215 nm (long) and 46–115 nm (short); even shorter particles are also associated with many strains. The particles sediment as two major components with sedimentation coefficients of c. 296 S (long particles) and 155–245 S (short particles) have buoyant densities in CsCl of c. 1.3 g cm^{-3} and

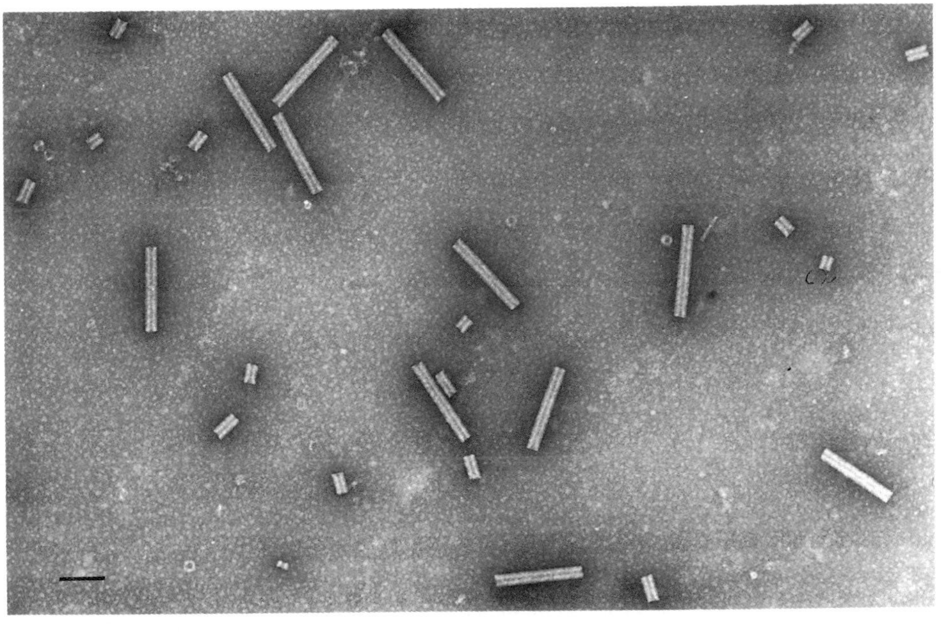

Fig. 5.22. Particles of tobacco rattle tobravirus; bar scale = 100 nm.

an M_r of c. 50×10^6 (long particles) and 11–29 (short particles). The capsid subunits consist of a single polypeptide of 24 kDa which encapsidate either linear positive-sense ssRNA of c. 6.8 kb (RNA 1; in long particles) or 1.8–4.5 kb (RNA-2; in short particles). The 3′ terminus of each is not polyadenylated and lacks a VPg, but the 5′ terminus is capped. RNA-1 codes for four non-structural proteins of 134 kDa, 194 kDa (both of which are probably associated with RNA replication) and (via subgenomic RNAs) of 29 kDa (movement protein) and 16 kDa (unknown function). RNA-2, via subgenomic RNAs, encodes the capsid protein of 24 kDa and that of some strains of another protein of c. 30 kDa (unknown function). Replication is probably associated with cytoplasmic ribosomes.

Tobacco rattle virus has a very wide natural host range, including several flowerbulb crops. The virus particles are readily detected in cytoplasm of hosts. It is transmitted by juveniles and adults of *Trichodorus* and *Paratrichodorus* nematodes, species of which are strain-specific vectors and can retain virus for many months if non-feeding. The virus is also seed-borne in some host species. The homologies in non-structural proteins indicates that tobacco rattle virus has wider affinities with tobamoviruses.

Type virus: tobacco rattle virus (TRV)

Diagnostic species: *Chenopodium amaranticolor*, necrotic lesions; *Cucumis sativus*, chlorotic or necrotic local lesions; *Nicotiana tabacum* cv. NN, necrotic spots or rings on isolated leaves and sporadic systemic distortion

Propagation species: *Nicotiana clevelandii*

Purification: Lister and Bracker (1969)

Serology: poorly immunogenic

Properties: A_{260}/A_{280} = 1.10-1.15
Specific absorbance at 260 nm=3.0

Selected Literature

Black, L.M. 1970. Potato yellow dwarf virus. CMI/AAB. Descriptions of Plant Viruses, No. 35.

Boccardo, G., Milne, R.G., Luisoni, E., Lisa, V. and Accotto, G.P. 1985. Three seedborne cryptic viruses containing double-stranded RNA isolated from white clover. Virology 147:29–40.

Doel, T.R. 1975. Comparative properties of type, nasturtium ringspot and petunia ringspot strains of broad bean wilt virus. J. Gen. Virol. 26:92–108.

Francki, R.I.B., Fauquet, C.M., Knudsen, D.L. and Brown, F. 1991. Classification and nomenclature of viruses. Fifth Report of the International Committee on Taxonomy of Viruses. Arch. Virol. Supplement 2.

Francki, R.I.B. and McLean, G.D. 1968. Purification of potato virus X and preparation of infectious ribonucleic acid by degradation with lithium chloride. Australian J. Biol. Sci. 21:1311–1318.

Fulton, R.W. 1967. Purification and some properties of tobacco streak and tulare apple mosaic viruses. Virology 32:153–162.

Gallitelli, D., Hull, R. and Koenig, R. 1985. Relationships among viruses in the tombusvirus group: Nucleic acid hybridization studies. J. Gen. Virol. 66:1523–1531.

Gardner, R.C. and Shepherd, R.J. 1980. A procedure for rapid isolation and analysis of cauliflower mosaic virus DNA. Virology 106:159–161.

Gooding, G.V. and Hebert, T.T. 1967. A simple technique for purification of tobacco mosaic virus in large quantities. Phytopathology 57:1285.

Goodman, R.M., Bird, J. and Thongmeearkom, P. 1977. An unusual virus-like particle associated with golden yellow mosaic of beans. Phytopathology 67:37–42.

Hollings, M. and Stone, O.M. 1970. Carnation mottle virus. CMI/AAB. Descriptions of Plant Viruses, No. 7.

Horne, R.W., Russell, G.E. and Trim, A.R. 1959. High resolution electron microscopy of beet-yellows virus filaments. J. Mol. Biol. 1:234–236.

Hull, R., Milne, R.G. and Regenmortel, M.H.V. van 1991. A list of proposed standard acronyms for plant viruses and viroids. Arch. Virol. 120:151–164.

Kassanis, B. 1970. Tobacco necrosis virus. CMI/AAB. Descriptions of Plant Viruses, No. 14.

Kurstak, E. 1981. Handbook of Plant Virus Infections and Comparative Diagnosis. Elsevier/North-Holland Biomedical Press, Amsterdam.

Lister, R.M. 1970. Apple stem grooving virus. CMI/AAB. Descriptions of Plant Viruses, No. 31.

Lister, R.M. and Bracker, C.E. 1969. Defectiveness and dependence in three related strains of tobacco rattle virus. Virology 37:262–275.

Lockhart, B.E.L. and Khaless, N. 1988. Commelina yellow mottle virus — a non-enveloped bacilliform virus containing double-stranded DNA. Phytopathol. 78:1548 (Abstract).

Lot, H., Marrou, J., Quiot, J.B. and Esvan, C. 1972. A contribution to the study on cucumber mosaic virus (CMV). II Quick method of purification. Ann. Phytopathol. 4:25–38.

Mahmood, K. and Peters, D. 1973. Purification of pea enation mosaic virus and the infectivty of its components. Neth. J. Plant Pathol. 79:138–147.

Martelli, G.P. 1992. Classification and nomenclature of plant viruses: State of the Art. Plant Dis. 76:436–442.

Matthews, R.E.F. 1960. Properties of nucleoprotein fractions isolated from turnip yellow mosaic virus preparations. Virology 12:521–539.

Mayo, M.A. and Martelli, G.P. 1993. New families and genera of plant viruses. Arch. Virol. 133:496–498.

Moghal, S.M. and Francki, R.I.B. 1976. Towards a system for the identification and classification of potyviruses. I. Serology and amino acid composition of six distinct viruses. Virology 73:350–362.

Rezaian, M.A. and Francki, R.I.B. 1973. Replication of tobacco rinspot virus. I. Detection of low molecular weight double-stranded RNA from infected plants. Virology 56: 238–249.

Scott, H.A. 1963. Purification of cucumber mosaic virus. Virology 20:103–106.

Tremaine, J.H., Ronald, W.P. and Valcic, A. 1976. Aggregation properties of carnation ringspot virus. Phytopatholgy 66:34–39.

van den Hurk, J., Tas, P.W.L. and Peters, D. 1977. The ribonucleic acid of tomato spotted wilt virus. J. Gen. Virol. 36:81–91.

Van Vloten-Doting, L. and Jaspars, E.M.J. 1972. The uncoating of alfalfa mosaic virus by its own RNA. Virology 48:690–708.

Wetter, C. and Paul, H.L. 1961. Die Darstellung des Carnation latent-Virus in hochgereinigter Form. Phytopath. Z. 43:207–212.

6

Viroids

Rosemarie Hammond
Molecular and Plant Pathology Laboratory, Bldg. 011A, ARS-BARC-W, Beltsville, MD 20705–2350, USA

Introduction

Unique among plant pathogens, viroids are small, covalently closed, circular, autonomously replicating, unencapsidated RNA molecules which apparently do not encode proteins (for reviews, see Diener, 1987). As such, they rely upon their host for both the replication and transport functions necessary to produce a systemic infection.

A number of plant diseases are now known to be caused by viroids, and hosts include both herbaceous and woody plants — agronomic as well as ornamental. Viroid infection can be characterized by mild to severe symptoms (dependent upon the host, the viroid strain, or both) or the infection may be latent, i.e., no obvious symptoms are visible on that host. In the case of latent infections, the inoculation of the viroid onto an alternate host can frequently result in severe symptoms. Examples of each of these situations are given below, particularly as they relate to ornamentals, as well as a brief summary of what is known to date about viroid replication and methods for diagnosis and control.

Economic Importance in Ornamentals

Several viroid-incited diseases are of considerable economic importance in crops (e.g., potato spindle tuber viroid (PSTVd) in potato, citrus exocortis viroid (CEVd) in citrus, and coconut cadang cadang viroid (CCCVd) in coconut palm). In ornamentals, the most important viroid disease is caused by the chrysanthemum stunt viroid (CSVd) in cultivated chrysanthemum, *Chrysanthemum morifolium* (for a review, see Lawson, 1987). A viroid which has not been fully characterized and which also poses a potentially serious threat to the chrysanthemum industry is chrysanthemum chlorotic mottle viroid (CCMVd) (Horst, 1987). Considerable economic losses are also attributed to the presence of a viroid-like small circular RNA that has been associated with a stunting syndrome observed in carnation (*Dianthus caryophyllus*) in central Italy (Lisa et al., 1985; Boccardo et al., 1988). Viroids infecting ornamental plants such as *Coleus blumei* (Fonseca et al., 1989; Spieker et al., 1990; Singh et al., 1991), *Columnea erythrophae* (Hammond et al., 1989), *Nematanthus wettsteinii* (Singh et al., 1992), and *Hibiscus* cvs (Sänger, 1988) have also been reported, but the infections appear to be symptomless in these hosts. Transfer of the *Columnea* and *Nematanthus* viroids to tomato and potato, however, resulted in symptoms typical of those incited by PSTVd (Owens et al., 1978; Singh et al., 1992). These ornamentals may form potential reservoirs from which spread to susceptible crops may occur with resulting serious diseases.

The experimental host range of PSTVd includes a number of ornamentals; and susceptible plants were found in the families Boraginaceae, Campanulaceae, Caryophyllaceae, Compositae, Convolvulaceae, Dipsaceae, Sapindaceae, Scrophulariaceae, Solanaceae, and Valerianaceae (Singh, 1973). Most of the infected plants were symptomless, e.g., *Dianthus barbatus*, but symptoms were observed with the severe strain of PSTVd in *Petunia hybrida* in the family Solanaceae.

In summary, viroid diseases and, more importantly, viroid infections in ornamentals are becoming increasingly evident as methods of detection have improved (see section on Detection and Control). Many of the viroids found in ornamentals do not cause *obvious* symptoms in their hosts, but these hosts may provide sources of infection for sensitive hosts.

Symptomatology and Transmission

Most of the *symptom* types observed with viral diseases also occur with one or more diseases incited by viroids; therefore, diagnosis of viroid diseases by symptomatology alone is not sufficient. For example, the chrysanthemum stunt disease was believed to be caused by a conventional virus disease from the time it was known to be infectious (Brierley and Smith, 1949) until it was identified as a viroid disease by Diener and Lawson (1973).

Environmental conditions play a role in symptom development in many viroid infections. Most viroids replicate best at high temperatures (30–33°C), and symptom expression is usually expressed with more intensity at higher temperatures (Diener, 1979). Chrysanthemum stunt, however, replicates and shows best symptoms at 22–29°C (Handley and Horst, 1988). High light intensities also favor replication and symptom production in viroid infected plants (Diener, 1987). Hollings and Stone (1973) found that CSVd-infected plants exhibited prominent symptoms at 21°C and high light intensity, but at temperatures above 30°C, symptoms were masked even with the same high light intensity.

Spread of viroid diseases most often occurs by mechanical transmission through vegetative propagation, foliar contact between neighboring plants, and handling during cultivation. Seed transmission has been demonstrated with many, but not all, viroids, including the *Coleus* viroid (Diener, 1987; Singh et al., 1991). Reports of seed transmission of CSVd in chrysanthemum have been contradictory, with Monsion et al. (1973) obtaining seed transmission, whereas Hollings and Stone (1973) concluded that the disease is not seed transmitted. Insect transmission has been reported for tomato planta macho viroid (Galindo et al., 1989). The aphid, *Myzus persicae*, was found to vector the viroid from the wild host, *Physalis* aff *foetens*, to the experimental tomato host.

Replication

As small circular RNAs of 246–463 nucleotides (Fig. 6.1), viroids represent the smallest known autonomously replicating RNAs. Viroid RNAs do not function as messenger RNAs and do not encode their own replicases. They rely upon their plant host for the replication machinery (i.e., RNA polymerases, transcription factors) that is either preexisting in the host or induced in the host following viroid infection. Viroids do not require a helper virus for replication and movement. Comparative analysis of a number of viroid sequences led Keese and Symons (1985) to propose a model of viroid structure in which five structural/functional domains are distinguished, as shown in Fig. 6.1. The five domains are: the central conseved region, the most highly conserved region among groups of viroids, e.g., within the PSTVd group; the pathogenicity domain, in which sequence variations

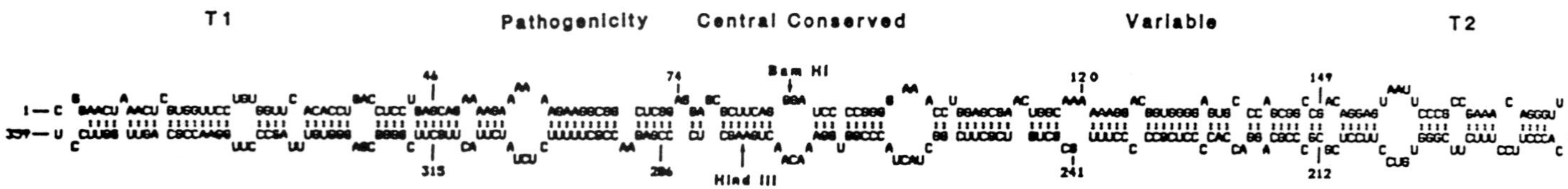

Fig. 6.1. The nucleotide sequence and proposed secondary structure of the intermediate strain of potato spindle tuber viroid (PSTVd). The five structural/functional domains are indicated, eg. pathogenicity, variable, central conserved, and terminal domains (T1,T2); and restriction enzymes sites that are used in cDNA cloning of viroid RNAs are indicated.

may be correlated with different symptoms when the viroid is propagated in the same host; the variable domain, which shows less than 50% homology between otherwise closely related viroids; and the T, or terminal sequence, domains. Assignment of evolutionary relationships among viroids has been based upon examination of the sequence similarities within these domains among the different viroids (Koltunow and Rezaian, 1989) as well as sequence similarities of the complete nucleotide sequences (Elena et al., 1991).

Various models for viroid replication have been proposed, but the consensus is that viroid replication proceeds in the nucleus via a rolling circle type of mechanism (for a review, see Diener, 1979). In the simplified scheme shown in Fig. 6.2, the mature, monomeric (one molecule length, e.g., in PSTVd= 359 nucleotides) enters the nucleus of injured cells and functions as a template for the synthesis of minus (−) sense oligomeric strands (A and B). The oligmeric (−) sense strands act as templates for the synthesis of plus (+) sense oligomeric strands (C) which are subsequently cleaved into unit length molecules (D and E) and circularized into mature (+) sense circular viroids (F). The mature viroid accumulates in the nucleolus and spreads to adjacent cells by an unknown mechanism to continue the infection process. Cell-to-cell spread is followed by long distance spread and results in a systemic infection. As mentioned above, viroid replication can occur without the formation of disease symptoms, and in some cases high titers of viroid can be detected in plants exhibiting little or no symptoms.

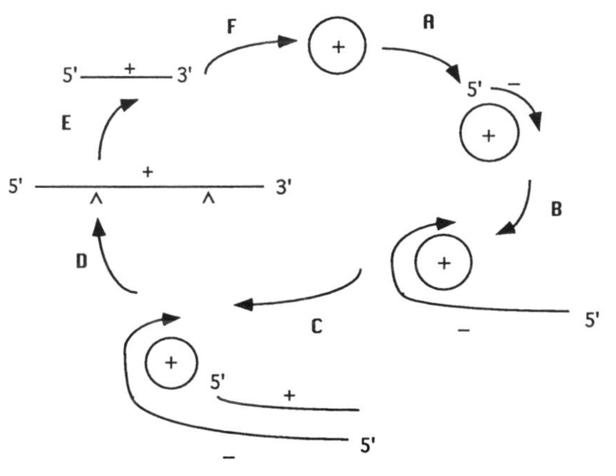

Fig. 6.2. Model for the rolling circle mechanism of viroid replication as proposed by Branch and Robertson (1984). A–F are explained in the text.

The enzymes that catalyze viroid RNA synthesis are those residing in the host cell. The structure of circular viroid RNA resembles that of double-stranded DNA but differs principally in that it contains single-stranded loops (Fig. 6.1). It is therefore possible that DNA-dependent RNA polymerases may play a role in viroid replication. Spiesmacher et al. (1985) and Tabler and Tsagris (1990) conclude that DNA-dependent RNA polymerase II of the host is responsible for the synthesis of both the (+) and (−) strands in viroid replication. The mechanisms for cleavage and ligation to form mature plus strand circles are still a matter of debate, and the reader is referred to Diener (1987) for further information.

Detection and Control

In the past, routine diagnosis of viroid diseases was made by mechanical inoculation to indicator plants, e.g., "Rutgers" tomato for PSTVd, Etrog citron for CEVd, and Mistletoe chrysanthemum for CSVd (Diener, 1979). *Bioassay* of CSVd can be achieved by chip-budding of Mistletoe chrysanthemeum plants with CSVd-infected material, and this results in symptoms being produced in the greenhouse within 30–35 days with the appearance of small chlorotic spots (Lawson, 1987). Symptom production on indicator plants, however, is not a good indicator of the identity of the viroid, as CEVd and PSTVd cause similar symptoms on their common tomato host (Singh and Clark, 1973), and Mistletoe chrysanthemum shows similar symptoms upon inoculation with CSVd and cucumber pale fruit viroid (Runia and Peters, 1980). Currently, detection and identification of viroids is based upon their nucleotide sequence and/or unusual migration in polyacrylamide gels.

Electrophoresis of partially purified nucleic acid extracts from infected plants can be used for rapid and sensitive detection of circular viroids without the need for knowledge of the nucleotide sequence of the viroid or the need for specialized nucleic acid probes or equipment. Schumacher et al. (1985) developed a two-dimensional electrophoretic technique that relies upon the unusual mobility of circular RNAs for the separation of viroids from host components, and Singh and Boucher (1987) simplified the technique so that the two dimensions can be run in the same gel (Fig. 6.3); silver-staining the gel increases the sensitivity of detection. For the analysis of large numbers of samples, *nucleic acid hybridization* techniques for viroid detection are the most rapid and sensitive. Owens and Diener (1981) developed a nucleic acid hybridization test for PSTVd in potato that relied upon highly radioactive, viroid-complementary DNA probes and was highly

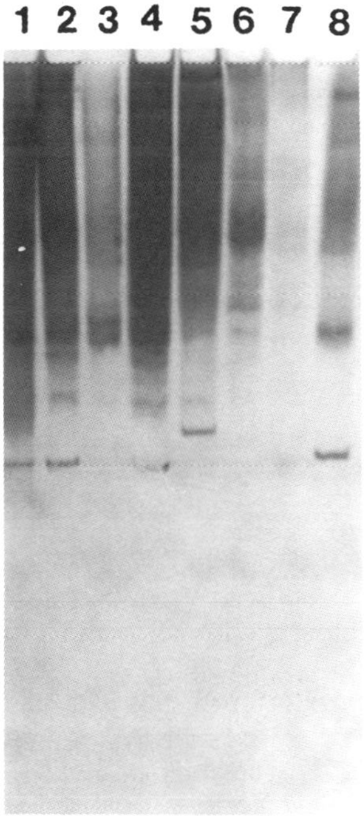

Fig. 6.3. "Return"-gel electrophoresis using conditions described by Singh and Boucher (1987) for the separation of viroids from host plant constituents. Lanes 1, 2, and 4 are nucleic acid extracts of Rutgers tomato plants inoculated with the *Columnea* latent viroid; lane 5 is tomato inoculated with the intermediate strain of PSTVd; lane 8 is tomato inoculated with tomato apical stunt viroid; lanes 3, 6, and 7 are mock-inoculated tomato plants. The gel was stained with silver prior to photography.

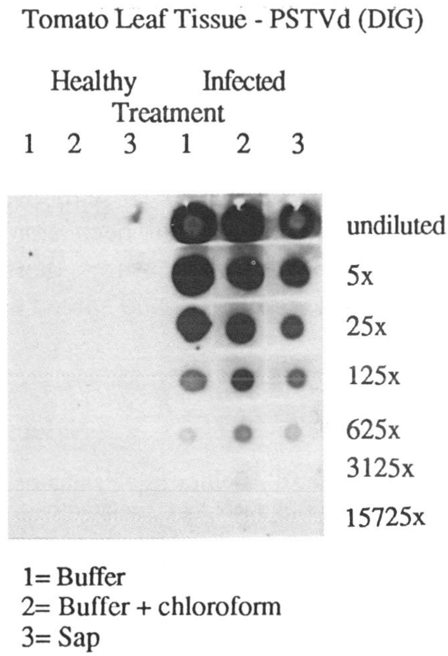

Fig. 6.4. Non-radioactive detection of PSTVd in tomato plants. Nylon membranes were prepared and hybridized with a digoxigenin (DIG)-labeled RNA probe for PSTVd as described by Podleckis et al. (1993). Fifty milligram samples of leaf tissue were ground in two volumes of buffer, and three microliter samples were applied to the membrane.

specific and ten times as sensitive as gel electrophoretic techniques. Since that time, sensitive and specific non-radioactive procedures for viroid detection have been developed (Roy et al., 1989; Podleckis et al., 1993) (Fig. 6.4) as well as polymerase chain reaction (PCR)-based techniques which require the use of sequence-specific oligonucleotide primers for the amplification of specific viroids (Puchta and Sänger, 1989; Hadidi and Yang, 1990).

Efforts to *control* viroid-incited plant diseases include the elimination of viroids from infected plants by low-temperature treatment and meristem tip culture of chrysanthemum (Paludan, 1985) and potato (Lizarraga et al., 1980). Most control measures, however, are based upon prevention instead of cure and rely upon the sensitive detection methods described above. Incorporation of resistance factors into host plants (e.g., pathogen-derived resistance) is on the horizon, and although cross-protection has been demonstrated for viroids (Niblett et al., 1978), its widespread use has not been employed in the field for the control of viroid diseases. Preventive measures, therefore, are by far the most practical and include measures to prevent viroid spread during propagation and planting of viroid-free seed.

Selected Literature

Boccardo, G., d'Aquilio, M., Palukaitis, P., Marzachi, C. and Lisa, V. 1988. Infectivity of the viroid-like RNAs associated with the carnation stunting syndrome. Acta Hortic. 234: 109–114.

Branch, A.D. and Robertson, H.D. 1984. A replication cycle for viroids and other small infectious RNAs. Science 223:450–455.

Brierley, P. and Smith, F.F. 1949. Chrysanthemum stunt. Phytopathology 39:501.

Diener, T.O. 1979. Viroids and Viroid Diseases. Wiley, New York.

Diener, T.O. (ed.). 1987. The Viroids. Plenum Press, New York.

Diener, T.O. and Lawson, R.H. 1973. Chrysanthemum stunt: a viroid disease. Virology 51: 94–101.

Elena, S.F., Dopazo, J., Flores, R., Diener, T.O. and Moya, A. 1991. Phylogeny of viroids, viroidlike satellite RNAs and the viroidlike domain of hepatitis δ virus RNA. Proc. Natl. Acad. Sci. USA. 88:5631–5634.

Fonseca, M.E.N., Boiteux, L.S., Singh, R.P. and Kitajima, E.W. 1989. A small viroid in *Coleus* species from Brazil. Fitopatol. Bras. 14:94–96.

Galindo, J., Lopez, C. and Aguilar, T. 1989. Discovery of the transmitting agent of tomato planta macho viroid. Revista Mexicana de Fitopatologia 7:61–65.

Hadidi, A., and Yang, X. 1990. Detection of pome fruit viroids by enzymatic cDNA amplification. J. Virol. Meth. 30:261–270.

Hammond, R., Smith, D.R. and Diener, T.O. 1989. Nucleotide sequence and proposed secondary structure of *Columnea* latent viroid: a natural mosaic of viroid sequences. Nucleic Acids Res. 23:10083–10094.

Handley, M.K. and Horst, R.K. 1988. The effect of temperature and light on chrysanthemum stunt viroid infections of florists chrysanthemum. Acta Hortic. 234:89–97.

Hollings, M. and Stone, O.M. 1973. Some properties of chrysanthemum stunt, a virus with the characteristics of an uncoated ribonucleic acid. Ann. Appl. Biol. 74:333–348.

Horst, R.K. 1987. Chrysanthemum chlorotic mottle. In: The Viroids (Diener, T.O., ed.) Plenum Press, New York, pp. 291–295.

Keese, P. and Symons, R.H. 1985. Domains in viroids: Evidence of intermolecular RNA rearrangements and their contribution to viroid evolution. Proc. Natl. Acad. Sci. USA 82:4582–4586.

Koltunow, A.M. and Rezaian, M.A. 1989. A scheme for viroid classification. Intervirology 30:194–201.

Lawson, R.H. 1987. Chrysanthemum stunt. In: The Viroids (Diener, T.O., ed.), Plenum Press, New York, pp. 247–259.

Lisa, V., D'Agostino, G. and Boccardo, G. 1985. A new disease causing stunting and shoot proliferation in carnation. Acta Hortic. 164:63–70.

Lizarraga, R.E., Salazar, L.F., Roca, W.M. and Schilde-Rentschler, L. 1980. Elimination of potato spindle tuber viroid by low temperature and meristem culture. Phytopathology 70:754–755.

Monsion, M., Bachelier, J.-C. and Dunez, J. 1973. Quelques propiétés d'un viroïde: Le rabougrissement du chrysantheme. Ann. Phytopathol. 5:467.

Niblett, C.L., Dickson, E., Fernow, K.H., Horst, R.K. and Zaitlin, M. 1978. Cross protection among four viroids. Virology 91:198–203.

Owens, R.A. and Diener, T.O. 1981. Sensitive and rapid diagnosis of potato spindle tuber viroid disease by nucleic acid hybridization. Science 213:670–672.

Owens, R.A., Smith, D.R. and Diener, T.O. 1978. Measurement of viroid sequence homology by hybridization with complementary DNA prepared *in vitro*. Virology 89: 388–394.

Paludan, N. 1985. Inactivation of viroids in chrysanthemum by low-temperature treatment and meristem-tip culture. Acta Hortic. 164: 181–186.

Podleckis, E.V., Hammond, R.W., Hurtt, S.S. and Hadidi, A. 1993. Chemiluminescent detection of potato and pome fruit viroids by digoxigenin-labeled dot blot and tissue blot hybridization. J. Virol. Meth. 43:147–158.

Puchta, H. and Sänger, H.L. 1989. Sequence analysis of minute amounts of viroid RNA using the polymerase chain reaction (PCR). Arch. Virol. 106:335–340.

Roy, B.P., Abouhaider, M.G. and Alexander, A. 1989. Biotinylated RNA probes for the detection of potato spindle tuber viroid (PSTV) in plants. J. Virol. Meth. 23:149–155.

Runia, W.Th. and Peters, D. 1980. The response of plant species used in agriculture and horticulture to viroid infections. Neth. J. Pl. Pathol. 86:135–146.

Sänger, H.L. 1988. Viroids and viroid diseases. Acta Hortic. 234:79–87.

Schumacher, J., Randles, J.W. and Riesner, D. 1985. A two-dimensional electrophoretic technique for the detection of circular viroids and viruses. Anal. Biochem. 135:288–295.

Singh, R.P. 1973. Experimental host range of the potato spindle tuber "virus". Am. Potato J. 50:111–123.

Singh, R.P. and Boucher, A. 1987. Electrophoretic separation of a severe from mild strains of potato spindle tuber viroid. Phytopathology 78:663–667.

Singh, R.P., Boucher, A. and Singh, A. 1991. High incidence of transmission and occurrence of a viroid in commercial seeds of *Coleus* in Canada. Plant Dis. 75:184–187.

Singh, R.P. and Clark, M.C. 1973. Similarity of host response to both potato spindle tuber and citrus exocortis viruses. FAO Plant Prot. Bull. 21:121–125.

Singh, R.P., Lakshman, D.K., Boucher, A. and Tavantzis, S.M. 1992. A viroid from *Nematanthus wettsteinii* plants closely related to the *Columnea* latent viroid. J. Gen. Virol. 73:2769–2774.

Spieker, R.L., Haas, C.Y., Freimüller, K. and Sänger, H.L. 1990. Primary and secondary structure of a new viroid "species" (CbVd 1) present in the *Coleus blumei* cultivar "Bienvenue". Nucleic Acids Res. 18:3998.

Spiesmacher, E., Mühlbach, H.-P., Tabler, M. and Sänger, H.L. 1985. Synthesis of (+) and (−) RNA molecules of potato spindle tuber viroid (PSTV) in isolated nuclei and its impairment by transcription inhibitors. Biosci. Rep. 5:251–265.

Tabler, M. and Tsagris, M. 1990. Viroid replication mechanisms. In: NATO ASI Series Vol. H 41, Recognition and Response in Plant-Virus Interactions, (Fraser, R.S.S., ed.), Springer-Verlag, Berlin, pp. 185–205.

7

Mycoplasma-like Organisms and Spiroplasmas in Ornamentals

Meir Klein
Department of Entomology, Agricultural Research Organization, The Volcani Center, Bet Dagan 50250, Israel

Mycoplasma-like organisms (MLOs) have been known to be the cause of plant diseases since 1967 (Doi et al., 1967). Marwitz (1990) recorded MLOs in some 400 species in approximately 100 different botanical families. The number of infected plant species has increased to more than 600 by 1994. Some MLOs, such as those causing the lethal coconut yellowing disease, are very destructive. Seventy MLO diseases of ornamentals belonging to 24 botanical families have been described. Apparently, some of the diseases are caused by the MLO which also infects diverse plant species of ornamentals. The family most affected is the Compositae, 12 species of which were infected by 27 diseases (Bellardi and

---

Contribution No. 1344-E, 1994 series from the Agricultural Research Organization, The Volcani Center, Bet Dagan, Israel.

Bertacinni, 1984). Some of these diseases (e.g., aster yellows in the US, virescence of gladiolus in Italy) are sometimes of great economic importance. A disease such as the phyllody of ranunculus in Europe, although not important *per se*, infects corms, and so an exported crop will be totally rejected when in quarantine even if the number of corms infected is small.

Citrus stubborn disease and corn stunt disease were considered to be of MLO etiology until about 1970. In 1971 and 1972, the microorganisms associated with these two diseases were categorized according to their structure and their movement in liquid phase and were named "spiroplasmas" (a new genus in the Order Mycoplasmatales). Recently a third spiroplasma was identified as causing yellows disease in periwinkle (*Catharanthus roseus*) in the Near East. Spiroplasmas not known to infect plants were cultured from the surface of certain flowers (McCoy et al., 1979).

Mycoplasmas, spiroplasmas and MLOs are structurally similar in many ways. In general they are considered to be the smallest free-living procaryotes, but whereas the first two microorganisms can be isolated from their hosts or from the environment and cultivated *in vitro* in artificial culture media, the MLOs cannot.

The morphology and structure of mycoplasmas from pure culture have been well described. MLOs and spiroplasmas were characterized mainly from phloem tissues of infected plants. MLOs observed in phloem cells of thin leaf sections by transmission electron microscope were pleomorphic; the rounded ones were 200–1100 nm in diameter and the filamentous ones 50–400 nm in diameter and 200–3000 nm in length. Like MLOs, spiroplasmas in phloem cells are mainly pleomorphic, but some helical bodies 3–25 μm in length and 100–200 nm in diameter also occur. Spiroplasmas are known to change their basic morphology during a different phase of their growth and become pleomorphic like MLOs as they age. These three microorganisms lack a rigid cell wall and are bound only by a triple-layer unit membrane, 8–10 nm thick. As a result, they are very pleomorphic. The organisms contain DNA, RNA and ribosomes and DNA strands (Fig. 7.2). Extra chromosmal DNA can also be isolated from these organisms. The genomes of the mycoplasmas and spiroplasmas are relatively small (500–1000 MDa), and it appears that the genome of the MLOs may be of the same size range. Multiplication of MLOs and spiroplasmas is believed to occur in a similar manner by binary fission. The multiplication of spiroplasmas was easier to study because they can be cultured *in vitro*. The smallest viable cell was found to be of a two-turn helix. During the logarithmic phase of growth, most of the bodies have two to four helices, and then cell divisions occur, yielding bodies of two helices.

The number of helices increases during the stationary phase.

MLOs and spiroplasmas occur mostly in the phloem tissue of plants. The overall concentration of these microorganisms is, therefore, low, and the microorganisms are not distributed evenly throughout the plants. Generally, the highest concentration of MLOs and spiroplasmas is found near meristematic tissues. Mycoplasmas, MLOs, and spiroplasmas are sensitive to tetracycline antibiotics but resistant to antibiotics such as penicillin, which are known to act on cell-wall constituents of bacteria. Thus, tetracycline antibiotic applications cause remission of symptoms in many MLO- and spiroplasma- infected plants.

Syndromes of MLO and spiroplasma infection indicate that toxins are involved in the disease process. In cases such as phyllody, virescence, and dwarfing, the mollicutes seem to induce also hormonal imbalance in the plant.

The known vectors for the MLOs and the plant pathogenic spiroplasmas are mostly auchenorrhynchans; very few are psyllids. The nature of their transmission resembles that of most circulative/propagative viruses spread by leafhoppers. This requires (1) a relatively long acquisition access feed on infectious sources; (2) a relatively long latent period in the vector which allows the microorganisms to multiply within its body and to circulate in it (no proof for transmission from parents to offspring via eggs); and (3) a comparatively short inoculation access feed to transmit the pathogen to new plants. The last characteristic causes some difficulties in the prevention of pathogen spread, as with field spread of viruses of a similar nature.

Until recently, confirmation of MLO etiology of a plant disease was based on the observation of mycoplasma-like bodies in ultrathin sections of the phloem tissue by electron microscopy and on the remission of symptoms following treatment with antibiotics of the tetracycline group. The classification was based on vector specificity, host range, and symptoms in certain hosts. Molecular techniques, such as nucleic acid probes for detecting homologous sequences by hybridization, provide a quicker diagnosis (Kirkpatrick et al., 1987). The existence of genetic relationships between groups of MLOs of various sources was thus proven. The southern hybridization, restriction fragment length polymorphism (RFLP) and polymerase chain reaction (PCR) techniques were also employed for finer analyses in the interrelationships of strains. For example, these techniques revealed that the many strains of aster yellows (AY) MLO could be arranged into three groups. These classifications were confirmed by using RFLP analyses of PCR-amplified 16S rDNA sequences from mycoplasmas, and from certain MLOs from infected plants (Kirkpatrick, 1989; Lee et al., 1993).

Aster Yellows MLO

The aster yellows disease has been identified in several locations in the world and in numerous plants of various botanical families. The disease occurs in ornamentals, vegetables, field crops, and trees. Until recently, only several strains of this MLO were known. During the last few years, this MLO has been thoroughly investigated, and the number of strains identified is increasing. Fourteen strains of the AY MLO were identified on the North American continent alone (Lee et al., 1992). These strains were classified into three genomic types: the eastern AY type (seven strains), the western AY type (six strains), and the clover phyllody type (one strain). The AY MLO has also been identified in Europe and Asia, thereby increasing the number of AY strains (Lee et al., 1992; Davis et al., 1993). Many more strains are likely to be identified worldwide in the future. Most of the AY MLO strains were transmitted and maintained in periwinkle (*Catharanthus roseus*) plants, which serve also as hosts for other MLOs (Fig. 7.1, see color plate; Fig. 7.2).

Kunkel first noticed in 1926 that the AY pathogen (then suspected of being a virus) is transmitted by leafhoppers. However, the specificity of the vector(s) has not yet been determined, as too many strains exist. It seems that in the US the aster (six-spotted) leafhopper, *Macrosteles fascifrons* (Stal), is the most important transmitter of the AY MLO (more than 20 species of leafhoppers). The AY MLO is transmitted by *M. fascifrons* in a circulative/propagative manner. The minimum acquisition feeding lasts about 10 minutes. The shortest recorded latent periods in the vector at 20, 25 and 30°C were 16, 12 and 11 days, respectively (Maramorosch, 1953). The MLO is retained in the leafhopper throughout its lifetime. Inoculation feeding was rather short.

Citrus stubborn spiroplasma, *Spiroplasma citri*

Plant species infected by *Spiroplasma citri* (Figs. 7.3, see color plate; Fig. 7.4) were found in nature in various botanical families which include trees, vegetables, ornamentals, and weeds. This pathogen was identified in the US and Mediterranean area. Every year and for reasons so far unknown, the number of plant species naturally infected is greater in the New than the Old World. However, laboratory inoculations showed that Old World isolates of *S. citri* and vectors have the same potential to infect different plant varieties as do the New World isolates. Ornamentals such as aster (*Callistephus chinensis*), stock (*Matthiola incana*) (Fig. 7.3, see color plate), and periwinkle (*Catharanthus roseus*) are naturally

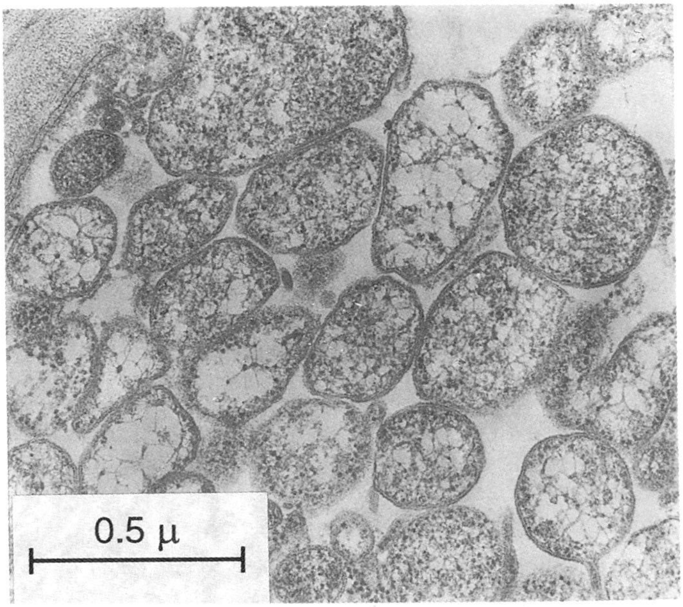

Fig. 7.2. Safflower phyllodys MLOs in phloem cells of a diseased periwinkle plant *(Catharanthus roseus)*. Courtesy of the Dept. of Virology, ARO, The Volcani Center, Bet Dagan.

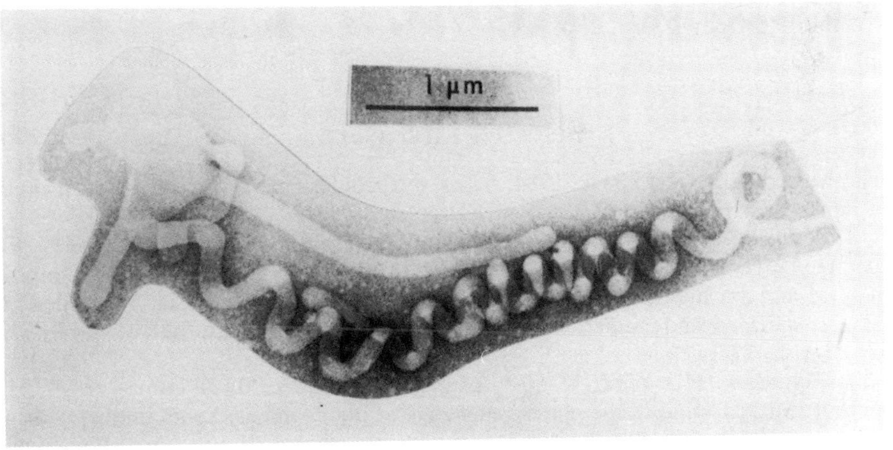

Fig. 7.4. *Spiroplasma citri* negatively stained. Courtesy of K. Maramorosch.

infected in the US and the Middle East. Strains of *S. citri* are known, but very little research has been conducted on this subject.

Leafhoppers of the genus *Circulifer* (Zakhvatkin) were found to be vectors of *S. citri* in the mid-1970s in the US and in the 1980s in the Middle East. In the US other leafhopper species from other genera, including *M. fascifrons*, proved to be vectors of *S. citri* but with very limited efficiency. The minimal conditions in the transmission process for the vector *Circulifer tenellus* tested in Israel were: an acquisition access period of at least 6 hours, latent period of 6 days, and inoculation access period of 60 minutes (at 25°C). The vector remains inoculative throughout its lifetime (Klein, 1992).

Insecticide applications were effective in preventing MLO inoculation only in certain cases. Too many applications are frequently needed and their effect is limited.

Preventive measures against the spread of yellows diseases should aim to interfere with or prevent the landing of the leafhoppers on susceptible plants during the most sensitive period of their inoculativity. Plant tissue culture of meristem tips and heat therapy, known as satisfactory methods, have not yet been used commercially to obtain vegetatively propagated crops free of mollicute infection. Tetracycline injections to infected trees are costly, involve repeated applications, and the treatment is generally not recommended. Breeding for varieties resistant to the pathogen is difficult and time-consuming, mainly because this has to be carried out with live vectors. Breeding for resistance to vectors may prove inadequate because even brief probing may be sufficient to permit transmission of the pathogen.

Selected Literature

Bellardi, M.G. and Bertacinni, A. 1984. Virosi e Micoplasmosi delle Piante Ornamentali e da Fiore Reciso. Edagricole–Edizioni Agricole, Bologna.

Davis, R.E., Dally, E.L., Bertaccini, A., Lee, I.-M., Credi, R., Osler, R., Savino, V., Carraro, L., Di Terlizzi, B. and Barba, M. 1993. Restriction fragment length polymorphism analyses and dot hybridizations distinguish mycoplasmalike organisms associated with *Flavescence doree* and southern European grapevine yellows disease in Italy. Phytopathology 83: 772–776.

Doi, Y., Teranaka, M., Yora, K. and Asuyama, H. 1967. Mycoplasma- or PLT-like microorganisms found in the phloem elements of plants infected with mulberry dwarf, potato witches' broom, aster yellows, or paulownia witches' broom. Ann. Phytopathol. Soc. Jpn. 33:259–266 (in Japanese).

Kirkpatrick, B.C. 1989. Strategies for characterizing plant pathogenic mycoplasma-like organisms and their effects on plants. In: Kasuze, T. and Wester, E.W. (eds.), Plant-Microbe Interactions, Vol. 3, Macmillan, New York, pp. 241–290.

Kirkpatrick, B.C., Stenger, D.C., Morris, T.J. and Purcell, A.H. 1987. Cloning and detection of DNA from a nonculturable plant pathogenic mycoplasma-like organism. Science 238: 197–200.

Klein, M. 1992. Role of *Circulifer/Neoaliturus* in the transmission of plant pathogens. In: Harris, K.F. (ed): Advances in Disease Vector Research, Vol. 9, Springer-Verlag, New York, pp. 152–193.

Kunkel, L.O. 1926. Incubation period of aster yellows in its insect host. Phytopathology 16: 67 (Abstr.).

Lee, I.-M., Davis, R.E., Chen, T.-A., Chiykowski, L.N., Fletcher, J., Hiruki, C. and Schaff, D.A. 1992. A genotype-based system for identification and classification of mycoplasma-like organisms (MLOs) in the aster yellows MLO strain cluster. Phytopathology 82: 977–986.

Lee, I.-M., Hammond, R., Davis, R.E. and Gundersen, D.E. 1993. Universal amplification and analysis of pathogen 16S rDNA for classification and identification of mycoplasma-like organisms. Phytopathology 83: 834–842.

Maramorosch, K. 1953. Incubation period of aster-yellows virus. Am. J. Bot. 40: 797–809.

Marwitz, R. 1990. Diversity of yellows disease agents in plant infections. Zentralbl. Bakteriol. 20 Suppl.: 431–434.

McCoy, R.E., Williams, D.S. and Thomas, D.L. 1979. Isolation of mycoplasmas from flowers. Proc. R.O.C.-U.S. CoopScience Seminar. Mycoplasma Diseases of Plants. NSC Symp. Ser. I. Taipai:National Council, pp. 75–80.

8

Symptoms of Virus and Virus-like Diseases

L. Bos
DLO Research Institute for Plant Protection (IPO-DLO), PO Box 9060, 6700 GW Wageningen, The Netherlands

Introduction

Infection of a plant by a virus implies multiplication of the virus in its cells. This often leads to malfunctioning of the plant that hosts the virus. The virus's total dependence on the host's cellular biochemistry involves initiation of abnormality at the cellular or even subcellular level. A subsequent *sequence of events* during *pathogenesis*, beginning with a period of inapparency (*incubation period*), may lead to externally visible *symptoms* and *disease*.

Whether — and to which extent — disease and damage result from virus infection depends upon the *pathogenicity* of the virus and on the *vulnerability* of its host. The pathogenicity of the virus is determined by its *aggressiveness* (the ease with which the virus attacks a plant and multiplies in it) and by its *virulence* (as expressed by the severity of the symptoms caused). The vulnerability of the plant is determined by its *susceptibility* or lack of *resistance* (the ease with which it contracts or resists infection) and by its *sensitivity* or lack of *tolerance* (the ease with which the plant reacts to or endures infection). Thus, symptoms and disease are determined by the complex interaction between virus and plant, and these are both highly variable genetically and phenotypically (environmentally conditioned) (Bos, 1983). Hence the tremendous variation in types and severity of symptoms (Bos, 1978).

Several infections do not lead to abnormalities visible to the naked eye. They are the so-called *latent* or *symptomless infections*. Such infections appear to be prevalent in wild vegetations (Bos, 1981). There, long-term association of virus with plant populations leads to selection of mild strains of the virus and of tolerant genotypes of the host, so that some sort of a balance ensues between virus and host, allowing optimal survival of both. New encounters of crops or of new crop genotypes with naturally occurring viruses, often hidden in wild vegetations, or introduction of a virus from elswhere at a new site, are more likely to result in disease and loss. The virus may then first of all strive for rapid establishment by maximal multiplication.

Whether an infection is considered latent or apparent depends on the precision of observation and on the tools used for examination. Plants that do not show external symptoms may be found to be abnormal internally in certain tissues or in certain cells or cellular organelles when viewed with the light or electron microscope. Thus, there is no clear-cut borderline between normal and diseased.

Theoretically, any virus infection should lead to some sort of abnormality. There are several cases where no external symptoms are visible, but counting of number and assessment of sizes of flowers and fruits and careful estimation of plant weight and final crop yield may reveal *growth* and *yield reduction*. Although such deviations are not specific for virus infections, they rank high in the range of symptoms of virus infection. Together with *quality reduction*, they are *the* symptoms that are of interest to the grower (Fig. 8.1) (Bos, 1982). By definition, disease refers to "any harmful deviation from normal functioning of physiological processes". But harmfulness to plants, or parts of them, is often judged by the grower of the plant. It is then interpreted in terms of economic damage.

In contrast to disease-inciting organisms such as fungi and bacteria that produce macroscopically visible molds, fruiting bodies or colonies on or in affected tissues, viruses were long considered not to contribute physically (corpuscularly) to the *syndrome* (characteristic complex of symptoms). However, a number of viruses are now known to aggregate within the cells into crystalline *inclusion bodies* that are visible with the light microscope. Other viruses that occur in lower concentration may be visible with the electron microscope in ultrathin sections of special cell organelles such as the nucleus or in endoplasmic reticulum. Several viruses produce virus-coded non-structural proteins (proteins not used as coat proteins for

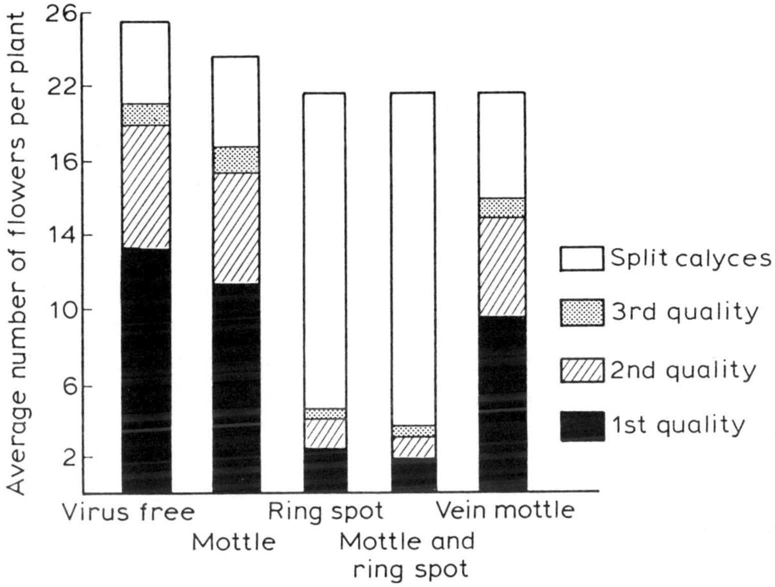

Fig. 8.1. Effect of carnation viruses on numbers and quality of flowers of carnation cv. William Sim (After Hakkaart, 1964; in Bos, 1982).

assembling virus particles), which aggregate into extensive *amorphous* and/or *crystalline inclusion bodies* (Fig. 8.2; see color plate). Shape and structure of virus-induced inclusion bodies are usually highly characteristic of the virus that has caused them, and they are formed irrespective of the plant species infected. Such inclusions are then part of the syndrome.

Pathogenesis

How viruses induce disease remains largely uncertain. Viruses multiply like normal host macromolecules, using the host's ribosomes and energy for replicating their own nucleic acid. They do so by redirecting the host's biochemistry to use its building blocks (amino acids and nucleotides) for assembling viral proteins and viral nucleic acids, respectively. First steps in disease development, therefore, are biochemical and physiological changes.

For a long time, much effort has been spent on detecting such aberrations as possible early signs of infection to be used in diagnosis. Decreases have indeed been found in rates of photosynthesis, increases in respiration rates, increases in activity of certain enzymes such as polyphenol oxidases, and changes in activity of growth regulators. However, such differences between virus diseased and normal are quantitative rather than qualitative. The effects are extremely variable, as in normal metabolism, and depend closely on leaf age, diurnal and seasonal variations, and on external growth conditions (Bos, 1978; Matthews, 1991).

Proof of infection is the presence of newly formed virus in infected tissues. Most present diagnostic methods are directed towards detection of the virus particles themselves or of their proteins and nucleic acids, with amino-acid and nucleotide sequences distinct from those of host components (Chapter 14). Virus multiplication often occurs "explosively" and in massive amounts to ensure maximum virus survival and spread. This then has tremendous consequences for the host. For example, with potyviruses the production of large quantities of viral coat protein is known to be paralleled by the amount of non-structural virus-encoded protein. In this virus group the viral genome is translated into a large polyprotein, which is later cleft into one molecule of coat protein and a number of other proteins (Dougherty and Carrington, 1988). Some of these proteins play a role in the multiplication of the virus, some in virus infection and transmission, and some possibly in symptom development. Some of these proteins are produced in such an excess that they aggregate into crystals and pinwheels in specific inclusion bodies (Fig. 8.2; see color plate; see also Internal Symptoms). Most host abnormalities that result from virus infection may be the result of a shortage of constituents for the plant's own essential enzymes and for its building materials.

Most virus symptoms in plants derive, in essence, from *growth reduction*, *premature ageing* (involving discolorations, which are general as in chlorosis and yellowing, or concern localized areas such as in variegations), and *premature death* of localized tissues (necrosis) or of entire plants. These phenomena largely resemble the basic physiological aberrations underlying other plant diseases. However, virus diseases fundamentally differ in that virus infections usually are systemic; that is, that most viruses spread internally throughout the plant together with assimilates, as if they were part of the host's own production and distribution system. That is why entire plants become involved so that *local symptoms* (developing at the site or on the leaf of virus entrance) are followed by *systemic symptoms*, and most virus symptoms show up in plant parts that develop after infection.

Symptoms may, and often do, gradually build up in severity when one symptom leads to another so that plants *decline* and often prematurely die, as, for example, with yellows diseases caused by luteoviruses and some mycoplasmas. They occur only in the phloem of vascular bundles, and infection by them primarily leads to *phloem degeneration* and often to *phloem necrosis* (Fig. 8.3).

Fig. 8.3. Vascular necrosis due to phloem degeneration or necrosis in corms of gladiolus, infected by mycoplasma. Left, healthy control. (After Groen and Van Slogteren, 1982).

In turn, this leads to transport problems and accumulation of starch in leaves and leafrolling, and to depletion of starch in the roots. This then results in poor root functioning so that nutrient uptake is poor and the ensuing mineral deficiency and chlorosis and yellowing in leaves (Fig. 8.11; see color plate) further contribute to *premature ageing* of the plant (*plant senescence*). With other viruses, particularly nepoviruses and ilarviruses, an initial *shock phase* may be gradually followed by a much less severe *recovery phase*. Symptoms may also be temporarily absent (*symptom masking*), usually under special climatic conditions, whereafter they may resurge.

Tissue injury often remains localized in the leaves where the virus has been introduced, but it may also occur in systemically infected leaves. Such localized tissue injuries are called *lesions*, and they are *local lesions* if they develop at the site of primary infection (Fig. 8.4; see color plate), and *systemic lesions* if they develop in secondarily infected leaves. Rapid development of local lesions may be an indication of high tissue sensitivity (a local shock reaction followed by some sort of recovery in surrounding tissues), which is called *hypersensitivity* (especially when necrosis is involved) if no further systemic infection ensues. In systemically infected leaves, localized injury may appear in *ringspots* or in attractive *line patterns* (Fig. 8.5; see color plate). Such localized reactions suggest host mechanisms to fight back by producing antiviral substances in reaction to infection. The plants concerned may also be more resistant when subject to later reinfection (*acquired resistance*). Zones of severely and mildly reacting tissues may also occur alternatingly (*concentric ringspots*, Fig. 8.5; see color plate) suggesting that the balance between the production of virus and of antiviral substances is swinging back and forth while systemic infection is spreading. Local lesions are especially important in virus research where they often show up after sap-inoculation, and their number may be used as an indication of virus concentration in the inoculum used and of the susceptibility of the host.

The systemic nature of virus symptoms, such as leaf variegation and plant dwarfing, may often make them resemble genetic aberrations, and systemic chlorosis may be hard to distinguish visually from mineral deficiency or shortage of water. Viral diseases, however, can always be discriminated by their transmissibility, especially by grafting if sap transmission fails and no vector is known or available.

Plant viruses have often been and are still named mainly after the most characteristic external symptom or symptoms of the disease they cause or have first been associated with. Hence the importance of unambiguous symptom terminology.

Types of Virus-induced Symptoms

Anatomical abnormalities

Anatomical abnormalities resulting from virus infections in plants have often been reviewed (Esau, 1967). Light microscopy may reveal that cells are abnormal in number (*hyperplasia* or *cell proliferation*, and *hypoplasia*) or abnormal in size (*hyper-* and *hypotrophy* up to *atrophy*) or become necrotic (as in *vascular necrosis*, especially *phloem necrosis* such as in plants with luteoviruses). The highly virus-specific intracellular *inclusion bodies*, consisting of virus particles in highly or nearly crystalline array (*crystalline or banded inclusion bodies*) or consisting of virus-coded non-structural proteins, aggregated either in crystalline array or together with abnormal cell organelles (*amorphous or granular inclusion bodies*), have attracted special attention. Several of these bodies can be easily observed in epidermal strips (Fig. 8.2; see color plate), or in free-hand tissue sections, after staining. They are of great — though often neglected — diagnostic value (Christie and Edwardson, 1977, 1986).

Electron microscopy has allowed the study of subcellular abnormalities in ultrathin sections of diseased tissues. The technique, supplemented by serological and radioactive labelling, has revealed the corpuscular fate of the virus particles in infected cells in relationship with cell organelles. It has also has shown the involvement of virus particles in inclusion bodies, and further molecular-biological studies have shown that the pinwheels and scrolls in inclusion bodies of potyviruses are composed of non-structural virus proteins (Christie and Edwardson, 1977; see also Internal Symptoms). Virus particles and virus-coded proteins are often formed in massive amounts and the nature and extent of structural cellular abnormality may be dramatic.

Growth reduction

Growth reduction is the most prevalent visible indication of virus disease and is of major economic importance. *Yield reduction* and even total *crop failure* may result. Growth reduction may be one of a series of symptoms of a syndrome. It may also be the only abnormality resulting from infection, and it may then be hard to assign to infection and be easily overlooked. A *reduction in vitality* may be associated with or lead to growth reduction, but it is also responsible for *increased susceptibility and sensitivity* to secondary pathogens or to adverse climatic conditions, such as summer drought or winter cold.

Color deviations

Color deviations are more easily ascribed to viruses, especially when they occur in sharply bordered angular patches alternating with normal green tissue. The descriptive term *mosaic,* as in "mosaic disease" (cf. tobacco mosaic and *Abutilon* mosaic; Figs. 2.1 and 2.3, respectively, in Chapter 2), has long stood for "virus infection". In discolored leaf tissue, chloroplasts change in shape, clump together, and degenerate in various ways. They change in color from green to colorless, and this results in *chlorosis* of the tissues concerned. In the dark green islands of tissue, which contain very little virus, chloroplasts appear normal (Chalcroft and Matthews, 1966).

There is a range of terms denoting various forms of *variegation* other than mosaic. The names *flecking* or *spotting*, for example, are used when the discolored parts are circular, and *mottling* when they are diffusely bordered. In fact, mottling is much more prevalent than mosaic, and the terms are often confused. Systemically infected leaves may also show conspicuous patterns known as *line pattern*

Fig. 8.6. Vein mosaic in *Wisteria floribunda*, naturally infected by *Wisteria* vein mosaic potyvirus. (After Bos, 1970.)

(Fig. 8.5; see color plate), *ring-spotting* (Fig. 8.5; see color plate), and *ring-flecking*. The terms *vein mosaic* (as of *Wisteria* with *Wisteria* vein-mosaic potyvirus; Fig. 8.6) and *vein-banding* refer to discolorations occurring in mosaic patterns and bands associated with leaf veins, and *striping* or *streaking* to a mosaic in leaves of monocots, such as the ornamental Liliiflorae, with parallel veins which force the discolored patches to become elongate (Fig. 8.7; see color plate).

An overall reduction in chlorophyll production in foliage leaves and stems leads to general *chlorosis*, evenly distributed throughout plant organs or entire plants. *Yellowing* (Fig. 8.11; see color plate) is at hand when yellow pigments dominate in discolored tissue or are increased in concentration, and *blanching* refers to disappearance of all pigments. There are several diseases, particularly those caused by phloem-limited aphid-transmitted luteoviruses and whitefly-transmitted geminiviruses, where yellowing or chlorosis, though occurring regularly, is limited to leaf tissue adjacent to and including the veins, and this is called *vein-yellowing* (Fig. 8.8; see color plate). *Vein chlorosis* often is a first (and transient) reaction to systemic invasion of leaf veins by a virus. Leaf *reddening* and *purpling* are physiologically related to chlorosis and yellowing, and they occur in plant species infected by luteoviruses or other phloem-limited virus-like pathogens where the disturbance of sugar metabolism leads to accumulation of anthocyanins. In such plants, reddening and yellowing are associated with starch accumulation in the leaves and with leaf curling and brittleness. *Browning* and *blackening* result from tissue necrosis (see below), and *bronzing* from such necrosis in epidermal tissue overlying normally green tissues. That is why tomato spotted wilt tospovirus, often causing this abnormality, is called "Tomatenbronzeflecken-Virus" in German and "tomatebronsvlekkenvirus" in Dutch.

Flower color-breaking is a mosaic or variegation of flower petals, at first sight an attractive feature in several species of ornamentals. Examples are tulips with tulip mosaic potyvirus (Chapter 2, Fig. 2.2; see color plate), iris with iris severe mosaic potyvirus, stock (*Matthiola incana*), common wallflower, and several other plant species with turnip mosaic potyvirus or other viruses, and gladiolus with bean yellow mosaic potyvirus and cucumber mosaic cucumovirus. Pigments then locally fade (*light breaking*), intensify, or accumulate (*dark breaking*) in the epidermal layer of petals. Such color-breaking cannot be observed in white or yellow cultivars since pigments are lacking in the epdermis of their petals. In tulips, dark breaking can also be caused by tobacco rattle tobravirus.

Water deficiency

Water deficiency caused by excessive transpiration or by impeded water supply leads to *wilting*, or more extremely, to *withering* which irreversibly entails death of tissue. *Etching* means desiccation of superficial tissues like the epidermis and outer parenchyma, as, for example, in leaves of carnations with carnation etched ring caulimovirus.

Necrosis

Necrosis covers a range of symptoms where tissue death remains localized, as in necrotic local lesions (Fig. 8.4; see color plate), and is usually associated with the production of melanin-like substances which lead to *blackening* or *browning*. In systemic infections, it may remain limited to the phloem of vascular bundles (*phloem necrosis*) or extend into adjacent tissues and become externally visible and develop into *vein necrosis* or *necrotic streaking* of petioles and stems (Fig. 8.9; see color plate). *Tip necrosis* and *stem necrosis* may extend and lead to whole-plant death.

Malformation

Malformation is very common in virus diseases. It usually results from imbalanced development of leaves, stems, and fruits when tissue growth is locally impeded by chlorosis (Chapter 2, Fig. 2.1) or completely arrested by necrosis (Fig. 8.9; see color plate). Some types are *leaf-curling* and *leaf-distortion*. *Leaf-rolling* is a more regular phenomenon. It is usually associated with accumulation of starch and leaf brittleness, as in plants infected by luteoviruses. *Leaf-narrowing* (restricted expansion of laminar tissue, up to complete absence so that the main veins remain: "shoe stringing"), *rugosity* (retarded growth of veinal tissue), and *rosetting* (impeded internodal expansion at stem tips) suggest hormonal imbalances. Infected plants may also branch more profusely, but this is mostly a matter of reduced apical dominance resulting from retarded development or necrosis of terminal buds.

New tissues or organs may also arise after virus infection when certain cells escape the host's regulatory mechanisms and proliferate independently into poorly differentiated tissues. Small outgrowths on leaves are called *enations* (Fig. 8.10; see color plate), and larger swellings on stems and roots are called *tumors*.

Symptoms Caused by Virus-like Pathogens

The symptoms caused by virus-like pathogens require special attention here because they are often confused with those of viruses. The diseases concerned also resemble those caused by viruses in their ecology and in various ways of control. Since the classical investigations in the US on peach yellows by Erwin Smith (1888) and later those of Kunkel (1926, 1931) on aster yellows — affecting an extremely wide range of plant species, including many ornamentals belonging to widely different plant families (Fig. 8.11; see color plate) — the diseases were often called "yellows-type diseases". They are spread by phloem-feeding vectors, mostly leafhoppers, and are now ascribed to the phloem-inhabiting mycoplasmas (see Chapter 7 on mycoplasmas). They are known to cause *phloem degeneration* as one of their primary symptoms, thereby often causing chlorosis and yellowing (Fig. 8.11) like the luteoviruses, which also are phloem-limited. Mycoplasma-infected plants may first react to systemic infection with vein clearing and vein chlorosis, as do many virus-infected plants. Several mycoplasmas cause more or less rapid death of phloem tissue, as in gladiolus with aster yellows (Fig. 8.3). Plant decline and plant death may then proceed more or less rapidly, as with witches' broom and decline of *Eucalyptus*, or be the predominant symptoms, as with elm phloem necrosis.

Most characteristic, however, of mycoplasma infections are the hormonal disturbances they cause. These are obvious from aberrations in growth-hormone dependent processes such as floral initiation, periodicity, and geotropy. In mycoplasma-diseased plants, flower development is disturbed or completely arrested, even when floral initiation had already proceeded partially before infection. That is why a variety of floral abnormalities may be observed, viz., a range of morphological transitions between normal flowers and vegetative leafy shoots. In plant teratology — the ancient science of growth abnormalities or "monstrosities" of plants (Bos, 1957b) — this phenomenon had long been called *antholysis*. The term refers to a metamorphosis which makes flower organs become more or less foliaceous. Antholysis consists of *virescence* (greening of the floral corolla, Figs. 8.11 and 8.12; see color plate), *phyllody* (the transition of all floral parts, including stamens and carpels, into foliage leaves), *apostasis* (the elongation of internodes between the toral nodes), and *prolification* or *proliferation* (the development of leafy sprouts at the top of the torus and in the axils of floral parts) (Fig. 8.13). Vegetative development is thus considerably enhanced, even of parts of the plant (flowers) that were meant for sexual reproduction, which are often completely suppressed (Fig. 8.13, Fig. 8.14; see color plate). Precocious development throughout the plant of all axillary buds that are potentially present results in systemic *witches' broom growth*. Localized witches' brooms are well known from birch trees and other tree species with localized infection by certain fungi or localized infestation by certain mites. Part of the mycoplasma syndrome is the erect, upright growth direction of newly formed branches and the upward position of stems of flowers and fruits that normally hang downwards (*enhanced negative geotropy*). Plants that have become infected early during development thus develop entirely into dwarfed, densely bushy structures. Suppression of sexual plant development renders affected plants completely sterile (*sterility*). Seeds that were formed in fruits prior to infection may prematurely develop into sprouts while still on plants, and this *disturbed periodicity* is another clear indication of hormonal disturbance.

Thus we are dealing with a complex of *witches' broom phenomena*. They comprise witches' broom growth in vegetative parts of the plants and a complex of floral transitions (antholysis) that contribute to the overall witches' broom development of affected plants. For mycoplasma diseases of plants the name *(systemic) witches' broom diseases* is more descriptive and appropriate than yellows-type diseases. Morphological abnormalities may not be prominent when

Fig. 8.13. Advanced stage of antholysis in garden nasturtium (*Tropaeolum majus*) caused by mycoplasma infection. Phylloid sepals, petals, and stamens in rosette at basis, followed by elongation of torus, phyllody of the three carpels (c) and proliferation (development of leafy sprouts) from axils of carpels and at the top of torus. (After Bos, 1957.)

phloem degeneration dominates and infected plants turn yellow and rapidly decline rather than show new outgrowths, as in gladiolus with aster yellows during the first season of infection (see also Fig. 8.3). Elm phloem necrosis in the US (Wilson et al., 1972) and Italy is even more extreme. Most characteristic and highly diagnostic of the vast majority of mycoplasma infections, however, are the morphological aberrations resulting from hormonal disturbance (Bos, 1957, 1978).

Mycoplasma diseases of plants have been reported in the literature under a wide variety of names (Bos, 1957; Grunewaldt-Stöcker and Nienhaus, 1977), which suggest non-existent differences. *Witches' broom*, as of hibiscus, lilac, and passionflower, or *witches' broom disease*, as of locust, is the best name covering the syndromes caused by mycoplasmas in plants. Some other names of mycoplasma diseases in ornamentals, referring to single symptoms only, are: *brooming disease* of black locust (*Robinia pseudacacia*); *decline* of lavender; *dieback* of eucalyptus (witches' broom and dieback) and *Sambucus*; *green flowering disease* of *Justicia gendarussa*; *little leaf* of *Eucalyptus*; *systemic brooming* of *Robinia*; *virescence* of *Hydrangea*; *witches' broom and decline* of *Eucalyptus* (Fig. 8.14; see color plate); and *yellows* of aster. Related names of similar diseases in other crops are *bunchy top*, *dwarf*, *false blossom*, *grassy shoot*, *green flowering disease*, *green petal*, *greening*, *phyllody*, *proliferation*, *rosette*, and *stunt*. The incitants of most of these diseases have meanwhile been confirmed to be mycoplasmas, but their distinction at the species level remains uncertain. Most of them were found to naturally or artificially infect wide ranges of plant species.

Selected Literature

Bos, L., 1957a. Heksenbezemverschijnselen, een pathologisch-morfologisch onderzoek (With a summary: Witches' broom phenomena, a patho-morphological study). Mededelingen van de Landbouwhogeschool te Wageningen 57 (1): 79 pp.

Bos, L., 1957b. Plant teratology and plant pathology. Tijdschrift over Plantenziekten 63: 222–231.

Bos, L., 1978. Symptoms of virus diseases in plants; with indexes of names of symptoms in English, Dutch, German, French, Italian, and Spanish, 3rd ed. Pudoc, Wageningen.

Bos, L., 1981. Wild plants in the ecology of virus diseases: In: K. Maramorosch and K.F. Harris (eds.), Plant Diseases and Vectors. Ecology and Epidemiology. Academic Press, New York, pp. 1–33.

Bos, L., 1982. Crop losses caused by viruses. Crop Protection 1:263–282.

Bos, L., 1983. Introduction to Plant Virology, Pudoc, Wageningen/Longman, London and New York.

Chalcroft, J.P. and Matthews, R.E.F., 1966. Cytological changes induced by turnip yellow mosaic virus infections. Virology 28:555–562.

Christie, R.G. and Edwardson, J.R., 1977. Light and electron microscopy of plant virus inclusions. Florida Agricultural Experiment Stations Monograph Series Nr. 8 (2nd printing 1982).

Christie, R.G. and Edwardson, J.R., 1986. Light microscopic techniques for detection of plant virus inclusions. Plant Dis. 70:273–279.

Dougherty, W.G. and Carrington, J.C., 1988. Expression and function of potyviral gene products. Annu. Rev. Phytopathol. 26:123–143.

Esau, K., 1967. Anatomy of virus infections. Annu. Rev. Phytopathol. 5:45–70.

Grunewaldt-Stöcker, G. and Nienhaus, F., 1977. Mycoplasma-ähnliche Organismen als Krankheitserreger in Pflanzen (With a summary: Mycoplasma-like organisms as plant pathogens). Acta Phytomedica 5:1–115.

Kunkel, L.O., 1926. Studies on aster yellows. Am. J. Bot. 13:646–705; Also in Contributions Boyce Thompson Institute 1:181–240.

Kunkel, L.O., 1931. Studies on aster yellows in some new host plants. Contributions Boyce Thompson Institute 3:85–123.

Matthews, R.E.F., 1991. Disease symptoms and effcts on metabolism. In: Matthews, R.E.F., Plant Virology, 3rd ed., Academic Press, San Diego, pp. 379–422.

Smith, E.F., 1888. Peach yellows, a preliminary report. Bull. Div. Bot., U.S. Dep. Agric. 9:1–254.

Wilson, C.L., Seliskar, C.E. and Krause, C.R., 1972. Mycoplasmalike bodies associated with elm phloem necrosis. Phytopathology 62:140–143.

9

Intracellular Symptoms

D.-E. Lesemann
Federal Biological Research Centre for Agriculture and Forestry, Institute of Biochemistry and Plant Virology, Braunschweig, Germany

Introduction

External symptoms of virus-infected plants as described herein are the result of a complex set of virus-host interactions. They express on the one hand the degree of damage in the host organism which is exerted by the virus. This can show up, for example, as chloroplast damage leading to chlorosis of photosynthetic tissues, as eventual fatal breakdown of cellular functions resulting in necrotic cells and tissues, or as an imbalance of growth regulating factors indicated by an altered anatomy of the host plant like deformation of leaves or stunting of the shoots. On the other hand, the symptoms are not uniformly spread over the host plant but often arranged in certain virus-specific patterns. These patterns of symptom distribution are an indication for the complex physiological regulation mechanisms within a virus-infected plant. External symptoms are, like the intracellular symptoms addressed now, not confined or specifically attributable to ornamental plants.

Virus-infected cells can show a variety of certain pathological alterations which are likely to be typical for the reactions of the host cells to disturbing effects in general. So the breakdown of thylakoid membranes and accumulation of osmiophilic lipid globules, starch, and phytoferritin in plastids of chlorotic tissues is a quite nonspecific indication for disturbed plastid functions. These alterations can also be caused by other damaging biotic or abiotic factors. Similarly, cell death is accompanied by structural changes which are more typical for the reaction system of the host cell to fatal influences in general than for the virus causing this reaction.

One virus can induce very severe symptoms in some plant species but can also cause latent infections in other host plant species or under certain environmental conditions. Two points should be kept in mind when studying cytological effects of virus infections: (1) cytopathological alterations indicating secondary effects which are due to disease- or symptom-provoking imperfect adaptations of virus and host have to be clearly distinguished from the functional virus-specific alterations; (2) the specific processes represented by the virus-specific structurally recognizable effects of the biochemical events during virus propagation may in their clearest expression be studied in latently infected tissues.

In the following the primarily virus-specific cytological alterations, as they can be vizualized in the electron microscope, will be preferentially demonstrated. The alterations comprise more or less massive cellular inclusions (traditionally designated inclusion bodies if big enough as to be vizualized in the light microscope) which represent either accumulations of viral gene products, i.e. of virions and nonstructural proteins or accumulations of hypertrophied cellular membranes, or complexes mixed of either components. Apart from the latter alterations of the cellular membrane system, specifically constructed and/or located dsRNA-containing vesicles which are newly formed under the influence of specific virus infections have to be recognized. Several comprehensive reviews of this field or of specific aspects are available: Martelli and Russo (1977); Christie and Edwardson (1977); Francki et al. (1985); Lesemann (1988, 1991). Due to space limitations, the citations in the following text preferentially refer to these reviews as sources for the vast number of original publications which are the basis for the following very short review.

Accumulations of Viral Gene Products

Virions and their aggregates

Non-enveloped viruses with isometric particles

The final products of the infection of host cells, i.e. the virions, are in the case of non-enveloped plant viruses, irrespective of their type of nucleic acid, located as a rule in the cytoplasmic compartment (cytoplasm or nuclei) of the host cells. In ultrathin sections low concentrations of small isometric virions (c. 30 nm in diameter) will very often not be distinguishable from ribosomes when they occur scattered throughout the cytoplasm of infected cells. Only if they are aggregated or localized in specific tissues (e.g. luteoviruses in phloem cells), may they be recognized under certain conditions or if they reach high concentrations and are especially well preserved and contrasted. Tymoviruses and tombusviruses are examples of virus groups whose virions appear relatively discrete from equally large appearing ribosomes (Fig. 9.1). Many small isometric viruses like alfamo-, carmo-, cucumo-, faba-, necro-, tombus- and tymoviruses may form fairly well organized crystal-like aggregates (Fig. 9.2). These virus crystals are very characteristic and easily recognized when they become large enough to be observed with a light microscope (Christie and Edwardson, 1977). The particles of some virus groups like nepo- and comoviruses are characteristically arranged in specific tubules which traverse the cell walls (Francki et al., 1985).

Virions of the 50 nm large isometric dsDNA containing caulimoviruses are mostly localized within viroplasm-like structures in the cytoplasm (Fig. 9.23) but can also occur free in the cytoplasm (Fig. 9.3) accompanied only by proliferated endoplasmic reticulum (ER).

Non-enveloped virions with helical structure

Elongated particles also are not very conspicous if present in low concentrations and scattered in the host cells. However, often such particles are not equally distributed throughout the cell but aggregated or confined to more or less clearly recognizable cytoplasmic areas which are occupied by complex accumulations of normal cytoplasmic organelles, proliferated endoplasmic reticulum (ER), virus particles, and accumulations of nonstructural virus proteins. Most groups of elongated viruses like tobamo-, tobra-, hordei-, furo-, potex-, carla-, and closteroviruses, except the potyviruses, show a tendency to form more or less highly

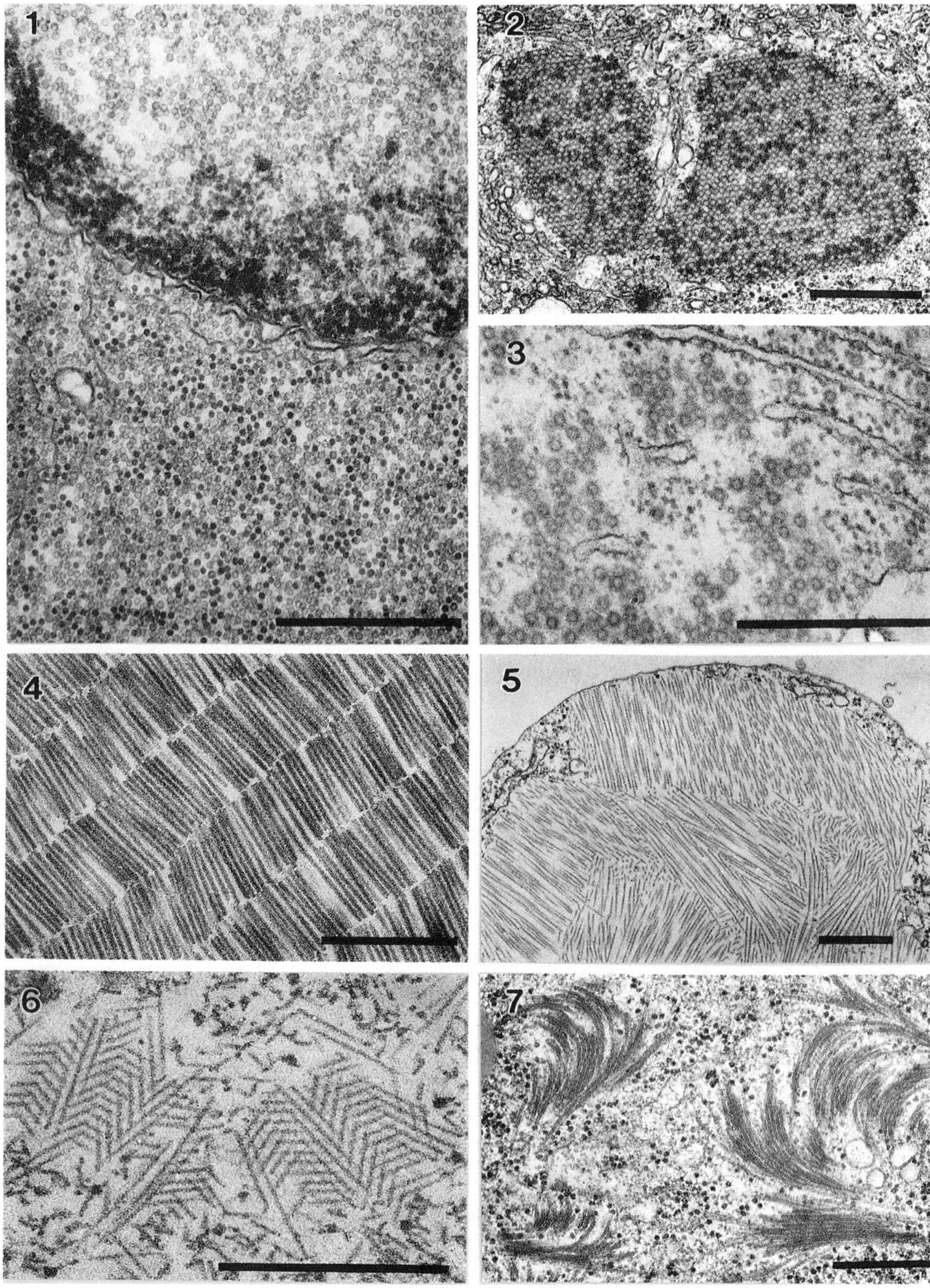
1
2
3
4
5
6
7

ordered particle aggregates, which mostly consist of side-to-side aggregates with the virion ends in register. Such aggregates appear as plates of the thickness of one (rarely of two) particle's lengths. These plates can occur singly but often are stacked together to form very massive paracrystalline or crystalline virion aggregates, as exemplified by tobamovirus crystals (Fig. 9.4). The crystals may be recognized *in vivo* in the light microscope, and their outer shape can be determined as hexagonal or rounded depending on the particular virus (Christie and Edwardson, 1977). The more flexible and longer the virions are, the more labile is the crystalline *in vivo* arrangement to disturbing influences during preparation of the tissues for microscopy. Thus, often masses composed of small virion aggregates in random orientation are the remnants of the originally highly regular structure of the aggregates. This can be found, for example, with tobamo- (Fig. 9.5), potex-, and closteroviruses.

Apart from the stacked plate aggregates, some tobamoviruses like odontoglossum ringspot virus and certain strains of tobacco mosaic virus (Christie and Edwardson, 1977; Francki et al., 1985) can form angle layer aggregates which consist of multiple monolayers of virions stacked onto each other in alternating orientations at angles of 60° (Fig. 9.6). A brush-like arrangement of virions is characteristic for virion aggregates of isolates of cowpea mild mottle virus (Fig. 9.7).

Virions of potyviruses may form massive bundles if produced in high concentrations. Those parts of the bundles which expose the virion ends are often in contact with cellular membranes or inclusions like pinwheels (Fig. 9.8). In many cases, however, potyvirus virions show the tendency to form monolayers, and these are usually attached to the tonoplast membrane. A common feature is the formation of cytoplasmic bridges through the vacuole which consist of a monolayer of virions lined only by two closely adjacent parts of the tonoplast (Fig. 9.9).

Opposite
Figs. 9.1–9.7. Arrangement of non-enveloped virions in cytoplasm of infected cells. 9.1. Nucleoprotein particles (black dots) and empty capsid protein shells (circles) of poinsettia mosaic (tymo?)virus in cytoplasm and vacuole (lower part) and empty shells in nucleus only (upper part). 9.2. Crystals of a mixture of nucleoprotein and empty shell particles of broad bean wilt fabavirus. 9.3. Clusters of 47 nm virions of petunia vein clearing (caulimo?)virus in the cytoplasm adjacent to proliferated ER. 9.4. Part of virion crystal of tobamovirus isolate Dorm. 9.5. Partially irregular virion arrangement in aggregate of tobamovirus isolate Aller, due to preparation artefacts. 9.6. Angle-layer virion aggregates of odontoglossum ringspot tobamovirus. 9.7. Brush-like virion aggregates of a Crotalaria isolate of cowpea mild mottle carlavirus. Bars=500 nm.

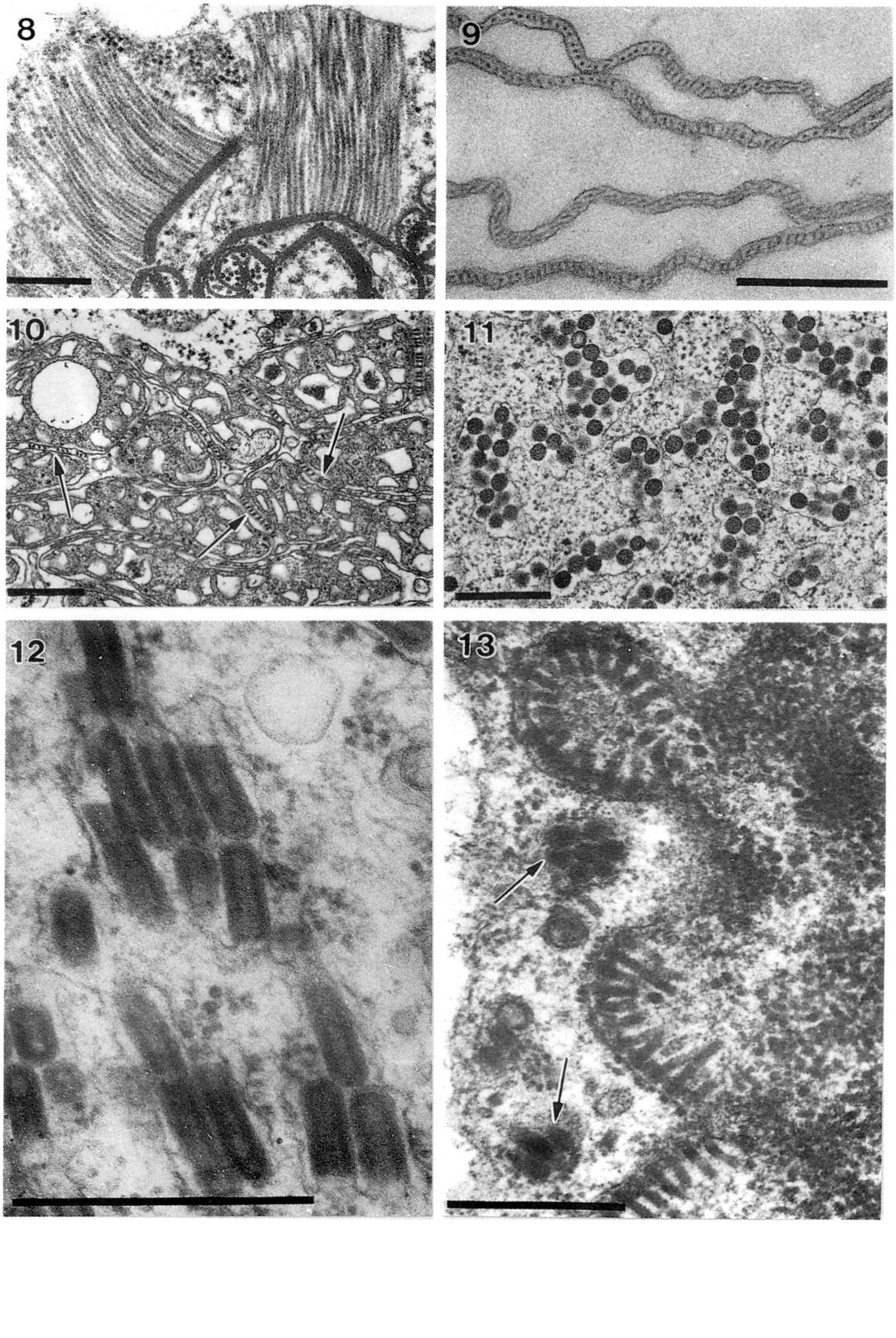
8
9
10
11
12
13

In other cases potyvirus particles may also exhibit a specific affinity to the outer membranes of mitochondria (Lesemann, 1988) or chloroplasts. The affinity of virion monolayers to the respective membranes is obviously strong enough to result in a conspicuous dense clumping of organelles like mitochondria or chloroplasts (Fig. 9.10).

Enveloped virions

Virions of the only two plant virus groups with enveloped particles, rhabdoviruses and tospoviruses, attain their host membrane-derived envelope by a budding process at ER or inner nuclear membrane sites. Membrane buds formed at these sites become filled with nucleocapsids synthesized in the cytoplasm or in viroplasm-like structures in the nucleus, break off from the budding sites, and thereby produce mature, membrane-enveloped virions in the extracytoplasmic compartments of ER (Figs. 9.11 and 9.12) and/or the perinuclear space. The maturation process of tospoviruses appears to be confined to cytoplasmic membranes. Here single particles or small clusters of mature particles are eventually found within ER cisternae (Fig. 9.11) (Francki et al., 1985). Fig. 9.12 shows mature rhabdovirus particles within ER-elements of *Phragmipedium* orchid. Immature particles (non-enveloped nucleocapsids) of particular types of rhabdoviruses, like orchid fleck virus which occurs in cultivated orchids, are produced in the cytoplasm or nucleoplasm and have a special affinity to the inner nuclear membrane and/or the ER where they can form characteristic spoke-wheel like arrangements (Fig. 9.13) (Francki et al., 1985).

Opposite

Figs. 9.8–9.13. Arrangement of virions of non-enveloped potyviruses and enveloped tospo- and rhabdoviruses. 9.8. Dense bundles of peanut stripe potyvirus associated with pinwheel structures (lower part) and tonoplast membrane (upper part). 9.9. Monolayers of turnip mosaic potyvirus (in cross section) within four bridges of cytoplasm traversing the vacuole of infected cell. 9.10. Monolayers of virions (arrows) of unidentified potyvirus between clumped mitochondria. 9.11. Groups of mature tomato spotted wilt virions within ER cisternae. 9.12. Groups of mature rhabdovirus particles from *Phragmipedium* orchid within ER cisternae. 9.13. Immature nucleocapsid particles of orchid fleck rhabdovirus at the inner nuclear membrane. Note spokewheel-like arrangement in the upper part and small bundles of mature particles within ER cisternae at left side (arrows). Bars=500 nm.

Inclusions consisting of nonstructural proteins

Potyviridae

Potyvirus-induced cytoplasmic and/or nuclear inclusions are outstanding examples of inclusions formed in infected cells additionally to virions and their aggregates. The formation of aggregates of nonstructural proteins may be favored by the genome organization of potyviruses which have only one large open reading frame (two for bymoviruses). This results in the primary production of one large polyprotein which is later processed into at least seven individual functional proteins. Consequently coat protein and the six nonstructural proteins (NSP) are produced in equimolar amounts. Thus, there will be high amounts of potyvirus NSP relative to NSP levels of viruses of other groups which individually regulate the expression of NSP genes. The relatively high amounts of NSP may favor their aggregation into massive proteinaceous inclusions. Four of the NSP have already been identified as forming specifically shaped inclusions, i.e. cylindical inclusion (CI) protein (homology to helicase and ATPase), nuclear inclusion proteins A and B (NI a and NI b) (RNA polymerase and protease plus Vpg, respectively), and helper component protein of aphid transmission (HC). Antisera to individual NSP could be produced using bacterially expressed fusion proteins and provide a tool for locating the proteins in inclusions on ultrathin sections (Hiebert and Dougherty, 1984; Brunt, 1988).

All members of the Potyviridae produce conspicuous cylindrical cytoplasmic inclusions (CI) (name referring to the outer shape), often also called pinwheel inclusions (PWI) with reference to their structure in cross section (Fig. 9.14). CI consist of CI proteins solely. Other NSP aggregates in inclusions are found with particular potyviruses only, although these NSP are produced by all potyviridae studied in this respect. Obviously their aggregation is not favored in all virus-host systems.

1. Cylindrical cytoplasmic inclusions — Pinwheel inclusions have long been known in ultrathin sections of potyvirus-infected cells (Edwardson, 1974; Edwardson and Christie, 1991). They all have the same principal construction but develop into differing final structures which result from the particular arrangement of their "pinwheel plates" into scrolls and/or laminated aggregates (Edwardson, 1974) or scrolls plus short curved laminated aggregates (Edwardson et al., 1984). On the basis of these pinwheel differentiations and their combinations four CI subdivisions have been defined by Edwardson and coworkers (Figs. 9.14–9.17), which appear very useful for obtaining a morphological classification of the CI types of

all well studied potyviruses. This is helpful in identifying and/or characterizing unknown potyviruses. CI types and their complex aggregates in infected cells can also be recognized in the light microscope and, with appropriate differential staining, used for virus identification (Christie and Edwardson, 1977).

2. Crystalline and amorphous nuclear inclusions (inclusions composed of NI) – Crystalline nuclear inclusions of tobacco etch potyvirus-infected cells have been long known, and more recent analyses have suggested that they are composed of equimolar amounts of Ni a and Ni b (Francki et al., 1985; Baunoch et al., 1991). A homologous composition has been shown for crystalline (also amorphous appearing) inclusions in nuclei (and also in the cytoplasm) of bean yellow mosaic potyvirus (Fig. 9.18) and plum pox potyvirus-infected cells (Hiebert et al., 1984; Martin et al., 1992). Recent unpublished results (L.D. Riedel and D.-E. Lesemann, 1994) indicate that Ni a and Ni b are also present in crystalline inclusions formed by henbane mosaic potyvirus in nuclei and cytoplasm (Fig. 9.19) and clover yellow vein potyvirus in nuclei (Fig. 9.20). Thus it might be suspected that these Ni proteins have a tendency to aggregate in the nuclear region of infected cells but can also be found aggregated in the cytoplasm.

3. Amorphous cytoplasmic inclusions containing HC — The HC protein is the fourth NSP of potyviruses identified in cytoplasmic inclusions. It occurs in massive amorphous cytoplasmic inclusions induced by the watermelon strain of papaya ringspot potyvirus (for references see Lesemann, 1988) and has also been detected in amorphous inclusions depicted in Fig. 9.21 induced by a potyvirus isolate from *Lagenaria* in Niger which is related to Moroccan watermelon mosaic potyvirus (D. Riedel and D.-E. Lesemann, unpublished results).

Examples of inclusions induced by other groups of ss-RNA viruses (tobamo- and potexviruses)

Tobamovirus-infected cells have long been known to contain "X-bodies" in addition to crystalline virion aggregates. For tobacco mosaic virus the "X-bodies" are known to consist of bundles of three to a few tubules (Francki et al., 1985). With other tobamoviruses X-body-like inclusions may consist of amorphous material (Fig. 9.22). For classical X-bodies of TMV and amorphous inclusions with a tobamovirus, it has been shown that they contain nonstructural proteins of the respective viruses (Hills et al., 1987; Saito et al., 1987). In the potexvirus group usually virion aggregates are the only virus-induced inclusions. However,

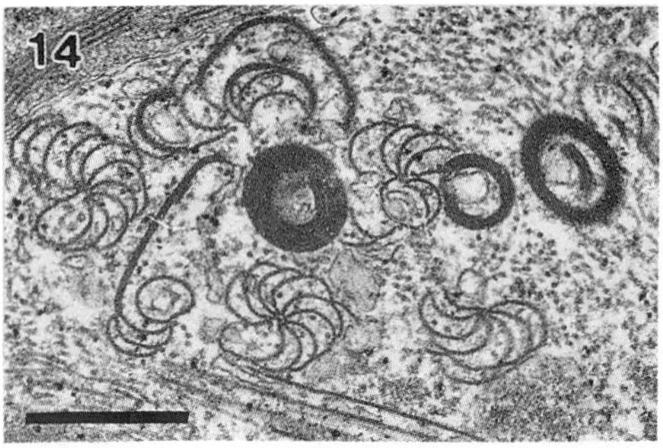
14

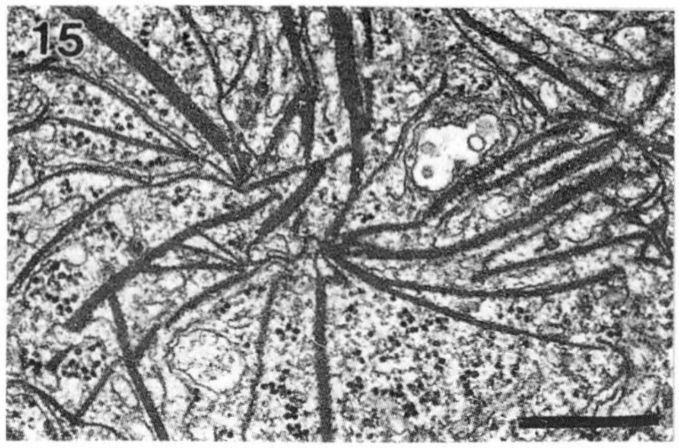
15

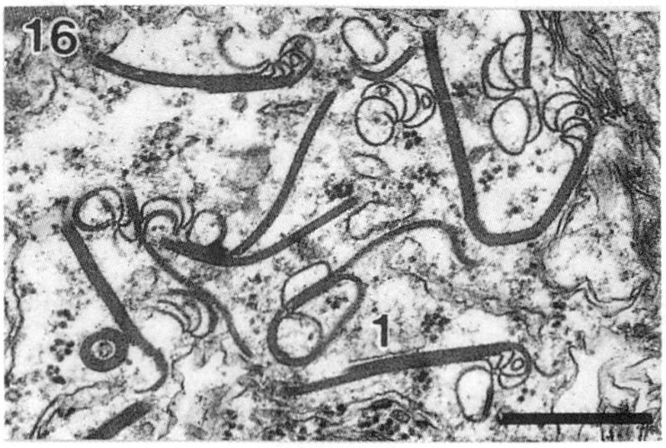
16
1

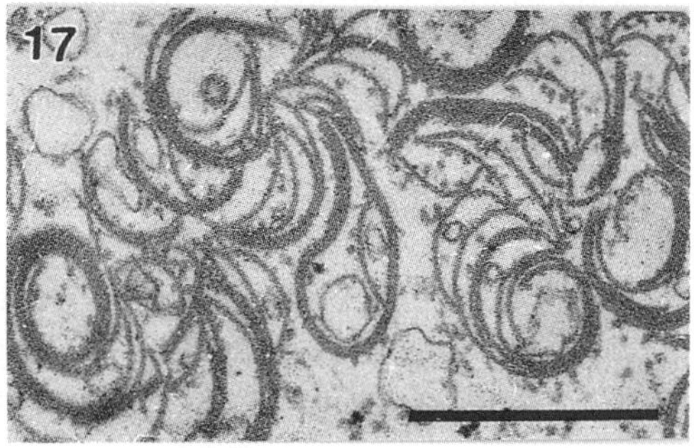
17

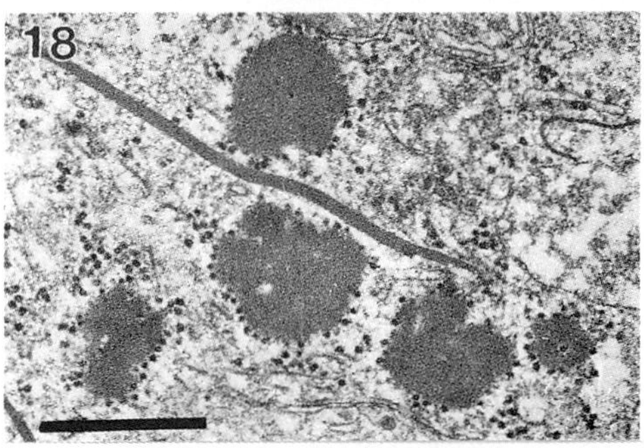
18

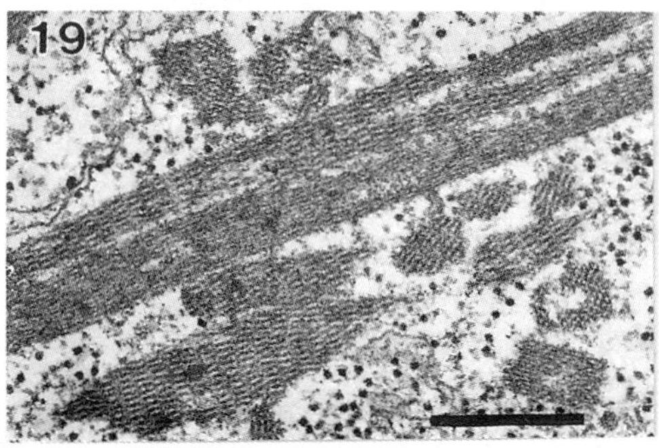
19

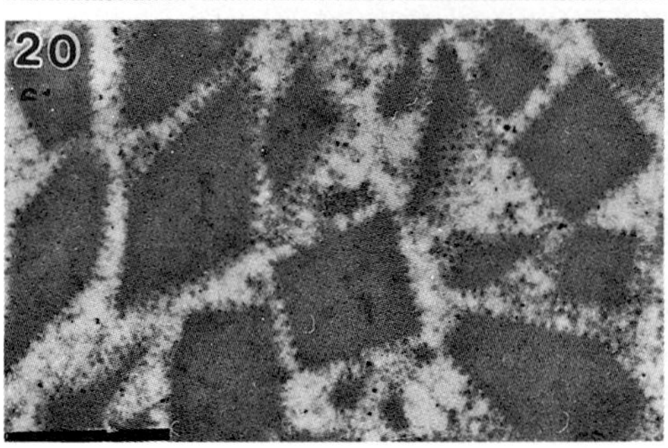
20

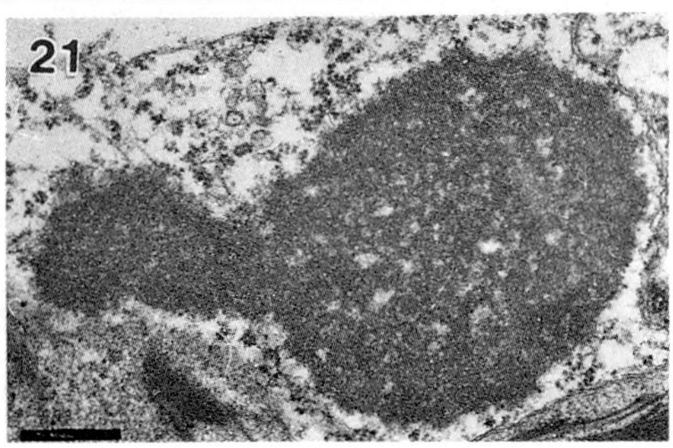
21

PVX and the virus cluster serologically related to boussingaultia mosaic virus produce very characteristic additional inclusions, apparently specific for these viruses. They produce laminated sheets densely covered by small beads (Francki et al., 1985) or numerous crystalline aggregates of unknown material in cytoplasm and nuclei (Lesemann, 1988), respectively. Both structures are suspected to be composed of NSP.

Inclusions of dsDNA viruses: caulimoviruses

This group of dsDNA viruses induces one or more types of amorphous cytoplasmic inclusions, also designated as viroplasms (Francki et al., 1985; Espinoza et al., 1991; Blanc et al., 1993). The mainly occurring viroplasms of caulimoviruses are oval to round, distinctly delimited, and dark staining masses, also known as "vacuolar viroplasms" because they contain lightly stained internal "lacunae". With most caulimoviruses the virions are exclusively found in these viroplasms (Fig. 9.23). But virions of petunia vein clearing virus, a possible member of the caulimovirus group (Richert, 1992), are always located outside the viroplasm (Fig. 9.3) (Lesemann and Casper, 1972). The viroplams of cauliflower mosaic virus have recently been shown to contain viral NSP (Espinoza et al., 1991; Blanc et al., 1993).

Rhabdo- and tospoviruses

Both these groups of enveloped plant viruses produce not only mature virion accumulations in extracytoplasmic compartments (see above) but also "viroplasm-like" accumulations of unknown material in the cytoplasm or the nucleus. The

---

Opposite

Figs. 9.14–9.21. Non-structural proteins forming inclusions of potyviruses. 9.14.–9.17. Cytoplasmic cylindrical inclusions in cross sections showing: 9.14. Typical pinwheel structures and scrolls of peanut stripe potyvirus. 9.15. Laminated aggregates of statice Y potyvirus. 9.16. Scrolls and laminated aggregates of turnip mosaic potyvirus. 9.17. Scrolls and short curved laminated aggregates of potato Y potyvirus. 9.18–9.20. Nuclear and cytoplasmic inclusions containing Ni a and Ni b proteins: 9.18. Amorphous-appearing cytoplasmic bodies formed by bean yellow mosaic potyvirus. 9.19. Crystalline cytoplasmic inclusions of henbane mosaic potyvirus. 9.20. Crystalline nucleolar inclusions of clover yellow vein potyvirus. 9.21. Amorphous cytoplasmic inclusion containing helper protein component of a virus isolate related to Moroccan watermelon mosaic potyvirus from *Lagenaria*. Bars=500 nm.

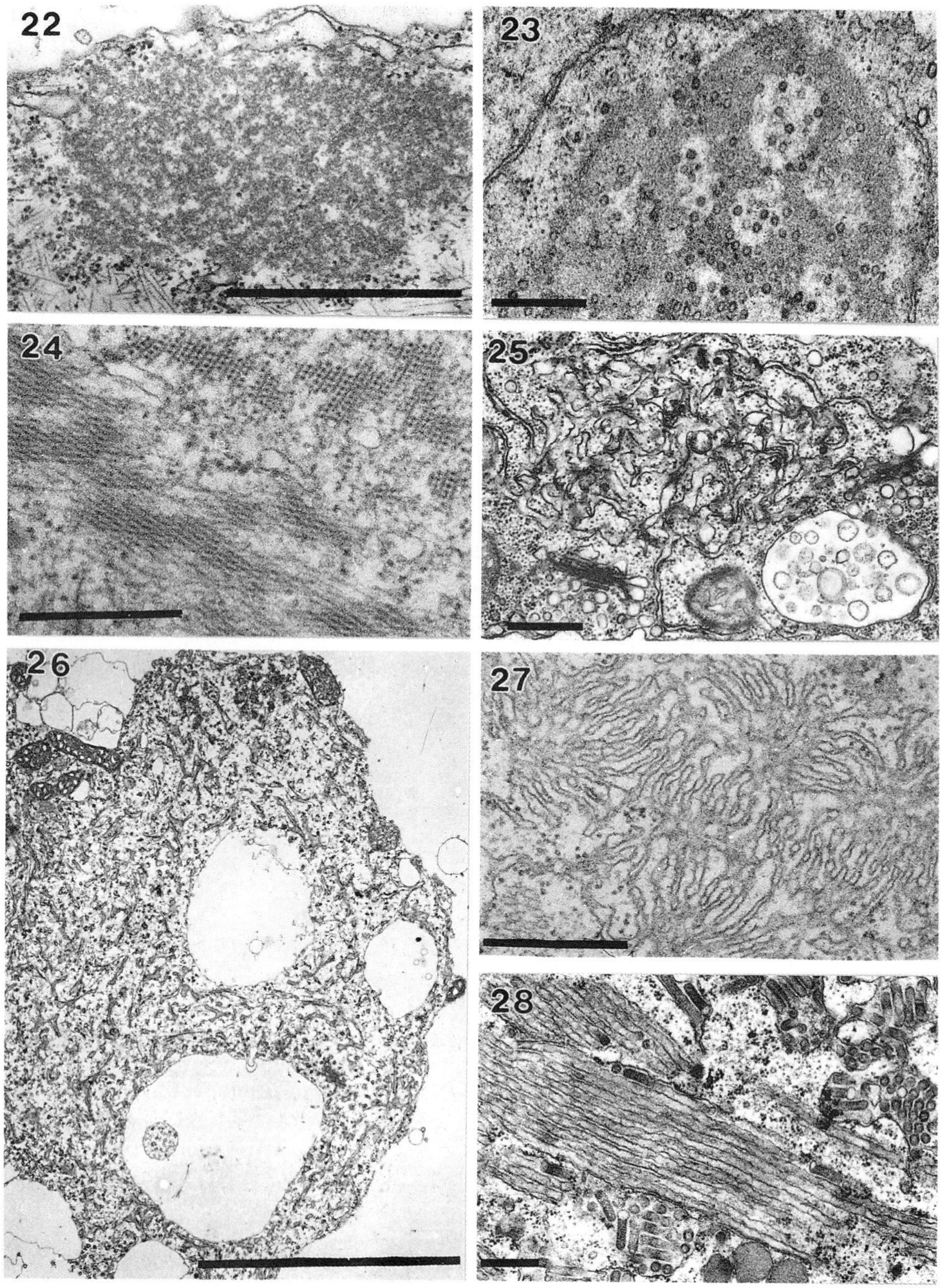
22
23
24
25
26
27
28

viroplasms may appear uniformly granular but can also contain convoluted tubular or thread-like material or nucleocapsids of immature virions in various stages of assembly (Francki et al., 1985).

Tomato spotted wilt (TSWV) and impatiens necrotic spot (INSV) tospoviruses both induce characteristic cytological features additional to virion accumulations. These may be viroplasm-like materials and/or accumulations of defective virions (dense bodies) lacking the membrane envelope. In the cytoplasmic viroplasms and in extended crystalline (INSV) or fibrillar (TSWV) proteinaceous aggregates NS_S, NSP has been detected (Fig. 9.24) (Kitajima et al., 1992; D. Riedel and D.-E. Lesemann, unpublished results).

Membrane Proliferations

The production of virions or NSP aggregates in the course of the viral infection is often accompanied by more or less conspicuous accumulations of proliferated ER-derived membranes. They may be an indication for the increased synthetic activity of virus-infected cells but may also be an expression of more specific influences of the viral infection on the cellular membrane system. While often barely recognizable, in some cases proliferated ER elements can loosely surround cytoplasmic areas in which virus production appears to be concentrated (e.g. petunia vein clearing virus, peanut clump furovirus) (Fig. 9.25); or they can accumulate to form more or less densely convoluted masses of ER-like membranes, including mitochondria, lipid droplets, and virus particles scattered or aggregated (Carla-, potex-, furo-, etc.) (Fig. 9.26). These complex accumulations can become big enough to be recognized as "amorphous granular inclusions" in the light microscope (Martelli and Russo, 1977). They can be differentially stained

---

Opposite

Figs. 9.22–9.28. Cytoplasmic aggregates of non-structural proteins of other plant viruses: 9.22. Amorphous X-body of maracuja mosaic tobamovirus. 9.23. Electron-dense vacuolate virion-containing viroplasm of cauliflower mosaic caulimovirus. 9.24. Paracrystalline aggregate of NS_S protein of impatiens necrotic spot tospovirus. Figs. 9.25–9.28. Virus-induced membrane proliferations: 9.25. Cluster of hypertrophied ER loosely surrounded by ER induced by peanut clump furovirus. 9.26. Large membraneous complex including small organelles and virions induced by kalanchoë latent carlavirus. 9.27. Finger-like ER-differentiation induced by bean necrosis mosaic potyvirus. 9.28. Excess production of membraneous tubules within ER cisternae induced by eggplant mottled dwarf rhabdovirus (Morocco strain). Bars= 500 nm.

by cytological dyes and used for diagnostic purposes (Christie and Edwardson, 1977). In some cases the membrane proliferations can attain a specific structure or arrangement, like the finger-like membrane complexes characteristically induced by bean necrosis mosaic virus isolates (Fig. 9.27) or the complex three-dimensional crystal-like arrays of ER-derived membranes ("coat of mail"-structures) induced by the members of the *Polymyxa*-transmitted bymovirus group (Lesemann, 1988). Some rhabdoviruses induce conspicuous vesicular or tubular membrane proliferations (Fig. 9.28) (El Maataoui et al., 1985).

Virus-specific dsRNA Containing Membrane-associated Vesicles and Multivesicular Bodies

Many of the groups of positive sense ssRNA viruses have been shown to induce the formation of particular kinds of small vesicles (for references see Francki et al., 1985; Lesemann, 1991). They are characterized by their content of fibrous structures proven in some cases to contain dsRNA. Such vesicles occur in two principal types. With picorna-like viruses they sometimes appear as large clusters of vesicles lying free in the cytoplasm, which are derived from vesicular buds formed at the ER (como-, nepo-, faba-, and potyvirus groups) (Fig. 9.29). However, with sindbis-like viruses to which most other groups of ssRNA viruses belong, c. 100 nm thick membrane-associated vesicles are formed at cytoplasmic membranes as flask-like evaginations from the cytoplasmic side into the adjacent extracytoplasmic compartments. The actual membrane sites where they are formed are characteristic for the inducing virus or the virus group to which it belongs. For example, they are associated with the tonoplast with some necro- (Fig. 9.30),

---

Opposite

Figs. 9.29–9.34. dsRNA containing vesicles induced by positive sense ssRNA viruses. 9.29. Cluster of free vesicles, some with dsRNA-like material (arrows), induced by azuki bean mosaic potyvirus as example of the vesicle-type characteristic of picorna-like viruses. 9.30– 9.34. Membrane-associated vesicles characteristic of sindbis-like viruses: 9.30. At tonoplast induced by the NFT-strain of tobacco necrosis necrovirus; inset: enlarged view of tonoplast-associated vesicle showing clear dsRNA-like fibrous material, induced by groundnut rosette virus. 9.31. At ER membrane (arrows) induced by another isolate of tobacco necrosis necrovirus. 9.32. At peroxisome induced by Moroccan pepper tombusvirus. 9.33. At mitochondrium induced by dendrobium vein necrosis clostero-like virus. 9.34. Double membrane-bounded invaginations induced by belladonna mottle tymovirus at the chloroplast envelope. Bars=500 nm (100 nm for inset of Fig. 26).

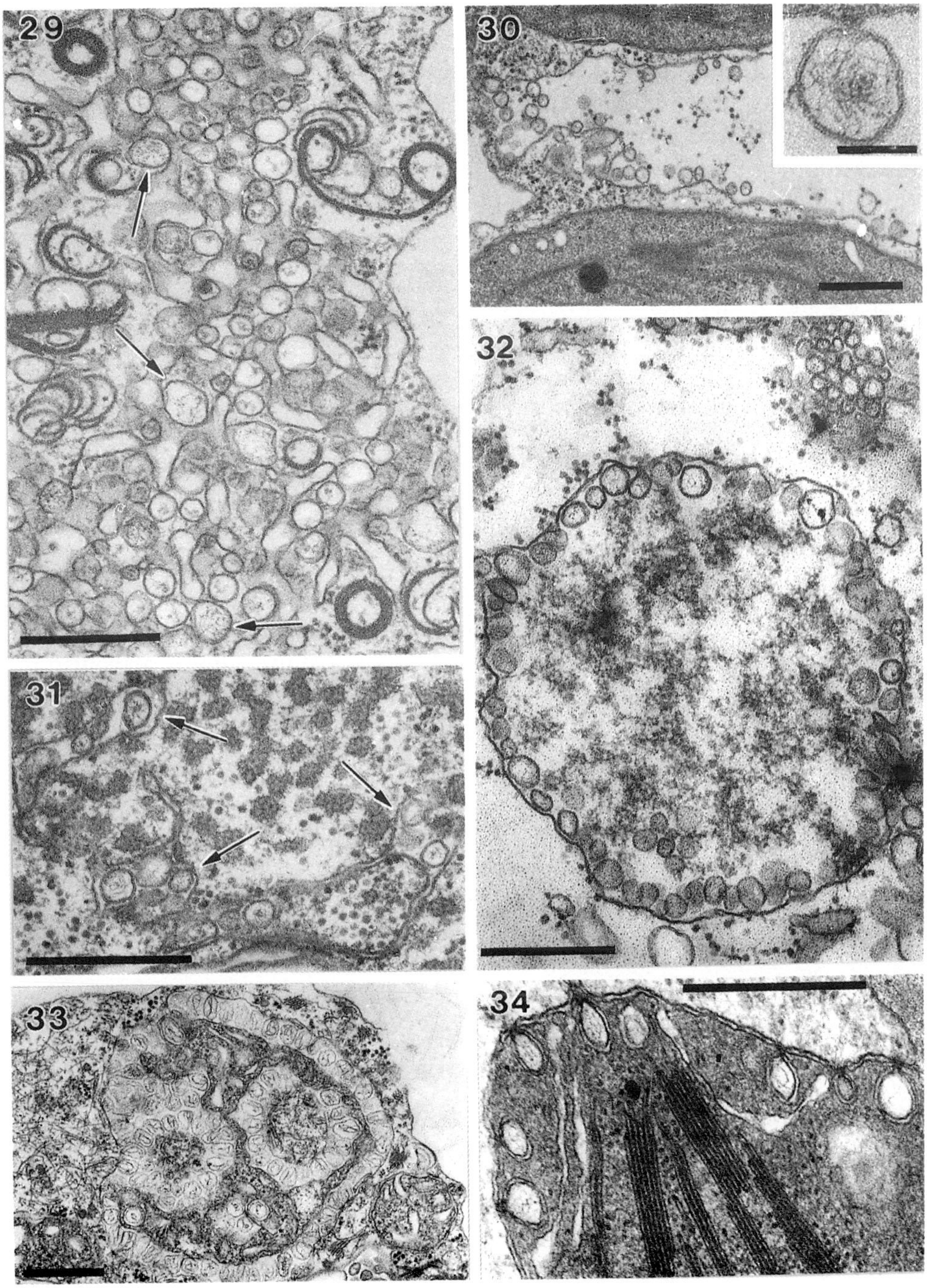
29
30
31
32
33
34

tobamo- and potexviruses, with ER with some necro- (Fig. 9.31), potex-, luteo- and carmoviruses, with the nuclear envelope with some luteo- and carmoviruses, with peroxisomes with most tombus- (Fig. 9.32), some carmo- and furoviruses, with mitochondria with some tombus- and a possible closterovirus (Fig. 9.33), and with chloroplasts with most tymoviruses (Fig. 9.34). With some virus groups like tombus- or tymoviruses the organelles (peroxisomes and chloroplasts, respectively), whose membranes carry associated vesicles, are eventually pathologically vacuolized and disorganized by extreme proliferation of membranes so that the resulting multivesicular bodies can completely obscure the cytological origin of the vesicles and the normal cytological organization of the host cells.

Complex Agglomerations of Virus-induced Cytological Alterations

The intracellular symptoms of virus-infected cells are very often not confined to one or several of the above-described features, but the infected cells contain cytoplasmic inclusion complexes (Martelli and Russo, 1977). Such a complex may include — for example, with carlaviruses virion aggregates — scattered virions, proliferated ER, cellular organelles such as peroxisomes and mitochondria, and lipid droplets (Fig. 9.26), or — for example, with potyviruses virion aggregates — virions associated with cylindrical inclusions, the distinct cylindrical inclusions, amorphous inclusions composed of HC, crystalline inclusions containing Ni a and Ni b proteins, and again small organelles and lipid droplets and clusters of free dsRNA containing vesicles. Such complexes may form masses larger than a nucleus and are often clearly differentiated by their morphology as inclusion masses within the infected cell, either near the nucleus or at other cytoplasmic places. In certain cases conspicuous complexes of clumped chloroplasts (tymoviruses, some potyviruses) or of mitochondria are formed and represent the most evident sign of an altered cytology.

The Use of Cytological Alterations in Virus Diagnosis and Classification

The internal symptoms described in this chapter do not represent sequences of pathological reactions to cell damage which would depend merely on the genetically fixed general reaction pattern of the cells to damaging outer or inner influences. What has been described here represents, however, specific effects of the respective virus in terms of alterations of the cellular membrane system or specifically structured accumulations of gene products of the infecting viruses. These features can be virus-specific or can be virus group-specific. Therefore, the observed specific internal symptoms can be very indicative of the identity of the infecting unknown virus or at least for its group relationship. The more complex the intracellular alterations caused by an unknown virus are, the more they can help in classifying and/or identifying this virus. Examples are (1) the CI and other NSP inclusions of potyvirus-infected cells which on the basis of the knowledge of characteristic inclusion combinations may lead to a very direct serological testing for the final identification (e.g. of beet mosaic or tobacco etch or potato Y viruses); (2) the typical localization of vesicles at outer chloroplast membranes pointing to tymoviruses or at peroxisomes pointing to either tombus-, carmo-, or very few other viruses; or (3) the "coat of mail" membrane aggregates suggesting bymoviruses (Lesemann, 1988).

The often feasible possibility to study intracellular symptoms by light microscopy, and more in detail by electron microscopy, offers a technically easy approach to a sound preliminary characterization or identification of viruses. This is especially true for those unknown or incompletely characterized viruses of which other, more intrinsic properties could not yet be determined because of technical problems in virus transmission, propagation, or purification.

Selected Literature

Baunoch, D.A., Das, P., Browning, M.E. and Hari, V. 1991. A temporal study of the expression of the capsid, cytoplasmic inclusion and nuclear inclusion proteins of tobacco etch potyvirus in infected plants. J. Gen. Virol. 72:487–492.

Blanc, S., Schmidt, I., Kuhl, G., Esperandieu, P., Lebeurieur, G., Hull, R., Cerutti, M. and Louis, C. 1993. Paracrystalline structure of cauliflower mosaic virus aphid transmission factor produced both in plants and in a heterologous system and relationship with a solubilized active form. Virology 197:283–292.

Brunt, A.A. 1988. Purification of filamentous viruses and virus-induced noncapsid proteins. In: Milne, R.G. (ed.), The Plant Viruses, Vol. 4. Plenum, New York, pp. 85–110.

Christie, R.G. and Edwardson, J.R. 1977. Light and electron microscopy of plant virus inclusions. Fla. Agric. Exp. Stn. Monogr. Ser. No. 9.

Edwardson, J.R. 1974. Some properties of the potato virus Y-group. Fla. Agric. Exp. Stn. Monogr. Ser. No. 4.

Edwardson, J.R., Christie, R.G. and Ko, N.J. 1984. Potyvirus cylindrical inclusions — subdivision IV. Phytopathology 74:1111–1114.

El Maataoui, M., Lockhart, B.E.L. and Lesemann, D.-E. 1985. Biological, serological and cytopathological properties of tomato vein-yellowing virus, a rhabdovirus occurring in tomato in Morocco. Phytopathology 75:109–115.

Espinoza, A.M., Medina, V., Hull, R. and Markham, P.G. 1991. Cauliflower mosaic virus gene II product forms distinct inclusion bodies in infected plant cells. Virology 185: 337–344.

Francki, R.I.B., Milne, R.G. and Hatta, I. 1985. Atlas of Plant viruses, Vols. I and II. CRC Press, Boca Raton.

Hiebert, E., Purcifull, D.E. and Christie, R.G. 1984. Purification and immunological analyses of plant viral inclusion bodies. Meth. Virol. 8:225–280.

Hiebert, E. and Dougherty, W.C. 1988. Organization and expression of the viral genomes. In: Milne, R.G. (ed.), The Plant Viruses, Vol. 4. Plenum, New York, pp. 159–178.

Hills, G.J., Plaskitt, K.A., Young, N.D., Dunnigan, D.D., Watts, J.W., Wilson, T.M.A. and Taitlin, M. 1987. Immunogold localization of the intracellular sites of structural and nonstructural tobacco mosaic virus proteins. Virology 161:488–496.

Kitajima, E.W., de Avila, A.C., Resende, O., Goldbach, R.W. and Peters, D. 1992. Comparative cytological and immunogold labelling studies on different isolates of tomato spotted wilt virus. J. Submicrosc. Cytol. Pathol. 24:1–14.

Lesemann, D.-E. 1988. Cytopathology. In: Milne, R.G. (ed.), The Plant Viruses, Vol. 4. Plenum, New York, pp. 179–235.

Lesemann, D.-E. 1991. Specific cytological alterations in virus-infected plant cells. In: Mendgen, K. and Lesemann, D.-E. (eds.), Electron Microscopy of Plant Pathogens. Springer, Berlin, pp. 147–159.

Lesemann, D.eE. and Casper, R. 1972. Electron microscopy of petunia vein clearing virus, an isometric plant virus associated with specific inclusions in petunia cells. Phytopathology 63:1118–1124.

Martelli, G.P. and Russo, M. 1977. Plant virus inclusions. Adv. Virus Res. 21:175–226.

Martin, M.T., Garcia, J.A., Cervera, M.T., Goldbach, R.W. and van Lent, J.W.M. 1992. Intracellular localization of three non-structural plum pox potyvirus proteins by immunogold labelling. Virus Res. 25:201–211.

Richert, K. 1992. Untersuchungen zur Charakterisierung des petunia vein-clearing virus (PVCV), ein samenübertragbares Pararetrovirus. Dissertation Göttingen.

Saito, T., Hosokawa, D., Meshi, T. and Okada, Y. 1987. Immunocytochemical localization of the 130K and 180K proteins (putative replicase components) of tobacco mosaic virus. Virology 160:477–481.

10

Transmission of Plant Viruses

Roger H. Lawson[1], A. Gera[2] and Hei-ti Hsu[1]
[1]*US Department of Agriculture, Agricultural Research Service, US National Arboretum, Floral and Nursery Plants Research Unit, Beltsville, MD 20705, USA*
[2]*Department of Virology, Agricultural Research Organization, The Volcani Center, Bet Dagan 50250, Israel*

Introduction

A characteristic common to plant viruses is their transmissibility. From the time viruses were first recognized as infectious agents, this common characteristic has been exploited for purposes of detection of viruses and virus-like agents. Viruses are obligate parasites and unaided are unable to pass through the cuticle of a plant. Viruses have evolved to avoid the need to penetrate the cuticle by passive transmission through vegetative propagation and seed transmission. Viruses penetrate the host through wounds as occurs in mechanical transmission or transmission by vectors. Understanding the nature of transmission for different viruses is essential in the development of nuclear stock programs and has been the cornerstone of research on this group of pathogens.

Plant viruses are transmitted in many different ways that include experimental transmission by mechanical rubbing, and naturally through grafting, pollen and seed, arthropods, nematodes, and even fungi. It is important to recognize these different modes of transmission for a particular virus or virus-like pathogen in order to understand its epidemiology. Knowledge of the factors that influence spread is essential for the development of control procedures.

Transmission of viruses is an essential aspect in studies of virus infectivity, genome composition, and structural morphology. Mechanical transmission of some viruses may also be important in plant-to-plant contact. Most field transmission of viruses in seed-produced crops probably results from specific insect or nematode vectors or through seed. In many vegetatively propagated ornamental crops, transmission occurs through passage of the disease agent through bulbs, corms, or rhizomes from underground or infected cuttings, scion wood, and rootstocks from diseased plants.

Mechanical Transmission

Transfer of a virus in sap from a diseased to a healthy plant was first recognized with the successful artificial inoculation of a healthy plant by injection of sap from a diseased plant. This classical experiment with TMV was repeated with many other viruses. Today, viruses in more than 20 of the 47 different groups are known to be mechanically transmitted.

Mechanical transmission of most viruses in natural field and greenhouse conditions is probably an uncommon event. There are, however, exceptions such as TMV in which the virus may spread by leaf contact. More likely, however, transmission occurs when plants are propagated by cutting divisions or flowers are harvested, as in the case of orchids infected with odonotoglossum ringspot virus, orchid tobacco mosaic (TMV-O), or clover yellow mosaic virus in sweetpeas. The chrysanthemum stunt viroid is an example of a pathogen that can be rapidly spread by leaf contact and handling plants.

Mechanical transmission of many viruses by leaf rubbing is a useful experimental tool as well as a valuable method for virus detection and disease diagnosis. Leaf rubbing with a finger, cheesecloth pad, or cotton swab is a common method of applying inoculum onto a leaf surface (Figs. 10.1 and 10.2). Inoculation efficiency is greatly increased with the use of celite or carborundum. The abrasive action of these materials creates micro wounds in surface cells that allow entry of the virus without causing sufficient mechanical damage to kill the cells. Abrasive materials may be dusted directly onto a leaf surface prior to rubbing or they may be added to the virus extract when the inoculum is prepared. The addition of a suitable abrasive is particularly important when the source virus is in low concentration. Abrasives may also be mixed with inocula and applied to a

Fig. 10.1. Application of carborundum to the leaves of a test plant prior to inoculation. Only a small quantity is applied that is barely visible.

Fig. 10.2. Sap extract prepared with a morter and pestle is rubbed onto a Carborundum-dusted leaf of an indicator plant.

susceptible host plant through a pressurized vessel such as an air brush. This method is particularly useful when a large population of plants is evaluated for resistance or mild strains are introduced to afford protection against more severe virus strains.

The addition of phosphate ions at the time tissue is extracted and inoculum is prepared often improves the infectivity of a preparation. Dipotassium phosphate, first used in a 1% concentration, was found to significantly increase the number of TMV local lesions on bean leaves. Phosphate buffer ranging from 0.05–0.1 M at pH 7.0–7.5 is now commonly used. Many different buffer systems have been developed for particular host-virus combinations with phosphate, citrate, and Tris buffers being most commonly used.

Although virus inoculum may be obtained from various parts of the infected donor plants, usually younger leaves contain a higher concentration of virus than older tissues. In some cases the roots of the infected plant may be a better source of inoculum, and occasionally the flower petals may be used.

Inhibitory plant constituents may reduce virus infection in extracts from many different species. Although these inhibitors may not inactivate virus particles, they prevent infection. For example, strong inhibitors occur in *Chenopodium* and *Dianthus*. These inhibitors may be overcome by diluting the inoculum. This method can be successful if the concentration of virus is high and the inhibitor is diluted more rapidly than the virus particles. In instances where virus titer is too low and the inoculum cannot be diluted, it may be possible to concentrate the virus by high-speed centrifugation to separate the particles from the inhibitors.

Tannins and other virus inactivators are present in many woody and herbaceous species. Infectivity can be retained by grinding infected leaves in solutions containing caffeine, nicotine, polyvinyl pyrrolidone (PVP), or at high pH (8–9) to minimize precipitation of the virus particles. Oxidative enzymes such as polyphenol oxidase may act as potent inhibitors of virus infection. Many different reducing agents such as mercaptoethanol, cysteine hydrochloride, sodium sulfite, and thioglycollic acid have been used to prevent oxidative inactivation of virus particles. Chelating agents such as sodium diethyldithiocarbamate (DIECA) have also been used to prevent inactivation.

A major factor influencing the success of mechanical inoculation is the type and condition of the host plant. Commonly used genera include different species of *Nicotiana* and *Chenopodium*. It should be noted that in some instances reproducibility of reported experimental results may depend on using the same seed source of the genotype reported. *Chenopodium amaranticolor* and *C. quinoa* are examples of two commonly used species that are susceptible to a wide range of viruses. Frequently, different species of host plants are required for studying characteristic symptoms for diagnostic purposes.

Growing conditions and plant age may greatly influence host susceptibility and/or symptom expression. For example, *Gomphrena globosa* inoculated with the TMV-O may show only small necrotic local lesions when test plants are grown at 32/29°C, day/night temperatures, but will show an abundance of large red lesions with necrotic centers on plants grown at 21/18°C, day/night temperatures (Lawson, 1990). A contrasting example is impatiens necrotic spot virus that produces more local lesions earlier on inoculated leaves of seedling impatiens grown at 27°C compared to 21°C.

The susceptibility of a plant depends on the light conditions as well as the temperature. Plants grown in lower light intensity and at lower temperatures are usually the most susceptible. Those grown in high light intensity and high temperature usually have very hard leaves that are less susceptible. If the environment for growing test plants cannot be easily controlled, it is possible to alter susceptibility by predarkening plants for 1 or 2 days prior to inoculation.

Age and stage of expansion of the inoculated leaves may also affect symptom expression. Generally, the most susceptible leaves are those that are only partially expanded when the inoculum is applied.

Washing leaves soon after inoculation is a common practice that may be beneficial with some virus-host combinations and may increase lesion expression. In other instances, however, this practice may actually reduce the number of local lesions. This procedure should be evaluated with each virus and host tested to empirically determine if infectivity is enhanced using this procedure.

Viruses transmitted in a persistent manner by leafhoppers or aphids are transmitted poorly or not at all by mechanical inoculation.

Graft Transmission

Grafting is a long established procedure used in propagation. Insertion of the cut surface of a bud or a scion into an understock of the same or related genotype will result in growth of a scion into an understock. This resulting union provides an avenue for movement of water and chemical elements from the rootstock into the scion or top portion of the graft.

Movement of viruses across a graft union has been recognized for many years. In the seventeenth century, Dutch growers found that tulip breaking virus could be transferred from bulb to bulb by grafting a diseased bulb onto a healthy one.

This secret was held by only a few growers who made great fortunes selling the bulbs with broken flower colors.

Grafts in dicotyledonous species are formed when there is a union between the cambium layer of the stock and scion and is usually most efficient when the two plants involved are from the same or related species. Virus transmission may occur, however, in the absence of such a union. Although a complete union is the most certain method of transmission, it is not the only one. Some viruses have been transmitted by wound contact along a graft line where some proliferation of parenchyma cells occurs in the absence of a vascular connection. Examples of successful transfer of virus from a diseased to a healthy plant include prunus necrotic ringspot in roses, strawberry latent ringspot in strawberry and some citrus, and grapevine viruses and viroids.

Grafting was at one time commonly used in virus testing and certification schemes for pathogens such as the chrysanthemum stunt viroid, but has been supplanted by other techniques such as nucleic acid hybridization probes. The method is still used, however, to detect viruses in woody ornamentals such as *Prunus* and continues to be a major component of international schemes for virus-indexing of nursery stocks. Grafting is also used to test for the presence of pathogens causing diseases of unknown etiology in woody plants.

Although grafting is usually considered the most definitive method of proof for the presence of a virus or viruslike agent, it may sometimes fail because of incomplete systemic invasion of the disease agent. In addition, that portion of the plant that is being grafted may not become infected with the scion material due to the erratic distribution of the virus such as plum pox virus. Failure to graft transmit may, therefore, result from the absence of the virus in the source tissue used for graft transmission. Despite its usefulness, graft inoculation is a laborious and time-consuming procedure, and in woody plants symptom development may take many months or sometime years.

Seed and Pollen Transmission

Although most plant viruses are not seed transmitted, the number has increased from about 30 in 1976 (Gibbs and Harrison, 1976) to 108 conventional viruses transmitted through seed or pollen in 1993 (Mink, 1993). Transmission rates may vary from a few hundredths of a percent to nearly 100% in some host-virus combinations. Some viruses, including several transmitted by nematodes, infect a high percentage of seed from in diseased mother plants.

Seed and pollen transmitted viruses occur in 25 taxonomic virus groups. Only in five of the 25 groups are all of the members known to be seed transmitted. Three of the groups — those containing alfalfa mosaic, maize chlorotic dwarf, and pea enation mosaic — are single-member groups. There are four members of the cucumovirus group, and they are all seed transmitted. At least half of the members of the comovirus, hordeivirus, ilarvirus, nepovirus, and tobravirus groups are seed transmitted.

Factors that influence the frequency of seed transmission include some that are host dependent and others that depend on the environment. These factors have been summarized (Mink, 1993). Seed-transmitted viruses are carried within the embryo with the exception of TMV which is a contaminant on the surface of tomato and pepper seeds. The virus can mechanically infect seedlings during germination and seedling development. Although a virus may infect the embryo, it is not necessarily seed transmitted. Also, seed transmissibility through the ovule parent depends upon the ability of a virus to infect floral tissue in early stages of development.

Frequency of seed transmission is determined primarily by two factors. First is the host-virus interaction and the second is the timing of infection. Temperature may also affect the frequency of transmission to some extent. Although the frequency of transmission may be as high as 100%, it rarely exceeds 50% for true viruses. Viruses transmitted through seed are also mechanically transmitted. These viruses infect parenchyma cells, in constrast to phloem-limited viruses that are not seed transmitted.

Virus infected pollen can result in seed infection through fertilization of a healthy mother plant. Frequency of transmission to seedlings is generally less than through ovules.

Although seed transmission is an important method for spreading viruses in several vegetable crops and fruit trees, it is not commonly reported for viruses in herbaceous ornamentals. In contrast, seed transmission in some woody ornamental shrubs and trees may be of great economic importance. For example, Prunus necrotic ringspot virus is pollen transmitted in *Prunus*. This virus and others in the ilarvirus group are present in pollen from infected plants. Cherry leaf roll nepovirus was transmitted to 100% and 0% of seeds from *N. rustica* plants grown at 20°C and 30°C, respectively. Tomato (TomRSV) and tobacco ringspot (TobRSV) viruses were transmitted from 1% to 30% of the progeny in plants experimentally infected with these viruses (Scarborough and Smith, 1977). TomRSV may become seedborne via infected female gametophytic tissue (ovule) or pollen, whereas tobacco ringspot transmission may occur only through the ovule. Spread of TobRSV in this manner is of little consequence in a clonally propagated geranium crop whereas pollen transmission of tomato ringspot within a geranium crop may represent a serious threat.

The presence of virus in pollen does not always mean the virus will be seed transmitted. For example, although cymbidium mosaic virus (CyMV) is present in orchid pollen, virus was not detected in progeny populations of cattleya seedlings of which either of both the female and male parents were virus infected (Lawson, unpublished). In one report, however, one of 123 *Dendrobium* seedlings from green pod culture was infected with CyMV when the female parent was infected (Yuen et al., 1979).

If seed transmission is to occur, the plant usually must be infected before the ovules are fertilized. The percentage of infected seed also depends on the time of flowering in relation to the time of infection.

The percentage of infected seed from a diseased plant also varies depending on the particular virus. For example, TobRSV may be transmitted in up to 100% in soybean. The extent of transmission also depends on the host genotype. Some viruses such as tomato black ring may infect many different species in several families while other viruses may be transmitted in one host but not in others.

Virus transmission in or on seed tissues outside the embryo may be important with some viruses. For example, in tomato seed, TMV transmission is mainly due to contamination of the seedling by mechanical means. The virus can be inactivated by treatment with trisodium phosphate. Several viruses have been demonstrated in immature seed coats as part of the mother plant. With the exception of TMV and tomato mosaic virus, most viruses do not persist and remain infectious in the seed coat. Although these viruses may be detected in the endosperm, they apparently never occur in the embryo.

Transmission through infected pollen may produce virus-infected seed on healthy mother plants. Infected pollen may, however, also be less viable and may not effectively transmit virus.

Tomato ringspot nepovirus and the related cherry leafroll virus may carry virus on the surface of the contaminated pollen grains. Infection may also occur if pollen carrying the tobacco streak ilarvirus rub together or wounds are created on the leaf surface by feeding insects or thrips (Sdoodee and Teakle, 1987). This process may explain the occurrence of closely related serotypes of cherry leaf roll virus in birch, cherry, and beech (Cooper, 1993). Ilarviruses naturally infect up to 50% of the wild cherries in North America, and prune dwarf was detected in 40% of the seed of *Prunus avium* analyzed in the UK (Cooper, 1993). Pollen transmitted viruses, which tend to be seed transmitted, are adapted to survive long periods of time in their woody perennial hosts. Interesting examples of the occurrence of ilarviruses include prunus necrotic ringspot, apple mosaic, and tobacco streak viruses in rose. Although these viruses are widely distributed in both roses and fruit trees, transmission through pollen has not been reported in rose.

Cryptoviruses

This group of viruses is unusual in that they are transmitted through seed or pollen but not by vectors, grafting, or mechanical methods (Boccardo, 1987). These viruses are not transported from cell to cell. They are propagated and distributed only through multiplying cells. Cryptic viruses are symptomless in their host plants. These viruses are often transmitted in a high percentage of plants grown from seed sources that contain them.

Insect Transmission

The ability of vectors to transmit plant viruses is very specific. Although TMV is considered the most stable and contagious plant virus, its transmission by an insect has not been established. Invertebrate vectors of viruses include the phyla Arthropoda and Nematoda. Although four families of Coleoptera or chewing insects have been implicated in virus transmission, it is the sucking insects in the order Hemiptera that are numerically the most important group of virus vectors. This order includes aphids, leafhoppers, planthoppers, treehoppers, mealybugs, psyllids, and whiteflies. Of all these vectors, the aphid *Myzus persicae* is probably the most important since it can transmit over 60 different viruses. With the exception of a few rhabdoviruses, no aphid-borne viruses replicate in their vectors. Many leafhopper/planthopper-borne plant viruses, however, multiply in their vectors before transmission occurs.

Virus vectors such as insects or nematodes have several features in common. The stages of transmission have been described as (1) acquisition access period and acquisition feeding period; (2) inoculation access period and inoculation feeding period; (3) latent period and (4) transmission threshold period (Gibbs and Harrison, 1976). The acquisition access period is the time that a non-viruliferous vector is allowed to access and feed on the source. The acquisition feeding period or the time the vector feeds on the source is followed by the inoculation access period or the time the viruliferous vector is allowed access to a virus-free plant. The inoculation feeding period is the time the virus-carrying vector is allowed access to the virus-free plant. There is usually a latent or pre-infective period from the end of an acquisition feeding period until the vector can infect a healthy plant. The transmission threshold period is the minimum time that a vector needs for acquisition of a virus and inoculation to a healthy plant. These different stages have been studied in detail for many different virus-vector systems. The pattern of acquisition and transmission is often complex and varies with the specific vector and host. Factors such as light, temperature, and nutritional status of the host may all affect the behavior of the vector and its ability to transmit the virus.

Transmission by arthropod vectors is a function of mouthpart structure and feeding characteristics. Insects like the Coleoptera (beetles) and Lepidoptera (caterpillars) have biting and chewing mouthparts that often produce significant wounds in foliage which generally hinder transmission, although there are some examples of specific transmission of viruses in the tymovirus, comovirus, bromovirus, and sobemovirus groups. In some instances a virus may persist in beetles for a long period of time.

Insects in the Hemiptera with piercing and sucking mouthparts have a feeding process that consists of a labium which supports a stylet which is the feeding structure. Stylets are present as extensions of the upper and lower jaws and form a bundle composed of an outer pair of mandibular stylets (or upper jaw) and an inner pair of maxillary stylets (lower jaw). Stylet penetration occurs with alternating movement of the two types of stylets.

An important aspect of aphid transmission is the nature of the interactions of the virus with the insect (Table 10.1). Transmission characteristics have been classified on this basis as stylet-borne, persistent, and semi-persistent transmission. Stylet borne viruses are those carried on the stylets after a very short acquisition feeding period of only a few seconds. The insect can transmit the virus immediately, but the ability to transmit is rapidly lost, usually from a few seconds up to 4 hours. Aphids transmit many different viruses in this manner. These viruses have also been referred to as non-persistent and include viruses in the potyvirus, carlavirus, cucumovirus, and alfamovirus groups. The rapid loss of infectivity of stylet-borne viruses is apparently the result of molting when the exoskeleton is shed. The process of transmission usually occurs when aphids feed during brief periods of probing.

Persistent transmission of viruses occurs when a virus is retained through the molt. In persistent aphid transmission, there is no increase in transmission efficiency when feeding is preceded by a starvation period. Persistent viruses are usually acquired after a minimum feeding period of 10–60 minutes. The feeding period is followed by a latent period of at least 12 hours before transmission

Table 10.1. Some properties and characteristics of plant virus-aphid vector relationships

	Persistent		Semi-persistent	Non-persistent
	Propagative	Non-propagative		
Vector specificity	Very specific	Very specific	Very specific	Less specific
Virus localization		Phloem limited	Phloem limited	Mesophyll
Mechanical transmission	No	No	With difficulty	Yes
Transovarial transmission	Demonstrated	No	No	No
Acquisition	Few minutes to 1 hour	Few minutes to 1 hour	Several minutes	Few seconds to none
Incubation period	Long (days)	Few hours	Few minutes	Few seconds to none
Retention of virus	Whole life	Not affected by molting	Few days or lost after molting	Few hours or lost after molting
Replication in vectors	Yes	No	No	No

can occur. The relationship between a virus and vector is very specific in persistent transmission. Some persistent viruses are efficiently transmitted by only one or a few aphid species. In a few cases, transmission is strain specific. Uptake of virus and inoculation occur after long periods of host penetration. Most viruses transmitted in this manner are associated with phloem and most are, therefore, not mechanically transmitted. In persistent virus transmission, the vector retains the ability to transmit virus as long as there is a supply of virus in the insect (Matthews, 1991).

Semi-persistent transmission is primarily a non-persistent mode of transmission but with some characteristics of persistent transmission. For example, most semi-persistent viruses can be mechanically transmitted with difficulty. Virus is acquired from phloem tissue. There is no latent period, and the virus may persist only a few days. Infectivity is not retained when the insect molts. The mechanism of virus uptake with food apparently involves absorption to the lining of the pharynx or anterior portion of the foregut. Virus is apparently released by a gradual elution from the foregut. This structure is lost when the insect molts.

In addition to the insect-specified nature of virus transmission that depends on the structure of the vector mouthparts and feeding characteristics, there is another virus-specified transmission factor. This factor, known as helper component, is a low molecular weight non-structural protein associated with transmission of some viruses by aphids. Helper component is coded by the viral genome and required for aphid transmission of non-persistent potyviruses and semi-persistent caulimoviruses. If an isolate of these viruses lacks this portion of the genome, the virus is not transmitted by aphids. Also, purified helper component-dependent aphid-borne viruses are highly infectious by mechanical inoculation. They are not transmissible by the vector when artificially acquired.

Transmission by Thrips

Thysanoptera (thrips) have previously been considered to have rasping and sucking mouthparts. The western flower thrips, *Frankliniella occidentalis* (Pergande), and several other thrips species have been shown to pierce plant cells and suck up the cell contents (Hunter and Ullman, 1992). During the feeding process the insect may take up nutrients, inject saliva, and also transmit viruses in the genus *tospovirus*. This group includes tomato spotted wilt (TSWV) and impatiens necrotic spot (INSV) viruses. Both of these viruses cause severe economic damage in many vegetable and ornamental crops. Thrips acquire virus only during the larval stage. Acquisition occurs in 15–30 minutes with an incubation period of 4 up to 18 days, depending on the genus and species of thrips. The viruses are transmitted by the adults that retain the ability to transmit for life. Recently the multiplication of TSWV in *Frankliniella occidentalis* has been demonstrated (Wijkamp et al., 1993). Although thrips are not regarded as highly efficient vectors of these viruses, both TSWV and INSV have become widespread in recent years because of the large thrips populations that have developed in many parts of the US as well as other areas of the world. Several species of thrips also reportedly transmit tobacco ringspot, tobacco streak, and prunus necrotic ringspot viruses. However, the mode of transmission of these viruses by thrips are different from tospoviruses.

Transmission by Mites

Eriophyid mites (Eriophyidae) are so small they cannot be seen without magnification. They have narrow host ranges and primarily colonize leaves, buds, and tender new plant growth. Mites are probably carried most efficiently plant-to-plant by wind. Members of this family feed by puncturing plant cells and sucking the contents. These mites reportedly transmit the cause of the rose rosette disease as well as diseases of wheat, peach, fig, and current. The spider mite (*Tetranychus telarius*) has been implicated in transmission of tobacco ringspot virus and potato virus Y.

Aceria tulipae is the most extensively studied mite-vector. It reportedly transmits wheat streak mosaic and wheat spot mosaic viruses, simultaneously. The virus can be acquired in 15 minutes. A similar minimum period is required to transmit the virus. The virus persists in the vector for up to 9 days in the greenhouse. The virus occurs in high concentration in the midgut, body cavity, and salivary glands of the mite-vector, suggesting that transmission is probably circulative.

Transmission by Whiteflies (Aleyrodidae)

Whitefly-transmitted virus diseases are of considerable economic importance. The insects are present mainly in tropical countries, but are also known to occur in sub-tropical and temperate agricultural areas. Their impact on such crops as cassava, cotton, sweetpotato, legumes, tomato, and recently ornamentals has been recognized.

Whitefly-transmitted viruses have been reported mostly on herbaceous plants and more rarely on shrubs and trees. The symptoms associated with these diseases include mosaic, leaf curl and yellowing.

Bemisia tabaci is the most important vector species. It is the vector of more than 60 different viral agents. They are transmitted by non-persistent, semi-persistent, and persistent mechanisms. The best studied group is the geminiviruses. Adult whiteflies generally acquire the virus in 10–30 minutes. A longer feeding time increases the transmission efficiency. A latent period of 4–20 hours is required before transmissions can occur. The minimum feeding period required to transmit a virus is often shorter than that required for acquisition. The vector may retain the virus for periods ranging from a few days to about 3 weeks. The viruses do not multiply within the vector. Transovarial passage of the virus to offspring has not been demonstrated.

Transmission in Soil

Nematode vectors of plant viruses are soil-inhabiting ectoparasites. They usually feed on young tissue near root tips using their stylets (Bos, 1983). Nematodes are important vectors of viruses in the *Nepo* and *Tobravirus* genera. Nepoviruses are transmitted by members of the family Dorylaimidae that includes the genera *Xiphinema* and *Longidorus*. Tobraviruses are transmitted by members of the Trichodoridae that includes the genera *Trichodorus* and *Paratrichodorus*. Viruses in these two groups have wide experimental host ranges including ornamental plants that are susceptible to some important pathogens.

Transmission by nematodes shows considerable vector specificity not only in the vector relationship to the virus group transmitted but also in the different species within a genus that transmit different but related viruses. For example, *Longidorus* transmits the nepoviruses tomato black ring and raspberry ringspot but not arabis mosaic and tobacco ringspot virus that are transmitted by *Xiphinema*. This specificity is apparently not related to ingestion of the virus since both transmitted and non-transmitted viruses occur in the same nematode species. Rather, transmission appears to be a function of absorption of virus particles to the cuticular lining of the esophagus.

It is apparent from these and other experiments that viruses transmitted by nematodes are similar to semi-persistent transmission of viruses by arthropods. Viruses in nematodes are retained in the vector in the lining of the guide-sheath of the odontostyle in *Longidorus* and in the lumen of the ondotophore and the esophagus in *Xiphinema*. In *Trichodorus* the virus is in the pharynx and esophagus. Despite this similarity, there is a strong vector specificity not only between different viruses but also between virus strains. Specific adsorption is apparently related to the structure of the coat protein that is correlated with the serological reactivity and vector specificity.

Although nematode transmitted viruses are apparently not retained through the molt, the viruses have been reported to persist for up to 12 weeks in *Longidorus* that are starved. In *Xiphinema* the viruses may persist for up to a year and for a longer period in *Trichodorus*. The virus does not multiply in the nematode vector, and there is no indication that virus is transmitted through the eggs of any of the genera that transmit viruses.

Nematode transmitted viruses are primarily a problem in field soils where the vectors are not controlled by soil treatment. The use of "bait" plants has been a common practice in detecting nematode transmitted viruses (Matthews, 1991). Using this method, a susceptible host such as cucumber is planted in soil infested with viruliferous nematodes where the nematodes can feed on the roots of the test plant. Infection is verified in the "bait" plants by mechanical inoculation to susceptible diagnostic species or by serological testing. Care must be taken in conducting these types of tests since the number of virus transmitting nematodes may be only a small proportion of the total population.

Transmission by Fungi

Viruses transmitted by soil-inhabiting fungi may cause serious diseases of high-value ornamental crops, root, and tuber crops worldwide. Root infecting endoparasite fungi in the family Olpidiaceae (chytrids) are the most important in transmission of some serious soilborne virus diseases (Matthews, 1991). *Olpidium* sp. transmits tobacco necrosis virus and freesia leaf necrosis agent that may pose a serious threat to susceptible bulb crops such as tulip and freesia, respectively. The fungus also transmits satellite and cucumber necrosis viruses as well as lettuce big vein agent and tobacco stunt virus. The lettuce big vein agent may persist for years in the fungal vector in the form of viable resting spores that persist for long periods of time.

Olpidium is a simple Phycomycete that infects its host by zoospores that move in films of water in soil from one root to another where they attach to the surface and produce a thin outer cyst wall. After about 2 hours the zoospore produces an infection canal that penetrates the root cell but leaves the cyst wall behind. A thallus subsequently grows in the cell and produces a new zoosporangium and zoospores that repeat the cycle. Virus particles attach to the surface of the zoospores but may also be associated with the cilium of the zoospore and are incorporated in the cytoplasm of the zoospore. This transfer of virus may result in movement of virus when the protoplasm of the zoospore is transferred to the protoplasm of the recipient plant host cell.

A second group of fungal root infecting vectors are the Plasmodiophorales that includes *Polymyxa graminis* as the vector of soil-borne wheat mosaic and other cereal viruses. *Spongospora subterranea*, a serious pathogen by its own, inducing powdery scab in potato, is the vector of potato mop-top virus and beet necrotic yellow vein viruses.

Our knowledge of the interaction between fungal vector and the viruses they transmit is limited, mainly due to the fact that fungal vectors are obligate parasite that are not easy to culture.

Virus transmission occurs by zoospores and, with the exception of TNV, appears to have a persistent relationship with their vectors. The virus is retained in the resting spores. These resting spores and the virus may remain infectious for years. TNV and its satellites are acquired from soil water and roots on the surface of zoospores. The virus appears to enter the zoospores, but not the resting spores (non-persistent). These transmission features have implication for the epidemiology of the viruses.

Transmission by Dodder

Dodder (*Cuscuta* spp.) is a vine-like parasitic plant belonging to the family Convolvulaceae. It is widely distributed in Europe and North America causing important yield losses on agricultural crops including those of several ornamentals.

Dodder has special value in transmitting viruses that are not mechanically transmissible and which have no known insect vectors or which have vectors with limited host range.

In nature dodder seed germinates and produces a slender yellowish shoot but no roots. When the dodder stem comes in contact with a susceptible host, the stem immediately encircles the hosts and sends haustoria which connect with the vascular tissue of two or more plants, establishing vascular bridges between them. The bridges may serve as a pipeline for transmission of viruses from virus-infected to virus-free plants.

Dodder species were experimentally used for virus transmission by early workers. It is now rarely used by plant virologists. The advantage of using dodder is the ability to transmit some viruses that for technical reasons cannot be transmitted by any other methods. Dodder growing on infected plants has proved to be a more readily available source of virus for insect transmission than the infected host plants. Dodder may also be useful for the separation of viruses from complexes.

Selected Literature

Boccardo, G., Lisa, V., Luisoni, E. and Milne, R.G. 1987. Cryptic plant viruses. Adv. Virus Res. 32:171–214.

Bos, L. 1983. Introduction to Plant Virology. Longman, New York.

Cooper, J.I. 1993. Virus Diseases of Trees and Shrubs, 2nd ed. Chapman and Hall, New York.

Gibbs, A.J. and Harrison, B. 1976. Plant Virology. The Principles. A. Halsted Press Book, Wiley, New York.

Hunter, W.B. and Ullman, D.E. 1992. Anatomy and ultrastructure of the piercing-sucking mouthparts and paraglossal sensilla of *Frankliniella occidentalis* (Pergande) (Thysanoptera: Thripidae). Intl. J. Insect. Morphol. and Embryol. 21:17–35.

Lawson, R.H. 1990. Orchid viruses and their control. In: Handbook on Orchid Pests and Diseases, A.M. Pridgeon and L.L. Tillman (eds.). Am. Orchid Soc., West Palm Beach, Florida, pp. 66–101.

Matthews, R.E.F. 1991. Plant Virology, 3rd ed., Academic Press, San Diego.

Mink, G.I. 1993. Pollen- and seed-transmitted viruses and viroids. Annu. Rev. Phytopathol. 31:375–402.

Scarborough, B.A. and Smith, S.H. 1977. Effects of tobacco and tomato ringspot viruses on the reproductive tissues of *Pelargonium* × *hortorum*. Phytopathology 67:292–297.

Sdoodee, R. and Teakle, D.S. 1987. Transmission of tobacco streak virus by *Thrips tabaci*: a new method of plant virus transmission. Plant Pathol. 36:377–180.

Wijkamp, I., van Lent, J., Kormelink, R., Goldbach, R. and Peters, D. 1993. Multiplication of tomato spotted wilt virus in its insect vector, *Frankliniella occidentalis*. J. Gen. Virol. 74:341–349.

Yuen, C.K.K.H., Kamemoto, H. and Ishii, M. 1979. Transmission of cymbidium mosaic virus through seed propagation in *Dendrobium*. Am. Orchid Soc. Bull. 48:1245–1247.

11

Epidemiological Aspects of Virus Spread

B. Raccah

Department of Virology, Agricultural Research Organization, The Volcani Center, Bet Dagan 50250, Israel

Introduction

Epidemiology describes the movement of diseases among healthy hosts. Accumulation of data on disease spread in many important crops and the availability of computers greatly added to this field in the last 20 years, a fact that is reflected in many chapters and several books (Thresh and Plumb, 1983; McLean et al., 1986; Madden et al., in press). The ultimate goal of epidemiological studies is to achieve a reliable prediction and allow to prevent and/or suppress epidemics. Prediction models may be used for disease management and the choice and timing of control measures.

Virus diseases may cause reduction in the market value. Ornamental and flower growers often tend to introduce new varieties and plant species. Frequent introduction of plant material and the readiness of air transportation may contribute to disease outbreaks especially if the new introduction may carry seed borne of vegetatively-borne viruses. If this is combined with the economic importance of ornamentals and flower production, it will explain why the epidemiology of viruses and virus-like diseases in ornamentals is so important for the economy of the industry.

Major Factors in Plant Virus Epidemiology

The major factors that are traditionally considered in epidemiological studies of plant diseases include the host, the pathogen, and the environment. Most plant viruses depend on vectors for the movement in the field. Therefore, the vector component should be included (Raccah, 1986). Epidemic outbreaks occurred whenever one of the above-mentioned factors is changed in a steady situation. Secondary effects may modify each of the major factors. Therefore, the understanding of the interrelationships between these factors is often complex and needs many years of data collection, thorough analysis, and validation of assumptions on the role of principal parameters. A short review of the role of each of the major factors is given below.

The host

The selection of crop is mostly dictated by economical considerations and by the preference and experience of the grower. Susceptibility of crops to virus infection is an outcome of its genotype and phenotype. Age and physiological condition may influence its susceptibility to infection.

The virus source

The quality of the virus source is affected by several factors, namely (1) the virus property (2) the type and physiological status of the host from which the virus is acquired, and (3) the location of the virus source in relation to the field.

1. Among the virus properties that affects spread, we may consider the ability to infect a range of hosts, a range of host tissues, the mode of transmission, the retention by the vector (non- semi- or persistent) and the susceptibility to adverse environmental conditions.

2. The host from where the virus is acquired and its physiological status may have profound effect on its ability to serve as a source. The host may have an indirect effect on attraction or deterrence of vectors.

3. The aptitude of the virus source is also affected by the position of the source plants in the field. The distance and arrangement of virus sources in relation to the plot may affect the rate and form of virus spread. Internal sources are found among plants infected with seed-borne viruses (such as soybean mosaic virus in soybean), among vegetatively propagated viruses (e.g. bean yellow mosaic and cucumber mosaic virus in gladioli corms), or among the remains of former crops.

External sources may be found in the vicinity or at a distance from the crop. Weeds are a special case which may contribute virus both internally and externally (Bos, 1981).

The vector

The incidence, efficiency, and mobility of the vector species involved in transmission have a direct effect on virus spread (Raccah, 1986; Raccah et al., 1988). In these studies, many aphid species were trapped over crop under investigation. However, few species were responsible for more than 70% of the infection. The behavior of vector species versus the target host (colonizing, transient, or visiting) influence the duration of stay and the probability to inoculate the host. In certain cases, aphids that are abundant in flight do not tend to land, and the relative importance of a specific vector is corrected by its landing rate (Raccah, 1986). The epidemiological role of vectors in virus spread was analyzed by Marcus and Raccah (1986) and by Madden et al. (1990). In these studies, an attempt has been made to quantify the contribution of each species to the infection pressure. In the work analyzed by Marcus and Raccah (1986), it was established that field transmission of CMV to peppers was mainly caused by non-colonizing species, *Aphis spiraecola* at the beginning of the season, and by the colonizing species *Myzus persicae* later in the season. Although the majority of viruses diseases in flowers and ornamentals are transmitted by aphids, several important viruses are vectored by whiteflies, e.g. tomato yellow leaf curl virus (TYLCV) in lisianthus (Cohen et al., 1993) and tomato spotted wilt virus (TSWV) in flowers and ornamentals (asters, chrysanthemum, ranunculus, kalanchoë, alstroemeria, cyclamen, pelargonium and others). Finally, leafhoppers (Cicadellidae, Auchenorrhyncha) and especially those that belong in the *Macrosteles* and *Euscellis* species were found to transmit mycoplasma-like-organisms (MLO) in ornamentals and flowers (for further details, readers are referred to Chapter 7).

Environmental factors

Temperature enhances propagation of viruses in the host up to the optimum and may cause reduction and even suppression of propagation at elevated temperature. Low humidity may cause desiccation thus desertion of vectors from infected hosts and migration to irrigated crops. Wind speed affect the direction and speed passive and active air-borne vectors. Jet stream may be responsible for long distance transport of vectors and for the dissemination of viruses to distant sites. Local environmental conditions may create the micro climate for the survival of virus and vectors in adverse conditions (Eastop and Raccah, 1998).

Methods for Monitoring Various Epidemiological Factors

Determination of virus infection in the field

Monitoring of virus infection in the field depends on the objectives of the study, the required degree of accuracy, and simplicity of use. The simplest method is visual inspection of virus symptoms in the plot to be studied. In most cases, this method is sufficient and accurate. However, in cases that the disease is not conspicuous and when more than one virus is likely to be present, costly and laborious methods are required. The reader is referred elsewhere for more on these methods (e.g. the recent review of Bar-Joseph and Lee, 1995).

Monitoring of infection is made at a predetermined time interval (according to the speed of the disease movement) and provides cumulative data on the number of plants infected over time. These records are especially required for the evaluation of the disease increase at the early stages of spread. In many cases, plants found to be infected are labeled by flags and on maps. A less laborious and less accurate approach is based on sampling of the field. It may provide the rate of disease increase but will not provide information about the distribution of infection.

When there is a need to relate the marginal infection to vector activity, different methods are reqiured. These include the use of bait plants or repetitive plantings of field plots and often require infectivity studies with field-captured aphids as potential vectors.

Bait plants are virus-free potted plants that are placed in field plots for a short periods of time. The time and incidence of infection in bait plants represent the infection pressure during the time of exposure (Raccah and Irwin, 1988). Bait plants have been used to relate the infection pressure of cucumber mosaic virus (CMV) and potato virus Y (PVY) to the incidence of live trapped aphids tested for transmission (Raccah, 1983). Indeed, the peaks of CMV of PVY infection in bait plants have correlated well with vector species that were active during bait plant exposure (Raccah, 1983, 1986). Exposure of potted bait plants requires irrigation and maintenance. This method is difficult when the test plots are distant. A useful alternative may provide similar information about infection pressure and vector activity and can be obtained by repetitive plantings in isolated plots (Schultz et al., 1985).

Determination of vectorial capacity and transmissibility of virus strains

The intrinsic capacity of an insect species to transmit a virus can be determined in the laboratory. For this objective, insect species that are frequent in the field are raised and tested in laboratory for the intrinsic ability to transmit the virus that is prevalent in plot. However, the intrinsic ability of the vector to transmit is complicated by the flight and landing behavior of vectors in nature. Landing and the duration of stay on the host reflect the degree of attraction or deterrence (for review, see Raccah, 1986). These factors explain discrepancies between data obtained in the laboratory and those seen in the field. Laboratory tests were therefore modified to examine the tendency of the aphid species to freely move from the infected to the healthy test — called the "arena test" — that attempts to take into consideration important vector behavioral components (Irwin and Ruesink 1986). These transmission tests provide actual vector propensity that takes into consideration natural preferences of the host and the virus. It also allows multiple acquisitions and multiple inoculations by each specimen. Although more realistic than tests involved with manual handling of vectors, this method may still be affected by the artificial conditions in the laboratory (such as caging and artificial light conditions).

Another approach is based on the sampling live insects in the field and then determining their vectorial capacity. Sampling is mostly made from the air or from vegetation. These methods have been developed for vectors of persistent viruses (Plumb, 1983). Non-persistent viruses that are retained by vectors for short times require immediate testing of the captured vectors (Raccah, 1986). Additional tests are required to determine the tendency of vectors flying in the field to land on the crop (Irwin, 1980).

The number and frequency of sampling for naturally occurring vectors depend on the amount of inoculum present in the field and on the vector propensity. This test becomes impractical when the number of insects in flight or inoculum level is low (Raccah and Pirone, unpublished). Data obtained by live-trapping are useful for predicting the disease progress as a function of vector activity (Marcus and Raccah, 1986; Ruesink and Irwin, 1986).

The above-mentioned methods were developed for aphid vectors. However, most of them are usable for other insect groups with appropriate changes. Unlike most aphid-borne non-persistent viruses, the specificity of the other viruses or MLOs for their vector is high. The nature of the transmission resembles the persistent or propagative mode (acquisition and inoculation access feeding times are long, a latent period in the vector is required and the pathogen persists in vector for long periods). Therefore, when virus sources are in excess, the incidence of vectors strongly suggests the potential for the spread of the disease.

Leafhoppers is the second largest group of known vectors. In ornamentals and flowers, they are mainly responsible for transmission of mycoplsama-like and spiroplasmas (see Chapter 7). Leafhoppers are sampled by portable suction traps of the D-Vac type. A model describing the relationships between leafhopper numbers and rice dwarf virus disease incidence was proposed by Kiritani and Sasaba (1978).

The best known thrips-borne virus is the tomato spotted wilt virus (TSWV). This virus is common in flowers and ornamentals. The TSWV is transmitted by the nymphal stages of several thripses species (Ullman et al., 1992). Nymphs of thrips are mostly hidden in the internal parts of plants. This explain the difficulties in sampling. Suction traps may provide information about adult presence. However, attempts to relate this information to disease spread have been futile until now (Ullman, personal communication).

Whiteflies are not common vectors in ornamentals and flowers. However, in view of the finding that the geminivirus tomato yellow leaf curl virus (TYLCV) severly affects lisianthus, the role of whiteflies should also be considered. Monitoring methods that were designed for *Bemisia tabaci* in vegetables grown in glasshouses and screenhouses should be suitable for flower and ornamentals (for review, see Cohen and Antignus, 1994). In most cases, horizontal sticky yellow traps were exposed, and the information was used to decide on timing and frequency of treatments against the vector.

A different approach is necessary for the nematode-borne viruses (Harrison, 1977). They are long-lived, have a wide host range, and may survive in absence of soils for long periods. On the other hand, the mobility of nematodes in undisturbed soil is limited. They move vertically in the soil as moisture and temperature fluctuate, thereby increasing their chances of survival under adverse conditions. Longer dispersion is passive with soil erosion or by winds. The nematode vectors are ectoparasitic and free-living species. Nepoviruses are transmitted by *Longidorus* or *Xiphinema* species, while the tobraviruses are transmitted by *Trichodorus* or *Paratrichodorus* species. Both the larvae and the adult stages can efficiently transmit viruses (Harrison, 1977). The limited mobility of the nematode vectors explains the slow movement of virus, mostly in "patches" from one plant to the next. Sampling is generally difficult, however; finding a single individual in 200 ml soil is equivalent to 29 million per hectare of top soil (Cotten, 1979). Preventive measures made in the past using methyl bromide fumigation are now banned in many countries, and loss due to new virus outbreaks may increase. Procedures for sampling nematodes were detailed by Barker and Campbell (1981).

Activity of vectors in the field

Sampling of vectors in the field provides information on the incidence in flight or on vegetation. Virus spread depends on the mobility and activity of the vector. Activity of vectors depends on the proportion of migrant forms in the population (alates in aphids, macropterous forms in planthoppers), on the migration urge, and on environmental conditions that favor take-off and landing. Host conditions and crowding were found to affect the proportion of alates in the population.

A number of activities are related to the ability of a vector to transmit virus: host seeking for food and ovi- or larviposition), landing, probing, feeding, take-off, and flight. Leafhoppers, planthoppers, and aphids can move great distances in the air [see several reviews in the book edited by Thresh (1981)]. For aphids, there is also a movement to shorter distances between and within fields. Taylor (1986) suggests a continuum between the two. Vectors can move within plants in the field by flying or walking if the canopies are closed.

Landing is probably the most common method to determine vector activity. The Moericke yellow water pan trap is the most widely used. Yellow sticky traps are also used to trap aphids, whiteflies, and leafhoppers. Colorless sticky and vertically oriented cylinder traps were also used (see review by Raccah and Irwin, 1988). Use of vertical traps may provide biased data when wind speeds vary (faster winds will result in higher catch). In an attempt to study the role of vector's landing on plants, Irwin (1980) used a trap that will attract aphids like the crop canopy on which the aphids were landing naturally. The trap was exposed in an horizontal attitude, close to the top of the canopy, and its color was of a spectral reflectivity and characteristics as the canopy (Irwin, 1980). Later, the capture of the traps consistently provided numbers that were closely related to the progress of viral epidemics (Schultz et al., 1985). While the horizontal mosaic green pan trap was designed specifically to collect alate aphids in soybean canopies, the differences in spectral characteristics among many chlorophyll producing plants are minimal, making the trap generally adaptable to most row crop and low growing agroecosystems.

Temporal and Spatial Distribution of Virus Spread

The distribution of viruses within a plot is described in several ways: showing the actual layout of infected and non-infected plants, progress curves in which the number of infections or the respective rate is plotted against the time or distance; and mathematical expressions which incorporate and quantify parameters which are involved in the spread in time and space are the major variables to which infection is related. Accordingly, the terms "temporal" and "spatial" distribution are in use.

Mapping the actual position of infected and non-infected plants

In the field, the infected plants are marked by flags, and data are then drawn on maps or inserted in spreadsheet maps using a simple software written for personal computers (Raccah and Raccah, unpublished). Overlaying maps of infection drawn following a sequence of field sampling allow the visualization of the actual shape of spread, the direction of secondary infection.

Progress curves

Progress curves are used to plot the incidence of infection (number, percent, or logits) on the vertical axis against time (mostly the Julian calendar) against the horizontal axis (for a review, see relevant chapters in McLean et al., 1986).

Temporal curves have three phases: the lag phase, the exponential, and plateau phases. The lag phase represents the first primary infections in the field. The exponential phase represents the fast secondary spread into the predominantly healthy plants. The plateau represent the phase where almost all the plants in the field are infected. Use of logit [log $(y/y-1)$ where y is the proportion of infected plants] serve to transform the sigmoid curve.

Models that describe the probability of new infection at a distance from former ones were presented in a number of studies. Analysis of spatial arrangement of infection assist in understanding the mechanisms which are the cause for the pattern of spread. The mathematical analysis of spatial distribution is beyond the scope of this chapter. The readers are referred to older and newer books (Thresh, 1981; Madden et al., in press). Mapped data for tobacco etch and tobacco vein mottling viruses in tobacco in Kentucky were used by Madden et al. (1987) to analyze the spatio-temporal autocorrelation approach. The subject of spatial analysis of spread was summarized recently by Hughes and Nelson (1995).

Conclusion

Epidemiological studies as a rule are based on field data. Therefore, assumptions to be tested require several seasons. Infection in one season is not necessarily the same in climate, in number of vectors, or in the intensity of infection pressure as in another. These limitations may explain why only few attempts were thus far made to construct prediction models for virus disease outbreaks. In the few that are available (Marcus and Raccah, 1986; Ruesink and Irwin, 1986; Madden et al., 1988), there are still many missing links that reflect the complexity and difficulty of selecting parameters with a direct effect on virus spread.

Moreover, most of the epidemiological studies were carried out in open fields. In the last two decades, intensification of agrotechnology is pushing many growers to use protected crops (tunnels, screen-, and glasshouses). Ornamental and flower growers are amongst the principal users of this technology. Evidently, epidemiology of viruses in the confined condition of covered crops is different, and little can be drawn from the information in the open field on the behavior of viruses and vectors under a closed and protected ambient. Knowledge on virus spread in protected crop conditions is still lacking, and it is stressed that such studies should be made. This chapter is intended to demonstrate the principal tools to be used and the attempts to be made to analyze epidemiological data. It is also intended to stimulate workers in ornamental and flower virology to add epidemiological parameters to work schedules which, when accumulated, will provide a return in better prediction of virus outbreaks and in more intelligent use of control measures.

Selected Literature

Bar-Joseph, M. and Lee, R.F. 1995. Virus detection methods. In: Epidemiology and Management of Virus Diseases, L.V. Madden, B. Raccah and J.M. Thresh, eds. Springer Verlag, New York.

Barker, K.R. and Campbell, C.L. 1981. Sampling nematode population. In: Plant Parasitic Nematodes, Vol. 3., B.M. Zuckerman and R.A. Rhode, eds. Academic Press, New York, pp. 451–454.

Bos, L. 1981. Wild plants in the ecology of virus diseases. In: Plant Diseases and Vectors: Ecology and Epidemiology, K. Maramorosch and K.F. Harris, eds. Academic Press, New York, pp. 1–33.

Cohen, J., Gera, A., Jokes, M., Aker, R. and Antignus, Y. (1993). Tomato yellow leaf-curl a new disease agent in lisianthus. Hassadeh 84:169 (in Hebrew).

Cohen, S. and Antignus, Y. (1994). Tomato yellow leaf curl virus, a whitely-borne geminivirus of tomatoes. Adv. Dis. Vector Res. 10:259–288.

Cotten, J. 1979. The effectiveness of soil sampling for virus–vectors nematodes in MAFF certification schemes for fruit and hops. Plant Pathol. 28:40–44.

Eastop, V.F. and Raccah, B. 1988. Aphid and host plant species in the Arava valley of Israel: Epidemiological aspects. Phytoparasitica, 16:23–32.

Harrison, B.D. 1977. Ecology and control of viruses with soil–inhabiting vectors. Annu. Rev. Phytopathol. 15:331–360.

Hughes, G. and Nelson, S.C. 1995. Spatial pattern analysis of plant virus diseases. In: Epidemiology and Management of Virus Diseases, L.V. Madden, B. Raccah and J.M. Thresh, eds., Springer Verlag, New York.

Irwin, M.E. 1980. Sampling aphids in soybean fields. In: Sampling Methods in Soybean Entomology, M. Kogan and D.C. Herzog, eds., Springer Verlag, New York, pp. 239–259.

Irwin, M.E. and Ruesink, W.G. 1986. Vector intensity: A product of propensity and activity. In: Plant Virus Epidemics: Monitoring, Modeling and Predicting Outbreaks, G.D. McLean, R.G. Garrett and W.G. Ruesink, eds., Academic Press, Sydney, Australia, pp. 13–33.

Kiritani, K. and Sasaba T. 1978. An experimental validation of the system model for the prediction of rice dwarf virus infection. Appl. Entomol. Zool. 13:209–214.

Madden, L.V., Pirone, T.P. and Raccah, B. 1987. Analysis of spatial patterns of virus diseased tobacco plants. Phytopathology 77:1413–1417.

Madden, L.V., Raccah, B. and Pirone, T.P. 1990. Modeling plant disease as a function of vector numbers: non-persistent viruses. Res. Popul. Ecol. 32:47–65.

Madden, L.V., Raccah, B. and Thresh, J.M. In press. Epidemiology and Management of Plant Virus Diseases. Springer-Verlag, New York.

Marcus, R. and Raccah, B. 1986. A model for spread of nonpersistent virus diseases. J. Appl. Stat. 13:167–175.

McLean, G.D., Garrett, R.G. and Ruesink, W.G. 1986. Plant virus epidemics: monitoring, modeling and predicting outbreaks. Academic Press, Sydney, Australia.

Plumb, R.T. 1983. Barley yellow dwarf virus a global problem. In: Plant Virus Epidemiology: The Spread and Control of Insect-borne Viruses, R.T. Plumb and J.M. Thresh, eds., Blackwell, Oxford, pp. 185–198.

Plumb, R.T. and Thresh, J.M. 1983. Plant Virus Epidemiology: The Spread and Control of Insect-borne Viruses, R.T. Plumb and J.M. Thresh, eds., Blackwell, Oxford.

Raccah, B. 1983. Monitoring insect vector populations and the detection of viruses in vectors. In: Plant Virus Epidemiology: The Spread and Control of Insect-borne Viruses, R.T. Plumb and J.M. Thresh, eds., Blackwell, Oxford, pp. 51–60.

Raccah, B. 1986. Nonpersistent viruses: epidemiology and control. Adv. Virus Res. 31: 387–429.

Ruesink, W.G. and Irwin, M.E. 1986. Soybean mosaic virus: A model and some implications. In: Plant Virus Epidemics: Monitoring, Modeling and Predicting Outbreaks, G.D. McLean, R.G. Garrett and W.G. Ruesink, eds. Academic Press, Sydney, Australia, pp. 295–313.

Schultz, G.A., Irwin, M.E. and Goodman, R.M. 1985. Relationship of aphid landing rates to the field spread of soybean mosaic virus. J. Econ. Entomol. 78:143–147.

Taylor, L.R. 1986. The distribution of virus disease and the migrant vector aphid. In: Plant Virus Epidemics: Monitoring, Modeling and Predicting Outbreaks, G.D. McLean, R.G. Garrett and W.G. Ruesink, Academic Press, Sydney, Australia, pp. 35–57.

Thresh, J.M. 1981. Pests, Pathogens and Vegetation, Pitman, London.

Ullman, D.E., Cho, J.J., Mau, R.F.L., Hunter, W.B., Westcot, D.M. and Custer, D.M. 1992. Thrips–tomato spotted wilt virus interactions: morphological, behavioral and cellular components influencing thrips transmission. Adv. Dis. Vector Res. 9:197–240.

12

Replication of Viruses and Movement in the Plant

John Hammond

Florist and Nursery Plants Research Unit, USDA–ARS, Beltsville, MD 20705-2350, USA

Viruses with Single-Stranded (+ and −) and Double-Stranded RNA Genomes

For viruses with (+), or messenger, sense RNA genomes, the first step of replication following entry to the cell is dissociation of at least some of the coat protein from the RNA to allow translational initiation, as the polymerase and associated enzymes must be translated from the viral genome before replication can begin. If the 5′ end of tobacco mosaic virus (TMV) genomic RNA is exposed by a brief alkaline treatment of virions, ribosomes are then able to associate with the exposed RNA *in vitro*; translation proceeds and drives further dissociation of the coat protein from the RNA. This "co-translational disassembly" is presumed to safeguard the RNA from ribonucleases at least until the replicase is available (Wilson, 1985). Turnip yellow mosaic tymovirus (TYMV), an isometric virus, appears to undergo a more specific particle degradation resulting in release of the RNA. A loss of 5–8 coat protein subunits allowed the RNA to be released *in vitro*; most of the virus inoculated to plants was uncoated in a similar manner within

2 minutes (reviewed in Matthews, 1991). Virus dissociation within the cell may occur at the surface of lipid-containing structures such as the plasma membrane.

Translation strategies differ significantly between plant virus groups; the 5′-proximal open reading frame (ORF) on each RNA is generally the most efficiently translated. Internal ORFs may be translated by ribosomes that remain attached after termination, and re-initiate. *Re-initiation*, or *internal initiation*, are potential means of regulating the quantity of a specific gene product. More often, internal ORFs are translated from *subgenomic RNAs* (sgRNAs; see Fig. 12.1a and 12.1b), allowing regulation both by the level of efficiency of the sgRNA promoter (transcriptional control) and the efficiency of translation of the sgRNA. *Multipartite viruses* have separated ORFs onto separately packaged and replicated genomic RNAs, with the coat protein typically produced from an sgRNA. This allows distinction between *early* genes, needed for replication, and *late* genes, required for other phases of the viral life cycle. Multipartite viruses also allow differential regulation at both translational and replicative levels. Some virus groups (e.g. *como-* and *potyviruses*) express each genome segment as a single *polyprotein*, which is then proteolytically cleaved to yield the individual gene products (see Fig. 12.1c). Regulation of "early" and "late" genes can be regulated by genome separation (e.g. comoviruses) as well as differential rates of processing; partially processed forms may have activities distinct from the fully processed products (reviewed in Riechmann et al., 1992); a disadvantage of the polyprotein processing is equimolar production of each gene product even though only a single molecule of VPg (genome-linked protein) is required per potyvirus virion, but about 2,000 coat protein monomers. *Readthrough* proteins are another means of regulating the level of a protein that is required in only minor amounts — the 183K readthrough protein of TMV is produced to about 10% of the amount of the shorter product, the 123K protein (Fig. 12.1a). *Frameshift* proteins also permit differential regulation of overlapping genes and the ability to yield two products from over-lapping sequence as in the luteovirus polymerase gene. *Luteoviruses* also have ORFs *over-lapping in different reading frames*. Combinations of these strategies may be found — luteoviruses utilize frameshifts, readthroughs, subgenomic RNAs, and an ORF contained within another functional ORF but in a different reading frame (Fig. 12.1b). Some viruses have RNAs of both (+) and (−) polarity; some proteins are translated from the genome-sense RNA, and some from the opposite strand. The *tospoviruses* use this *ambisense* strategy.

Once translation has occurred, the viral polymerase interacts with the 3′ end of the viral RNA(s) to initiate (−) strand production. The (−) strand will then serve as a template for production of progeny (+) strand molecules. Replicative

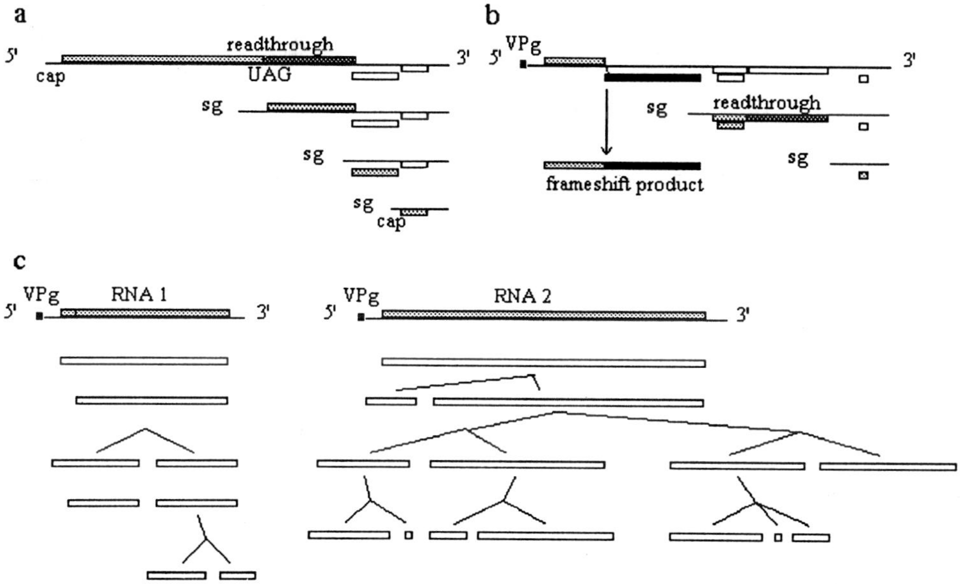

Fig. 12.1 Translation strategy of (a) TMV, (b) barley yellow dwarf luteovirus, and (c) cowpea mosaic comovirus (CPMV). = protein translated from the RNA segment shown; = readthrough protein resulting from suppression of a UAG codon; = frameshift protein resulting in translational fusion; sg, subgenomic RNA; cap, capped mRNA; VPg, genome linked protein. Note (a) that a shorter version of the readthrough domain of TMV is apparently translated from an sgRNA; (b) that the ORF in a separate reading frame, internal to the luteovirus coat protein gene, is apparently translated by internal initiation on the same sgRNA; and (c) there are two functional initiation sites on RNA 1 of CPMV in the same reading frame, and alternative proteolytic processing routes for the products of CPMV RNA 2.

intermediate consists of a single full-length (−) strand from which multiple copies of the (+) strand are transcribed at once (Fig. 12.2), resulting in an excess of (+) over (−) strands. Subgenomic RNAs are presumably transcribed at the same time, apparently only from the genome length (−) strand. Typically the ratio of (+) to (−) strand production will be quite large; the regulatory mechanisms are not clear, although the replicase recognition signals may differ in sequence and structure. VPg present at the 5′ end of *cowpea mosaic comovirus* (CPMV) RNA after translation may act as a primer of RNA synthesis; the release of VPg from the precursor protein may be coupled to initiation of RNA synthesis.

For *brome mosaic bromovirus* (BMV; reviewed by Ahlquist, 1992) it has been shown that the 3′ tRNA-like structure need not be amino-acylated for replication

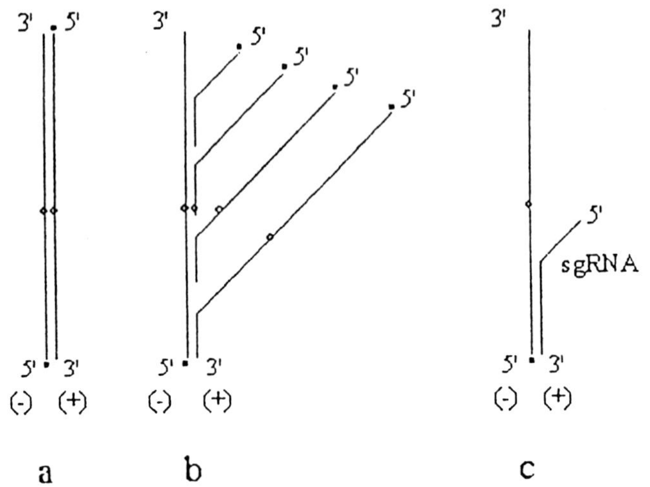

Fig. 12.2. (a) Replicative form dsRNA, with full-length (+) and (−) strands; (b) replicative intermediate, a partially dsRNA, with multiple (+) strands being transcribed from a single (−) strand template; and (c) production of sgRNA from the sgRNA promoter (represented by the open circle) on the genome-length (−) strand template.

to occur. This structure may act like a telomere in protection of the coding sequences from degradation; the (−) strand promoter is contained within the tRNA-like region and dependent upon its tertiary structure. There is greater homology in the 3′ end of the genome segments than the 5′ ends — indicative perhaps of the necessity for production of (−) strand of all genome segments for successful replication. Regulation of the genome segments is by differential production of the (+) strands, which are transcribed in much greater amounts than the (−) strands. Whether the (+) strand promoter (in the 3′ end of the (−) strand) is recognized by the same site of the replicase as the (−) strand promoter, or one that is modified by interaction with another viral gene product or host protein is not yet clear. The BMV RNA 5′ non-coding region resembles internal control regions of tRNA genes and has some homology to a consensus element in cellular RNA polymerase III promoters. It is not known if cellular factors interact with these BMV sequences.

For *TMV* (reviewed by Dawson, 1992), (−) strand production ceases at 6–8 hours after infection of protoplasts when there is already a 100-fold excess of (+) RNA. Production of genomic RNA and the coat protein sgRNA continues to accumulate until 14–18 hours post infection. TMV has three sgRNAs, and replicase differentially recognizes each sgRNA promoter; overall structure, rather than specific sequence, seem to determine replicase binding.

Alfalfa mosaic virus and the similar *ilarviruses* require *genome activation* by the coat protein (reviewed by Matthews, 1991) to establish replication; activation may occur either through the presence of the protein itself or of the sgRNA 4 (mRNA for the coat protein) in the inoculum. Bound protein may alter the conformation of the RNA to act as the replicase recognition signal.

The enzymes involved in replication include, minimally, the viral encoded RNA-dependent RNA polymerase, or *replicase*, to transcribe the (+) and (−) strand templates, and a *helicase* activity to unwind the replicative form RNA. *Methyltransferase* activity probably functions to cap RNAs for additional stability.

The replicase complex of BMV has been shown to include a host protein, closely related to a eukaryotic translation initiation factor; host factors of similar molecular weight have been identified in the cucumber mosaic cucumovirus and TYMV replicase complexes.

Viruses with (−) polarity genomes encapsidate molecules of the polymerase or replicase in the virion. Some of these viruses (*rhabdoviruses* and *tospoviruses*) have larger and more complex membrane-bound virions. Upon entry to the cell the particle dissociates and replicase transcribes mRNAs for each gene product, which are then translated. Replication is presumed to occur via a genome-length (+) RNA template. Some *rhabdoviruses* mature from the nucleus through the inner lamella into the perinuclear space; others bud through endoplasmic reticulum (ER) and accumulate in vesicles on the ER, while a third group form large viroplasms in the cytoplasm.

The *tospoviruses* include tomato spotted wilt (TSWV) and impatiens necrotic spot (INSV) viruses; these viruses have three virion RNAs — L (large), M (middle) and S (small) — each of which is found in a circular form in the particle as a result of terminal inverted repeat sequences which are probably also involved in replication. The S RNA of TSWV has an ambisense nature; one protein is translated in the genome sense, and the other from the virion-complementary strand; both are translated from sgRNAs. Replication occurs in the cytoplasm.

Viruses with dsRNA genomes include *phytoreoviruses* (12 genome segments), *fijiviruses* (10 genome segments), and *cryptoviruses* (2 genome segments); the isometric particles include RNA transcriptase to produce mRNAs from the genomic dsRNA. Wound tumour *phytoreovirus* (reviewed by Nuss and Dall, 1989) encapsidates methyltransferase, resulting in capped mRNAs; each genome segment is translated to a single protein product. Each dsRNA has common 5′ and 3′ sequences, internal to which are inverted repeats specific for each genome segment. Replication occurs in viroplasms in the cytoplasm. Each particle contains one copy of each genome segment; the segment-specific inverted repeats may act as signals for the encapsidation of a single copy of each.

Viruses with Single-stranded and Double-stranded DNA Genomes

Geminiviruses have single-stranded DNA genomes and are differentiated into three subgroups according to vector type (whitefly or leafhopper), host type (monocot versus dicot), and whether the genome is monopartite or bipartite; all have a "common region" of about 200 nucleotides of similar sequence. This forms a stem-loop structure that acts as the origin of replication and probably of RNA transcription (reviewed by Lazarowitz, 1992). All have genes transcribed from both the virion-sense (+) and virion-complementary (−) strands. The coat protein gene of the bipartite whitefly-transmitted viruses is encoded by the (+) strand of DNA A. Three ORFs in the (−) sense of DNA A overlap in different reading frames, but the ORFs of the (+) strand do not overlap with the ORFs in the (−) sense in either DNA A or B. There are individual mRNAs for each expressed gene.

In the monopartite geminiviruses infecting monocots, three ORFs are apparently translated from two 3′ co-terminal RNA transcripts in the virion sense. Its DNA sequence suggests that it contains two partially overlapping ORFs in the (−) strand, but an intron in the single RNA transcript is spliced out to yield a single product similar to the unspliced replicase ORF of bipartite geminiviruses, the only gene essential for replication. The (+) and (−) ORFs do not overlap, but in the monopartite viruses of monocots there is a second intergenic region. The monopartite viruses of dicots have at least four ORFs corresponding to ORFs on DNA A of the bipartite viruses, with one in the virion sense and three in the (−) strand.

Replication probably occurs via a rolling circle mechanism, with the dsDNA intermediate localized in nucleosomes where transcription of ssDNA is initiated. The replicase has been shown to interact with the stem loop formed by the common region. The other genes in the DNA A (−) strand are thought to be a transactivator of the movement protein gene, and a replication enhancer.

The *caulimoviruses* and *badnaviruses* are pararetroviruses, and their circular dsDNA genomes are replicated through an RNA intermediate. Cauliflower mosaic caulimovirus (CaMV) is the best-studied (reviewed by Mason et al., 1987); one strand of virion DNA has a single discontinuity of 1-2 nucleotides and is the strand used for transcription. The complementary strand has one, two or three gaps at which there are overlaps of a few nucleotides. All of the gene products are transcribed from the (+) strand from two RNAs; the 35S RNA extends around the whole genome and contains a direct terminal repeat of 180 nucleotides, while the 19S RNA is 3′-coterminal with the 35S RNA and carries the coding region only of gene VI. The other six or seven genes are translated from the 35S RNA. There are either gaps, or overlaps, of a few nucleotides between the termination codon of one gene and the initiation codon of the next, except for ORF VIII (which lies within ORF IV) and the intergenic region where the 35S RNA initiates and both 35S and 19S RNAs terminate. ORF V is the viral replicase/reverse transcriptase. ORF VI, found in viroplasms, is also a transactivator of other viral genes.

Replication occurs when the infecting virion dsDNA reaches the nucleus of the cell. There the overlaps at the (−) strand gaps are removed, and gaps in both (+) and (−) strands repaired, forming a minichromosome. Both 35S and 19S RNAs are transcribed from the minichromosome by host DNA-dependent RNA polymerase II. Both RNAs are polyadenylated and migrate to the cytoplasm; translation from the 19S RNA forms the gene VI inclusion protein, the matrix of the viroplasm, while all of the other proteins are translated from the 35S RNA. The 35S RNA is unusual in that at least five and possibly seven proteins are translated from a single RNA, and each gene has its own start and stop codons. Virion DNA is reverse-transcribed from the 35S RNA, the RNA template degraded, and the second DNA strand synthesized. Both virion DNA synthesis and particle maturation take place in the viroplasm, and most particles are retained in viroplasms; encapsidation may be closely linked to DNA synthesis.

Replication Sites

Most plant viruses with RNA genomes are replicated essentially in the cytoplasm; exact sites have not been determined for many viruses. Altered membranous structures are characteristic cytopathic effects of many viral infections, typically including endoplasmic reticulum and attached ribosomes. Accumulations of viral proteins are not necessarily sites of virus synthesis. *Potyviruses* (reviewed by Riechmann et al., 1992) provide ample evidence that viral protein accumulation is not always associated with sites of replication, and some of the accumulations may be a means of removing excess protein from circulation. Early in infection the cytoplasmic inclusion protein (CIP, helicase) accumulates at the cell wall, over plasmodesmata; later in the infection cycle masses of cytoplasmic inclusions are found free of the cell wall in the cytoplasm. The nuclear inclusion proteins of some potyviruses (VPg, protease and RNA polymerase functions) accumulate in highly ordered crystals in the nucleus; no such accumulations are found in infections of other potyviruses in the same hosts.

Potexviruses form beaded sheets in the cytoplasm of infected cells with parallel arrays of virus particles in close proximity; these structures are not proven to be the site of virus synthesis. *Tombusviruses* induce membranous, multivesiculated bodies, which contain dsRNA, in the cytoplasm. Nearby chloroplasts develop opposed, narrow-necked invaginations that are probably also sites of RNA synthesis; separate granular inclusions are largely accumulations of coat protein. *Tymoviruses* also induce peripheral vesicles on chloroplast membranes, which contain fibrillar material presumed to be dsRNA; particles may be assembled at the mouths of these vesicles. *TMV* induces viroplasms — also known as amorphous inclusions or X-bodies — in the cytoplasm. These inclusions consist of a mixture of endoplasmic reticulum, virus particles, ribosomes and a fibrillar material (or tubules). Antiserum to the 126K and 183K replication-associated TMV proteins labels these viroplasms, and especially the tubules. *Luteoviruses* are replicated only in the phloem and associated tissues; mixed infections with certain other viruses allow limited spread into the mesophyll. It has been suggested recently that luteoviruses and some other viruses largely limited to the phloem have phloem-specific promoters. *Geminiviruses* replicate in the nucleus, and spherical fibrillar structures are sometimes found in nuclei. Replication is thought to be largely coordinated with cell division, in particular to S-phase.

Movement of Viruses in the Plant

The ability of a virus to establish infection is dependent upon multiple interactions, but it has been shown for a number of plant viruses that replication can occur in protoplasts of species not infected to a detectable level in intact tissues. Movement occurs in at least two phases — *cell-to-cell* (or local) and *long distance* — which are quite distinct. Virus movement has been recently reviewed by Leisner and Turgeon (1993) and Hull (1989). *Cell-to-cell* movement of many, if not all, plant viruses involves a specific transport protein; that of TMV is probably the best studied. It has been shown to localize to, and alter the structure of, plasmodesmata. Normal plasmodesmata have a size exclusion limit of 800–1000 Da — allowing photosynthate and small molecules through, but not complex macromolecules like virus particles, or even large proteins. The size exclusion limit is increased in virus-infected plants; that this effect is due to the viral transport protein was demonstrated in transgenic plants expressing the TMV 30K transport protein in the absence of TMV infection. The 30K protein also acts as a single-stranded nucleic acid binding protein, unfolding secondary structure to form a thread-like ribonucleoprotein only 1.5–2 nm in diameter, compared to about 17 nm for a TMV virion.

Most other plant virus groups probably also have transport proteins that act to facilitate movement of a non-virion ribonucleoprotein complex through plasmodesmata — proteins with some homology to the TMV 30K protein have been identified in several virus groups. Transport proteins of different virus groups may function quite differently. It has been suggested that the *potyvirus* CIP has a role in intercellular transport because of its association with plasmodesmata, but the only demonstrated activity of this protein is as a helicase, solely a replicative function unless unwinding of dsRNA is linked to ssRNA passage through plasmodesmata. The transport function of *comoviruses* and *nepoviruses* also differs; in these virus groups virions can be observed by electron microscopy in tubules connecting adjacent cells. Similar tubules have been observed on the exterior of protoplasts infected with CPMV, suggesting that these structures are not modified plasmodesmata, which are not observed in protoplasts.

The *cryptoviruses* may not encode a movement function; their distribution within plants is irregular, and they are apparently not transmitted by inoculation either mechanically or by any vectors examined. These viruses may spread solely by cell division.

Cell-to-cell transport in local lesion hosts may result in activation of host defenses — in some cases preventing further spread and systemic infection. However, access to the vascular system typically allows long-distance movement, via either encapsidated RNA or a specific ribonucleoprotein complex. Spread into the phloem may require a mechanism separate from normal cell-to-cell movement, as there are differences in intercellular connections between phloem companion cells and sieve elements compared to those between the companion cells and mesophyll cells; there are specialized cell types with different numbers and types of connection in different plant families. Thus plant anatomy has a strong influence on virus access to the phloem — and probably on viral egress from the phloem as well. TMV can apparently spread throughout the mesophyll of a susceptible leaf within six hours after inoculation, but not move out of the leaf until two days later, demonstrating the presence of a barrier to transport that may occur in vascular loading or unloading. Once entry to the phloem has been achieved, virus moves at rates comparable to photosynthate, which will depend upon the sites of sources and sinks, and the way the vascular system is interconnected. Symptom appearance in partially-expanded leaves is often restricted to the basal portion of the leaf, reflecting the reduced ability of these leaves to act as sinks for assimilates — indeed such leaves soon become net exporters of photosynthates, and virus will either not spread into the tips of these leaves or not replicate there to levels sufficient to induce symptoms.

Experiments with temperature-sensitive mutants of TMV show that particle formation is not necessary for efficient long distance movement; ribonucleoprotein complexes containing both genomic and sgRNAs as well as both viral and host proteins were isolated from plants at the permissive temperature. Mutants defective in virion assembly were also capable of efficient long distance movement, indicating that virion assembly is not necessary, but that coat protein may contribute to transport. The coat proteins of some other virus groups have also been shown to contribute to long-distance transport. Whether any additional viral protein is required for egress from the vascular system has not yet been proven, but the normally phloem-limited potato leafroll luteovirus (PLRV) can spread into mesophyll cells in plants also infected with potyviruses, potexviruses or tobraviruses; dual infections with bromoviruses, cucumoviruses or nepoviruses did not facilitate spread of PLRV out of the phloem. PLRV also spreads systemically in plants expressing TMV movement protein, suggesting that movement protein is responsible in this instance. However, it may not be a normal factor for many viruses, as translation of the movement protein in adequate quantity in the phloem is unlikely — at least without prior egress from the sieve tubes into companion cells, which raises the problem of exiting the sieve tube afresh.

Selected Literature

Ahlquist, P. 1992. Bromovirus RNA replication and transcription. Curr. Opin. Genet. Dev. 2:71–76.

Dawson, W.O. 1992. Tobamovirus-plant interactions. Virology 186:359–367.

Hull, R. 1989. The movement of viruses in plants. Ann. Rev. Phytopathol. 27:213–240.

Lazarowitz, S.G. 1992. Geminiviruses: Genome structure and gene function. Crit. Rev. Plant Sci. 11:327–349.

Leisner, S.M., and Turgeon, R. 1993. Movement of virus and photoassimilate in the phloem: A comparative analysis. BioEssays 15:741–748.

Mason, W.S., Taylor, J.M., and Hull, R. 1987. Retroid virus genome replication. Adv. Virus Res. 32:35–96.

Matthews, R.E.F. 1991. Plant Virology, 3rd ed. Academic Press, San Diego.

Nuss, D.L., and Dall, D.J. 1989. Structural and functional properties of plant reovirus genomes. Adv. Virus Res. 38:249–306.

Riechmann, J.L., Lain, S., and Garcia, J.A. 1992. Highlights and prospects of potyvirus molecular biology. J. Gen. Virol. 73:1–16.

Wilson, T.M.A. 1985. Nucleocapsid disassembly and early gene expression by positive-strand RNA viruses. J. Gen. Virol. 66:1201–1207.

13

Isolation and Purification of Viruses

Alan A. Brunt[1] and Roger H. Lawson[2]

[1]*Horticulture Research International, Littlehampton, Sussex, UK*

[2]*US Department of Agriculture, Agricultural Research Service, US National Arboretum, Floral and Nursery Plants Research Unit, Beltsville, MD 20705, USA*

Introduction

Purified virus preparations are mainly required for the physico-chemical characterization of individual viruses and their protein and nucleic acid components. In addition, purified viruses are needed to produce reagents (especially antisera and c-DNA probes) for rapid virus detection and identification. Purified and concentrated preparations also provide highly infective inocula, especially of viruses that occur naturally in hosts containing virus inhibitors and/or virus inactivators.

Different viruses usually require different procedures for purification, and methods for well characterized viruses are described in original publications and reviews. Appropriate methods for the purification of uncharacterized or newly recognized viruses usually need to be developed experimentally. Here we discuss the principles of virus purification, but with emphasis on procedures that are especially applicable to viruses infecting ornamental plants.

The essential stages of virus purification, each of which is discussed below, are the establishment of a "pure" culture of the relevant virus, propagation of the virus, if possible, in a convenient host (often an unrelated herbaceous species), preparation of virus-containing plant extracts, clarification of the extracts, sedimentation of virus particles and, finally, the removal of any contaminants from the concentrated virus preparations.

Virus Isolation and Propagation

Naturally-infected plants, especially those of vegetatively propagated species such as flowerbulbs, trees, and shrubs, frequently contain two or more viruses. It is thus an important pre-requisite that a "pure" culture be established of each virus to be investigated. Viruses that induce lesions in inoculated leaves of differential hosts can often be obtained in "pure" culture by repeated sub-culture of single lesions. Finally, inoculum provided by a single lesion is used to establish cultures in a suitable propagation host. Uncontaminated cultures of other viruses initially occurring in mixtures can sometimes be obtained by differences in their host range, stability *in vitro*, vector specificity or, if antisera are available, by the serological absorption and removal of the contaminating viruses; the use of monoclonal antibodies in affinity chromatography is especially useful for this purpose (e.g. Diaco et al., 1986). If viruses differ sufficiently in their sedimentation characteristics, it is sometimes possible to separate them by density gradient centrifugation (Brakke, 1960) or, if they differ in electrophoretic behavior, by density gradient electrophoresis (Regenmortel, 1964). Chromatography on controlled glass beads is also useful for separating viruses of different sizes (Barton, 1977).

Although it is sometimes necessary to purify viruses with restricted host ranges from naturally infected hosts, it is more usual to extract virus from an experimental host species to which the virus is readily transmissible in which it replicates faster and/or from which it is more readily extracted. For example, commonly occurring strains of carnation necrotic fleck, lilac chlorotic leafspot, narcissus mosaic, and tomato spotted wilt viruses are more readily purified from *Dianthus barbatus*, *Chenopodium quinoa*, *Nicotiana clevelandii*, and *N. rustica*, respectively, than from their natural hosts (Bar-Joseph and Smookler, 1976; Brunt, 1988). *N. benthamiana* and *N. clevelandii* are very convenient propagation hosts for a wide range of plant viruses (Hollings and Brunt, 1981; Hollings et al., 1981), but *N. bigelowii* and *N. acuminata* are even better hosts of plum pox potyvirus (Kerlan et al., 1975). Viruses with restricted host ranges that are not transmissible to unrelated herbaceous plant species, such as some infecting flowerbulbs, need to be purified from their primary host species. However, such viruses occur in higher concentration and/or are more easily isolated from intolerant crop cultivars; for example, lily symptomless virus is extracted more readily from lily cvs Enchantment and Dunkirk than other cultivars (Beijersbergen and Hulst, 1980; A.A. Brunt and R.S. Phillips, unpublished information).

It is also sometimes beneficial to seek an isolate of a virus which, because of its greater stability, reduced virulence, faster replication or for other unknown reasons, is more readily purified than other isolates. For example, although nerine X potexvirus from *Nerine sarniensis* is purified with difficulty (Maat, 1976), another isolate from *Agapanthus praecox* is purified easily (Phillips and Brunt, 1980).

To recover maximum amounts of virus, infected plants should be grown in conditions that favor virus replication. Infected material should be collected for processing when the virus has attained a relatively high concentration (Yarwood, 1971; Francki, 1972; Matthews, 1991). The optimum interval between inoculation of plants and the collection of infected material can vary greatly with different virus:host combinations. The optimal conditions need to be determined experimentally. The virus concentration in such source plants, as well as in preparations at various stages of the purification procedure, can now be rapidly determined by immunosorbent electron microscopy, enzyme-linked immunosorbent assay and/or dot-blot cDNA hybridization (e.g. Hull, 1985).

Preparation of Virus-Containing Plant Extracts

After the establishment of a pure virus culture in a suitable host and after growing conditions for maximum virus replication have been defined, it is necessary to determine the optimal conditions for making and clarifying plant extracts and for the concentration and further purification of virus particles.

Viruses are usually most effectively extracted from fresh plant material, although satisfactory yields of some can be obtained from frozen (see Brunt, 1988) or lyophilized (Derks et al., 1982) leaves. Although leaves of infected plants are usually used as sources of virus, roots of pea are better than aerial plant parts as a source of clover yellow vein potyvirus and bark of Etrog citron or pericarp of navel oranges as a source of citrus tristeza trichovirus (see Brunt, 1988).

It is sometimes advantageous, especially with viruses that are purified with difficulty, to extract virus only from parts of leaves with conspicuous symptoms and/or to reduce their dry matter by keeping plants in the dark for 24–48 h before removal of leaves (Lawson et al., 1985).

Viruses are usually released from cells by disrupting infected tissues in a suitable extractant using a pestle and mortar, food blender, juice extractor, or a mechanical press such as that manufactured in Germany by Pollähne. However, to minimize shearing forces that disrupt long filamentous virus particles, carnation necrotic fleck closterovirus-infected material is best disintegrated with a pestle and mortar (Bar-Joseph and Smookler, 1976).

Viruses are usually extracted by disrupting infected plant tissues in a buffer containing one or more additives, although buffer or even water alone has been used for the successful extraction of some potyviruses (Hollings and Brunt, 1981). A buffer is often used to stabilize virus particles *in vitro* and, in plant extracts that are highly acidic, to prevent their iso-electric precipitation. Commonly used buffers are phosphate, borate, citrate, TRIS or HEPES at 0.01–0.5 M, and pH 7.2–9.0. The choice of buffer, molarity, and pH is often critical as viruses, or even virus strains, differ greatly in their stability in different buffers. For example, citrate buffer is superior to any other for the purification of many strains of cucumber mosaic cucumovirus (Francki, 1972).

A buffer of high molarity (0.25–0.5 M) and relatively high pH (8.5–9.0) can often be used successfully to minimize particle aggregation of some filamentous viruses. The solubilization of particles is also often increased by extracting infected tissues in large volumes (3–20 ml/g tissue) of extractant, especially if the host species contains much mucilaginous material. Higher yields of tulip breaking potyvirus are obtainable from tulips by first heating leaves to 50°C for 15 min, a beneficial effect possibly attributable to heat-induced breakdown of membranes and dispersal of particle aggregates (van Slogteren and de Vos, 1966). Higher yields of other viruses are obtained by first storing crude leaf extracts for 24–48 h, a period during which virus particles are released from plant debris or inclusions (R.S. Phillips, personal communication); this procedure is especially useful for facilitating the separation of caulimovirus particles, such as those of dahlia mosaic and carnation etched ring viruses, from virus-containing inclusions (e.g. Brunt, 1971). Some additives to extractants also sometimes facilitate the dispersal and solubilization of particles; the addition of hemicellulase, cellulase, and pectinase reduces the viscosity of lily leaf extracts and may thus facilitate the purification of lily symptomless carlavirus (Beijersbergen and Hulst, 1980). Macerating enzymes such as Driselase (a mixture of pectinase and cellulase) or Celluclast (an industrial grade of cellulase) are also effective in facilitating the release of closteroviruses from phloem cells. Urea at 0.5–1.0 M in extractants prevents aggregation and/or facilitates dissociation of aggregates of some filamentous viruses (e.g. Brunt, 1988). Detergents such as Tween-20 and Triton X-100 and the chelating agent EDTA have also been used for this purpose (Hull, 1985; Matthews, 1991). EDTA is not recommended for viruses such as closteroviruses, which require divalent cations for particle integrity, although it is especially beneficial in disrupting ribosomes and phytoferritin that would otherwise occur as contaminants in preparations of stable viruses.

Some plant species contain substances which abolish the infectivity of viruses *in vitro*, and others have constituents which merely inhibit the mechanical transmission of viruses. The latter include species such as those of *Capsicum* and many of those included in the Chenopodiaceae and closely related families. Inhibitors of infection characteristically inhibit the transmission of such viruses from their primary host to unrelated species but have no adverse effect on the transmission of viruses from infected to healthy plants of the same species. Such inhibitors have relatively low molecular weights (usually less than 30 kDa), so inhibitor-free virus preparations are easily obtained by subjecting leaf extracts to ultracentrifugation, during which the denser virus particles are sedimented and so separated from the inhibitors remaining in the supernatant which is discarded; after resuspension of the precipitate in an appropriate buffer, the preparation usually provides a highly infective inoculum.

Virus inactivators are often o-quinones which are formed by the enzymatic oxidation of polyphenols. Such inactivators react with virus particles to form insoluble non-infective virus:tannin complexes, the latter being commonly formed in extracts of plants containing active copper-dependent polyphenoloxidases. Members of the Compositae, such as dahlia and chrysanthemum, contain such systems. As the formation of non-infective virus:tannin complexes is irreversible, it is often necessary to minimize their formation by including in the extractant a reducing agent such as sodium bisulphite, ascorbate, cysteine, 2-mercaptoethanol or dithiothreitol (to minimize oxidation), and/or copper-chelating agents such as sodium diethyldithiocarbamate (to inactivate polyphenoloxidases by removing copper ions), and making extracts at low temperature or even with liquid nitrogen. Preformed o-quinones in plant extracts can be removed by including in extractants additives such as 0.2% (w/v) polyvinyl pyrrolidone or alumina; the addition of 2–4% (w/v) protein, or even milk, is also used occasionally to remove preformed o-quinones (e.g. Brunt, 1988).

Clarification of Plant Extracts

Low-speed centrifugation (10–15,000 g for 10–20 min) alone is usually inadequate to clarify plant extracts, although commonly used to remove course plant debris. Extracts containing less stable viruses can be clarified by centrifugation through a 20–30% (w/v) sucrose cushion, during which the larger plant debris is sedimented and the virus particles concentrated at the sugar:extractant interface or, after prolonged centrifugation, in the sugar cushion. If the virus to be purified is sufficiently stable, it is sometimes possible to aid clarification of extracts by using simple physical treatments to denature normal plant constituents. Heating extracts to 40–50°C is sometimes a useful step in the purification of some potyviruses but can result in substantial or even total loss of other viruses. Freezing extracts — although this facilitates clarification without inactivation of some potex-, carla-, closteroviruses, and other viruses — can result in near total loss of many potyviruses. Lowering the pH of extracts briefly to 4.5 denatures some proteins and permits the clarification of leaf extracts. This is a useful procedure for viruses such as prunus necrotic ringspot and tobacco streak ilarviruses which are unaffected by the treatment.

Various additives to extractants also sometimes facilitate clarification of crude extracts. Adsorbents such as activated charcoal, DEAE cellulose, Celite, or calcium phosphate have been used effectively in the purification of potex-, ilar-, and carlaviruses (Francki, 1972; Hull, 1985; Brunt, 1988). Bentonite (40 mg/ml) has been used in the purification of some closteroviruses (Hull, 1985; Brunt, 1988). Silver nitrate has been used occasionally to clarify extracts containing potex- and potyviruses (e.g. Lesemann et al., 1983).

Organic solvents, however, are widely used to effect clarification of extracts. Although 10–20% ethyl alcohol and freon are occasionally effective (Lawson et al., 1985), other solvents (3–50%) such as n-butanol, chloroform, diethyl ether, and carbon tetrachloride — either alone or in mixtures — have been used in the purification of a wide range of viruses. The non-ionic detergent Triton X-100, by disrupting lipid membranes, has greatly facilitated the purification of some viruses. Whichever procedure or additive is used to denature plant components, the latter are usually removed from extracts by brief low speed centrifugation (10–20 min at 10,000 g).

Partial Purification of Viruses

Viruses are sometimes precipitated from clarified leaf extracts by the addition of ammonium sulphate (to one-third saturation). However, polyethylene glycol (PEG; M_r 6,000–20,000) is now commonly used to precipitate viruses. In this procedure virus particles are precipitated from clarified leaf extracts by adding sodium chloride to 0.1 M and PEG to 2–4% (w/v) for anisometric viruses or to 8–10% for isometric viruses; after stirring the mixture for 1–2 h at 2–4°C, the precipitated virus is collected by brief low speed centrifugation (10,000 g for 10 min), then resuspended in a small volume of an appropriate buffer and finally clarified by low speed centrifugation (e.g. Hull, 1985; Matthews, 1991). The universal use of this procedure is undoubtedly attributable to its simplicity, efficacy, and requirement for only simple equipment (the major item of which is a low-speed centrifuge).

Partially purified preparations of many viruses, however, are more conveniently obtained by two or more cycles of low- and high-speed centrifugation. In the first cycle of differential centrifugation, crude plant extracts are clarified by low-speed centrifugation (10–20 min at 10,000–15,000 g), and virus then sedimented from the clarified solution by high speed centrifugation (1–3 h at 70,000–240,000 g). After redispersion of the sedimented virus in a small volume of buffer, the cycle of differential centrifugation is repeated. Partially purified virus is resuspended from the final precipitate.

Further Purification of Viruses

Repeated cycles of differential centrifugation or further precipitation with PEG or ammonium sulphate usually fail to remove all plant contaminants from preparations. Such extended procedures often result in loss of virus. Similarly, although plant proteins can be removed by the addition (if available) of homologous antiserum, this procedure can also result in unacceptable losses of virus.

Of the alternative procedures available for removing contaminants, the most widely used is density gradient centrifugation. Rate-zonal density gradient centrifugation in which macromolecules are separated by differences in their sedimentation velocities is most commonly used. Isopycnic or equilibrium centrifugation in which components are separated by their different specific density is also, however, of considerable value (Brakke, 1960). Sucrose, because of its low cost and general availability, is often used for the formation of gradients, usually of 10–40% (w/v). However, caesium chloride and caesium sulphate are also commonly used; other gradient media (sodium bromide, sodium iodide, potassium tartrate, Renagrafin, and Nycodenz), although used less frequently, have also proved to be efficacious (e.g. Francki, 1972; Hull, 1985; Lawson et al., 1985; Gugerli, 1985). Centrifugation of partially purified virus preparations through a cushion of sucrose is a relatively gentle and effective procedure for removal of non-viral contaminants (e.g. Lawson et al., 1985).

Whatever material is used for gradient formation, density gradient centrifugation of partially purified preparations results in the separation of virus particles from contaminants which have different densities. After collection of the virus-containing zones from the gradients, either with a syringe or by passage of the tube contents through an ultraviolet monitoring device set at a wavelength of c. 260 nm with an attached fraction collector, the gradient material can be removed from the virus preparation by dialysis and the particles then concentrated by high-speed centrifugation.

Chromatography, using adsorbents such as calcium phosphate, Sephadex, or cellulose, can also be used for the further purification of impure virus preparations. Affinity chromatography using antibodies coupled covalently to agarose or other medium is particularly useful for the further purification of labile viruses (e.g. Diaco et al., 1986). Molecular permeation chromatography using controlled pore glass beads (70 nm pore size, 120–200 mesh) has also been used for a wide range of plant viruses (Barton, 1977). Although not now widely used, zonal density gradient electrophoresis is a very valuable technique which not only permits separation of viruses from contaminants of viruses in mixtures but also those which have different electrophoretic mobilities (Regenmortel, 1964).

Quality of Purified Preparations

Each preparation should be assessed for infectivity, concentration, and purity. Infectivity is best determined by inoculation, if available, of a sensitive local lesion assay host. Concentration is measured from the specific absorption at 260 nm by estimating the amount of virus present using the known extinction coefficient. The purity of a preparation can be assessed by electron microscopy or, better still, by electrophoresis of a sodium lauryl sulfate (SDS)-dissociated virus preparation in polyacrylamide gels (Alper et al., 1984); the latter procedure is especially useful in indicating the presence of contaminating plant proteins which should be removed to produce highly purified virus preparations.

Storage and Use of Virus Preparations

It is preferable to use purified virus as soon as possible after its preparation. Nevertheless, some viruses can be stored at 2°C after the addition of chloroform, glycerol, sodium azide, or chloroform-butanol which inhibit contamination by micro-organisms. Some viruses can be stored at −70°C, and others such as tulip breaking virus at −20°C after the addition of 30% (w/v) sucrose (Derks et al., 1982). Preparations of many viruses, especially those to be used for antiserum production, can be safely lyophilized and stored.

Selected Literature

Alper, M., Salomon, R. and Loebenstein, G. 1984. Gel electrophoresis of virus-associated polypeptides for detecting viruses in bulbous irises. Phytopathology 74:960–962.

Bar-Joseph, M. and Smookler, M. 1976. Purification, properties and serology of carnation yellow fleck virus. Virology 66:835–838.

Barton, R.J. 1977. An examination of permeation chromatography on columns of controlled pore glass for routine purification of plant viruses. J. Gen. Virol. 35:77–87.

Beijersbergen, J.C.M. and Hulst, C.T.C. van der. 1980. Application of enzymes during bulb tissue extraction for detection of lily symptomless virus by ELISA in *Lilium* spp. Neth. J. Plant Pathol. 86:277–283.

Brakke, M.K. 1960. Density gradient centrifugation and its application to plant viruses. Adv. Virus Res. 7:193–224.

Brunt, A.A. 1971. Some hosts and properties of dahlia mosaic virus. Ann. Appl. Biol. 67: 357–368.

Brunt, A.A. 1988. Purification of filamentous viruses and virus-induced non-capsid proteins. In: The Plant Viruses, Vol. 4. The Filamentous Plant Viruses, R.G. Milne (ed.), Plenum, New York, pp. 85–110.

Derks, A.F.L.M., Vink-Van den Abeele, J.L. and Schadewijk, A.R. van. 1982. Purification of tulip breaking virus and production of antisera for use in ELISA. Neth. J. Plant Pathol. 88:87–98.

Diaco, R., Hill, J.H. and Durand, D.P. 1986. Purification of soybean mosaic virus by affinity chromatography using monoclonal antibodies. J. Gen. Virol. 67:345–351.

Francki, R.I.B. 1972. Purification of viruses. In: Principles and Techniques in Plant Virology, C.I. Kado and H.O. Agrawal (eds.), Van Nostrand-Reinhold, New York, pp. 295–335.

Gugerli, P. 1984. Isopycnic centrifugation of plant viruses in Nycodenz density gradients. J. Virol. Methods 9:249–258.

Hollings, M. and Brunt, A.A. 1981. Potyviruses. In: Handbook of Plant Virus Infections and Comparative Diagnosis, E. Kurstak (ed.), Elsevier/North Holland, Amsterdam, p. 738.

Hollings, M., Pawley, R.R. and Atkey, P.T. 1981. Host ranges of potyviruses. Report of the Glasshouse Crops Research Institute for 1979, p. 160.

Hull, R. 1985. Purification, biophysical and biochemical characterisation of viruses. In: Virology: A Practical Approach, B.W.J. Mahy (ed.), IRL Press, Oxford and Washington DC, pp. 1–24.

Kerlan, C., Dunez, J. and Bellet, F. 1975. Methodes simples de purification du virus de la sharka (plum pox virus). Ann. Phytopathol. 7:287–297.

Lawson, R.H., Brannigan, M.D. and Foster, J. 1985. Clover yellow vein virus in *Limonium sinuatum*. Phytopathology 75:899–906.

Lesemann, D.-E., Makkouk, K.M., Koenig, R. and Natafji Sammon, E.N. 1983. Natural infection of cucumbers by zucchini yellow mosaic virus in Lebanon. Phytopathol. Z. 108: 304.

Maat, D.Z. 1976. Two potex viruses in Nerine. Acta Hortic. 59:81–82.

Matthews, R.E.F. 1991. Isolation. In: Plant Virology, 3rd ed., Academic Press, New York, pp. 53–67.

Phillips, S. and Brunt, A.A. 1980. Some hosts and properties of an isolate of nerine virus X from *Agapanthus praecox* sub. sp. *orientalis*. Acta Hortic. 110:65–70.

Regenmortel, M.H.V. van 1964. Purification of plant viruses by zone electrophoresis. Virology 23:495–502.

Van Slogteren, D.H.M. and de Vos, N.P. 1966. Tulip breaking, its serological behaviour and serological relationship to a virus isolated from lily. In: Viruses of Plants, A.B.R. Beemster and J. Dijkstra (eds.), North Holland, Amsterdam, pp. 320–323.

Yarwood, C.E. 1971. Procedures to increase virus yield from infected plants. In: Methods in Virology, Vol. 5, K. Maramorosch and H. Koprowski (eds.), Academic Press, New York, pp. 451–479.

14

Identification and Assay

A. Gera[1], Roger H. Lawson[2] and Hei-Ti Hsu[2]

[1]*Department of Virology, Agricultural Research Organization, The Volcani Center, Bet Dagan 50250, Israel*

[2]*US Department of Agriculture, Agricultural Research Service, US National Arboretum, Floral and Nursery Plants Research Unit, Beltsville, MD 20705, USA*

Introduction

The early and accurate diagnosis of plant diseases is a crucial component of any crop-management system. Once the cause of a disease has been established as a virus, a series of tests is necessary to determine its identity. Symptoms are of major importance because they are the main means by which a virus disease is diagnosed. Transmission of a virus from infected to healthy tissue is a fundamental procedure used to isolate viruses from diseased plants, to study viral symptoms and to assay for infectivity. Electron microscopy is the only direct method used for determining the shape and size of the virus particle. In many cases it serves as a rapid method for placing a virus in its taxonomic group. Serological procedures represent the most reliable method for identification and quantitative assay of viruses.

Contribution No. 1376-E, 1994 series from the Agricultural Research Organization, The Volcani Center, Bet Dagan, Israel.

Recent developments in molecular biology have resulted in more convenient, effective, and specific assays for the detection and diagnosis of viruses. The application of nucleic acid probes and related techniques based on properties of viral nucleic acid is becoming common in the detection of plant viruses. A major breakthrough in the detection of viruses which are present at very low concentrations in the infected tissue is the recent development of the polymerase chain reaction (PCR) technology.

In this chapter the various techniques commonly used for assay and identification of plant viruses are described.

Biological Tests

Assay of infectivity is an essential aspect of plant viruses. Since infectivity of a virus depends on the susceptibility of the host plant, the same basic procedure may simultaneously measure the infectivity of the virus and the susceptibility of the host.

Visual inspection for symptoms is often not adequate, since the symptom expression can be highly variable. Application of more than one assay method depends on different properties of the virus. The use of multiple assays is usually essential for studying new viruses. Host symptoms are very important to the applied plant virologist. In the field, symptoms give the first clue to the identity of a virus. In the laboratory and greenhouse, symptoms produced in a range of test plants may be of considerable diagnostic value.

Experimental host plants, under standardized conditions, will produce consistent and distinguishing disease symptoms when infected with a particular virus. Determining whether infection occurs and if it is local or becomes systemic remains a simple and most useful tool in plant virology. This procedure is usually essential for studying new viruses or virus strains. For the characterization of an unknown virus, transmission by specific means and the induction of particular symptoms in plants may provide useful clues to the identity of the virus. Sap transmission of many viruses by leaf rubbing is a valuable method for virus detection and disease diagnosis. In general, viruses transmitted in a persistent manner by vectors are transmitted poorly or not at all by mechanical inoculation. Grafting, dodder, and transmission by insect vector to indicator plants are transmission procedures used for phloem-limited viruses, especially of woody plants.

Several herbaceous plants are susceptible to a large number of viruses. Some plant species react to mechanical inoculation with several viruses by producing

Table 14.1. Examples of plant species that react by producing local lesions after inoculation with plant viruses

Chenopodium amaranticolor Cost and Reyn.
Chenopodium quinoa Willd.
Datura stramonium L.
Gomphrena globosa L.
Nicotiana glutinosa L.
Petunia hybrida Vilm.
Phaseolus vulgaris L.
Tetragonia expansa Murr.
Vigna unguiculata Endl.

local lesions that are useful in quantitative virus assay. The most widely used local-lesion hosts for plant viruses are listed in Table 14.1. Indicator hosts react diagnostically with certain viruses, whereas plants that can distinguish between certain viruses are called differential hosts.

Biological techniques for virus diagnosis and detection are usually highly accurate but too slow and not amenable to large-scale application. The test is routinely used for diagnosis of cucumber mosaic virus (CMV), tobacco mosaic virus (TMV), tomato spotted wilt virus (TSWV), and other viruses in ornamental crops. The advent of serological and molecular techniques has reduced the use of transmission for virus identification.

Electron Microscopy

The shape and size of a virus particle is a basic criterion for virus identification. The electron microscope was introduced in 1932. Subsequent development and improvements have provided a very powerful tool capable of visualizing viruses both *in vitro* and *in situ*.

Electron microscopy is commonly used to examine purified preparations of virus as well as crude extracts from infected plants. The development of the negative staining technique allowed simple, rapid, and direct examination of crude extracts of infected plants. Crude sap preparation may be made by grinding or chopping infected leaf, stem, or root tissue in a buffer or directly in a negative stain. Alternatively, a fragment of epidermis can be stripped from the leaf and passed through a drop of water or stain on a specimen grid. The technique provides information on virus morphology within minutes.

Many kinds of heavy metal salts have been tested, but none has proved to possess all desired qualities. Phosphotungstic acid, uranyl acetate, and ammonium molybdate are the most commonly used. These stains may also penetrate into the particles, revealing the internal structure. Rod-shaped viruses can be easily distinguished from host components in crude sap. However, small icosahedral viruses that occur in low concentrations may be confused with host components.

The use of electron microscopy in combination with serology for detecting and identifying plant viruses will be discussed later in this chapter.

Serology

Antibodies are animal host proteins produced in response to the presence of foreign molecules in the body. Any molecule that can bind to an antibody is known as an antigen. The molecule used to induce the immune response is called an immunogen. Immunogenicity is used to describe the ability of a molecule to induce the immune system to produce antibodies, and antigenicity is the ability of the foreign molecule to react with a specific antibody.

Antibodies are synthesized primarily by plasma cells. They belong to a large family of glycoproteins that share key structural and functional features. Structurally, antibodies are composed of one or more copies of a characteristic unit forming a **Y**-shape. Each unit contains four polypeptides — two heavy and two light chains. Functionally, they can be characterized by their ability to bind to specific antigens. Antibodies are divided into five classes of immunoglobulins known as IgG, IgM, IgA, IgE, and IgD. The first two classes are relevant to plant virology. IgG molecules have three protein domains. Two of the domains are identical (approx. 55 kD) and form the arms of the **Y**; each arm contains an antigen-binding site, making the IgGs bivalent. The third domain, composed of two identical light chains (approx. 25 kD), forms the base of the **Y**; IgM antibodies are pentamers of the **Y**-shaped units.

Most plant viruses are strong immunogens. When they are injected into an experimental animal, they stimulate the production of specific antibodies that can be used in various serological tests. Under suitable conditions the reactions between antibodies and the antigen can be demonstrated *in vitro*. Visible precipitates are formed as a result of the reaction between the bivalent antibodies and the multivalent viral-antigens. A prerequisite for the formation of precipitates is that both antibodies and antigen are present within a certain range of concentrations. Precipitation will not occur if one of the reactants is not present in the optimal concentration.

Production of Antisera

The production of virus specific antisera is the basis for a successful serological procedure. Antisera are produced by immunizing animals with a highly purified virus or viral coat protein. Although various animal species may be used to raise the antiserum, it is likely that rabbits will continue to be the preferred animal for most applications. Purified viral antigens at a concentration of 0.5–1 mg are usually administered in complete Freund's adjuvant, followed by subsequent immunization in complete adjuvant. Good-quality antisera of adequate titer can be obtained by injecting animals with as little as 50–100 μg of nucleoprotein with some viruses. The antigen is usually introduced intramuscularly or intravenously into the animal. There is considerable diversity in the literature concerning the most appropriate schedule of immunization. The production of high-titered antisera depends on the amount of virus injected and viral antigenicity. The number of injections will also affect the specificity of antisera. When few injections are given, highly specific antisera are produced. Multiple injections may produce broad-spectrum antisera without the specificity required for strain differentiation.

Most plant viruses are stable in the animal immunized and will stimulate the production of antibodies specific for the quaternary structure of the virions. However, certain viruses such as cucumoviruses that tend to degrade during the immunization process can be stabilized by adding a low concentration of formaldehyde before the virus is injected.

In our laboratory antisera are routinely produced by four–six biweekly intramuscular injections of 1 ml of virus preparation (0.5–2 mg) emulsified with an equal volume of Freund's complete or incomplete adjuvant. Bleedings are taken 10–20 days after the last injection when serum titer reaches its maximum. Blood samples are allowed to coagulate for 2 hours at 37°C and then overnight at 4°C. The serum is centrifuged at low speed to remove any remaining blood cells. Antisera are usually stored frozen at −20°C or mixed with glycerol and stored at −20°C. Non-specific antibodies can be removed from the antiserum by cross-absorption with any contaminating host antigens.

For some serological applications it is necessary to use purified immunoglobulins. These may be prepared from whole antiserum by ammonium sulfate precipitation and DEAE-cellulose filtration or protein A-Sepharose affinity adsorption.

Polyclonal antisera specific to viral coat protein treated with sodium dodecyl sulfate (SDS) and separated on polyacrylamide gel have been successfully produced by immunizing rabbits with protein bands cut from gels.

Monoclonal Antibodies

In the animal, antibodies are synthesized primarily by plasma cells, a type of terminally differentiated B lymphocytes. These cells, unfortunately, do not divide. However, when a B lymphocyte cell is fused with a mouse myeloma cell, a hybrid cell (hybridoma) is produced that can grow in culture indefinitely and continue to produce antibodies with a defined specificity. Antibodies produced by hybridoma cell cultures are known as monoclonal antibodies (Mabs). Monoclonal anti-virus-antibodies can be selected from hybridoma clones which bind to cryptotopes (epitopes exposed only on the inside of the capsomers), metatopes (epitopes present on the capsomers and on the nucleocapsid), or neotopes (epitopes present only on the nucleocapsid). Highly specific antibodies can be identified during the selection process with purified antigen, even though the immunogen was not highly purified. Hybridomas may be maintained in culture and produce 10–100 μg immunoglobulins/ml. However, when grown as ascites tumors, hybridoma cells may produce as much as 20 mg immunoglobulins/ml of ascites fluid.

Since Mabs against TMV were first described, these have been prepared against 60 different plant viruses of economically important crops. These antibodies are valuable for detecting viruses that are difficult to purify. In practice, however, production of Mab is often labor-intensive.

For diagnosis work Mabs that recognize a wide range of different plant virus species are recommended. Mab PTY-1 developed by Jordan and Hamond (1991) has been found to react with 48 distinct aphid-transmitted potyviruses in antigen-captured (ACP)-ELISA. Mabs against apple mosaic, CMV, bean yellow mosaic (BYMV), TSWV, carnation mottle, carnation etched ring, carnation necrotic fleck, cymbidium mosaic, impatiens necrotic spot, prunus necrotic ringspot, and tulip breaking viruses are commonly used for virus detection in ornamental crops.

Immunological Tests

Agar immunodiffusion

In this method antigen and antibody are placed in opposite wells cut into a gel medium. The antiserum is usually placed in the central well and the antigen in the surrounding wells. As the antigen diffuses into the medium, a precipitation zone is formed where the antigen and antibodies meet at optimal concentrations. Under standardized conditions the precipitation lines form a characteristic pattern (Fig. 14.1), and conclusions may be drawn on various antigen-antibody relationships. The size, shape, position, and density of the precipitation lines depend upon the characteristics of the antigen-antibody interactions.

The pattern of precipitation lines (fusion, crossing, or spur formation) indicates whether the isolates are identical, serologically related, or unrelated viruses. Spur formation indicates that the two antigens being compared possess common as well as distinct epitopes. Fusion and crossing precipitin lines are interpreted in terms of serological identity or non-relatedness of the two antigens.

Accurate information on the presence or absence of common epitopes can be obtained by comparing the two antigens and their respective homologous antisera. With isometric viruses, the interpretations of the results by this method are satisfactory. However, most elongated rod-shaped viruses do not move into the diffusion medium and thus need to be degraded with detergent such as SDS or pyrrolidine. Although an agar- or agarose-based medium has been the most commonly used for double gel diffusion recently, the use of the polysaccharide from *Pseudomonas* (Gelrite) enabled a 10–100 fold increase in sensitivity, which made possible the detection of 2–20 μg/ml of antigen.

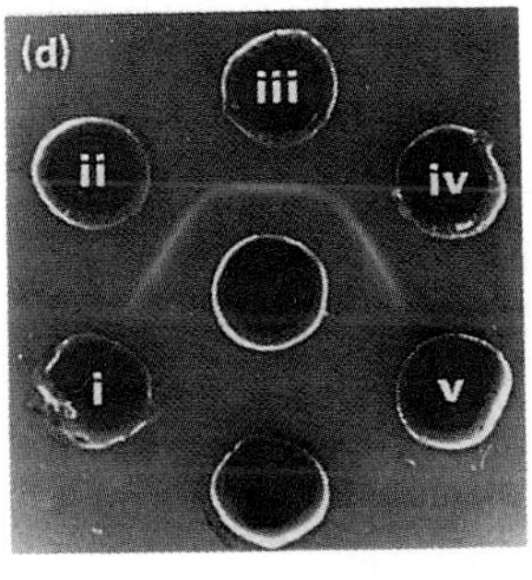

Fig. 14.1. Agar gel diffusion test with antiserum to cucumber mosaic virus (CMV) in the center well and crude sap from (1) healthy lisianthus, (2) CMV-infected lisianthus, (3) partially purified CMV isolate homologous to the antiserum, (4) *Nicotiana glutinosa* infected by the lisianthus isolate of CMV, and (5) healthy *N. glutinosa*.

Tests with labeled antibodies

In order to increase the sensitivity of serological tests, several labels such as enzymes, fluorescent molecules, and radioisotopes have been attached to antibody molecules. In this chapter only methods which are commonly used for plant virus detection will be discussed.

Enzyme-linked immunosorbent assay (ELISA)

ELISA was first applied to plant virus detection by Clark and Adams (1977). The method offers many advantages over the more traditional immunoprecipitation-based procedures. The method has been effectively applied for the detection and assay of plant viruses and has significantly increased the ability of plant virologists to detect and study plant viruses. In the last decade different variations of ELISA have become the preferred serological test for plant viruses because of their simplicity, adaptability, rapidity, sensitivity, and accuracy. The method has been applied to viruses of various morphological types in both purified preparations and crude plant extracts.

The double antibody sandwich (DAS-ELISA) test is the most commonly used. Virus is first selectively trapped by a specific antibody adsorbed on the surface of polystyrene plates. A specific enzyme-labeled antibody (conjugate) is then added to the immobilized virus, and the reaction is measured visually or spectrophotometrically after adding a suitable enzyme substrate. Alkaline phosphatase is the enzyme most frequently used for the antibody-conjugate, and this is normally detected using the substrate p-nitrophenylphosphate.

A variation of the above method is the indirect ELISA in which plates are coated with antigen, and the primary antiviral antibody of one animal species (e.g. rabbit) is then added. A secondary commercial antibody conjugate (e.g. goat anti-rabbit) is then added, which reacts with the first antibody. Following the addition of a substrate, a color is formed. The color reactions produced by unknown samples are compared with known positive and negative control wells.

DAS-ELISA is especially useful for detecting antigens in complex mixtures. This is because the bound antibody specifically traps the antigens of interest, whereas non-specific antigens are removed in the wash step. Although the indirect-ELISA is considered less strain-specific, the test is simple to perform. Enzyme antibody conjugates which can be used for detecting many different viruses are available commercially (Table 14.2).

Table 14.2

Examples of commercially available ELISA assay kits for ornamental viruses

Virus	Host plant
Apple mosaic	Rose
Arabis mosaic	Gladiolus, rose
Bean yellow mosaic	Gladiolus, orchid
Carnation viruses (mottle, latent, etched ring, necrotic fleck)	Carnation
Chrysanthemum virus B	Chrysanthemum
Cucumber mosaic	Aster, begonia, chrysanthemum, geranium, lily
Dahlia mosaic	Dahlia
Dasheen mosaic	Dasheen and other aroids
Hibiscus chlorotic ringspot	Hibiscus
Hosta virus X	Hosta
Impatiens necrotic spot	Aster, begonia, chrysanthemum, gladiolus, impatiens, orchid
Kalanchoë latent virus	Kalanchoë
Lily symptomless	Lily
Orchid viruses (cymbidium mosaic, cymbidium ringspot, and odontoglossum ringspot viruses)	Orchids (many genera)
Pelargonium flower break	Geranium
Poinsettia mosaic	Poinsettia
Prunus necrotic ringspot	Rose
Strawberry latent ringspot	Gladiolus, lily, rose
Tobacco mosaic	Aster, kalanchoë, geranium
Tobacco ringspot	Aster, begonia, chrysanthemum, geranium, gladiolus, lily, rose
Tomato spotted wilt	Aster, begonia, chrysanthemum, geranium, gladiolus, impatiens, orchid
Tomato aspermy	Aster, chrysanthemum, impatient, lily

Dot ELISA

Assays in which antibodies or antigens are bound to nitrocellulose or nylon membranes have been used to detect plant viruses. In principle, solutions containing purified antigen or crude sap from infected plants are applied directly to a membrane and air-dried. The membrane surface is then blocked with bovine serum albumin (BSA) and a specific viral antibody is added. Finally, a second antibody conjugate and substrate are added. The enzyme reacts with a soluble substrate to form an insoluble colored product at the site of the reaction (see Chapter 36, Orchids).

The method has been routinely used for the diagnosis of several potato viruses and for the detection of viruses in seed samples. The method was nearly eight times more sensitive than ELISA for the detection of tomato spotted wilt virus when biotinylated mouse Mabs were used (Hsu and Lawson, 1991).

Electroblot immunoassay

In this method, proteins are first separated by SDS-polyacrylamide gel electrophoresis before being transferred to a nitrocellulose membrane. After saturation of the membrane surface with BSA, the blots are incubated successively with specific antibody and a detection reagent. The method is more often used as an analytical tool than for virus diagnosis.

Tissue squashes and tissue blotting

DNA and RNA of plant viruses were detected specifically and sensitively by hybridization of infected plant tissues squashed onto a nylon membrane (squash blot) with a specific radioactive probe. Tomato yellow leaf curl (TYLCV) DNA was detected in squash blots of tomato leaves, roots, stems, and fruits (Navot et al., 1989). The method provides a specific, rapid, and simple tool for large-scale diagnostics of TYLCV and other viruses.

A similar technique based on tissue blotting on nitrocellulose and immunological detection of plant viruses was described recently (Lin et al., 1990). Tissue blots were made by gently pressing the freshly cut tissue surface on nitrocellulose membrane. Antigens in tissue blots were detected by enzyme-labeled probes (Fig. 14.2). The technique was applied to detect viruses among the cucumovirus, luteovirus, potexvirus, potyvirus, and TSWV groups in blots of ornamental plants showing virus-like symptoms (Hsu and Lawson, 1991).

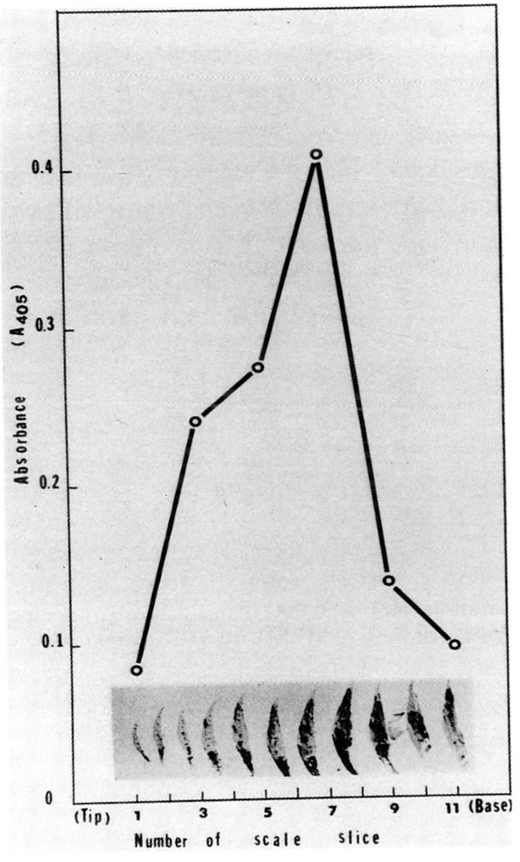

Fig. 14.2. Comparative sensitivity of detection methods and distribution of lily symptomless carlavirus in scales of lily bulbs by enzyme-linked immunosorbent assay and by tissue blot immunoassay.

Immunosorbent electron microscopy

Techniques involving the detection and identification of plant viruses by combining electron microscopy and serology are highly sensitive. The method was introduced by Derrick (1973). In this procedure an electron microscope grid is coated with specific antisera for the virus being investigated. Grids are floated on purified virus or crude sap from an infected plant, and antibodies on the grid surface trap virus particles from solution. If virus particles are in the preparation that is analyzed, more virus particles become attached to a grid coated with specific antibody than on one coated with normal serum. The sensitivity of this method may exceed that achieved by ELISA. Improved detection was achieved by combining ISEM with a decoration step (Fig. 14.3) whereby the virus particles are coated with a second layer of antibodies following virus attached to a grid

a b

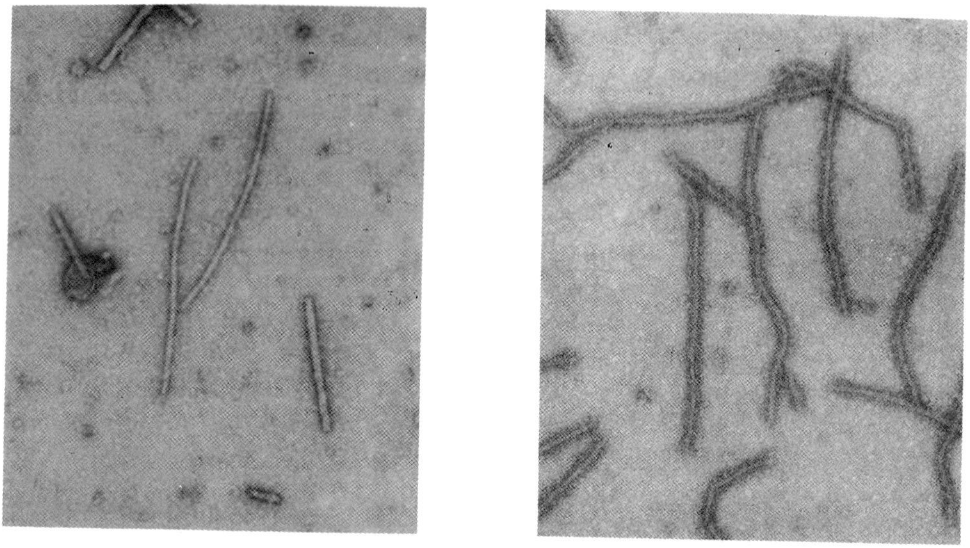

Fig. 14.3. Electron micrograph of a dip preparation from lisianthus plants naturally infected by bean yellow mosaic virus (BYMV) (a) and decoration of BYMV from lisianthus leaves using BYMV-specific antiserum (b).

(Milne, 1992). When the number of samples to be tested is small, ISEM is very useful. The application of ISEM for diagnosis is, however, very limited.

Molecular Tests

Nucleic acid hybridization

The rapid advances in the molecular biology of plant viruses and recombinant DNA led to the development of new techniques for virus diagnosis. These techniques involve hybridization between a specific nucleic acid probe and viral nucleic acid immobilized on filters. The hybridization is based on the interaction of a single-stranded polynucleotide chain with complementary base sequences to form a double-stranded structure. Depending on the initial reagents, the products of the reaction can be RNA:RNA, RNA:DNA, or DNA:DNA duplexes.

For plant virus diagnosis, the mixed-phase hybridization is the most commonly used. Viral nucleic acid is first denatured and immobilized on filter papers (nitrocellulose or nylon), which are immersed in solutions of labeled probe under conditions which favor hybridization. If the solution contains any labeled nucleic acid sequences which are complementary to the viral nucleic acid immobilized on the filter, reannealing occurs. Autoradiography is then used to detect the sample which shows a positive reaction. The nucleic acid probe can be DNA or RNA. Although the use of cDNA probes is more common, RNA probes (riboprobes) have the advantage of forming stronger duplexes with the target and thus are more sensitive. A cRNA probe is made by cloning the cDNA of interest into a transcription vector to synthesize cRNA *in vitro*. cRNA probes were shown to be more sensitive than cDNA probes (Robinson and Romero, 1991).

Due to the problems associated with the use of radioactivity, there is an increasing number of commercially available nonradioactive systems. The method has been used successfully for routine virus diagnosis, especially for viruses present in low concentrations and for phloem-limited viruses.

Polymerase chain reaction (PCR)

Recently, a new enzymatic assay, the PCR, was described. The method allows the amplification of very low concentrations of specific target sequences of DNA and RNA. A prerequisite for the use of PCR for virus detection is the availability of sequence data on the viral genome. The process involves the use of two oligonucleotide primers that flank the target DNA and repeated cycles of heat denaturation of the DNA, annealing of the primers to their complementary sequences, and extension of the annealed primers with thermostable DNA polymerase. The newly synthesized DNA strands serve as a target for subsequent DNA synthesis.

The genome of most plant viruses is made of RNA. Detection of an RNA sequence by this method requires some modifications. RNA has to be reverse-transcribed into cDNA before amplification (RT-PCR).

The specificity of the method depends on the choice of the synthetic oligonucleotide primers derived from the sequence of the target DNA. The use of highly conserved sequences for primer design is recommended for detecting all individuals of a particular virus. Alternatively, degenerate primers derived from a conserved amino acid sequence of viral coat protein may be used for the detection of all members of a virus group.

PCR was used successfully to detect very low amounts of poty-, nepo-, and luteoviruses and viroids. The method was also used for the detection of viruses in ornamental crops (Vunsh et al., 1990; Dekker et al., 1993). Recently a single PCR tube method which permits RNA denaturation, reverse transcription, and amplification was described for plant viruses (Wetzel et al., 1992).

Although highly sensitive PCR is too laborious to allow processing of only a limited number of samples, it was reported that an immunocapture step directly performed in the tube prior to PCR was a highly sensitive assay. This step allows the concentration of the particles from very diluted samples and increases a hundred-fold the sensitivity when compared with conventional PCR (Nolasco et al., 1993). Recently an immuno-PCR system using a specific DNA molecule as a marker has been reported, enabling the detection of fg antigen (Sano et al., 1992).

Conclusions

Serological test methods which are widely used for the diagnosis of ornamental viruses include ELISA, dot blot, tissue blotting, immunodiffusion, and ISEM.

In a comparison of ELISA, dot-blot immunoassay (DBIA), and tissue blot immunoassay (TBIA), lily symptomless virus (LSV) was detected in bulb scales by TBIA in samples where sap extracts from the same scale pieces were negative in both DBIA and ELISA (Table 14.3). When 26 bulbs were analyzed for LSV infection, ELISA showed 13 samples were infected with LSV, whereas DBIA revealed 19 samples, including those tested positive by ELISA, were infected with LSV. Analysis by TBIA revealed the presence of LSV antigens in tissue of 20 samples.

Table 14.3

Detection of lily symptomless (LSV) carlavirus in lily bulb scales by enzyme-linked immunosorbent assay (ELISA), dot blot immunoassay, and direct tissue blot immunoassay

Reaction	ELISA	Dot blots	Tissue blots
Positive	8	17	20
Weak positive	5	2	0
Negative	13	7	6
Total	26	26	26

One scale from each of 26 bulbs was cut twice near the middle portion and blotted onto nitrocellulose membranes. A 1:20 dilution of tissue extract was prepared. Extracts were assayed for LSV by avidin-biotin mediated ELISA at 1:20 dilutions and by dot blot immunoassay at 1:200 dilutions. ELISA readings (A_{405}) greater than 0.1 were considered positive, whereas those less than 0.1 were considered weak positive. Readings for healthy lily bulb scales ranged from 0–0.05.

The most important features of techniques for virus detection that attract much attention are reliability and sensitivity. Undoubtedly, ELISA has been the most widely used method for the diagnosis of plant virus diseases over the last 15 years. Although ISEM is very sensitive, it is essentially a research procedure because electron microscopes are not often available for routine diagnosis. The need for an expensive instrument limits the application of ISEM in diagnostic work.

The application of nucleic acid probes is becoming more widespread for the detection of plant viruses. The recently developed chemiluminescent assays are claimed to be able to detect less than fentamole of target DNA. Theoretically, PCR can detect a single molecule of target nucleic acid. A positive diagnosis should involve the application of two or more of the available techniques. Biological assays on sensitive indicator plants will remain vital for the diagnosis of plant virus diseases.

Selected Literature

Abad, J.A. and Moyer, J.W. 1992. Detection and distribution of sweet potato feathery mottle virus in sweet potato by *in vitro* transcribed RNA probes (riboprobes), membrane immunobinding assay, and direct blotting. Phytopathology 82:300–305.

Clark, M.F. and Adams, A.N. 1977. Characteristics of the microplate method of enzyme-linked immunosorbent assay (ELISA) for the detection of plant viruses. J. Gen. Virol. 34:475–483.

Colinet, D., Kummert, J., Lepoivre, P. and Semal, J. 1994. Identification of distinct potyviruses in mixedly-infected sweet potato by the polymerase chain reaction with degenerate primers. Phytopathology 84:65–69.

Crosslin, J.M., Hamond, R.W. and Hammerschlag, F.A. 1992. Detection of prunus necrotic ringspot virus serotypes in herbaceous and prunus hosts with a complementary RNA probe. Plant Dis. 76:1132–1136.

Dekker, E. L., Derks, A.F.L.M, Asjes, C.J., Lemmers, E.C., Bol, J.F. and Langeveld, S.A. 1993. Characterization of potyviruses from tulip and lily which cause flower-breaking. J. Gen. Virol. 74:881–887.

Derrick, K.S. 1973. Quantitative assay for plant viruses using serologically specific electron microscopy. Virology 56:652–658.

Hampton, R., Ball, E. and DeBoer, S. Eds. 1990. Serological Methods for Detection and Identification of Viral and Bacterial Plant Pathogens. APS Press, St. Paul, Minnesota.

Henson, J.M. and French, R. 1993. The polymerase chain reaction and plant disease diagnosis. Annu. Rev. Phytopathol. 31:81–109.

Hsu, H.T. and Lawson R.H. 1991. Direct tissue blotting for detection of tomato spotted wilt virus in impatiens. Plant Dis. 75:292–295.

Hsu, H.T., Vongsasitorn, D. and Lawson, R. 1992. An improved method for serological detection of cymbidium mosaic potexvirus infection in orchids. Phytopathology 82: 491–495.

Huguenot, C., van den Dobbelsteen, G., Dehan, P., Wagemaker, C.A.M., Drost, G.A., Osterhaus, A.D.M.E. and Peters, D. 1990. Detection of tomato spotted wilt virus using monoclonal antibodies and riboprobes. Arch. Virol. 110:47–62.

Joens, R.A.C. and Torrance, L. 1986. Developments and Applications in Virus Testing. AAB. Wellesbourne, Warwick, UK.

Jordan, R. and Hamond, J. 1991. Comparison and differentiation of potyvirus isolates and identification of strain-, virus-, subgroup-specific and potyvirus group-common epitopes using monoclonal antibodies. J. Gen. Virol. 72:25-32.

Langeveld, S.A., Dore, J.M., Memlink, J., Derks, A.F.L.M., van der Vlugt, C.I.M., Asjes, C.J. and Bol, J.F. 1991. Identification of potyviruses using the polymerase chain reaction with degenerate primers. J. Gen. Virol. 72:1531–1541.

Lawson, R.H. 1986. Pathogen detection and elimination. In: Tissue Culture as a Plant Production System for Horticultural Crops, R.H. Zimmerman (ed)., Kluwer Academic Publishers Group, Dordrecht, The Netherlands, pp. 97–117.

Lin, N.S. Hsu, Y.H. and Hsu, H.T. 1990. Immunological detection of viruses and mycoplasma-like organisms by direct tissue blotting on nitrocellulose membranes. Phytopathology 80:824–828.

Mathews, R.E.F. (ed.). 1993. Diagnosis of Plant Virus Diseases. CRC Press, Boca Raton, FL.

Miller, S.A. and Martin, R.R. 1988. Molecular diagnosis of plant diseases. Annu. Rev. Phytopathol. 26:409–432.

Milne, R.G. 1992. Immunoelectron microscopy of plant viruses and mycoplasmas. In: Advances in Disease Vector Research, Vol. 9, K.S. Harris (ed.), Springer-Verlag, New York, pp. 283–297.

Navot, N., Ber, R. and Czosnek, H. 1989. Rapid detection of tomato yellow leaf curl virus in squashes of plants and insect vectors. Phytopathology 79:562–568.

Nolasco, G., de Blas, C., Torres, V. and Ponz, F. 1993. A method combining immunocapture and PCR amplification in a microtiter plate for the detection of plant viruses and subviral pathogens. J. Virol. Meth. 45:201–218.

Robinson, D.J. and Romero, J. 1991. Sensitivity and specificity of nucleic acid probes for potato leafroll luteovirus detection. J. Virol. Meth. 34:209–214.

Sano, T., Smith, C.L. and Cantor, C.R. 1992. Immuno-PCR: Very sensitive antigen detection by means of specific antibody-DNA conjugates. Science 258:120–122.

Smith, T.D. and Bantarri, E.E. 1987. Dot-ELISA on nitrocellulose membranes for detection of potato leafroll virus. Plant Dis. 71:795–799.

Van Regenmortel, M.H.V. 1982. Serology and Immunochemistry of Plant Viruses. Academic Press, New York.

Vunsh, R., Rosner, A. and Stein, A. 1990. The use of polymerase chain reaction (PCR) for the detection of bean yellow mosaic virus in gladiolus. Ann. Appl. Biol. 117:561:569.

Vunsh, R., Rosner, A. and Stein, A. 1991. Detection of bean yellow mosaic virus in gladioli corms by the polymerase chain reaction. Ann. Appl. Biol. 119:289–294.

Wetzel, T., Candresse, T., Ravelonandro, M. and Dunez, J. 1992. A highly sensitive immunocapture polymerase chain reaction method for plum pox potyvirus detection. J. Virol. Methods 39:27–37.

15

Viruses with a Multipartite Genome

Ron Gafny
Department of Virology, Agricultural Research Organization, The Volcani Center Bet Dagan 50250, Israel

Introduction

Members of several plant virus groups have a genome divided into several molecules. Segmented genomes of single-stranded RNA viruses are divided into either two (bipartite) or three (tripartite) segments. Genomic RNAs of multipartite plant viruses are encapsidated in separate particles. In contrast, all the RNA segments of animal viruses with segmented genomes are encapsidated in the same virus particle. Plant viruses that have a segmented genome are found also among double-stranded RNA viruses (phytoreoviruses and fijiviruses) and single-stranded DNA viruses (some members of the geminivirus group have a bipartite genome). In this chapter we refer only to multipartite viruses with a positive sense, single-stranded RNA genome.

Multipartite Genome as a Strategy for Gene Expression

The messenger RNA of eukaryotes usually contains the coding sequence for the translation of a single protein. The accepted model for the initiation of protein translation is the "scanning mechanism model". According to this model, the ribosome which binds to the 5′ end of the messenger RNA moves along the RNA until the first AUG triplet where translation commences. As a consequence of the scanning mechanism, open reading frames (ORFs) beyond the first ORF on the ssRNA viral genome will not be translated. Plant viruses have evolved several mechanisms to overcome this constraint on gene translation. The division of the viral genome into a multipartite genome is one strategy by which ssRNA viruses ensure that all of their genes are accessible to the protein synthesizing machinery. The multipartite arrangement ensures that the 5′ gene on each RNA segment can be translated. In all multipartite ssRNA viruses, this strategy occurs together with a second strategy for gene expression. The latter involves the synthesis of subgenomic RNA, translation of a polyprotein, or the synthesis of a read-through (or frame-shift) protein. The additional mechanisms provide a further possibility to control the level of expression of particular genes. For example, the coat protein (CP) of many viruses with a segmented genome is translated from a subgenomic RNA. The multiple copies of the subgenomic mRNA allow the production of a relatively high level of CP required for virus encapsidation. Examples of the mechanisms involved in gene expression by some of the multipartite viruses are presented below. Genome structure and gene expression strategies of multipartite viruses have been comprehensively described by Matthews (1991).

Additional Advantages of Multipartite Genome Organization

Several possible advantages of segmentation of the viral genome were discussed by Goldbach (1986).

1. Genome segmentation allows the exchange of genetic information between related strains during mixed infection. This seems to be a more rapid route than RNA recombination for exchanging genetic information. Pseudorecombination has been well documented with the segmented genome of influenza virus. Naturally occurring pseudorecombination has not been reported for plant viruses. However, there are many studies involving the *in vitro* construction of plant virus pseudorecombinants. These have been experimentally constructed with bromoviruses, cucumoviruses, dianthoviruses, and nepoviruses.

2. Encapsidation of the segmented genome pieces in separate virions enables an increase in the genome size of spherical viruses, thereby overcoming the problem of encapsidating a large RNA in a small spherical capsid.

3. Segmentation of the genome reduces the lethality of progeny RNA caused by unfavorable mutations introduced either by mutagenic agents or by the high error level of RNA replication.

4. A divided genome encapsidated separately in smaller capsids may facilitate more efficient cell-to-cell movement and systemic spread of the virus in the infected plant.

The major disadvantage of segmented genomes may be the reduced efficiency in host-to-host transmission. In order to initiate infection, at least one particle of each of the separately encapsidated segments should enter a cell. It was suggested that the efficient transmission of plant viruses by invertebrate vectors or transmission by seeds enabled the evolution of multipartite viruses in plant viruses but not in animal viruses (Reanney, 1984).

Virus Genera with a Multipartite Genome

Ten genera of single-stranded RNA (ssRNA) have a multipartite genome. Table 15.1 lists these genera along with some of the characteristics of their genomes. Detailed descriptions of each genus can be found in Francki et al. (1985) and Matthews (1991). Some characteristics and the genome strategy of two bipartite and two tripartite virus genera are discussed below.

Bipartite Genome

Tobravirus genus

Tobacco rattle virus (TRV), the type member of this genus, was the first virus shown to have a genome divided between two particles. In addition to TRV, the two other distinct members are pea early browning virus and pepper ringspot virus. The viruses in this genus have a genome consisting of two single-strand RNA molecules. The two RNAs of TRV — 6791 and 1905 nucleotide long — are encapsidated separately in long and short rigid rod particles of 185–195 and 75–100 nm, depending on the particular strain.

Table 15.1. Properties of multipartite virus genera

Genus	Type member	Capsid structure	Size of RNA segments (nucleotides)	Additional gene expression strategy	5′-end group	3′-end structure
Comovirus	Cowpea mosaic	Isometric	5889, 3481	Polyprotein	VPg	Poly(A)
Nepovirus	Tobacco ringspot	Isometric	[a]7359, 4662	Polyprotein	VPg	Poly(A)
Bromovirus	Brome mosaic	Isometric	3234, 2865, 2144	Subgenomic RNA	Cap	$tRNA^{tyr}$
Cucumovirus	Cucumber mosaic	Isometric	3357, 3050, 2216	Subgenomic RNA	Cap	$tRNA^{tyr}$
Alfamovirus	Alfalfa mosaic	Bacilliform	3644, 2593, 2037	Subgenomic RNA	Cap	-AUGC-OH, no tRNA like, no poly(A)
Ilarvirus	Tobacco streak	Isometric	3.8Kb, 3.3Kb, 2.2Kb	Subgenomic RNA	?	-AUGC-OH, no tRNA like, no poly(A)
Hordeivirus	Barley stripe mosaic	Rod	3768, 3289, 3164	Subgenomic RNA	Cap	$tRNA^{tyr}$
Tobravirus	Tobacco rattle	Rod	6791, 1905	Subgenomic RNA and read-through protein	Cap	tRNA like
Furovirus	Soil borne wheat mosaic	Rod	[b]6746, 4612, 1775, 1431	Subgenomic RNA and read-through protein	Cap	Poly(A)
Dianthovirus	Carnation ringspot	Isometric	3756, 1394	Subgenomic RNA and read-through protein	Cap	no tRNA like no poly(A)

[a]The sizes of RNAs 1 and 2 of tomato black ring nepovirus. [b]The sizes are for beet necrotic yellow vein furovirus.

The viruses of the tobravirus genus are soil-borne and transmitted by nematodes of the genera *Trichodorus* and *Paratrichodorus*. The transmissibility persists in the nematodes for over a year but is lost when the nematode molts. Using a pseudorecombinant produced from efficiently transmitted and non-transmitted isolates of TRV, the factor determining the nematode's transmissibility was mapped to RNA 2, and it was suggested that the coat protein is involved in the transmission process. TRV causes diseases in a number of crops including potatoes and bulbous ornamentals.

The genes encoded by the two RNAs of TRV and the subgenomic RNAs are presented schematically in Fig. 15.1. There are four ORFs found in RNA 1 and one ORF in RNA 2. The size and nucleotide sequence of RNA 2 vary considerably between strains and may contain additional ORFs in specific strains.

RNA 1 of TRV encodes a protein of 134K and a read-through protein of 194K. The 194K protein shows sequence similarities with putative replicase genes of other ssRNA viruses. The 29K protein has sequence similarity with the tobacco mosaic virus movement protein and is probably the movement protein of TRV. The 16K protein encoded by RNA 1 has no suggested function. This protein has a duplicated motif of cysteine and histidine residues. This resembles the "zinc finger" motif which is a cysteine-rich sequence motif found in several zinc-binding proteins involved in nucleic acid binding and gene regulation. The 29K and 16K proteins are translated from their respective subgenomic RNAs.

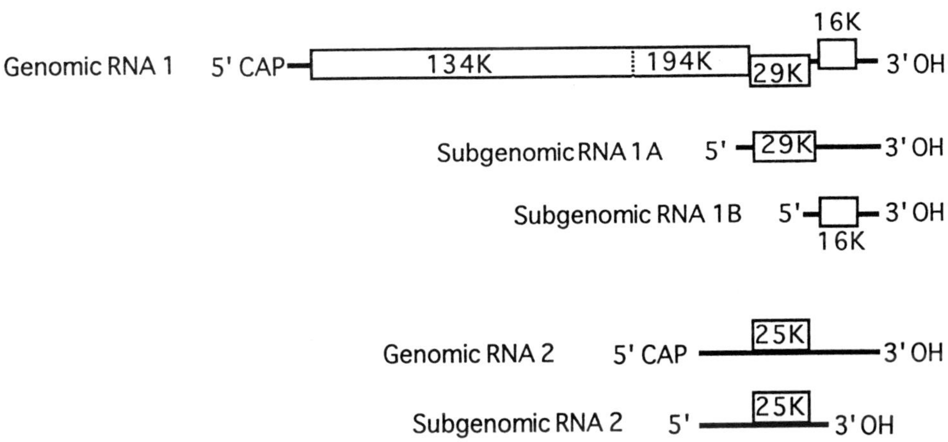

Fig. 15.1. Organization and expression of the tobacco rattle tobravirus genome, a bipartite virus that uses subgenomic RNA as a second strategy for gene expression.

RNA 2 encodes the TRV CP with a MW of 25K. There is a subgenomic RNA produced from RNA 2 which serves as the mRNA for the CP. It is not clear why a messenger RNA is needed for the CP which is coded by the 5′ ORF of RNA 2. The TCM strain of TRV contains an ORF for 29K protein. No protein product has been found for this ORF.

The genes encoded by RNA 1 are sufficient for replication and cell-to-cell movement of TRV. RNA 2 is required for CP synthesis and thus the production of complete virions.

Nepovirus genus

The name nepovirus is an acronym for **ne**matode-transmitted **po**lyhedral virus. The particle has a diameter of 28–30 nm. Nepoviruses have a wide host range. The viruses are transmitted mechanically and by nematodes of the genera *Xiphinema* and *Longidorus*. Many of the nepoviruses are also seed- and pollen-transmitted.

The genome is bipartite, and the two segments are encapsidated separately. RNA 1 has approximately 8110–8400 nucleotides. The size of RNA 2 is more variable: 3400 to 7200 nucleotides long. Both RNAs have a VPg linked to the 5′ end and a 3′ poly(A) tail. The two RNAs of tomato black ring virus have been sequenced. Both RNAs contain a single long ORF. RNA 1 codes for a 254K protein and RNA 2 for a 150 K protein. The two large polypeptides are proteolitically processed to smaller functional polypeptides as described in Fig. 15.2. RNA 1 which carries the RNA replicase and the VPg genes can replicate in the absence of RNA 2. The polyprotein translated from RNA 2 is processed to produce the cell-to-cell movement protein and the CP.

Tripartite Genome

Cucumovirus genus

The type member of the cucumovirus genus is cucumber mosaic virus (CMV). The two other definitive members of the genus are tomato aspermy virus (TAV) and peanut stunt virus (PSV). The virus particles are isomeric, approximately 29 nm in diameter.

CMV has the largest host range of any plant virus, exceeding 800 species. The virus is found in most countries in the world, predominantly in temperate zones.

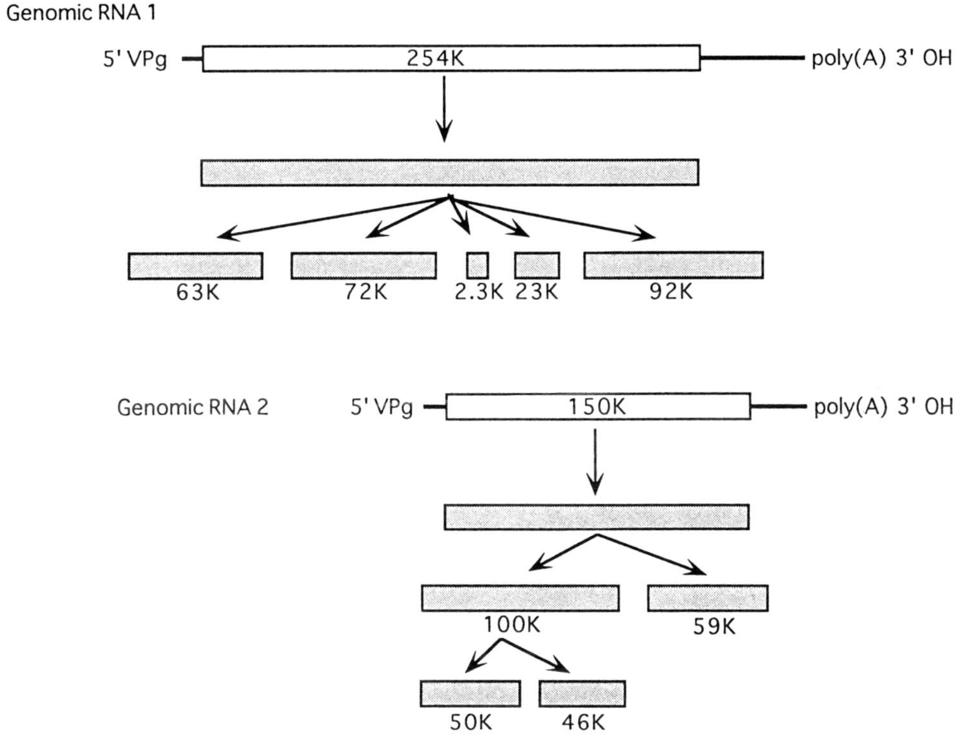

Fig. 15.2. Organization and expression of the tomato black ring nepovirus genome, a bipartite virus that uses polyprotein as a second strategy for gene expression. Open boxes represent ORFs, shaded boxes represent protein products.

CMV is a common pathogen of many food and ornamental crops. There have been reports on epidemics of CMV in vegetable crops such as tomato, cucurbits, and peppers. Other crops reported to be prone to CMV epidemics are banana, sweet potato, and several species of pasteur legumes. The virus is transmitted by aphids, mechanical inoculation, and is also transmitted by seeds in a number of plant species. A comprehensive review by Palukaitis et al. (1992) describes recent findings on the structure, molecular genetics, and ecology of CMV.

The CMV genome consists of three single stranded RNAs: RNA 1, 2 and 3 of 3357, 3050, and 2216 nucleotides, respectively (for the Fny strain of CMV). RNA 1 and 2 contain the coding sequence for the a 111K and a 97K proteins, respectively. These proteins are directly translated from their respective RNAs and are associated with the viral RNA replication. RNA 3 contains the genes for both the CP and the viral movement protein (3a protein). The CP is translated from a subgenomic mRNA. The subgenomic mRNA is also encapsidated in the CMV virion.

The construction *in vitro* of pseudorecombinants of CMV has allowed the mapping of phenotypic traits to specific RNAs. Thus, the CP gene and aphid transmissibility have been mapped to RNA 3. Several different host reactions have been mapped to RNA 1, 2 and 3. RNA 1 and 2 are exchangeable, *in vitro*, among strains of CMV but cannot be exchanged between CMV and TAV.

Some isolates of CMV contain small encapsidated satellite RNA. The satellite RNA of CMV was also named CARNA5 for CMV-associated RNA 5. The CMV satellite is a small ssRNA of 335 to 411 nucleotides long. The presence of satellite RNA in a CMV isolate can either ameliorate or acerbate the disease symptoms. The particular effect is the result of an interaction involving the satellite, the helper virus, and the host genome. Inoculation with a CMV isolate containing a symptom-attenuating satellite RNA is used for protection of vegetable crops on a commercial scale in China and in large-scale experiments in the US and Europe (Tien and Wu, 1991; Kaper, 1993).

Ilarvirus genus and alfamo genus

Most ilarviruses are labile *in vitro* and often produce ringspot-like symptoms. The particles are quasi-spherical and range in size from 20–30 nm. The group name stands for **i**sometric **la**bile **r**ingspot. Ilarviruses frequently infect woody hosts, including roses, but also herbaceous species. Alfalfa mosaic virus (AMV) has labile anisometric particles similar to the ilarvirus particles. Moreover, as discussed below, the genomic organization and the requirement for CP to initiate replication are similar in alfalfa mosaic and the ilarviruses. Although AMV is the sole member of a monotypic genus, it has been proposed that it should be considered as a "super group" together with the ilarviruses. AMV is aphid-transmitted, whereas no natural insect vector has been definitely identified for ilarviruses. Ilarviruses are often transmitted by seeds and pollen.

Ilarviruses have a tripartite genome. RNA-1 and -2 are monocistronic and encode the replicase proteins P1 and P2 respectively. RNA 3 is bicistronic, the two ORFs encoding the movement protein (MP or P3) and the CP. The MP gene is located at the 5′ half of RNA 3 and is translated directly from this RNA. The CP gene is translated from a subgenomic mRNA. This genomic organization and expression strategy are common to alfalfa mosaic virus, the bromoviruses and, as described above, the cucumoviruses. Fig. 15.3 describes schematically the organization of the AMV genome.

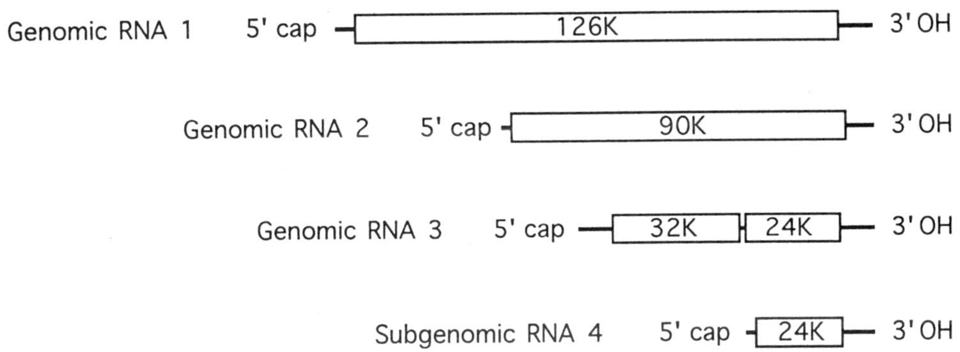

Fig. 15.3. Organization and expression strategy of the alfalfa mosaic virus genome.

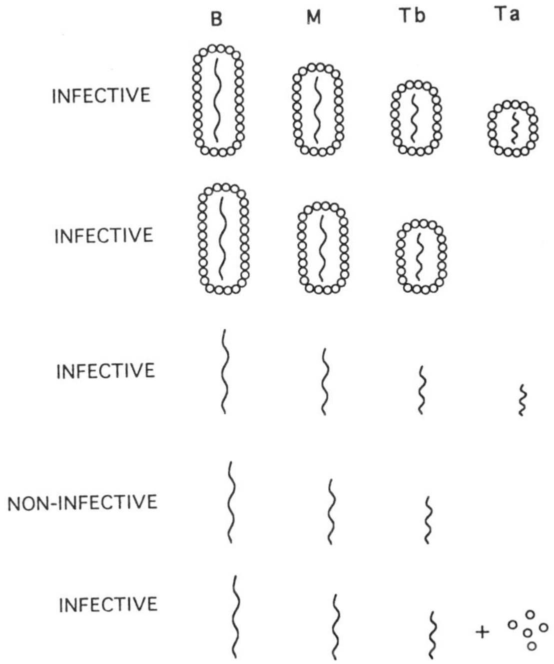

Fig. 15.4. Schematic representation of the "genome activation" phenomenon by the coat protein of the alfalfa mosaic and ilarviruses. Some coat protein molecules or their subgenomic messenger RNA must be present together with the three genomic RNAs to initiate infection. B, M, Tb, and Ta represent the bottom, middle, and top components of the alfalfa mosaic virus. Circles represent coat protein molecules.

Ilarviruses and AMV share the requirement for the presence of the capsid protein to initiate replication (see Jaspars, 1985; Matthews, 1991). Some CP molecules or their subgenomic messenger RNA must be present together with the three genomic RNAs to initiate infection (see Fig. 15.4). The CPs of some ilarviruses are capable of activating replication of the genome of serologicaly unrelated members of the group, as well as that of AMV. The 3′-termini of the RNAs of AMV and of the ilarviruses do not contain a tRNA-like structure. Instead, these RNAs contain a 3′-terminal, high-affinity binding site for the CP. The binding site is composed of several stem/loop structures, each flanked by the repeated sequence AUGC. When viral RNA is used as inoculum, each of the three genomic RNAs of AMV has to be complexed with a few CP molecules to permit initiation of infection. Several lines of evidence suggest that the binding of the CP to the 3′-termini of the RNA is necessary for the recognition of the RNAs by the viral replicase (Houwing and Jaspars, 1987). Recent work involving transgenic plants expressing genomic cDNAs of AMV and inoculation of plants with cDNAs suggests that the role of the CP in the "genome activation" phenomenon is to protect of the parental RNAs from degradation until the replication complex is assembled (Neeleman et al., 1993).

Selected Literature

Francki, R.I.B., Milne, R.G. and Hatta, T. 1985. Atlas of Plant Virus, Vols. I and II. CRC Press, Boca Raton, FL.

Goldbach, R.W. 1986. Molecular evolution of plant RNA viruses. Annu. Rev. Phytopathology 24:29–310.

Houwing, C.J. and Jaspars, E.M.J. 1987. *In vitro* evidence that the coat protein is the programming factor in alfalfa mosaic virus-induced RNA synthesis. FEBS Lett. 221: 337–342.

Jaspars, E.M.J. 1985. Interaction of alfalfa mosaic virus nucleic acid and protein. In: Molecular Plant Virology, Vol. I, J.W. Davies (ed.), CRC Press, Boca Raton, FL, pp. 155–221.

Kaper, J.M. 1993. Viral satellites, molecular parasites for plant protection. In: Pest Management: Biologically Based Technologies, R.D. Lumsden and J.L. Vaughn (eds.), American Chemical Society, Washington, DC, pp. 134–143.

Palukaitis, P., Roossinck, M.J., Dietzgen, R.G. and Francki, R.I.B. 1992. Cucumber mosaic virus. Adv. Virus Res. 41:281–348.

Matthews, R.E.F. 1991. Plant Virology, 3rd ed. Academic Press, San Diego, CA.

Reanney, D.C. 1984. The molecular evolution of viruses. In: The Microbe 1984, Part I, B.W.J. Mahy and J.R. Pattison (eds.), Cambridge Univ. Press, London, pp. 175–196.

Neeleman, L., Van der Vossen, E.A.G. and Bol, J. 1993. Infection of tobacco with alfalfa mosaic virus cDNAs sheds light on the early function of the coat protein. Virology 196: 883–887.

Tien, P. and Wu, G-S. 1991. Satellite RNA for the biological control of plant disease. Adv. Virus Res. 39:321–339.

16

Resistance and Interference

G. Loebenstein, E. Teverovsky and A. Gera
Department of Virology, Agricultural Research Organization, The Volcani Center, Bet Dagan 50250, Israel

The interaction of a virus with a plant may vary between two extremes — from complete susceptibility to immunity. Extreme susceptibility resulting in plant viral epidemics is, however, not the norm, and resistance — especially in wild plants — seems to be the rule, and that due to natural selection during evolution. Modern agriculture, where new hosts or pathogens are introduced into a different ecosystem, may disturb this balance, resulting in epidemics. Nowadays, when many crops are grown in monocultured areas, from cultivars which were selected in breeding nurseries and not in field populations, even minor epidemics can cause severe damage.

Contribution No. 1372-E, 1994 series from the Agricultural Research Organization, The Volcani Center, Bet Dagan, Israel.

Resistance may be defined as the ability of the plant to prevent or restrict virus multiplication. We may distinguish between *passive preformed defenses* prepared in advance of infections, such as mechanical barriers, lack of compatible infection sites or metabolites required for virus multiplication; and *active resistance* phenomena, which require metabolic processes of the host. The resistance of most plants to viruses probably falls within the first category, whereby escape from the disease is due to the inability of the pathogen to establish itself, or where after introduction into the host, the virus does not find the metabolic prerequisites for multiplication. On the other hand, the restriction of virus movement from the local lesion to adjacent tissues in hypersensitive hosts — *localization* — is one of the most efficient active resistance mechanisms, requiring metabolism of host cells.

Passive Preformed Defenses

Resistance to infection

Cuticle, epidermis and ectodesmata

Although direct evidence as to the role of the cuticle and epidermis in resistance is limited, it is nevertheless obvious that both are of major importance in preventing introduction of the virus into the plant. Monocotyledonous plants in particular have a hard epidermis, which may account for resistance to mechanical transmission. Thus, gladiolus plants are resistant to mechanical infection by cucumber mosaic virus (CMV) whether in the field or by intentional mechanical inoculation. However, rubbing the leaves with a detergent solution prior to inoculation markedly increased rates of transmission, probably by removing the waxy leaf surface that acts as a mechanical barrier to infection (Sushak, 1977).

A tendency to escape disease due to resistance to infection by mechanical inoculation has been observed in lines of tobacco for tobacco mosaic virus (TMV) and CMV. Thomas and Fulton (1968) found that differences in susceptibility to TMV and CMV among tobacco varieties were correlated with numbers of ectodesmata in the outer walls of epidermal cells. If indeed ectodesmata serve as infection sites, then the correlation between resistance and fewer ectodesmata observed in this system would be a good example of preformed passive resistance. The possibility that ectodesmata serve as infection sites has, however, been questioned (De Zoeten et al., 1972). Furthermore, the effective resistance of some of these tobaccos also to aphid transmission cannot be explained on the basis of limited numbers of ectodesmata. Thomas and Fulton (1968) suggested, therefore, that resistance to aphid transmission in this tobacco might be explained if, concurrently with fewer ectodesmata, there were also fewer plasmodesmata at the junctures between epidermal cells where aphids deposit virus during short probes.

Replicases

Analogies to bacterial systems might lead to an answer about preformed factors in the plant cell responsible for its susceptibility or resistance to infection. The Qβ RNA-dependent RNA polymerase (replicase) has been shown to consist of one virus-encoded polypeptide and three host polypeptides (Blumenthal and Carmichael, 1979). Recently it was shown that a CMV replicase obtained from CMV-infected tobacco contained two virus-encoded polypeptides and one host polypeptide (Hayes and Buck, 1990). The absence or inability of the host-coded polypeptide to replicate virus RNA may therefore represent one of the most important preformed resistance mechanisms.

Resistance to vector transmission

Use of vector-resistant cultivars is attractive in principle, although only a few actual cases of its use in practice are known (Loebenstein et al., 1984) Thus, preformed resistance to vector inoculation of maize dwarf mosaic virus by the aphid *Schizaphis gramineum* was observed in corn lines which were susceptible when inoculated mechanically. Some plants escape infection because of an adverse effect on vector behavior — nonpreference, resulting in decreased colonization or probing. For example, nonpreference to the vector has been observed in a melon line (Lecoq et al., 1980). This melon line, which is resistant to CMV and watermelon mosaic virus 1 and 2 transmission by *Aphis gossypii*, was susceptible to these viruses following mechanical inoculation. However, vector nonpreference of this melon line was not observed in transmission tests of CMV with three other aphid species: *Aphis citricola, A. fabae* and *Myzus persicae*.

Among the Solanaceae, the tomato and some wild potato species, such as *Solanum berthaultii*, bear trichomes with a glandular four-lobed head which ruptures when touched by aphids and glues the latter to the plant. Thus, wild tomato accession lines that exudate sticky material from their trichomes were used in breeding programs for resistance against tomato yellow leaf curl virus transmitted by the whitefly *Bemisia tabaci* (Berlinger and Dahan, 1987). Morphological analysis by light and scanning electron microscopy showed that resistant *Lycopersicon* accession lines have a specific type of glandular trichome (Channarayappa et al., 1992).

Variation in susceptibility to aphid infection with CMV has been observed in gladiolus cultivars (Aly et al., 1986). In confined access feedings on five cultivars, "Spic and Span" and "Peter Pears" were susceptible, whereas "Trader Horn" was highly resistant. These findings were corroborated in field experiments in Israel where CMV infection rates in gladiolus plots planted with virus-tested corms were monitored. "Commando", "Peter Pears" and "Spic and Span" were the most susceptible (infection rates reaching 22–80%), while "Eurovision" and "Trader Horn" did not become infected by CMV (Aly et al., 1986). The basis for the resistance is not yet understood.

Differences in virus resistance are known also in tulips (Asjes, 1981; Romanow et al., 1991) and lilies (Derks and Abbink, 1988). In tulips, two types of resistance were observed: resistance to infection, as for example in *Tulipa fosteriana* cv. "Princeps"; and a lower multiplication rate of tulip breaking virus, as in cv. "Madame Lefeber". A breeding program was started to introduce these types of virus resistance into tulip hybrids (Van Eijk et al., 1988).

Resistance to vectors could therefore become an important tool to overcome viral diseases.

Reduced translocation

Reduced translocation of virus in the plant is probably another type of preformed resistance. Thus, resistance to TMV in tomato lines, conferred by the gene TM-2^a, is due to limited cell-to-cell movement of the virus. This resistance mechanism functions only in the intact plant and not in isolated mesophyll protoplasts, which produce as much virus as protoplasts from susceptible lines (Motoyoshi and Oshima, 1975). Also, the spread of potato leafroll virus infection within the phloem of resistant potato cultivars appears to be impaired, leading to less efficient acquisition of the virus by the aphid vector (Barker and Harrison, 1986).

Active Resistance

Active resistance can be defined as a mechanism that is activated by the pathogen or another agent, and requires transcription of cellular DNA. Resistance in cucumber cultivars to cucumber mosaic virus is an active mechanism, developing in the plant after inoculation. Resistance depends on three genes giving a stable type of resistance whereby virus multiplication is markedly reduced. Injection of leaves with actinomycin D within 48 hours after inoculation or ultraviolet irradiation led to a substantial increase in virus multiplication, indicating that the resistance mechanism is activated by the virus and depends on transcription from cellular DNA (Loebenstein et al., 1984).

In several systemic infections, the virus is unevenly distributed, forming in several virus-host combinations "green islands" that contain extremely low concentrations of virus and are resistant to further infection. Especially in tobaccos infected with CMV strain 6, marked green islands are observed (Loebenstein et al., 1977). When protoplasts were prepared from these green islands, a substance that inhibits virus replication similar to IVR (the inhibitor of virus replication; see below), was found to be released from them during incubation (Gera and Loebenstein, 1988). It seems, therefore, that in green islands an active resistance mechanism is operative, which may be associated with an IVR.

Local lesion response

The local lesion response is probably the most notable active resistance phenomenon, whereby following inoculation the virus invades and multiplies in several hundred cells but does not spread to other tissues. The following types of localized infections are known: necrotic lesions of the nonlimited type, self- limiting necrotic local lesions, chlorotic local lesions, ring-like patterns or ringspots, starch lesions, microlesions, and subliminal infections. Localization in most cases depends on one dominant gene.

Several hypotheses have been raised to explain localization such as: death of cells, barrier substances, PR-proteins, inactivation of the virus transport protein, or substances that inhibit virus multiplication. For detailed reviews, see Loebenstein et al. (1984) and Loebenstein and Gera (1990, 1993). If indeed an inhibitory substance such as the IVR (Loebenstein and Gera, 1981) is responsible for localization, then this gene product could be used to isolate the gene. Subsequently, if plants transformed with the IVR gene become resistant to a spectrum of viruses (IVR inhibits replication of several viruses), this gene would then broaden markedly the battery of resistance sources available to plant breeders.

Apparent recovery

Some virus-infected plants, after showing initially acute symptoms, pass later to a chronic stage of infection where symptoms decrease in severity and the plants seem to "recover". These features are especially characteristic of diseases with primary symptoms of the ringspot type. Thus, tobacco plants infected with tobacco ringspot virus after showing the ringspot patterns ("shock phase") seem to recover. New leaves which form 2–3 weeks after inoculation show no typical ringspot patterns. Virus concentration decreases markedly in such plants and recovery is followed by an "acquired immunity" to reinfection. After reinoculation, these plants do not again pass through the "shock phase". Recovery is well known in fruit trees and has also been observed in roses infected with rose mosaic (Prunus necrotic ringspot virus) (personal observation). Recovery is often associated with a reduction in virus concentration, but the virus is not eliminated. Apparently the small amount of virus still present protects the plant from reinfection (cross protection).

Cross protection

A plant already infected with one strain of virus is commonly protected from infection by other, related strains. This phenomenon, known as cross protection, can be used to determine relationships between viruses and, in some cases, also to control virus diseases.

The classic procedure uses a combination of two strains, one which spreads systemically and another which produces local lesions (Fig. 16.1). This offers the advantage that quantitative data on the degree of protection may be obtained, which is not possible when two systemic strains are used. It is generally accepted that a high level of or complete inhibition occurs between closely related stains, provided that the interfering strain is systemically established in the plant before challenge inoculation with the local-lesion-forming strain. Unrelated strains generally are not cross-protective. Thus, relationships between tobacco mosaic virus stains may be determined on *Nicotiana sylvestris*, where the type strain is systemic and the aucuba mosaic strain produces local lesions. CMV strains can be identified on *Zinnia elegans* with the aid of an indicator local-lesion-forming strain, while most other strains of CMV cause mottle mosaic symptoms on this species. It is important to challenge-inoculate young leaves because local-lesion formation on older leaves is erratic.

Relationships may also be determined when both strains are systemic, as long as symptoms between the two differ; and even between intracellular symptoms as with the specific "pinwheel" inclusion bodies caused by potyviruses.

Several hypotheses have been proposed to explain cross protection. Here they will be mentioned only briefly and the reader is referred to specific reviews, as for example that by Urban et al. (1990).

Fig. 16.1. Cross protection in a leaf of *Nicotiana sylvestris*. The common (systemic) strain of tobacco mosaic virus inoculated on the right half of the leaf protected it from infection by the U2 (necrotic) strain, inoculated on the entire leaf three days later.

1. *Depletion of host metabolites and structures*. The initial virus could sequester the ribosomes, leaving the challenge virus nucleic acid unbound and susceptible to degradation. Another suggestion is that the host-coded domain of the viral replicase has been depleted by the protecting strains. These theories assume that different viruses utilize different host components.

2. *Capture of viral nucleic acid*. The (−) strand of the challenge virus is captured by excess (+) RNA of the protecting strain.

3. *Involvement of coat protein*. Coat protein of the protecting strain could prevent uncoating of the challenge strain or block attachment sites on the replicase.

The use of cross protection for disease control in practice is limited. Thus, mild strains of TMV have been used to protect glasshouse-grown tomatoes from severe isolates causing tomato mosaic; mild strains of citrus tristeza virus are widely used in Brazil to protect sweet orange trees; and cross protection of papaya from papaya ringspot disease under field conditions is under investigation in several countries.

Protection with CMV satellite (designated CARNA 5, for CMV-Associated RNA 5) is used in China where CMV infestations are endemic (Tien and Wu, 1991). Viral satellites are small nucleic acid molecules associated with a helper virus to which they are sequence-unrelated but which they need for replication support (Kaper, 1993). Satellites generally inhibit replication of the helper viral RNA. Satellites associated with CMV modulate symptoms; and some attenuate symptoms of CMV in a variety of crops. Controlled release by spraying of seedlings with mild variants protects the crop from epidemics caused by severe isolates of satellite.

Applying cross protection in the field should be done with extreme caution. Cross protection might be considered if the virus is widespread, most of the plants are naturally infected, and the virus does not endanger other crops. It should also be considered that mild strains can, under certain ecological (temperature) conditions, become severe. Due to these problems, cross protection should be considered only as a last resort in disease control.

Viral Genes

Plant genetic engineering provides new methods for developing virus-resistant plants, based on the integration of viral genes and sequences into plant genomes.

Coat protein (CP)-mediated virus resistance

At present the most promising way to produce virus-resistant plants by genetic engineering is to insert and express a viral CP gene(s). One of the reasons for the attractiveness of this approach is the relative accessibility of these genes. The cistron encoding the coat protein is often located at the 3' end of the encoding RNA. This simplifies the synthesis of cDNA molecules once a suitable primer sequence is identified.

Plants are then transformed with the CP gene linked to a suitable promoter and selected for expression of the protein and resistance. Thus, resistance to several viruses such as TMV, potato virus X and potato virus Y was obtained in CP-transformed plants (see reviews by Hemenway et al., 1990; and by Hull and Davies, 1992). However, in some cases CP-transformed plants are not resistant to infection by viral RNA. There seems to be some analogy between CP-mediated resistance and cross protection. When employing CP-mediated resistance, it should be kept in mind that transformation of plants with CP of a vector-transmitted virus (Cucumo-, Poty- or Alfamo-viruses) may result in heteroencapsidation of the nucleic acid of another strain of virus not naturally vector-transmitted (Bourdin and Lecoq, 1991; Candelier-Harvey and Hull, 1993).

Viral replicases

Tobacco plants transformed with the 54 kDa gene of TMV, which is derived from a portion of the replicase complex, were found to be immune to inoculation with high concentrations of TMV or TMV RNA (Golemboski et al., 1990). Recently, MacFarlane and Davies (1992) demonstrated that a sequence of pea early browning tobravirus, which is analogous to the 54 kDa gene of TMV, confers resistance to this virus in transformed *Nicotiana benthamiana* plants. Also extreme resistance to potato virus X was obtained in plants expressing a modified component of the viral replicase (Longstaff et al., 1993).

Satellite RNA

As mentioned above, certain isolates of CMV satellites (CARNA 5) attenuate symptoms and protect the crop from epidemics. Transgenic plants expressing CMV satellite sequences showed resistance to CMV. In these plants viral RNA synthesis and symptoms were reduced (Harrison et al., 1987; Tien and Wu, 1991).

Other approaches presently employed in investigations using genetic engineering to obtain resistance include the expression of viral antisense RNA (Hemenway et al., 1990), ribozymes (Haseloff and Gerlach, 1988), and expression of antibodies against viral proteins (Tavladoraki et al., 1993).

The Breeding Process

The breeder's work starts with the selection of plants which, after inoculation, express resistance in the form of absence of symptoms, mild symptoms, or local lesion formation. Resistant plants can be selected from wild species or from cultivar populations. The selected plants have to be retested by both mechanical and vector inoculation. Each plant is tested after inoculation for symptom expression, virus concentration, and horticultural properties. If the selected resistant plant is to be used in a breeding program, the above-mentioned tests have to be repeated in each generation until a uniform line is obtained. With vegetatively propagated plants, the retesting should be done in the offspring obtained during two or three additional growing periods. It is of advantage if the resistance to a specific virus is multigenic (two or three genes) to delay breakdown of resistance due to the appearance of new virus strains. This, of course, will delay the release of a resistant cultivar, but it is worthwhile, providing additional resistance genes are available. During the selection process, care also has to be taken that other important traits, such as resistance to other diseases, are not lost. Therefore, resistance breeding is a relatively long process — generally at least seven generations, and is often combined with producing hybrids. It is hoped that recent advances using molecular markers like restriction fragment length polymorphisms (RFLPs) (Tanksley et al., 1989) and random amplified polymorphic DNA (RAPD) (Welsh et al., 1991) may shorten the selection process.

Acknowledgement

This research was supported in part by the United States-Israel (Binational) Agricultural Research and Development Fund (BARD).

Selected Literature

Aly, R., Stein, A., Levy, S., Raccah, B. and Loebenstein, G. 1986. Spread and control of cucumber mosaic virus in gladiolus. Phytoparasitica 14:205–217.

Asjes, C.J. 1981. Verspreiding van het tulpemozaiekvirus in tulpen. Bloembollencultuur 91:1230–1232.

Barker, H. and Harrison, B.D. 1986. Restricted distribution of potato leafroll virus antigen in resistant potato genotypes and its effect on transmission of the virus by aphids. Ann. Appl. Biol. 109:595–604.

Berlinger, M.J. and Dahan, R. 1987. Breeding for resistance to virus transmission by whiteflies in tomatoes. Insect Sci. Appl. 8:783–784.

Blumenthal, T. and Carmichael, G.G. 1979. RNA replication: function and structure of Qβ-replicase. Annu. Rev. Biochem. 48:525–548.

Bourdin, D. and Lecoq, H. 1991. Evidence that heteroencapsidation between two potyviruses is involved in aphid transmission of a non-aphid-transmissible isolate from mixed infections. Phytopathology 81:1459–1464.

Candelier-Harvey, P. and Hull, R. 1993. Cucumber mosaic virus genome is encapsidated in alfalfa mosaic virus coat protein expressed in transgenic tobacco plants. Transgenic Res. 2:277–285.

Channarayappa, Shivashankar, G., Muniyappa, V. and First, R.H. 1992. Resistance of *Lycopersicon* species to *Bemisia tabaci*, a tomato leaf curl virus vector. Can. J. Bot. 70:2184–2192.

Derks, A.F.L.M. and Abbink, A. 1988. Symptomen van tulpemozaiekvirus in lelie varieren. Vakbl. Bloemisterij 43:60–63.

De Zoeten, G.A., Gaard, G. and Merkeus, W.S.W. 1972. Observations on ectodesmata and the virus infection process. Phytopathology 62:754.

Gera, A. and Loebenstein, G. 1988. An inhibitor of virus replication associated with green island tissue of tobacco infected with cucumber mosaic virus. Physiol. Mol. Plant Pathol. 32:373–386.

Golemboski, D.B., Lomonosoff, G.P. and Zaitlin, M. 1990. Plants transformed with a tobacco mosaic virus nonstructural gene sequence are resistant to the virus. Proc. Natl. Acad. Sci. USA 87:6311–6315.

Harrison, B.D., Mayo, M.A. and Baulcombe, D.C. 1987. Virus resistance in transgenic plants that express cucumber mosaic virus satellite RNA. Nature 328:799–802.

Haseloff, J.S. and Gerlach, W.L. 1988. Simple RNA enzymes with new and highly specific endoribonuclease activities. Nature 334:585–591.

Hayes, R.J. and Buck, K.W. 1990. Complete replication of a eukaryotic virus RNA *in vitro* by a purified RNA-dependent RNA polymerase. Cell 63:363–368.

Hemenway, C., Haley, L., Kaniewski, W.K., Lawson, E.C., O'Connell, K.M., Sanders, P.R., Thomas, P.E. and Tumer, N.E. 1990. Genetically engineered resistance: transgenic plants. In: Plant Viruses, C.L. Mandahar (ed.), Vol. 2, Pathology. CRC Press, Boca Raton, FL, pp. 347–363.

Hull, R. and Davies, J.W. 1992. Approaches to nonconventional control of plant virus diseases. Crit. Rev. Plant Sci. 11:17–33.

Kaper, J.M. 1993. Satellite-mediated symptom modulation: an emerging technology for the biological control of viral crop disease. Microb. Releases 2:1–9.

Lecoq, H., Labonne, G. and Pitrat, M. 1980. Specificity of resistance to virus transmission by aphids. Ann. Phytopathol. 12:139–144.

Loebenstein, G., Cohen, J., Shabtai, S., Coutts, R.H.A. and Wood, K.R. 1977. Distribution of cucumber mosaic virus in systemically infected tobacco leaves. Virology 81:117–125.

Loebenstein, G. and Gera, A. 1981. Inhibitor of virus replication released from tobacco mosaic virus infected protoplasts of a local lesion-responding tobacco cultivar. Virology 114:132–139.

Loebenstein, G. and Gera, A. 1990. Inhibitor of virus replication associated with resistance responses, In: Recognition and Response in Plant-Virus Interactions, R.S.S. Fraser (ed.), Springer-Verlag, Berlin, pp. 395–403.

Loebenstein, G. and Gera, A. 1993. The local lesion response to viruses: possibilities for engineering resistant plants, In: Biotechnology in Plant Disease Control, I. Chet (ed.), Wiley-Liss, New York, pp. 105–113.

Loebenstein, G., Gera, A. and Stein, A. 1984. Plant defense mechanisms of viral infections, In: Control of Virus Diseases, E. Kurstak (ed.), Marcel Dekker, New York, pp. 375–391.

Longstaff, M., Brigneti, G., Boccard, F., Chapman, S. and Baulcombe, D.C. 1993. Extreme resistance to potato virus X infection in plants expressing a modified component of the putative viral replicase. EMBO J. 12:379–386.

MacFarlane, S.A. and Davies, J.W. 1992. Plants transformed with a region of the 201K replicase gene from pea early browning virus RNA1 are resistant to virus infection. Proc. Natl. Acad. Sci. USA 89:5829–5833.

Motoyoshi, F. and Oshima, N. 1975. Infection with tobacco mosaic virus of leaf mesophyll protoplasts from susceptible and resistant lines of tomato. J. Gen. Virol. 29:81–91.

Romanow, L.R., Van Eijk, J.P., Eikelboom, W., Van Schadewijk, A.R. and Peters, D. 1991. Determining levels of resistance to tulip breaking virus (TBV) in tulip (*Tulipa* L.) cultivars. Euphytica 51:273–280.

Sushak, R.M. 1977. Gladiolus viruses transmitted in the seed. Gladio Grams 26:2–7.

Tanksley, S.D., Young, N.D., Paterson, A.H. and Bonierbale, M.W. 1989. RFLP mapping in plant breeding: new tools for an old science. Bio/Technology 7:257–264.

Tavladoraki, P., Benvenuto, E., Trinca, S., De Martinis, D., Cattaneo, A. and Galeffi, P. 1993. Transgenic plants expressing a functional single-chain Fv antibody are specifically protected from virus attack. Nature 366:469–472.

Thomas, P.E. and Fulton, R.W. 1968. Correlation of ectodesmata number with nonspecific resistance to initial virus infection. Virology 34:459–469.

Tien, P. and Wu, G.-S. 1991. Satellite RNA for the biological control of plant disease. Adv. Virus Res. 39:321–339.

Urban, L.A., Sherwood, J.L., Rezende, J.A.M. and Melcher, U. 1990. Examination of mechanisms of cross protection with non-transgenic plants, In: Recognition and Response in Plant-Virus Interactions, R.S.S. Fraser (ed.), Springer-Verlag, Berlin, pp. 415–426.

Van Eijk, J.P., Romanow, L.R. and Eikelboom, W. 1988. Componentenanalyse bij het onderzoek naar resistentie tegen mozaiekvirus in de tulpe. Prophyta 42:61–62.

Welsh, J. Peterson, C. and McClelland, M. 1991. Polymorphisms generated by arbitrarily primed PCR in the mouse: application to strain identification and genetic mapping. Nucleic Acids Res. 19:303–306.

17

Control of Virus Diseases

S. Spiegel and G. Loebenstein
Department of Virology, Agricultural Research Organization, The Volcani Center, Bet Dagan 50250, Israel

Prevention of Pathogen Movement in International Trade

The production and worldwide commercial distribution of ornamental plant material consisting of cut flowers, foliage or flowering potted plants, seeds, and vegetative propagants have become integral parts of domestic and international agricultural trade. One of the greatest concerns regarding movement of vegetatively propagated plant material from one area to another is the accidental introduction and distribution of pests, including virus and virus-like pathogens, into areas where they were not previously found. Viruses and virus-like agents that are often symptomless pose a special risk. For convenience, the term "viruses" will be used here to include virus-like agents, viroids, and mycoplasma-like bodies (MLOs).

Contribution No. 1935-E, 1994 series from the Agricultural Research Organization, The Volcani Center, Bet Dagan, Israel.

Phytosanitary regulations and policies established by individual countries and increasingly by organizations formed by groups of countries (i.e. the European and Mediterranean Plant Protection Organization, EPPO; North American Plant Protection Organization, NAPPO; and the European Community, EC, now recognized as the European Union, EU) are designed to prevent the entry and spread of pests, including viruses, over natural barriers. These regulations are enforced by quarantine and other authorised inspection agencies.

International trade of plant material for propagation purposes follows regulations and/or phytosanitary requirements that have been specified by the Plant Protection/Quarantine authorities in the importing country. In the US and elsewhere, quarantine regulations prohibit the importation of plant parts within genera that have been determined to have a high risk of carrying hazardous foreign pests, including viruses (Kahn, 1977).

In Europe, the EU council established in 1976 the first directive on protective measures against the introduction into member states of organisms harmful to plants or plant products and published specific directives regarding the introduction and movement of plant products from non-EU countries into and within member countries (*Official Journal of the European Communities, 1976-1993*). Recently, a summary of the complete EU phytosanitary regulations was prepared by EPPO (1995). These requirements are tailored to target known pests of potential economic importance (Lovisolo, 1988). The directives are mandatory for all member countries and introduce a degree of standardization within the EU.

Movement of plant material originating in a EU country within the EU has been facilitated by introducing an accompanying document known as a "plant passport". This consists of an official label which provides evidence that the provisions and restrictions detailed in the relevant directives and annexes, such as freedom from listed pests in the area of origin, have been fulfilled. The plant passport replaces the phytosanitary certificate in intra-community trade and allows free circulation of plant material within the single market of EU countries. A plant passport is generally issued at the place of production and after phytosanitary checks by the responsible official body in a Member State. Plants and plant products imported from non-EU countries require a phytosanitary certificate. If the import concerns plant material for which a plant passport is needed in domestic and intra-community trade, this document is also necessary for further transport within the EU, and in this case is issued at the EU frontier. Specific plants or plant parts listed in the directives are prohibited from entering the entire EU, e.g. *Euphorbia* infected with euphorbia mosaic virus, transmitted by the whitefly *Bemisia tabaci*, whereas plants or plant parts infected with tomato spotted wilt virus, known to occur and cause severe damage in Europe and other parts of the world, are prohibited only in protected zones of the EU in which it does not occur (e.g. Denmark).

The growing demand for high quality and disease-free planting material, the high sanitary standards imposed by many countries for importation of plant material, and the increasing amount of international trade have highlighted the significance of virus-free plant material. Since no practical treatments to cure virus-infected plants in the field are available, elimination of viruses from propagation material is currently the method used to obtain virus-free material.

Virus Elimination Procedures

Production of vegetatively propagated plant material is based on building up stocks from source plants. Viruses usually enter a propagation system at the breeding or germplasm level and are transmitted vertically down the propagation chain. The ultimate outcome of this process is virus-infected progeny. In many cases plants infected with a single virus tend to acquire a complex of viruses which results in severe distortion of entire plants or plant parts, e.g. flowers.

Elimination of viruses at the beginning of the propagation chain will generally result in virus-free regenerated plants which will serve as source plants for propagation (Stace-Smith, 1985; Walkey, 1991). Subsequent reliable testing for viruses, preferably over two growing seasons, and evaluation of trueness-to-type of propagation material are mandatory following a therapy step.

Tissue culture methods, thermotherapy, and chemotherapy have been used — either alone or in combination — to eliminate viruses from plants or plant parts that will be regenerated subsequently into plants. Traditionally, these methods have been applied to fully developed plants grown in pots. Recently, therapeutic procedures have also been applied to plant material maintained *in vitro*.

Tissue culture techniques

Shoot-tip culture has been the most commonly used tissue culture technique for production of virus-free plants from infected clones since the pioneering work of Morel and Martin (1952), who successfully regenerated virus-free plants from infected dahlias. The technique involves the excision of a shoot tip (0.3–0.8 mm in length) consisting of the meristematic dome and at least one set of leaf primordia from plants. The required minimum size of the excised tips varies with the plant species and virus. The shoot-tips are regenerated into plants on a nutrient medium *in vitro* and subsequently grown out in soil. This procedure has been used to establish virus-free plants of dahlia, geranium, lilium, freesia, and others (Walkey, 1991). Although shoot-tip culture is the method of choice for obtaining virus-free plants wherever possible, it has not been useful for several plant species. The reasons for this include failure to eliminate viruses in explants, lack of differentiation of small shoot tips during tissue culture, and failure to induce rooting of plantlets *in vitro*. To overcome these problems, variations on the basic shoot-tip culture method have been developed, two of which will be discussed here briefly.

Shoot-tip-grafting is a technique in which shoot-tips isolated aseptically from infected plants are grafted onto decapitated, young, etiolated rootstock seedlings. Growing plants at an elevated temperature prior to excising shoot-tips improved graft survival and recovery of virus-free plants, i.e. the success rate. The method has been adopted for various woody species including *Prunus salicina* L. (Japanese plum) (Navarro, 1988). However, success rates vary greatly among different viruses, plant species, and research groups. Recently, *in vitro* cultures of thin sections prepared from orchid plantlet shoots and protocorm-like bodies regenerated plantlets free of cymbidium mosaic virus and odontoglossum ringspot virus (ORSV) (Lim et al., 1993).

Thermotherapy

Replication of many plant viruses is significantly reduced when the host plant is grown at elevated temperatures. The effects of prolonged high temperatures on virus and host are not clearly understood. Direct and immediate suppressive effects of exposure to 40°C on the synthesis of host and viral RNAs and on the activity of viral replicases were demonstrated in cowpea and tobacco plants infected by various viruses. However, when the plants were transferred after a few hours at 40°C to an optimal temperature, host RNA synthesis was resumed immediately, while virus synthesis was delayed (Dawson and Lozoya-Saldana, 1984).

A basic thermotherapy procedure involves maintaining plants at a constant high temperature (usually 36–38°C), within the limits of their physiological tolerance, for periods of time ranging from several weeks to months. During this time, parts of or entire plants may become virus-free. In general, the higher the temperature and the longer the exposure time, the greater the reduction in virus titer. In practice, the selected temperature is a compromise between virus elimination and plant survival. Alternating high and low temperatures were found to be less stressful for plants than constant high temperature and effectively eliminated viruses in many cases. In procedures combining thermotherapy and tissue culture techniques, shoot tips (0.3–0.8 mm in length) may be excised from new growth produced during the heat treatment and established *in vitro* to regenerate new plants. The resulting plants have a high probability of being virus-free, but this has to be verified in each case. The combined thermotherapy-shoot tip procedure has been used to establish virus-free plants of carnation (Stone, 1968), narcissus (Mowat, 1980), chrysanthemum, and others (Quak, 1977; Walkey, 1991).

When thermotherapy is applied to plants grown in pots, preconditioning greatly improves their survival. This is achieved by growing plants under optimal conditions in large pots to allow a large root system to develop. In addition, the temperature in the heat chamber should be raised gradually over several days to the desired one for treatment. In spite of proper preconditioning, the exposure of several cultivated woody species (e.g. *Prunus*) to elevated temperatures resulted in very poor plant survival. Such sensitivity can be overcome by either increasing CO_2 concentration in the heat chamber to approximately 1,000 ppm and/or reducing O_2 to approximately 5%.

Whereas viruses can generally be eliminated from plants by heat therapy, viroids have been eliminated from plants by prolonged exposure to cool temperatures. Viroid-free chrysanthemum plants were obtained by exposing plants infected with chrysanthemum stunt viroid plants to 5°C, followed by meristem tip culture (Paduch-Cichal and Kryczynski, 1987).

Selecting an effective heat treatment for a given plant material requires an empirical approach if no information is available regarding heat tolerance of the host plant.

Chemotherapy

Producing virus-free plants from infected stock by applying chemicals that inhibit or interfere with virus replication has been defined as plant virus chemotherapy. Among the three main techniques available for eradication of plant viruses, this is the newest and so far the least utilized. A variety of natural and synthetic compounds, many of which are of interest in animal and human virology, have been tested for their potential to eliminate both RNA and DNA plant viruses. Research has been concentrated on two synthetic nucleotide analogues: ribavirin (Virazole) and 5-dihydroazauracil (DHT) (Schuster, 1988). The chemicals may be sprayed on, injected into infected plants, or incorporated into a nutrient medium on which plant material is grown. Shoot-tips are subsequently excised and regenerated into plants (Schuster, 1988; Hansen, 1989).

The biochemical events leading to virus eradication by chemicals are not understood. Ribavirin may be active in its triphosphate form, which inhibits the capping of viral RNAs (Dawson and Lozoya-Saldana, 1984). Schuster and Huber (1991) concluded that both ribavirin and DHT inhibit potato virus X replication at the early stage by impairing synthesis of RNA-dependent RNA polymerase and at a later stage by impairing synthesis of the coat protein.

Viruses of various groups and in different hosts have been successfully eradicated in young shoots treated with ribavirin and DHT. With ribavirin, the best examples are in potato (Griffiths et al., 1990) and apple (Hansen, 1989). However, in some cases, chemotherapeutants reduce virus only at phytotoxic levels, e.g. tobacco mosaic virus in systemically infected tobacco plants (Schuster, 1988).

Chemotherapy was recently applied to a few host-virus combinations by adding the chemotherapeutant to the *in vitro* medium on which plant material is grown. The presence of ribavirin in a meristem-inducing medium for lily bulblets infected with lily symptomless virus (LSV) and tulip breaking virus (TBV) increased virus eradication in cv. "Arai" but was ineffective for cv. "Enchantment" (Bloom-Barnhoorn and van Aartrijk, 1985). Similar experiments performed with hybrid lilies resulted in a marked reduction in LSV titer but not in TBV (Cohen et al., 1985). Ribavirin had no effect on the number of virus-free hyacinth plants infected with hyacinth mosaic virus (Bloom-Barnhoorn et al., 1985). Recently, Toussaint et al. (1993) reported that incorporation of ribavirin into a culture medium on which orchid protocorms infected with ORSV were grown enabled regeneration of virus-free plantlets. Effective eradication of the virus during subcultures depended on accurate isolation of new growth from initial tissues, ribavirin concentration, and frequency of transplanting into new media.

The potential of *in vitro* therapy is obvious. Large numbers of plants can be easily treated in a relatively small space; tissue culture procedures that increase survival can be easily applied (Stein et al., 1991); chemotherapeutants can be applied to plantlets in a contained medium, minimizing their possible environmental effects. As is the case with all therapeutic procedures, the genetic integrity of the resulting clones should be evaluated.

Certification Schemes for Virus-tested Ornamentals

Once viruses are eliminated and healthy plant material becomes available, it should be protected from reinfection by viruses and other pathogens. Schemes for the production of virus-tested, certified, vegetatively propagated planting stock are operational in many countries for carnation, chrysanthemum, pelargonium, narcissus, and others, either on a compulsory or a voluntary basis (Mowat, 1980; Elhanan, 1984; van Ruiten, 1988). A typical scheme is based on three consecutive stages of propagation:

- "Nuclear stock" is composed of absolutely healthy clone plants, derived from selected pre-nuclear plants after therapy, if necessary, and tested extensively for pathogens — preferably for two growing seasons — and trueness to type. Nuclear stock plants are grown in individual pots on raised benches under strict conditions minimizing the risks of reinfection by vectors and/or during handling by humans. Indexing for diseases is repeated several times throughout the year (zero tolerance for diseases). From nuclear stock, plant material is taken for growing "foundation stock".
- "Foundation (propagation) stock" is maintained under conditions minimizing reinfection as above. Plants are indexed for diseases on a clonal basis and not necessarily for each individual plant.
- "Certified stock" is propagated from foundation stock and used by propagators for mass production of planting material to be used by growers. This stock and its progeny are randomly checked for quality standards and diseases for which maximum tolerance levels are defined. It is recommended but not mandatory to grow certified stock under minimum reinfection conditions.

A scheme outlining the carnation certification scheme which is operating successfully in Israel on a compulsory basis is presented in Fig. 17.1.

Control of Seed-transmitted Viruses

Seed transmission of viruses in ornamental crops is of minor economic importance, as many of these crops are vegetatively propagated. Furthermore, although seed transmission occurs more frequently and in a wider range of viruses than was once thought, seed transmission in nature tends to be confined to a few plant virus groups: *inter alia* Crypto, Ilar, Nepo, Hordei, Poty, Tobra, and viroids. Also,

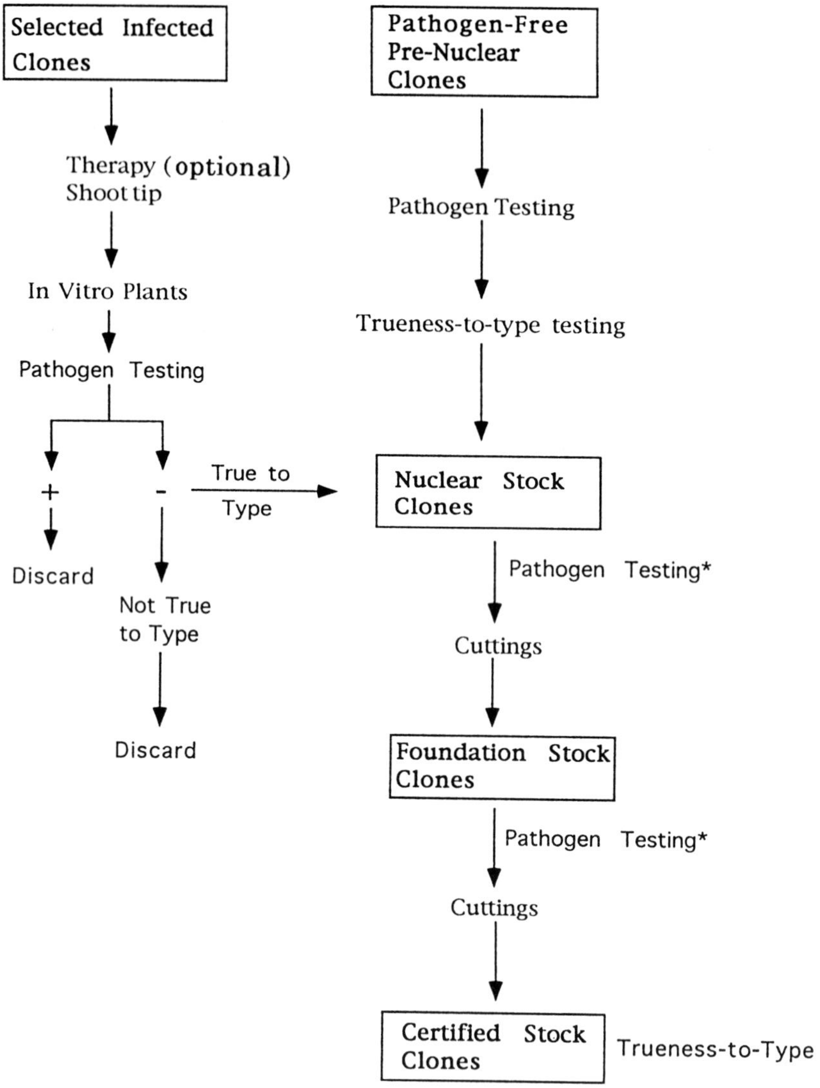

Fig. 17.1. Scheme for production of certified carnations in Israel.
*Pathogen testing of nuclear stock is done on individual plants; of foundation stocks on each clone and in certified stocks at random. Testing for carnation mottle virus is by ELISA; for other viruses by sap inoculation onto *Chenopodium quinoa* or *C. amaranticolor*, at least three times every year for two growing seasons; and testing for *Fusarium*, *Alternaria* spp. and *Pseudomonas wadsii*. Infected plants are discarded except for infection with *P. wadsii* which requires eradication of the entire nuclear stock. Nuclear and foundation stock plants are grown individually in pots, maintained on raised inert surfaces, in a strict insect-proof facility.

in many cases only a small percentage of the seeds developing on an infected plant carry virus inside the seed. Therefore, many of the embryo-infecting viruses cannot survive solely by seed transmission; each also uses another method of spread, mainly insect or nematode transmission. Seed transmission creating loci of infection can, however, cause severe damage in the field if an efficient vector is present. Thus, for example, lettuce mosaic potyvirus is transmitted through lettuce seeds at about 1–8%. This can result in a complete infection of the crop due to rapid spread by aphids.

Data on seed transmission in ornamental crops are limited (see Chapter 11). Carnation cryptic virus is 90–100% seed-transmitted in Mediterranean carnation cultivars (Lisa et al., 1980). Cherry leaf roll and arabis mosaic nepoviruses are transmitted at low rates through seeds of *Viola tricolor* and *Petunia hybrida*, respectively, and tobacco rattle through seeds of *Viola arvensis* (Lister and Murant, 1967). Hippeastrum mosaic potyvirus has been reported to be transmitted through seeds of *Hippeastrum hybridum* (Brants and van den Heuvel, 1965) and 1% transmission of cucumber mosaic virus (CMV) was observed in gladiolus seeds (Sushak, 1977). Seed transmission of tobacco ringspot and tomato ringspot nepovirus has been reported to occur in pelargonium (Scarborough and Smith, 1977). Seed transmission in this case could be of economic importance in seed-propagated pelargonium nurseries. Tobacco ringspot is also transmitted in seeds of *P. hybrida* and *Zinnia elegans*, at 20% and 5%, respectively (see Mandahar, 1981). In other crops such as lisianthus, the possibility was raised that lisianthus line pattern ilarvirus might be seed-transmitted (see Chapter 39); and lettuce mosaic potyvirus, which infects *Dimorphotheca sinuata,* might also be seed transmitted, though no data are available at present. Transmission of viroids through seeds of ornamentals has been reported for chrysanthemum stunt viroid, but this still has to be confirmed (see Chapter 6).

The best way to control seed-transmitted viruses is to grow virus-free plants under insect-proof screens. As with lettuce mosaic virus (LMV) in lettuce, these plants are tested individually and infected plants are removed. Nurseries are surveyed for early detection and infected plants are rogued. Furthermore, with LMV in lettuce and CMV in lupin seeds, seed lots are tested by enzyme-linked immunobsorbent assay (ELISA) or by polymerase chain reaction (PCR) (Wylie et al., 1993) to ensure that infection is below the allowed tolerance level.

At present, seed transmission of viruses in ornamentals is of little importance. If and when such problems become acute, the above-mentioned practices will have to be implemented. It is nevertheless advisable to choose only virus-free plants for breeding programs.

Control of Virus Dissemination by Vectors

At present no antiviral pesticides that affect the virus without harming the host plant are available. It is, however, possible to reduce the dissemination of viruses by interfering with their spread by vectors. In general, it is easier to prevent by insecticides the spread of persistently transmitted viruses than of nonpersistently transmitted ones. Nonpersistent viruses can be acquired and inoculated within minutes, and before pesticides have time to act, whereas persistent viruses require hours or days to be acquired and inoculated efficiently. Thus, dissemination of the persistently transmitted potato leaf roll luteovirus can be decreased markedly by using insecticides. Conversely, attempts to reduce spread of nonpersistently transmitted viruses through vector control by insecticides have in general been ineffective. In certain cases infection rates may even increase as a result of activity of alate aphids before being killed by the insecticides. As many of the viruses in ornamentals are nonpersistent, care should be taken to use insecticides only when infestation by the vector damages the plant or against persistent viruses.

Control of nematode vectors can be achieved by chemical treatments in the field or by heat. The difficulty with nematode treatment is in achieving complete kill of the nematodes because the latter frequently occur at a soil depth below the nematicide penetration. Thus, DD (dichloropropane-dichloropropene), 1,3-dichloropropene or methylbromide reduced spread of certain nepo- and tobraviruses in some crops, as for example arabis mosaic virus in *Lilium tigrinum* (Asjes and Segers, 1983). Steam or chemical treatments such as methylbromide, formalin, etc., of the growth medium in greenhouses are now common practice in ornamental crops.

Recently, soil pasteurization by solar heating, in combination with low dosages of fumigant, has been attempted in reducing nematode populations and subsequent virus spread (Katan, 1987).

Nonfumigant systemic nematicides, such as the oxime carbamates aldicarb and oxamyl, can also be used to control virus spread. Thus, oxamyl decreased the incidence of tobacco rattle virus in potato tubers from 19% to 3–4% in field trials in Scotland. Pentachloronitrobenzene, a fungicide, is also used to control nematode-borne viruses. A single application before planting reduced the numbers of *Longidorus elongatus* in the soil by 98% as assessed 15 months later, and protected strawberry plants from infection with raspberry ringspot and tomato black ring viruses for 4 years (Harrison, 1984).

One disadvantage of nematicides is their cost, which restricts their use to relatively valuable crops. Also, they rarely, if ever, eradicate vector nematodes completely; usually a few individuals survive the treatment, and populations may return to normal in 2–6 years. Furthermore, some of the chemicals still in use are already banned in certain countries, and such bans are likely to be extended.

Among the nematode-transmitted viruses in ornamentals, the following have been reported: strawberry latent ringspot virus in roses; tobacco ringspot virus in gladiolus, pelargonium, and zinnia; tomato ringspot virus in pelargonium; and tobacco rattle virus in bulb and corm crops (Hussey and McGuire, 1987). Control of nematodes is justified in high-value crops, especially in nuclear stock and foundation nurseries, taking care also to remove all virus source plants.

Oil sprays

Oil sprays to reduce dissemination of stylet-borne and some semi-persistent viruses have been attempted in various crops since Bradley (1963) showed that transmission of potato Y virus (PVY) by *Myzus persicae* was impeded by coating either the source plant or the test plant with mineral oil. Mineral oil sprays have proved useful in controlling viruses in bulb and corm crops, such as tulip breaking, lily symptomless, and brown ring in lilies (Asjes, 1974, 1975); iris yellow mosaic in iris (Deutsch and Loebenstein, 1967); cucumber mosaic in gladiolus (Aly et al., 1986); and mosaic in hyacinths (Asjes, 1980). Oil sprays are practiced on a larger scale in lilies in Israel and tulips in the Netherlands.

Spraying with oils should be started before the virus becomes established in the field in order to reduce primary spread. Oils should be applied at regular intervals (once a week) to keep the foliage covered with oil droplets. Oil sprays are efficient especially when virus dissemination occurs during relatively short periods and plants are still small (Loebenstein and Raccah, 1980). Various oils differ in their effectiveness, and viscosity was studied relative to their inhibitory effect on aphid transmission (De Wijs et al., 1979). Thus, Sunoco 7E, with a viscosity of 14 centi-Stokes at 37°C, was more effective than Bayol 50, with a viscosity of 8 centi-Stokes. De Wijs et al. (1979) also showed that the heavier fractions of Bayol 50, with a viscosity above 10, were more effective. They recommended using oils with a viscosity above 12 centi-Stokes. An oil (JMS Stylet oil) has been developed in Florida which collects selectively in the grooves over the anticlinal walls of the epidermal cells, where aphids appear to start their probe into the leaf epidermis. In Israel no differences between JMS stylet oil and "Virol", a white medium-light oil, were observed (Aly et al., 1986). The effectiveness of an oil spray is also dependent on the choice of emulsifier, type of sprayer nozzle, and spray pressure (Simons, 1982) so that small droplets of oil cover all the foliage. Combining pyrethroids with oils improved control of PVY and beet mosaic (Gibson and Rice, 1986) and of several viruses in lilies (Asjes, 1991).

Care must be taken to test each oil, with or without additives, on the specific crop, on a small scale first to ensure that there are no phytotoxic damages in the particular growth environment.

As to the mode of action whereby oil interferes with aphid transmission of stylet-borne viruses, it has been shown that brief contacts between the labium and oil lower both acquisition and inoculation (Bradley, 1963). The presence of oil on the aphid's labium is apparently involved in this process, as absorbing oil from the aphid's labium with clay restores transmission, although the site of action is not necessarily on the labium. A change in probing behavior of aphids on oil-sprayed plants was reported (Simons et al., 1977), although previous studies had not shown differences in behavior of aphids on oil-coated leaves (Bradley, 1963). Thus, by electronic monitoring Simons et al. (1977) showed that preprobing times of aphids on oil-sprayed plants increased markedly, while ingestion was reduced. These results are of major importance and should be verified in other virus–aphid systems.

Aphids are apparently unable to acquire virus from oil- sprayed plants, as virus content measured by ELISA of aphids decreased markedly.

It seems, therefore, that oil prevents virus attachment to the stylets, concomitantly, perhaps, with a reduction in liquid ingestion from epidermal cells. Oils also inhibit infection during mechanical inoculation, and it may well be that they also have an effect during the early stages of infection, after inoculation by aphids.

Reflective mulches and white nets

Repelling aphids with a reflective surface covering the ground between plants, thereby reducing virus dissemination, has been attempted in several parts of the world with variable success (Loebenstein and Raccah, 1980). Johnson et al. (1967) found that this procedure decreased vector numbers to less than 10% and the spread of CMV in gladiolus to less than 25% of the values for control plots. In Israel mulches of grey plastic sheeting decreased aphid numbers in peppers by more than 80% and the spread of PVY and CMV by more than 90% (Loebenstein et al., 1975). In experiments with gladiolus in Israel, however, grey plastic mulch was less effective, reducing CMV incidence to between 50–80% of the control

(Aly et al., 1986). Apparently, if virus spread occurs over longer periods, it is expected that infection rates will increase markedly, due to the reduced reflectance of the mulch.

Coarse white nets (2–8 mesh) were found to be effective for controlling the spread of PVY and CMV in peppers in Israel (Cohen, 1981). These nets are placed at least 0.5 m above and around the plots. As aphid vision is apparently limited to less than 50 cm, these nets seem to obscure the plants from the aphids. In experiments with gladiolus, white nets in combination with oil sprays reduced CMV incidence by 80% or more compared with nontreated plots (Aly et al., 1986). It seems, therefore, that loose nets in combination with oil sprays could be of value in producing high-value propagation material.

Breeding for Resistance

The best way to overcome diseases in general — and viruses in particular — is to breed resistant or tolerant plants. Major achievements, based on various resistance mechanisms (see Chapter 16) have been obtained with seed-propagated crops and potatoes. Thus, virus-resistant vegetable cultivars — tomato, cucurbits, and sweet pepper — and cereal crops were obtained using conventional breeding procedures. So far only limited work has been done in ornamental crops as most of them are propagated vegetatively and are therefore not uniform genetic lines. New lines resulting from crosses are immediately vegetatively propagated without screening for resistance to viruses. In recent years several breeding programs for resistance to viruses in ornamental plants have been started, such as introduction of virus resistance into tulips (Van Eijk et al., 1988). Differences observed in lily species to isolates of tulip breaking virus (Derks et al., 1989) could also serve as a basis for a breeding program.

A promising approach to obtain resistance in vegetatively propagated crops is through genetic engineering (see Chapter 16). Thus, basic work to obtain resistance to potyviruses derived from anti-sense RNA and coat protein genes gave resistance to bean yellow mosaic virus (BYMV) in *Nicotiana benthamiama* (Hammond and Kamo, 1992). The final aim is to transform bulb and corm crops with these genes.

With the rapid development of the ornamental industry, breeding for resistance to viruses, both by conventional and novel techniques, will probably receive more attention.

Selected Literature

Aly, R., Stein, A., Levy, S., Raccah, B. and Loebenstein, G. 1986. Spread and control of cucumber mosaic virus in gladiolus. Phytoparasitica 14:206–217.

Asjes, C.J. 1974. Control of the spread of the brown ring formation virus disease in the lily midcentury hybrid "Enchantment" by mineral oil sprays. Acta Hortic. 36:85–92.

Asjes, C.J. 1975. Control of the spread of tulip breaking virus in tulips with mineral oil sprays. Neth. J. Plant Pathol. 81:64–70.

Asjes, C.J. 1980. Toepassing van minerale olie om verspreiding van hyacintemozaikvirus in hyacinten tegen te gaan. Bloemobollencultuur. 90:1396–1397.

Asjes, C.J. 1991. Control of air-borne field spread of tulip breaking virus, lily symptomless virus and lily virus X in lilies by mineral oils, synthetic pyrethroids, and a nematicide in the Netherlands. Neth. J. Plant Pathol. 97:129–138.

Asjes, C.J. and Segers, L.C. 1983. Incidence and control of necrotic leaf mosaic caused by arabis mosaic virus in *Lilium tigrinum splendens* in the Netherlands. Phytopathol. Z. 106:115–126.

Bloom-Barnhoorn, G.J. and van Aartrijk, J. 1985. The regeneration of plants free of LSV and TBV from infected lilium bulb scale explants in the presence of virazole. Acta Hortic. 164:163-168.

Bloom-Barnhoorn, G.J., van Aartrijk, J. and van der Linde, P.C.G. 1985. Effect of virazole on the production of hyacinth plants free from hyacinth mosaic virus by meristem culture. Acta Hortic. 177:571-574.

Bradley, R.H.E. 1963. Some ways in which a paraffin oil impedes aphid transmission of potato virus Y. Can. J. Microbiol. 9:369–380.

Brants, D.H. and van den Heuvel, J. 1965. Investigation of Hippeastrum mosaic virus in *Hippeastrum hybridum*. Neth. J. Plant Pathol. 71:145–151.

Cohen, D., Milne, K.S. and Hyland, M.J. 1985. *In vitro* manipulations of virus concentrations in hybrid lilies. Acta Hortic. 164:319-324.

Cohen, S. 1981. Reducing the spread of aphid-transmitted viruses in peppers by coarse-net cover. Phytoparasitica 9:69–76.

Dawson, W.O. and Lozoya-Saldaña, H. 1984. Examination of the mode of action of ribavirin against tobacco mosaic virus. Intervirology 22: 77-84.

Derks, A.F.L.M., Hollinger, Th.C. and Lemmers, M.E.C. 1989. Differences between lily species and cultivars in their resistance to various isolates of tulip breaking virus. (abstr.). Fourth International Plant Virus Epidemiology Workshop, Montpellier, France, p. 35.

Deutsch, M. and Loebenstein, G. 1967. Field experiments with oil sprays to prevent yellow mosaic virus in irises. Plant Dis. Rep. 51:318–319.

De Wijs, J.J., Sturm, E. and Schwinn, F.J. 1979. The viscosity of mineral oils in relation to their ability to inhibit the transmission of stylet-borne viruses. Neth. J. Plant Pathol. 85:19–22.

EC. 1976–1993. Council directives on protective measures against the introduction into the Member States of organisms harmful to plants or plant products. Official Journal of the European Communities.

Elhanan, S. 1984. Certification schemes and their benefit to plant protection. EPPO Bull. 14:377-379.

EPPO. 1995. EPPO Summaries of Phytosanitary Regulations. EPPO Technical Documents (in press).

Gibson, R.W. and Rice, A.D. 1986. The combined use of mineral oils and pyrethroids to control plant viruses transmitted non- and semipersistently by *Myzus persicae*. Ann. Appl. Biol. 109:465–472.

Griffiths, H.M., Slack, S.A. and Dodds, J.H. 1990. Effect of chemical and heat therapy on virus concentrations in *in vitro* potato plantlets. Can. J. Bot. 68:1515-1521.

Hammond, J. and Kamo, K.K. 1992. Potyvirus resistance derived from antisense RNA and native or chimeric coat protein genes. (abstr.). Ninth International Symposium on Virus Diseases of Ornamental Plants (Prague, Czech Republic), p. 19.

Hansen, A.J. 1989. Antiviral chemicals for disease control. Crit. Rev. Plant Sci. 8:45-88.

Harrison, B.D. 1984. Progress and problems in the control of arthropod-, nematode-, and seedborne plant viruses. In Kurstak, E. (ed.), Control of Virus Diseases. Marcel Dekker, New York, pp. 265–299.

Hussey, R.S. and McGuire, J.M. 1987. Interaction with other organisms. In: Brown, R.H. and Kerry, B.R. (eds.), Principles and Practices of Nematode Control in Crops. Academic Press, Sydney, pp. 293–328.

Johnson, G.V., Bing, A. and Smith, F.F. 1967. Reflective surfaces used to repel dispersing aphids and reduce spread of aphid-borne cucumber mosaic virus in gladiolus plantings. J. Econ. Entomol. 60:16–18.

Kahn, R.P. 1977. Plant quarantine: principles, methodology and suggested approaches. In: Hewitt, W.B. and Chiarappa, L. (eds.), Plant Health and Quarantine in International Transfer of Genetic Resources. CRC Press, Cleveland, pp. 289-307.

Katan, J. 1987. Soil solarization. In: Chet, I. (ed.), Innovative Approaches to Plant Disease Control. Wiley, New York, pp. 77–105.

Lim, S.T., Wong, S.M. and Goh, C.J. 1993. Elimination of cymbidium mosaic virus and odontoglossum ringspot virus from orchids by meristem culture and thin section culture with chemotherapy. Ann. Appl. Biol. 122:289-297.

Lisa, V., Luisoni, E. and Milne, R.G. 1980. Carnation cryptic virus. Acta Hortic. 110:175–176.

Lister, R.M. and Murant, A.F. 1967. Seed-transmission of nematode-borne viruses. Ann. Appl. Biol. 59:49–62.

Loebenstein, G., Alper, M., Levy, S., Palevitch, D. and Menagem, E. 1975. Protecting peppers from aphid-borne viruses with aluminum foil or plastic mulch. Phytoparasitica 3:43–53.

Loebenstein, G. and Raccah, B. 1980. Control of non -persistently transmitted aphid-borne viruses. Phytoparasitica 8:221–235.

Lovisolo, O. 1988. New potential quarantine problems for ornamentals in Europe. Acta Hortic. 234:331-344.

Mandahar, C.L. 1981. Virus transmission through seed and pollen. In: Maramorosch, K. and Harris, K.F. (eds.), Plant Diseases and Vectors: Ecology and Epidemiology. Academic Press, New York, pp. 241–292.

Morel, G. and Martin, C. 1952. Guerison de dahlias atteints d'une maladie à virus. C.R. Hebd. Seances Acad. Sci. 235:1324-1325.

Mowat, W.P. 1980. The production of virus-free narcissus stocks in Scotland. Acta Hortic. 109:513-521.

Mowat, W.P. and Woodford, J.A.T. 1975. Control of field spread of tulip breaking (TBV) and lily symptomless virus (LSV). Ann. Rep. Scott. Hortic. Res. Sta. 71.

Navarro, L. 1988. Application of shoot-tip grafting *in vitro* to woody species. Acta Hortic. 227:43-55.

Paduch-Cichal, E. and Kryczynski, S. 1987. A low temperature therapy and meristem tip culture for eliminating four viroids from infected plants. J. Phytopathol. 118:341-346.

Quak, F. 1977. Meristem culture and virus-free plants. In: Reinert, J. and Bajaj, Y.P.S. (eds.), Plant Cell, Tissue and Organ Culture. Springer-Verlag, Berlin. pp. 298-615.

Scarborough, B.A. and Smith, S.H. 1977. Effects of tobacco- and tomato ringspot viruses on the reproductive tissues of *Pelargonium* × *hortorum*. Phytopathology 67:292–297.

Schuster, G. 1988. Synthetic antiphytoviral substances. In: Kurstak, E., Marusyk, R.G., Murphy, F.A. and Van Regenmortel, M.H.V. (eds.), Applied Virology Research, Vol. 1. New Vaccines and Chemotherapy. Plenum Press, New York, pp. 265-284.

Schuster, G. and Huber, S. 1991. Evidence for the inhibition of potato virus X replication at two stages dependent on the concentration of ribavirine, 5-azadihydro-uracil as well as 1,5-diacetyl-5-azadihydrouracil. Biochem. Physiol. Pflanz. 187:429-438.

Simons, J.N. 1982. Use of oil sprays and reflective surfaces for control of insect-transmitted plant viruses. In: Harris, K.F. and Maramorosch, K. (eds.), Pathogens, Vectors and Plant Diseases: Approaches to Control. Academic Press, New York, pp. 71–93.

Simons, J.N., McLean, D.L. and Kinsey, M.G. 1977. Effects of mineral oil on probing behaviour and transmission of stylet-borne viruses by *Myzus persicae*. J. Econ. Entomol. 70:309–315.

Stace-Smith, R. 1985. Virus-free clones through plant tissue culture. In: Moo-Young, M. (ed.), Comprehensive Biotechnology: The Principles, Applications and Regulations of Biotechnology in Industry, Agriculture, and Medicine. Pergamon Press, Oxford, England, pp. 69-179.

Stein, A., Spiegel, S., Faingersh, G. and Levy, S. 1991. Responses of micropropagated peach cultivars to thermotherapy for the elimination of prunus necrotic ringspot virus. Ann. Appl. Biol. 119:265-271.

Stone, O.M. 1968. The elimination of four viruses from carnation and sweet william by meristem-tip culture. Ann. Appl. Biol. 62:119-122.

Sushak, R.M. 1977. Gladiolus viruses transmitted in the seed. Gladio Grams 26:2–7.

Toussaint, A., Kummert, J., Maroquin, C., Lebrun, A. and Roggemans, J. 1993. Use of virazole to eradicate odontoglossum ringspot virus from *in vitro* cultures of Cymbidium Sw. Plant Cell, Tissue and Organ Culture 32:303-309.

Van Eijk, J.P., Romanow, L.R. and Eikelboom, W. 1988. Componentenanalyse bij het onderzoek naar resistentie tegen mozaiekvirus in de tulpe. Prophyta 42:61–62.

van Ruiten, J.E.M. 1988. Production and trade of healthy propagating material: a need for a quality control system. Acta Hortic. 234:523-528.

Walkey, D.G.A. 1991. Production of virus-free plants. In: Applied Plant Virology, Wiley New York, pp. 270-292.

Wylie, S., Wilson, C.R., Jones, R.A.C. and Jones, M.G.K. 1993. A polymerase chain reaction assay for cucumber mosaic virus in lupin seeds. Aust. J. Agric. Res. 44:41–51.

Fig. 2.2. Tulip with color breaking. Many such tulips were depicted by Dutch painters in the 17th century. Reproduction of a painting by P. Cos in 1637 (Agricultural University, Wageningen). At that time bulbs were sold for fabulous prices; one of the series painted by Cos was sold for 4200 guilders.

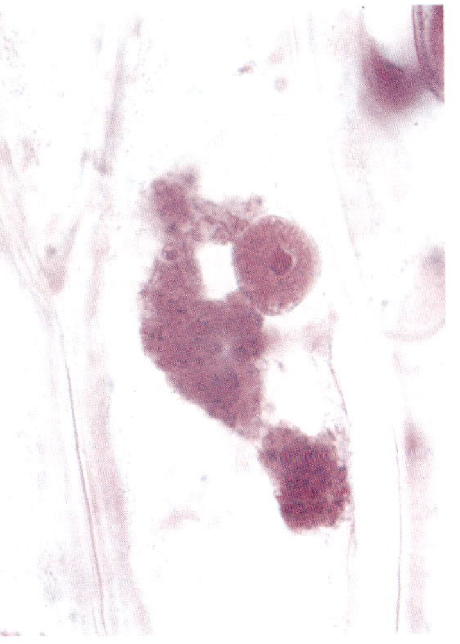

Fig. 8.2. Granular inclusion bodies in the cytoplasm and abnormal enlargement of nucleolus caused by bean yellow mosaic potyvirus — i.a. infecting sweet pea, gladiolus and some flower bulb crops — in epidermal cells of faba bean after staining with phloxine and methylene blue (mag. ×300).

Fig. 7.1. Aster yellows in periwinkle (*Catharanthus roseus*): left, several flowers suffering from phyllody; right, a healthy flower. Courtesy of K. Maramorosch.

Fig. 7.3. Stock plant (*Matthiola incana*) infected with *Spiroplasma citri* (left), and healthy plant of the same age (right).

Fig. 8.4. Desiccated and necrotic local lesions in *Nicotiana clevelandii*, a widely used test plant and artificial host for many plant viruses, after sap inoculation with tobacco necrosis virus from *Primula*. (Photo courtesy of Ms. I. Bouwen, IPO-DLO, Wageningen.)

Fig. 8.5. Ringspots (often concentric) and line patterns in *Dianthus barbatus*, systemically infected with carnation ringspot nepovirus. (Photo courtesy of Ms. I. Bouwen, IPO-DLO, Wageningen.)

Fig. 8.7. Striping on leaf of *Galtonia candidens* with *Ornithogalum* mosaic potyvirus. (Photo courtesy of Bulb Research Center, Lisse.)

Fig. 8.8. Vein yellowing in *Lonicera japonica* var. *aureoreticulata*, ascribed to tobacco leafcurl geminivirus.

Fig. 8.10. Enations on leaf veins of *Nicotiana rustica* infected by tomato aspermy cucumovirus.

Fig. 8.9. Systemic leaf and stem necrosis and secondary malformation in tulip, naturally infected by tobacco necrosis virus.

Fig. 8.11. Foliar yellowing and floral virescence and phyllody in periwinkle (*Catharanthus roseus*) with aster yellows-like agent from periwinkle. At right, flower with early stage of virescence.

Fig. 8.12. Greening and early stages of phyllody in garden nasturtium (*Tropaeolum majus*) caused by mycoplasma infection: left, normal flower. (After Bos, 1957.)

Fig. 8.14. Abnormal branches and inflorescences of *Eucalyptus* with witches' broom and decline; right, flowering healthy branch.

Fig. 18.1. Alstroemeria plants just before planting in a glasshouse for a comparative trial with virus-infected plants (right) and virus-free plants from meristems (left).

Fig. 18.2. Flower color break symptoms (right) in alstroemeria. Left: healthy control.

Fig. 21.2. Faint mosaic on leaf bases of *C. flavus* caused by bean yellow mosaic virus (BYMV) (right). Healthy plant (left).

Fig. 22.4. Concentric rings becoming necrotic in leaf infected with tomato spotted wilt virus (TSWV).

Fig. 22.5. Chlorotic vein-banding, mosaic and leaf distortion in plant infected with dahlia mosaic virus (DaMV).

Fig. 24.1. Color break associated with infection by cucumber mosaic virus (CMV) — cv. Peter Pears. Right, CMV and bean yellow mosaic virus (BYMV) infected. Left, symptomless infection by BYMV.

Fig. 24.3. Leaf spots suspected to be associated with tobacco streak virus.

Fig. 25.1. Irregular light and dark green mosaic pattern on the leaf of a *Hippeastrum* hybrid caused by hippeastrum mosaic virus.

Fig. 26.2. Green ring patterns on flower stalk of cv. King of the Blues caused by hyacinth mosaic virus.

Fig. 27.1. Faint mosaic symptoms caused by iris mild mosaic virus (IMMV) in iris cv. Ideal (left) and a virus free plant (right).

Fig. 28.4. Yellowing and browning of leaves and veins in the stem of cv. Stargazer infected with lily mottle virus (LMoV) and lily symptomless virus (LSV).

Fig. 28.2. Yellow and brown spots on the leaves of cv. Connecticut King infected with lily symptomless virus (LSV).

Fig. 32.1. Ranunculus plant infected by cucumber mosaic virus (CMV) showing flower breaking, stunting, and leaf malformation.

Fig. 32.2. On ranunculus plants tobacco necrosis virus (TNV) alone induces yellowing, vein chlorosis, and necrotic areas on the leaves.

Fig. 32.4. In Ligurian (Italy) ranunculus cultivation tospoviruses induce severe damage destroying entire crops; necrotic spots on the leaves are diagnostic symptoms.

Fig. 32.5. The presence of MLOs is associated with virescence, malformations, and phyllody of ranunculus flowers.

Fig. 33.3. Tulip cv. Rose Copland: (left) tepal from healthy plant; (right) tepal, with pigmented veinal streaks, from plant experimentally infected with lily symptomless virus.

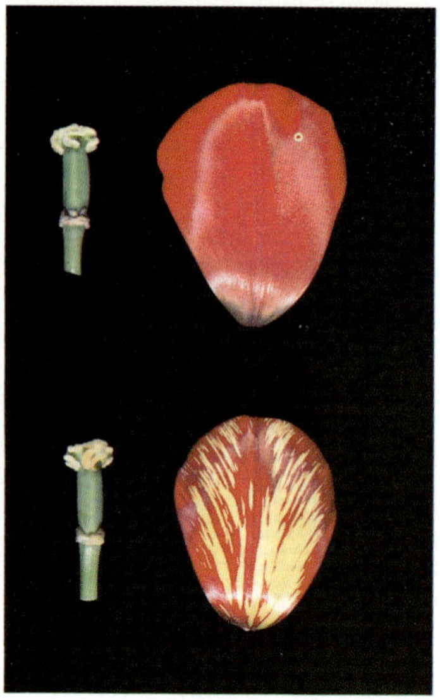

Fig. 33.5. Pistil and tepal of tulip cv. Paul Richter: healthy plant (top); plant infected with tulip breaking virus showing discoloration of stigma (bottom).

Fig. 33.4. Tulip cv. Orange Marvel infected with (left) tulip chlorotic blotch virus and (right) tulip breaking virus; center flower uninfected.

Fig. 34.1. Flower breaking on a carnation flower infected by carnation vein mottle, carnation mottle and carnation etched ring viruses, field infection. Reproduced with permission from "Colture Protette" No. 5, 1976, Edagricole, Bologna, Italy.

Fig. 34.2. Carnation infected by carnation necrotic fleck, carnation mottle and carnation etched ring viruses, field infection. The leaf necrosis is caused mainly by carnation necrotic fleck virus. Reproduced with permission from "Colture Protette" No. 7/8, 1992, Edagricole, Bologna, Italy.

Fig. 34.3. Flowers of Mediterranean carnation affected by the carnation stunting syndrome. Reproduced with permission from "Inf. Fitopatol." No. 3, 1985, Edagricole, Bologna, Italy.

Fig. 34.4. Proliferation of carnation shoots associated with the carnation stunting syndrome. Photo Istituto Fitovirologia Applicata CNR, Torino.

Fig. 35.4. Two stunted plants infected with chrysanthemum stunt viroid (left). The two plants on the right are not infected. Photograph: Research Station for Floriculture, Aalsmeer.

Fig. 35.3. Tomato spotted wilt virus induced symptoms showing necrotic spots surrounded by chlorosis on a leaf, and necrotic streaks on a stem. Photograph: Glasshouse Crops Research Station, Naaldwijk.

Fig. 36.1. Flower necrosis induced by cymbidium mosaic virus in *Cattleya*. Necrosis normally first appears in flowers 10-14 days after the tight buds break open.

Fig. 36.2. Severe color break in a *Cattleya* flower induced by orchid tobacco mosaic virus. The increased pigment intensity appearing as streaks and patches parallel to the midvein is typical of this disease reaction.

Fig. 40.1. Lisianthus shoot infected by lisianthus necrosis virus, field infection. Photo courtesy of Dr. M. Iwaki, Yamaguchi University, Yamaguchi, Japan.

Fig. 40.2. Lisianthus shoot infected by lisianthus line pattern and tomato mosaic viruses, field infection. The necrotic symptoms are associated with lisianthus line pattern virus. Photo courtesy of Istituto Fitovirologia Applicata CNR, Torino.

Fig. 40.3. Lisianthus leaves infected by tobacco mosaic virus (right) and healthy control (left). Photo courtesy of Institute of Plant Protection, Bet Dagan.

Fig. 40.4. Lisianthus plants infected by tomato yellow leaf curl virus, field infection. Photo courtesy of Institute of Plant Protection, Bet Dagan.

Fig. 41.2. Necrotic rings on leaf of plant infected with impatiens necrotic spot virus (INSV).

Fig. 42.1. Cyclamen plant infected by tomato spotted wilt virus (TSWV) showing necrotic spots on the leaves.

Fig. 44.1. Three *Dieffenbachia maculata* "Exotica" leaves with chronic systemic symptoms of dasheen mosaic virus (right) and an asymptomatic leaf (left).

Fig. 44.2. A plant of *Dieffenbachia maculata* "Rudolph Roehrs" with ringspot symptoms induced by dasheen mosaic virus.

Fig. 44.3. Unfurling leaf of *Caladium hortulanum* with mosaic symptoms of dasheen mosaic virus.

Fig. 44.4. Dasheen mosaic virus-infected (left) and healthy (right) leaves of *Zantedeschia elliottiana*.

Fig. 45.2. *Dimorphotheca sinuata* plants infected by lettuce mosaic virus. Photo courtesy of Dr. T. Natsuaki, Utsunomia University, Utsunomia, Japan.

Fig. 45.3. *Dimorphotheca sinuata* plant infected by tomato spotted wilt virus, field infection. Photo courtesy of Istituto Fitovirologia Applicata CNR, Torino, Italy.

Fig. 46.3. Leaves of *Hibiscus rosa-sinensis* infected with hibiscus chlorotic ringspot virus showing chlorotic spots and ringspots. Photo courtesy of B.C. Raju.

Fig. 47.4. Hydrangea virescence associated with a mycoplasma-like organism.

Fig. 47.5. Healthy hydrangea cv. Strafford with normal cymes.

Fig. 48.1. Seed produced impatiens naturally infected with impatiens necrotic spot virus. Note the severe necrosis and spotting in the upper foliage.

Fig. 48.4. Stem necrosis in seedling impatiens infected with a severe isolate of impatiens necrotic spot virus. Left: "Dazzler Orange". Center: "Accent Lilac". Right: "Accent Salmon". Top row: plants incubated at 27/24 °C day/night temperature. Bottom row: Plants incubated at 21/18 °C day/night temperature.

Fig. 49.1a. *Kalanchoë blossfeldiana* naturally infected with tomato spotted wilt virus (TSWV). (Photo: Verhoeven, 1991.)

Fig. 50.2. Flower break symptoms on petals of pelargonium cv. Palais.

Fig. 50.3. Yellow spots, partial rings and ringspots in *Pelargonium peltatum* infected with pelargonium ringspot disease (courtesy of Drs. N. Paludan and J. Begtrup, Denmark).

Fig. 50.4. Symptoms caused by tomato spotted wilt virus on *Pelargonium peltatum* (courtesy of Dr. J. Th. J. Verhoeven, Plant Protection Service, Wageningen, Netherlands).

Part II

Virus Diseases of Major Ornamental Crops

Bulb and Corm Crops

Cut Flower Crops (vegetatively propagated)

Cut Flower Crops (seed propagated)

Pot and Bedding Plants

18

Alstroemeria

Annemarie van Zaayen
Inspection Service for Floriculture and Arboriculture (NAKB), Johan de Wittlaan 12, 2517 JR Den Haag, The Netherlands

Introduction

Recently the ornamental crop alstroemeria has become more and more important all over the world. Many desirable cultivars are now available, and flowers with a long vase-life are produced under conditions requiring low energy. About 600 million flower-stems, grown on c. 300 ha, have been sold world-wide in 1993. In the past ten years sales at the Dutch auction showed more than a threefold increase from 65 million cut-flowers in 1983 (Hakkaart and Versluijs, 1985) to a total of 222 million in 1993 (Produktschap voor Siergewassen en Produktschap voor Groenten en Fruit, 1994). The total cropping area under glass in the Netherlands increased from c. 15 ha in 1980 to 103 ha in 1993. At the Dutch auction in 1993, the turnover of alstroemerias produced in the Netherlands was 65.3 million Dutch guilders (Produktschap voor Siergewassen en Produktschap voor Groenten en Fruit, 1994). Most of the starting material is produced in this country.

The genus *Alstroemeria* belongs to the Amaryllidaceae, but there are also close relationships to the Liliaceae. Various authors differ in their opinions on this subject. The genus was named after Linnaeus' friend C. Alstroemer, who drew his attention to one of the species (Linnaeus, 1762). The monocotyledonous plants originate from South American countries, primarily Chile, but also Argentina, Bolivia, Brazil, Ecuador, Paraguay, and Peru. Bayer (1987) described the genus *Alstroemeria* in Chile: it comprises at least 31 species. During the last decades, a number of varying botanical species have been introduced in Europe and especially in the Netherlands for breeding purposes, yielding different groups of products, e.g. the so-called Butterfly-group, the Aurea-group and the Orchid-group. These breeding products are grown in glasshouses for cut-flower production. The vegetative propagation is through rhizomes in the soil. Reproduction through seed is used only in breeding.

For many years the crop has been multiplied vegetatively by splitting mature plants. Consequently, viruses occurring in the crop — and probably originating in part from parental lines used in breeding — have also been multiplied. Several viruses have been described by authors in Denmark (Rønde Kristensen et al., 1962), England (Brunt and Phillips, 1981), the Netherlands (Hakkaart and Versluijs, 1985), and Italy (Bellardi and Bertaccini, 1991). Alstroemeria of the Butterfly-type nearly always contains alstroemeria mosaic potyvirus (AlMV; Brunt and Phillips, 1981; Hakkaart and Versluijs, 1985; Schönfelder et al., 1993). The Aurea-type alstroemerias usually do not contain AlMV, but they occasionally do contain alstroemeria carla virus (AlCV) and cucumber mosaic virus (CMV). AlCV and CMV sometimes occur in the other alstroemeria types.

Recently, another potyvirus was detected and tentatively named alstroemeria streak virus (AlSV; Wong et al., 1992).

The Viruses

Alstroemeria mosaic potyvirus (AlMV)

Symptoms

Symptoms induced by AlMV, presumably the most frequently occurring virus, are streaking of the leaf (chlorosis), which shows like light green and darker spots parallel to the midrib. Considerable variation in symptom expression can be noticed, probably dependent on cultivar, growing conditions, and time of the year. As there still is uncertainty about the number of (poty)viruses involved, and as symptoms caused by individual and combined viruses are not yet fully known, future research should elucidate these gaps.

Sometimes a few dark stripes are visible on the flower petals. Occasional "flower-break" symptoms have been mentioned by Brunt and Phillips (1981) and by Phillips and Brunt (1986), but Hakkaart and Versluijs (1985) reported such symptoms to be unknown in the Netherlands. Some years later, however, flower color break symptoms were supposed to be caused by an uncharacterized virus in the Netherlands (van Zaayen et al., 1992). Subsequently such symptoms were correlated with a second potyvirus (van Zaayen et al., 1994), which appeared to be identical to AlSV (Wong et al., 1992; van Zaayen et al., 1994). The recent finding that AlMV and AlSV frequently occur together in virus-infected alstroemeria plants (van Zaayen et al., 1994) clearly indicates that all data published on AlMV, including the description of symptoms, should be revised.

Description of the virus

AlMV is a member of the potyvirus group (Phillips and Brunt, 1986; Schönfelder et al., 1993; Milne, 1988); see also Chapter 5, potyvirus.

AlMV preparations, partially purified from systemically infected *Nicotiana clevelandii* plants, contained numerous flexuous filamentous particles, c. 750 nm in length (Phillips and Brunt, 1986). Such particles had a sedimentation coefficient of 145 S and comprised c. 5% ss-RNA and 95% protein with a single polypeptide with a MW 32,500.

Typical of potyviruses, AlMV induces conical or cylindrical cytoplasmic inclusions in cells of infected plants (Phillips and Brunt, 1986). According to Schönfelder et al. (1993), the cytoplasmic inclusions found in infected cells comprised cylindrical inclusions type III.

In tube precipitin tests, AlMV showed a close serological relationship to iris severe mosaic virus, to another potyvirus from iris, and to a virus isolated from narcissus (Phillips and Brunt, 1986). These results should be considered with care, however, since the finding that AlMV often is not the only potyvirus in infected alstroemeria plants.

Geographical distribution and damage caused

AlMV occurs widespread wherever alstroemeria's are grown. van Zaayen et al. (1992) tried to obtain data on the supposed detrimental effects to flower quality and yield caused by alstroemeria viruses. An experiment was set up with 20 virus-free plants from meristems of the variety Snow Queen, a Butterfly-type, planted in a glasshouse next to 20 plants of the same variety containing AlMV and AlCV

(Fig. 18.1; see color plate). The results showed that the virus-free "Snow Queen" had in this trial, in comparison with infected plants, a 50% increase in yield, more first quality flower stems, better quality of the leaves, and a longer vase-life as the leaves did not yellow. The 50% increase in yield may be an exception and may have resulted from the presence of two viruses in the infected plants. Moreover, the presence of AlSV, which had not yet been detected at the time of this experiment, could not be excluded.

Methods of assay

AlMV is mechanically transmissible to a limited range of indicator plants. It induced occasional faint chlorotic lesions in inoculated leaves followed by mild systemic veinal chlorosis in *Nicotiana benthamiana*, *N. clevelandii*, and *N. megalosiphon* (Phillips and Brunt, 1986); and it caused a systemic chlorotic mottle, which later became necrotic, in *N. hesperis*-67A (De Blank et al., 1994). Local lesions were obtained in several *Chenopodium* species and in *Tetragonia expansa* (Phillips and Brunt, 1986), in *N. occidentalis*-37B and cv. Japan, and in *Hyoscyamus niger* (De Blank et al., 1994). The AlMV-containing sap used in all of these experiments, however, could have been contaminated with AlSV.

AlMV can be readily detected in alstroemeria sap by electron microscopy and even better by immunosorbent electron microscopy (ISEM). ISEM was applied parallel to ELISA to examine the reliability of ELISA (De Blank et al., 1994). ISEM-studies revealed the presence of other potyviruses in alstroemeria plants with or without flower colour breaking (De Blank et al., 1994). Large-scale testing with ISEM, however, is time-consuming, and differences in quality between grids may cause variable results.

For detection of AlMV, also in large-scale routine testing, ELISA is the obvious assay. A double antibody sandwich (DAS)-ELISA with a polyclonal antiserum (Maat, 1983, 1989) has been used in the Netherlands for several years to detect the virus. ELISA is also applied in Britain for detection of AlMV (Phillips and Brunt, 1986). Although the distribution of AlMV in the alstroemeria plant can be erratic, sampling of leaves for routine testing by ELISA is reliable provided that leaves from different shoots of a plant are used and lower leaves are avoided (De Blank et al., 1994). Antisera to AlMV, prepared in England and in the Netherlands, later appeared to be able to detect AlSV in addition to AlMV (van Zaayen et al., 1994). Consequently, the sensitivity of ELISA for the detection of AlMV (and of AlSV) could be increased by the preparation and application of separate specific antisera.

AlMV reacts positively to a commercially available monoclonal potyvirus group specific antiserum (Agdia potyvirus group test; van Zaayen, unpublished results).

Natural and experimental transmission

Mechanical inoculation experiments on single plants did result in virus transmission, not only by trimming rather young plants but also through harvesting flowers from older plants (De Blank et al., 1994). There was, however, a huge variability in sensitivity of cultivars: those belonging to the Butterfly-group are most sensitive, whereas a cv. of the Orchid-group could not be infected (De Blank et al., 1994).

The virus was not apparently seed-borne in *N. clevelandii*, and was occasionally transmitted by the aphid *Myzus persicae* but not by *Macrosiphoniella sanborni* (Phillips and Brunt, 1986). Although De Blank et al. (1994) did agree with the statement about seed transmission, which had not been observed in an experiment with alstroemeria seeds, their transmission trials with aphids were not conclusive. These authors concluded that pruning and cutting of plants most probably is the main cause of virus transmission. In addition, splitting of plants for propagation may also play an important role.

The discrepancy in the results with aphid transmission may have been due to the presence of mixed infections of AlMV and AlSV.

Control methods

Several varieties have been freed from AlMV and AlCV by meristem (tip) culture (Hakkaart and Versluijs, 1985, 1988). The meristems are located in the rhizomes in the soil, implying that these have to be disinfected thoroughly before the meristems can be isolated. Plantlets growing from the meristems should form a rhizome, or else they will die when transferred to soil.

Meristem culture is nowadays practiced in commercial companies (van Zaayen et al., 1992). In the case of alstroemeria, virus tests of the *in vitro* material grown from meristems gave satisfactory results. The *in vitro* alstroemeria plantlets were tested twice for the virus or viruses present in the starting material with an interval of at least one month (preferably two or more). The most efficient way is to collect redundant plant parts (e.g. top shoots, leaves), which are cut off during the regular transfers of the *in vitro* plantlets to fresh nutrient media, and to test this material. As the virus(es) in question may occur in the young plantlets in a low concentration, repeated testing is required. A test after weaning of the plantlets in the glasshouse, as required for lily, is not necessary as it does not yield additional information (van Zaayen et al., 1992).

"Butterflies" are usually difficult in tissue culture, in contrast with, for instance, "Orchids".

Unfortunately, once pathogens have been eliminated, crops do not automatically remain free from these pathogens. Many measures must be taken to prevent re-infection. The introduction of infected propagation material must be prevented. Goals can be realized to improve the quality of the propagating material by the introduction of certification schemes (Van Ruiten, 1988 and 1993; OEPP/EPPO, 1991; Lopian, 1993). The worldwide trade in starting material implies the necessity of a high quality product, which resulted in the start of a certification program for alstroemeria in the Netherlands from September 1991 by the Inspection Service for Floriculture and Arboriculture (NAKB). Companies may participate voluntarily in such programs. The certification system implies that there should be separate glasshouse sections, which meet given isolation (e.g. insect-proof) and hygiene standards. Consequently, there should be separate glasshouses for

- entry and pre-testing of *candidate material*
- maintenance of *nuclear stock*
- propagation of basic material for *propagation stock*
- production of *certified material*.

Requirements for glasshouses have been described elsewhere (OEPP/EPPO, 1993).

Obviously, sanitary measures should be taken to prevent virus spread including removal of obviously diseased plants, disinfection of knives, insect control, etc. Furthermore, regular visual inspections and testing for viruses in the first production stages and visual inspections and random testing in later stages are indispensable.

Alstroemeria streak potyvirus (AlSV)

Symptoms

Wong et al. (1992) reported streak symptoms of alstroemeria leaves in a glasshouse in Ithaca, New York. A virus-free alstroemeria plant inoculated with AlSV obtained from S.M. Wong showed small chlorotic streaks parallel to the midrib (van Zaayen et al., 1994). Symptoms caused by AlSV probably are very similar to those of AlMV, although somewhat fainter. Again, as with AlMV, cultivar, conditions, time of the year, and mixed infections may influence symptom variability.

Flower color break symptoms (Fig. 18.2; see color plate) are almost certainly correlated with AlSV (or with AlSV in combination with AlMV), but not with AlMV alone (see above: symptoms caused by AlMV). Such symptoms, however, can only be noticed in purple, pink, and red flowers.

Description of the virus

AlSV is a tentative member of the potyvirus group (Wong et al., 1992; van Zaayen et al., 1994; Milne, 1988); see also Chapter 5, potyvirus.

Purified preparations from alstroemeria (Wong et al., 1992), and crude preparations from symptomatic leaves of indicator plants (van Zaayen et al., 1994) contained flexuous filamentous particles resembling a potyvirus, 780 nm long × 12 nm wide (Wong et al., 1992).

Potyviruses other than AlMV have been noticed earlier in alstroemeria by Hakkaart and Versluijs (1985), J. Dijkstra (pers. comm., 1985) and by Bellardi et al. (1992). Probably because AlMV and AlSV frequently occur together in infected alstroemeria plants (van Zaayen et al., 1994), the separate isolation and characterization of AlSV was hampered. Wong et al. (1992) apparently had the disposal of alstroemeria plants naturally infected with AlSV only. Although such plants yielded a rather small amount of virus, the authors still performed some purification trials and observed both pinwheel and laminated inclusions in the cytoplasm of leaf tissues. This virus later appeared to be one of the potyviruses earlier noticed but different from AlMV (van Zaayen et al., 1994).

Decoration experiments in the electron microscope with sap from virus-infected alstroemeria plants and various antisera showed very close relationships between AlSV and freesia mosaic virus (FreMV). ELISA, however, gave a negative result when the alstroemeria sap was tested for FreMV (van Zaayen et al., 1994).

Geographical distribution and damage caused

Probably: see AlMV, but little knowledge is available about this recently detected virus.

Methods of assay

Although Wong et al. (1992) could not obtain symptoms on test plants upon inoculation, AlSV appeared to be mechanically transmissible, after a slow start with only 6 μg of AlSV purified and provided by S.M. Wong, to a limited range of indicator plants (van Zaayen et al., 1994). The symptoms caused by AlSV differed from and were fainter than those caused by different AlMV-isolates (possibly contaminated with AlSV), but there was a resemblance and a similar host range was involved (De Blank et al., 1994; van Zaayen et al., 1994).

S.M. Wong (pers. comm., 1992) found no reactions with an antiserum to AlMV and with Agdia's potyvirus group test, but a positive reaction with a polyclonal potyvirus group-specific antiserum provided by D.D. Shukla. Decoration experiments in the electron microscope and experiments with DAS-ELISA showed that AlSV reacted strongly with the commonly used antisera to AlMV (van Zaayen et al., 1994). Although AlMV reacted with the monoclonal potyvirus group-specific antibodies, the negative result in this test of sap from leaves infected with AlSV, as reported by S.M. Wong (pers. comm., 1992), could be confirmed. Consequently, detection of AlSV and distinction from AlMV is possible by applying an ELISA with antiserum to AlMV (Maat, 1983, 1989; Phillips and Brunt, 1986), followed by an ELISA with potyvirus group-specific monoclonal antibodies (van Zaayen et al., 1994). The currently available antisera to AlMV obviously have been prepared from infected plants with both viruses. In the future, the preparation of specific antisera and an improved ELISA for both potyviruses may be feasible.

Natural and experimental transmission

Partial transmission through seed was supposed by S.M. Wong (pers. comm., 1992). Apart from mechanical transmission as purified virus in buffer or with sap from infected plants (van Zaayen et al., 1994), ways of spreading this virus have not yet been investigated.

Control methods

See AlMV for meristem culture. As AlSV has most probably been present for a long time and as it can be detected with antiserum to AlMV, it is almost certain that it has been included in earlier experiments (Hakkaart and Versluijs, 1985, 1988) and in commercial practices with meristem culture. As this virus probably is seed-borne, an additional measure would be to free the breeding parents from AlSV by the application of meristem culture.

Although all modes of transmission of this virus are not yet known, it would be advisable to apply the sanitary measures and other recommendations mentioned for AlMV.

Alstroemeria carla virus (AlCV)

AlCV was reported to be closely related to lily symptomless virus (LSV) and was considered to be a host-adapted strain of LSV (Phillips and Brunt, 1986). Recently it was found to be identical to LSV (A.F.L.M. Derks, pers. comm., 1994). LSV is described in Chapter 26 herein.

Cucumber mosaic cucumovirus (CMV)

CMV was first reported from alstroemeria by Hakkaart (1986). Because of its occasional occurrence in alstroemeria, very little is reported on CMV-infection of this crop. (See Chapter 5 for a fuller description.)

Symptoms caused in alstroemeria are not described and are also not clear, as CMV, if it is present, usually occurs together with other viruses. For control, see under AlMV, although it is not known whether alstroemeria plants can be freed from the virus through meristem culture. It is also not known whether CMV is seed-borne and is transmitted by aphids in alstroemeria.

The alstroemeria strain of CMV can be detected in ELISA since good antisera are available (e.g. Maat, 1980).

Tomato spotted wilt (TSWV)/Impatiens necrotic spot (INSV) tospoviruses

Symptoms

Alstroemeria plants can be systemically infected with TSWV or, although very seldom, with INSV. The first symptoms of TSWV are large (5–15 mm), oval to round chlorotic spots on the leaves. These spots later become necrotic. In a later stage very characteristic symptoms may appear: necrotic streaks, starting from the leaves, are visible on the petioles and extend through these to the stem, which becomes necrotic and breaks. As such symptoms usually occur in the top of the plant, the upper end topples over. Infected plants are sometimes stunted, and leaves and flowers may be deformed. There is a great variability in sensitivity for TSWV between different cultivars.

INSV causes other symptoms. On leaves of a botanical variety large (5–20 mm), oval to round chlorotic spots consisting of many parallel concentric circles (Fig. 18.3) were visible. Other leaves showed, in addition to such spots, many necrotic streaks and spots. Stem necrosis and toppling upper ends of plants were not observed.

Fig. 18.3. A botanical alstroemeria species with symptoms of Impatiens necrotic spot virus (INSV). Photograph: IPO-DLO, Wageningen.

Description of the viruses

See Chapter 5, tospovirus. Hakkaart and Versluijs (1985) reported the isolation of TSWV from alstroemeria as early as 1977.

Geographical distribution and damage caused

TSWV occurs widespread and may occur in alstroemeria crops wherever these are grown adjacent to, for instance, infected vegetable crops like tomato and cucumber, or where alstroemerias are grown outside.

The damage caused is strongly dependent on the cultivar grown, but is in most cases limited to focuses of infection, provided that effective measures are taken.

Methods of assay

TSWV is mechanically transmissible from alstroemeria to indicator plants (Hakkaart and Versluijs, 1985). For (large-scale) detection with ELISA of tomato spotted wilt virus, a polyclonal antiserum to its nucleocapsid protein is used (BR-01; De Avila et al., 1990). TSWV and INSV are currently detected in the Netherlands with a combined antiserum prepared by D. Peters.

As both viruses cannot be detected in ELISA long before the symptoms are visible, visual inspection is most important.

Natural and experimental transmission

Although TSWV is mechanically transmissible to indicator plants (Hakkaart and Versluijs, 1985), trials to inoculate alstroemeria with sap from diseased plants failed (F.A. Hakkaart, pers. comm., 1979). Thrips, however, will readily transmit the virus, probably especially the Western flower thrips, *Frankliniella occidentalis*. This would explain the characteristic toppling over of the upper ends of the plants, as the thrips starts its campaign in the flowers at the top. Accurate data, however, e.g. on thrips species involved, are not known.

Control methods

The crop should be regularly subjected to visual inspection for symptoms and for thrips damage. Thrips should be adequately prevented. If an infection is noticed, infected and adjacent plants should be rogued, and thrips should be controlled and prevented. Weeds outside the glasshouse can be host plants for TSWV; these should be removed. Wherever possible, resistant cultivars should be grown.

The application of meristem culture to TSWV-infected plants can result in TSWV-free plants (J.M.A. Versluijs and A. van Zaayen, unpublished results).

Final Remarks

Alstroemeria may harbor other viruses, some of which are not yet characterized, but which should be mentioned here as their importance and the damage caused is not yet known:

- an ilarvirus, which was isolated from field-grown "ligtu" hybrids in Essex with faint leaf chlorosis (Phillips and Brunt, 1986). This is the only report of such a virus. "Ligtu" hybrids are hardly grown nowadays;
- tobacco rattle virus (TRV), reported by Hakkaart and Versluijs (1985) and by Phillips and Brunt (1986). The symptoms, which were more severe than those of AlMV, were malformed leaves and yellow flecks and rings. TRV, which is

nematode-transmitted (see Chapter 5, tobravirus), will be limited mainly to outdoor grown alstroemeria;

- arabis mosaic virus (F.A. Hakkaart, pers. comm., 1980). Symptoms were not described. This virus is also nematode-transmitted, so it will be limited to outdoor grown alstroemeria like TRV;
- freesia mosaic virus (FreMV), which was reported to occur naturally in *Alstroemeria* sp. by Bellardi et al. (1992). The report was mainly based on decoration experiments. As mentioned above, AlSV showed a very strong relationship with FreMV in decoration experiments with antiserum to this virus;
- a third potyvirus: in experiments with sap of an infected alstroemeria plant, decoration was occasionally obtained of potyviruslike particles with antiserum to Ornithogalum mosaic virus. Such particles did not react with antiserum to AlMV. Young virus-free *Ornithogalum* plants could not be infected by inoculation through aphids from an alstroemeria plant containing such virus particles (A.F.L.M. Derks and A. van Zaayen, unpublished results);
- a virus 1300–1800 nm in length, which has been found in some plants with flower color breaking (De Blank et al., 1994); a closterovirus has earlier been reported by Bellardi and Bertaccini (1991);
- a rhabdovirus, which may give very serious stunting, chlorosis, and leaf curl upon inoculation of *Nicotiana benthamiana*, was detected in *Alstroemeria caryophylla* with small necrotic spots on the leaves (A. van Zaayen, unpublished results).

Selected Literature

Bayer, E. 1987. Die Gattung Alstroemeria in Chile. Mitteilungen der botanischen Staatssammlung München 24:1–362.

Bellardi, M.G. and Bertaccini, A. 1991. Indagine preliminare sulle virosi dell'Alstroemeria in Italia. Atti "Giornata di studio Alstroemeria", San Remo, pp. 115–124.

Bellardi, M.G., Vibio, M. and Bertaccini, A. 1992. Natural occurrence of freesia mosaic virus in *Alstroemeria* sp. Plant Dis. 76:643 (abstract).

Brunt, A.A. and Phillips, S. 1981. Alstroemeria. Annual Report of the Glasshouse Crops Research Institute, Littlehampton 1979, pp. 151–152.

De Avila, A.C., Huguenot, C., Resende, R. de O., Kitajima, E.W., Goldbach, R.W. and Peters, D. 1990. Serological differentiation of 20 isolates of tomato spotted wilt virus. J. Gen. Virol. 71:2801–2807.

De Blank, C.M., van Zaayen, A. and Bouwen, I. 1994. Towards a reliable detection of Alstroemeria mosaic virus. Acta Hortic. 377:199–208.

Hakkaart, F.A. 1986. Virusziekten van Alstroemeria. Jaarverslag Instituut voor Plantenziektenkundig Onderzoek 1985:96.

Hakkaart, F.A. and Versluijs, J.M.A. 1985. Viruses of Alstroemeria and preliminary results of meristem culture. Acta Hortic. 164:71–75.

Hakkaart, F.A. and Versluijs, J.M.A. 1988. Virus elimination by meristem-tip culture from a range of Alstroemeria cultivars. Neth. J. Plant Pathol. 94:49–56.

Linnaeus, C. 1762. Planta *Alströmeria* . . . (J.P. Falck) Amoenitates Academicae 6: 247–262, Dissertationes CXIV. Stockholm, Leipzig. (from: Mitteilungen der Botanischen Staatssammlung München, 1987)

Lopian, R. 1993. EPPO's contribution to a harmonized Europe: certification schemes to improve the health status of ornamental plants. In: BCPC Monograph No. 54, Plant Health and the European Single Market, D. Ebbels (ed.). BCPC/AAB/BSPP, UK, pp. 273–278.

Maat, D.Z. 1980. Serologie van virussen als ziekteverwekkers in land- en tuinbouwgewassen; Alstroemeria. Jaarverslag Instituut voor Plantenziektenkundig Onderzoek 1979: 77.

Maat, D.Z. 1983. Serologie van virussen als ziekteverwekkers in land- en tuinbouwgewassen; Alstroemeria. Jaarverslag Instituut voor Plantenziektenkundig Onderzoek 1982: 80.

Maat, D.Z. 1989. Detection and biotechnology department. Annu. Rep. Res. Inst. Plant Prot. 1989:44.

Milne, R.G. (Ed.) 1988. The Plant Viruses. Volume 4: The Filamentous Plant Viruses. Plenum Press, New York and London.

OEPP/EPPO. 1991. General scheme for the production of certified pathogen-tested vegetatively propagated ornamental plants. Bull. OEPP/EPPO Bull. 21:757.

OEPP/EPPO. 1993. Certification scheme. Nursery requirements-recommended requirements for establishments participating in certification of fruit or ornamental crops. Bull. OEPP/EPPO Bull. 23:249–252.

Phillips, S. and Brunt, A.A. 1986. Four viruses of Alstroemeria in Britain. Acta Hortic. 177: 227–233.

Produktschap Voor Siergewassen en Produktschap voor Groenten en Fruit, 1994. Tuinbouwstatistiek 93. The Hague, the Netherlands.

Rønde Kristensen, H., Jørgensen, H.A. and Jørgensen, J. 1962. New attacks of virus diseases, fungi and pests 1961. Tiddskr. for Planteavl. 66: 605–607.

Schönfelder, M., Bürstenbinder, U., Maiss, E., Frötschl, R., Lesemann, D.-E. and Adam, G. 1993. Sequence analysis, serology and cytoplasmic inclusions identify Alstroemeria mosaic virus (AlMV) as a distinct potyvirus species. Abstracts IXth Int. Congr. Virol., Glasgow, p. 369.

Van Ruiten, J.E.M. 1988. Production and trade of healthy propagating material: a need for a quality control system. Acta Hortic. 234:523–528.

Van Ruiten, J.E.M. 1993. Quality inspection and certification schemes for ornamental plants and fruit crops in the Netherlands. In: BCPC Monograph No. 54, Plant Health and the European Single Market, D. Ebbels (ed.). BCPC/AAB/BSPP, UK, pp. 83–88.

van Zaayen, A., De Blank, C.M. and Bouwen, I. 1994. Differentiation between two potyviruses in Alstroemeria. Eur. J. Plant Pathol. 100:85–90.

van Zaayen, A., Van Eijk, C. and Versluijs, J.M.A. 1992. Production of high quality, healthy ornamental crops through meristem culture. Acta Bot. Neerlandica 41:425–433.

Wong, S.M., Reiser, R.A. and Horst, R.K. 1992. Characterization of a new potyvirus from *Alstroemeria* in the USA. Abstracts 8th International Symposium on Virus Diseases of Ornamental Plants, Prague, p. 42.

Bulb and Corm Crops

19

Anemone

Alan A. Brunt
Horticulture Research International, Littlehampton, Sussex, UK

Introduction

Anemones are popular ornamental plants, mainly because individual species can be grown in different environmental conditions. Of the many species that are cultivated, *A. coronaria* is by far the most important as it can be grown either as a garden plant or forced for cut flower production. Indeed, most commercial cultivars (including cvs Wicarbri, San Piran, Tetramone, and Jerusalem) originate from *A. coronaria* cv. de Caen.

Anemones are grown commercially mainly in western Europe (especially the Netherlands, France, Italy, and England), the US, and Israel. Of the estimated total world production of 150–170 ha/annum, approximately 80 ha are grown in Italy, 60 ha in France, and 26 ha in England (Meynet, 1993). Approximately half of the tubers are produced for garden use and half for commercial flower production.

The Viruses

The following six viruses — three with insect vectors and three with soil-borne vectors — are known to occur in naturally-infected anemones.

Insect-transmitted viruses

Turnip mosaic potyvirus (TuMV) (aphid-transmitted virus)

A disease of anemones first found in southwest England in 1953 was shown to be induced by a virus which, although then designated anemone mosaic virus (Hollings, 1957), is now recognized as a strain of TuMV. Infected plants, most of which originated from infected corms imported from the Netherlands, were usually stunted and had slightly mottled and distorted leaves and "broken" flowers; the leaf chlorosis in some plants, however, was very conspicuous. In winter the petioles of infected plants developed necrotic pitting and streaking which eventually resulted in dieback and death of some plants. In some years the virus occurred in epidemic proportions. In the affected areas the virus was also found to occur naturally in *Senecio vulgaris*, *Capsella bursa-pastoris*, *Mattiola incarna*, and *Ranunculus* sp. The virus has also been found to occur naturally in anemones in Israel (Alper et al., 1987); due to the past international sale and exchange of anemone corms and the wide geographical distribution of the virus, it probably also occurs in anemones elsewhere. It is interesting, however, that the virus (*Ranunculus* mottle potyvirus) infecting *R. asiaticus* in the US is distinct from TuMV (Laird and Dickson, 1965, 1966; Elliott et al., 1988).

TuMV and cucumber mosaic virus together were shown by Hollings (1957) to be the probable cause of a disease previously described as anemone "crinkle" or "parsley leaf" (Smith, 1954). Plants infected with the two viruses in complex are stunted and have leaves which, in addition to being mottled, are curled and rugose (Smith, 1954; Hollings, 1957).

TuMV is transmitted from infected to healthy plants in the non-persistent manner by aphids. Although transmissible by numerous aphid species, its most important vectors are *Myzus persicae*, *Aphis fabae*, *Macrosiphum euphorbiae*, and *Acyrthosiphon pisum*. Infection in partially infected stocks is usually attributable to within-crop spread but, as TuMV has an extensive natural host range, it can also originate from infected but unrelated weed and cultivated species. Conversely, infected anemones can also act as sources of infection for any of numerous other susceptible crop species.

Identification — Like the other five viruses, TuMV is mechanically transmissible without difficulty from infected anemones to herbaceous diagnostic hosts. It induces conspicuous chlorosis in systemically infected leaves of *Brassica pekinesis* (and many other *Brassica* spp.), *Spinacia oleracea*, and *Cichorium endiva* and local lesions in *Chenopodium quinoa*. It is, however, more quickly identified by serological procedures, of which the most useful are ELISA and immunosorbent electron microscopy (ISEM).

Cucumber mosaic cucumovirus (CMV) (aphid-transmitted virus)

This virus has long been known to infect anemones in the UK (Smith, 1954, 1955) and to be present in corms imported into the UK from the Netherlands (Hollings, 1957). More recently, it has been reported also in anemones grown in Italy (Lisa and Conti, 1981; Lisa et al., 1990) and Israel (Alper et al., 1987). Plants infected with this virus alone have leaves which are conspicuously chlorotic and "broken" flowers with fine white or pale lines and streaks in red and blue petals (Hollings, 1957). In some years the incidence of infection in the UK has been very high.

The virus together with TuMV, as described above, causes the "crinkle" or "parsley leaf" disease (Hollings, 1957).

CMV, like turnip mosaic virus, has a very extensive natural host range and wide geographical distribution and is readily transmissible in the non-persistent manner by numerous aphid species. It is also seed transmitted in some host species, although probably not in *A. coronaria*.

Identification — The virus is mechanically transmissible without difficulty from anemones to diagnostic hosts. It induces local lesions only in some species such as *Chenopodium amaranticolor* and *C. quinoa*, and systemic leaf chlorosis and distortion in many cucurbitaceous (such as cucumber and melon) and solanaceous hosts (especially *Nicotiana* species). It is, however, quickly identified by serological procedures such as ELISA and, although less reliably with some strains, by ISEM.

Tomato spotted wilt tospovirus (TSWV) (thrips-transmitted virus)

It has long been known that TSWV is experimentally transmissible to anemone (Hollings, 1965). More recently, however, it has been found to occur naturally in anemone crops grown in several countries including France, Italy, the Netherlands, and Germany (Lisa et al., 1990; Marchoux et al., 1991; Adam and Lesemann, 1994; Verhoeven and Roenhorst, 1994). These recent incidences of the virus were all correlated with the occurrence in high populations of the efficient vector *Frankliniella occidentalis*. The virus induces vein clearing and necrotic spots in severe stunting of plants.

Identification — TSWV induces local necrotic lesions and systemic chlorosis and necrosis in a range of solanaceous species of which *Nicotiana rustica*, *N. tabacum*, *N. glutinosa*, and *N. clevelandii* are the most useful. Diagnostic antisera to strains of the virus are available, and their use permits rapid identification by ELISA of the virus, both in infected plants and viruliferous thrips.

Soil-borne viruses

Tobacco necrosis necrovirus (TNV)

This virus apparently occurs symptomlessly in anemones and is often detected in roots but more rarely in leaves (Smith, 1955; Hollings, 1957). In addition to its occurrence in field-grown anemones in England, it has also been isolated from corms imported from the Netherlands (Hollings, 1965) and in crops in Israel (Alper et al., 1987) and Italy (Lisa et al., 1990).

Identification — TNV induces characteristic local necrotic lesions but usually no subsequent systemic infection in mechanically inoculated plants of *Phaseolus vulgaris*, *N. tabacum*, and *N. glutinosa*. It is, however, easily and quickly identified by ELISA or any other serological procedure.

Tobacco ringspot nepovirus (TobRSV)

A necrotic disease was first found in 1957 in anemones grown in southwest England (Hollings and Stone, 1963). It was shown to be induced by a virus which, although originally designated anemone necrosis virus (Hollings and Stone, 1963), was later identified as a strain of tobacco ringspot nepovirus (Hollings, 1965).

Field-infected plants usually occurred in small discrete patches approximately 1–2 meters in diameter although single plants were found scattered throughout the crop. The young leaves and then the petioles of infected plants became necrotic, and such plants failed to flower and often died prematurely; some, however, partially recovered and had less severe symptoms which included slight leaf mottling, necrosis, and dwarfing.

The virus is mechanically transmissible to anemones with difficulty; although some experimentally infected plants developed very faint leaf chlorosis and were stunted, others developed severe leaf chlorosis and necrosis and later died (Hollings, 1965).

TobRSV is transmitted from infected to healthy plants by nematodes (mainly *Xiphinema americanum*). Unlike TuMV and CMV which, under favorable conditions for aphid reproduction and movement, can reach epidemic proportions, TobRSV occurs in anemones only locally in southwest England and is usually of such a low incidence to be economically unimportant.

Identification — The virus, which is readily isolated from infected anemones, induces chlorotic local lesions and subsequent systemic infection in *Chenopodium amaranticolor* and *C. quinoa*. Serological procedures, of which ELISA and ISEM are the most frequently used, allow its rapid identification.

Raspberry ringspot nepovirus (RRSV)

This virus was detected in anemone corms imported into Japan from the Netherlands (Nishio et al., 1980). Like TobRSV, its vectors are nematodes (mainly tje *Longidorus* and *Xiphinema* species).

Identification — This virus induces symptoms in herbaceous hosts similar to those of TobRSV. Like the latter, it is also readily identified by serological procedures, especially ELISA and ISEM.

Control

With the possible exception of RRSV, the viruses have extensive natural host ranges. Initial infection, therefore, can originate from any of their numerous foci. However, most subsequent spread probably occurs within anemone crops.

The viruses are best controlled by using virus-free planting stock and promptly removing any anemone plants that become infected; as most of the viruses (especially the two aphid-borne viruses), TSWV, TNV, TobRSV have extensive natural host ranges, infected weeds or other foci of infection should be removed. It is also prudent to cultivate anemones as far as possible from cruciferous crops in which TuMV is likely to occur. Infection by the soil-borne viruses can be minimized by growing crops on land with low populations of their respective fungal and nematode vectors.

Other Virus-like Diseases

A disease described as alloiophylly was described in Germany almost 70 years ago (Klebahn, 1926) and, although sometimes attributed to virus-infection, might well be induced by a *Mycoplasma*-like organism.

A disease described as anemone brown ring was found in 1958 in southwest England; although then attributed to virus infection (Hollings, 1958), the putative virus has yet to be isolated and characterized.

Selected Literature

Adam, G. and Lesemann, D.-E. 1994. Identification and characterisation of a new tospovirus isolate from *Anemone* coronaria by TAS-ELISA and electron microscopy. Acta Hortic. 377: 129–139.

Alper, M., Levy, S. and Loebenstein, G. 1989. Virus diseases in anemones. Hasadeh 65:105 (in Hebrew).

Elliott, M.S., Zettler, F.W., Gallegati, M. and Ko, N.-J. 1988. A potyvirus infecting *Ranunculus asiaticus*. Acta Hortic. 234:39–43.

Hollings, M. 1957. Anemone mosaic — a virus disease. Ann. Appl. Biol. 45:44–61.

Hollings, M. 1958. Anemone brown ring — a virus disease. Plant Pathol. 7:95–98.

Hollings, M. 1965. Anemone necrosis, a disease caused by a strain of tobacco ringspot virus. Ann. Appl. Biol. 55:447–457.

Hollings, M. and Stone, O.M. 1963. Anemone necrosis virus (ANV). Rep. Glasshouse Crops Res. Inst. for 1962, p. 89.

Klebahn, M. 1926. Die Alloiophyllie der *Anemone nemorosa* und ihre vermutliche Ursache. Planta 1:419–440.

Laird, E.F. and Dickson, R.C. 1965. Certain biological properties of a virus from *Ranunculus asiaticus*. Plant Dis. Rep. 49:449.

Laird, E.F. and Dickson, R.C. 1967. Ranunculus mottle virus: field incidence and insect relationships at Ponto, California. J. Econ. Entomol. 60:10–13.

Lisa, V. and Conti, M. 1981. Isolation of the necrosis virus from tobacco and the cucumber mosaic virus from *Anemone coronaria* L. cultivated in Liguria. Colture Protette 10:53–56.

Lisa, V., Vaira, A.M., Milne, R.G., Luisoni, E. and Rapetti, S. 1990. Tomato spotted wilt virus in five crops cultivated in Liguria. Informatore Fitopatologico 40:34–41.

Marchoux, G., Gebre Selassie, K. and Villevielle, M. 1991. Detection of tomato spotted wilt virus and transmission by *Frankliniella occidentalis* in France. Plant Pathol. 40:347–351.

Meynet, J. 1993. Anemone. In: The Physiology of Flowerbulbs, A. de Hertogh and M. Le Nard (eds.), Elsevier, Amsterdam.

Nishio, T., Kawai, A. and Kobayashi, T. 1980. Raspberry ringspot virus transmitted through seed of *Spinacia oleracea* detected in anemone imported into Japan from Holland. Res. Bull. Plant Prot. Serv. Jpn. 17:35–42.

Smith, K.M. 1954. Recent work on plant viruses. Scientific Hortic. 11:100–103.

Smith, K.M. 1955. Recent work on plant viruses. NAAS Quarterly Review 28:153–161.

Verhoeven, T.Th. J. and Roenhorst, J.W. 1994. Tomato spotted wilt virus: ecological aspects in ornamental crops in the Netherlands from 1989 up to 1991. Acta Hortic. 377:175–182.

20

Aster

A. Gera and J. Cohen
Department of Virology, Agricultural Research Organization, The Volcani Center, Bet Dagan 50250, Israel

Introduction

The genus *Aster* is a member of the Compositae family (Asteraceae). It consists of 500 species of herbaceous perennials, sometimes woody at their bases. A very few are annuals. The genus name comes from the Latin *aster*, meaning a star. Most species are natives of North America; others occur in Europe, Asia, and South Africa (Schwabe, 1985). Several species and their hybrids are commonly grown in gardens, others are now grown commercially for cut flower production.

Under commercial conditions, Aster is propagated by division of the rootstock with its spreading basal shoots and rhizomes. It is also possible to use stem cuttings and even detached leaves. The continuous vegetative propagation of

Contribution No. 1461-E, 1994 series from the Agricultural Research Organization, The Volcani Center, Bet Dagan, Israel.

infected Aster may result in the spread of viruses through propagules. The most common viral diseases are caused by tomato spotted wilt virus (TSWV), cucumber mosaic virus (CMV) and a potexvirus.

The Viruses

Tomato spotted wilt tospovirus (TSWV)

On Aster, TSWV causes necrotic line patterns, ringspots on leaves, malformation, and flower distortion (Fig. 20.1). Infected plants are stunted and exhibit systemic chlorosis. The necrotic lines persist as long as the affected leaves live; other symptoms may disappear. The cultivar "Painted Lady" is the most severely damaged. Symptoms are more severe in cool weather. TSWV in Aster has been reported in Europe (Marchoux and Gebre-Selassie, 1991) and recently in Israel (Gera, unpublished).

Cucumber mosaic cucumovirus (CMV)

The virus has been repeatedly reported in Korea. Naturally infected *Aster yomena* plants show leaf mottle and severe mosaic (Choi et al., 1990). In aphid transmis-

Fig. 20.1. Leaves of Aster naturally infected with tomato spotted wilt virus (TSWV) showing mild mosaic and necrotic spots.

sion tests conducted in Korea, *Myzus persicae* transmitted the virus with a frequency of 30%.

Argentine plantago potexvirus (APlaV)

A potexvirus with particles 538 nm long was isolated from *Plantago major*, *Taraxacum officinale*, and *Callistephus chinensis* (Aster) in Argentina. The virus, provisionally named APlaV, has a wide host range and produces systemic symptoms in several species of Chenopodiaceae, Compositae, and Solanaceae. Infected Aster exhibits severe mosaic and leaf malformations. The coat protein of APlaV has a molecular weight of 21.6 kDa. The virus is serologically related to papaya mosaic, plantago severe mottle, and Boussingaultia mosaic viruses (Gracia et al., 1983). It has been reported only from Argentina.

Aster chlorotic stunt carlavirus (AsCSV)

A carlavirus with particles 590–640 nm long and 13 nm in diameter was isolated from *Aster novae-angliae* in Ontario (Stobbs and van Schagen, 1991). It was found in one commercial greenhouse containing cultivars of *Aster* spp. The virus was tentatively named AsCSV. It is readily transmitted mechanically and by several aphid species between Aster and Chenopodium spp. but is not seedborne. Infected plants are stunted and show systemic chlorosis and rosetting. Apical leaves are distorted, often showing necrotic flecking. The RNA is single-stranded (Mr 2.78 MDa). The coat protein was found to migrate in PAGE as a single component of 35.4 kDa. The virus did not react with antisera prepared against helinum virus S and carnation latent virus. A similar carlavirus with a narrow host range was found in Israel (Gera, unpublished).

Tobacco rattle tobravirus (TRV)

Symptoms consist of retarded growth, wrinkled curly leaves, thickened chlorotic rootlets, and deformed flowers (Komuro et al., 1971). The virus is easily transmitted by mechanical means to a wide host range. In nature, it is transmitted by nematodes (*Trichodorus* spp.).

Bidens mottle, oat blue dwarf, aster ringspot, and potato virus N have occasionally been detected in *Aster* spp.

Identification and Assay

All viruses infecting Aster spp. are readily transmissible by mechanical means to herbaceous indicator plants. *Chenopodium* spp. are reliable local lesion host for APlaV, AsCSV, CMV, and TRV. AsCSV has a narrow host range, infecting *C. amaranticolo* and *C. quinoa* only. TSWV induces necrotic local lesions on *Petunia hybrida* and *Nicotiana rustica* 2–4 days after inoculation. All five viruses can be observed in negatively stained leaf-dip preparations. Antisera against TSWV, CMV, and TRV are commercially available and are routinely used by conventional ELISA.

Control

Cultivated aster is propagated vegetatively by division of the rootstocks with the basal shoots and rhizomes. Methods of controlling aster viruses consist of growing healthy mother plants. Propagation material should be collected from virus-free plants. Visual observation for viral symptoms and application of serology for screening of mother plants should be considered.

Control of TRV can be achieved through soil sterilization prior to planting. Control of CMV and the carlavirus, both of which are transmitted non-persistently by aphids, is difficult. Control of aphid populations using nets and chemicals, together with rouging of infected plants, is recommended. Similar precautions should be adapted for TSWV.

Selected Literature

Choi, J.K., Kwon, S.B., Lee, S.Y. and Park, W.M. 1990. Some properties of two isolates of cucumber mosaic virus isolated from *Aster yomena* Makino and *Commelina communis* L. Korean J. Plant Pathol. 6:139–143.

Gracia, O., Koenig, R. and Leseman, D.E. 1983. Properties and classification of a potexvirus isolated from three plant species in Argentina. Phytopathology 73:1488–1492.

Komuro, Y., Yoshino, M. and Ichinoche, M. 1971. Tobacco rattle virus from aster showing ringspot syndrome and its transmission by *Trichodorus minor* Colbran. Rev. Plant Prot. Res. 4:130–132.

Marchoux, G. and Gebre-Selassie, K. 1991. Detection of tomato spotted wilt virus and transmission by *Frankliniella occidentalis* in France. Plant Pathol. 40:347–351.

Schwabe, W.W. 1985. *Aster novae-belgii*. In: A.H. Halevy (ed.), Handbook of Flowering, IV. CRC Press, Boca Raton, FL, pp. 20–41.

Stobbs, L.W. and van Schagen, J.G. 1991. Characterization of a carlavirus isolated from *Aster novae-angliae*. Plant Dis. 75:421–424.

BULB AND CORM CROPS

21

Crocus

A.F.L.M. Derks
Bulb Research Centre, PO Box 85, 2160 AB Lisse, The Netherlands

Introduction

The genus *Crocus* belongs to the family of the *Iridaceae*. About 80 species originate in the Balkans and Turkey, four in the Iberian Peninsula, and three in Russia (Mathew, 1982). The species of horticultural importance are *C. vernus* (in the Netherlands: 279 ha), *C. chrysanthus* (112 ha), and *C. flavus* (75 ha). Crocuses grown in the Netherlands occupied 514 ha in 1993 (PVS/BKD, 1994). Other countries where crocuses are grown are: England, USA, and Japan (acreages are not known). Crocuses are used as pot plants to a limited extent.

C. sativus grown in Mediterranean countries, India, and Pakistan is a source of saffran, which is used to dye and flavor food, in perfumes, and as source of vitamin B2 (Mathew, 1982). The main viruses are discussed below.

The Viruses

Iris severe mosaic potyvirus (ISMV)

After flowering a mosaic of light green, small stripes are visible, particularly on the basal parts of the leaves. The flowers of blue-colored cultivars show dark blue stripes (Fig. 21.1). The virus occurs naturally in *C. vernus* and *C. flavus* (Van Slogteren, 1956; Brunt et al., 1988). ISMV is well detectable with ELISA in leaves. The virus is not reliably detected in the corms directly after harvesting. At the end of the storage period, the virus is detectable in the sprouts but not in the corm itself (Derks and Vink-Van den Abeele, 1982). For more details about the virus, see Chapter 27.

Bean yellow mosaic potyvirus (BYMV)

After flowering, a faint mosaic is visible on the leaves of *C. flavus* which is indistinguishable from the symptoms caused by ISMV (Fig. 21.2; see color plate). Later in the season leaf symptoms disappear (Derks and Vink-Van den Abeele, 1982). *C. sativus* can be latently infected (Russo et al., 1979). The crocus isolate of BYMV is serologically more closely related to the pea strain than to the

Fig. 21.1. Dark blue stripes in flowers of cv. Remembrance infected with iris severe mosaic virus (ISMV).

gladiolus strain (Kaneshige et al., 1991). The virus induces cylindrical inclusions in *C. sativus* (Russo et al., 1979). For a further description of the virus, see Chapter 24.

Tobacco rattle tobravirus (TRV)

Leaves show yellowish green to white spots and stripes (Fig. 21.3) and can be notched. The flowers may show dark spots (Van Slogteren, 1958; Asjes, 1974). TRV is described in Chapter 5. The virus is detectable in the leaves by electron-microscopy and serologically (Van Slogteren, 1958). The vector, *Trichodorus* spp., occurs in sandy and light loamy soil types where most crocuses are grown. Nevertheless, TRV occasionally occurs in crocus (Asjes, 1974). The virus can be gradually lost during vegetative propagation (Asjes, 1983).

Fig. 21.3 Yellow spots and rings on the leaves of *C. flavus* caused by tobacco rattle virus (TRV).

Fig. 21.4. White streaks at the edge of the petals of *C. vernus* infected with arabis mosaic virus (ArMV).

Arabis mosaic nepovirus (ArMV)

The petals of blue-colored cultivars show white streaks extending towards the edge of the petals (Fig. 21.4). The petals are crinkled and wavy, but the leaves remain symptomless (Rydén, 1974). The disease is caused by arabis mosaic virus (Asjes, 1973, 1974). The virus is easily isolated from the petals of infected crocus when sap is rubbed on leaves of *Nicotiana tabacum* cv. Samsun (Rydén, 1974). ArMV occurs occasionally in susceptible cultivars (Asjes, 1974). The virus description is found in Chapter 5.

Control

Symptom-bearing plants have to be rogued as early in the season as possible. This, together with other cultural methods such as keeping soils weed-free, make the rattle disease in crocus controllable (Asjes, 1974). Stocks are field-inspected in the Netherlands by the Bulb Inspection Service (BKD). The maximum virus percentages allowed vary from 0.1% (class I) to 0.3% (class II) for *C. vernus* and *C. flavus*.

Final Remarks

Viruses incidentally detected in crocus are: cucumber mosaic cucumovirus (Bellardi and Pisi, 1987; Brunt et al., 1988) and tobacco ringspot nepovirus (Van Slogteren, 1956; Waterworth and Povish, 1972).

Selected Literature

Asjes, C.J. 1973. Jversl. Lab. Bloembollen Onderz., Lisse 1972:45.

Asjes, C.J. 1974. Soil-borne virus diseases in ornamental bulbous crops and their control in the Netherlands. Agric. Environ. 1:303–315.

Asjes, C.J. 1983. Virus diseases of ornamental bulbs and strategies for control. In: Exotic Plant Quarantine Pests and Procedures for Introduction of Plant Materials. K.G. Singh (ed.), ASEAN Plant Quarantine Centre and Training Institute, Malaysia, pp. 243–271.

Bellardi, M.G. and Pisi, A. 1987. Indagine sulle virosi del Crocus sp. in Italia. Inf. tore fitopatol. 37 (9):33–38.

Brunt, A.A., Derks, A.F.L.M. and Barnett, O.W. 1988. Iris severe mosaic virus. A.A.B. Descriptions of Plant Viruses, No. 338.

Derks, A.F.L.M. and Vink-Van den Abeele, J.L. 1982. Jversl. Lab. Bloembollen Onderz., Lisse 1981:51–53.

Kaneshige, H., Maeda, T. and Inouye, N. 1991. Host range and properties of bean yellow mosaic virus (BYMV) infecting Crocus, and serological relationships among three strains of BYMV. Nogaku Kenkyu 62:225–240.

Mathew, B. 1982. The Crocus. B.T. Batsford, London.

PVS/BKD, 1994. Beplante oppervlakten Bloembollen 1990/1991–1993/1994: Voorjaarsbloeiers. Ed. Produktschap voor Siergewassen, The Hague, The Netherlands.

Russo, M., Martelli, G.P., Cresti, M. and Ciampolini, F. 1979. Bean yellow mosaic virus in Saffron. Phytopathol. Mediterr. 18:189–191.

Rydén, K. 1974. Crocus breaking, a disease caused by an isometric virus. Phytopathol. Z. 80:361–365.

Van Slogteren, D.H.M. 1956. Jversl. Lab. Bloembollen Onderz., Lisse 1955:82.

Van Slogteren, D.H.M. 1958. Ratelvirus als oorzaak van ziekten in bloembolgewassen en de mogelijkheden de infectie door middel van grondontsmetting te bestrijden. Tijdschr. Pl. Ziekten 64:452–462.

Waterworth, H.E. and Povish, W.R. 1972. Tobacco ringspot virus from naturally infected dogwood, autumn crocus and forsythia. Plant Dis. Rep. 56:336–337.

22

Dahlia

Josette Albouy
Station de Pathologie Végétale, INRA, Versailles, France

Introduction

Dahlia is an important bulb crop in Europe, the US, and Japan. It is classified according to its use, size, and flower shape. The cactus and decorative types remain the most widely grown while the acreage of dwarf and semi-dwarf cultivars is increasing. Dahlia plants have been popular since their first introduction from America to Europe at the beginning of the 19th century. Their easy propagation by cuttings, tubers, and seeds and the ease of producing hybrids has led to a rapid increase in the number of cultivars (300–400).

Dahlias are mainly used in gardens but also by florists for cut flowers. Annual dahlia tuber production in Europe runs from February until May in greenhouses for cutting and rooting. Then they are planted in the open field until tubers are harvested.

Diseases of dahlia consist mainly of viral pathogens due to the use of traditional and continuous vegetative propagation of infected plant material and to activity of vectors. The most important viral diseases of dahlia are caused by dahlia mosaic virus (DaMV), tobacco streak virus (TSV), tomato spotted wilt virus (TSWV), and cucumber mosaic virus (CMV). These viruses infect commercial cultivars and cause severe symptoms such as foliar mosaic and necrosis stunting.

The Viruses

Dahlia mosaic caulimovirus (DaMV)

The occurrence of stunting of dahlias in Massachusetts (US) was recorded as early as 1911. Dahlia mosaic virus has been described by Brierley and Smith (1950) and Brunt (1971).

DaMV belongs to the caulimovirus group. Caulimoviruses have isometric particles of 50 nm in diameter. The particles contain a single molecule of circular double-stranded DNA. The particles occur in characteristic cytoplasmic inclusion bodies. Caulimoviruses are transmitted by inoculation with sap, and their natural mode of transmission is by aphids in a non-persistent or semi-persistent manner. A virus encoded protein facilitates aphid transmission. DaMV is efficiently transmitted by many aphid species that commonly infest dahlias.

DaMV is generally considered to be the most prevalent and damaging of the four viruses that infect dahlias; it is a limiting factor in production. It has been found all over the world and probably occurs wherever dahlias are grown. Although it is geographically widespread, it is very limited in its natural host range; natural infection is found only in dahlia species. Other members of the *Compositae*, *Solanaceae*, *Chenopodiacae*, and *Amaranthacae* can be infected by mechanical sap inoculation (Brunt, 1971). Many cultivars are seriously affected. DaMV also often occurs in combination with cucumber mosaic, tobacco streak, or tomato spotted wilt viruses.

Symptoms vary considerably according to the cultivar infected. The different types may be described as tolerant, semi-tolerant, and intolerant (Brierley, 1933). The most characteristic symptoms are mosaic and vein banding; chlorotic bands adjacent to the midrib or veinlets appear on the affected leaf. Shortening of the internodes is also very common in susceptible cultivars, giving a short bushy appearance known as stunting (see Fig. 22.1). Some cultivars also show marked leaf distortion and rugosity. Flowers may be smaller than normal. No change in the coloring of the petals occurs, but a slight decrease in the number of blooms can be observed in diseased plants. Some more tolerant cultivars are only slightly

Fig. 22.1. General stunting (right) in plants infected by dahlia mosaic virus (DaMV).

shortened. The chlorotic symptoms of DaMV frequently become masked during the growing season and are more pronounced on cuttings in the greenhouse. Many cultivars remain almost symptomless throughout the vegetative period.

Cucumber mosaic cucumovirus (CMV)

The second damaging virus affecting dahlias is CMV. It has an extremely wide host range and a worldwide distribution. CMV is one of the widespread plant viruses recorded in more than 700 species of 85 families. It is transmitted by more than 60 species of aphids in a non-persistent manner. Reservoirs of CMV are ornamental and vegetable crops and weeds.

CMV currently infects dahlias but reaches epidemic proportions in some years. Symptoms vary in severity depending on the growth stage of the plant. CMV causes a light mosaic with an extreme narrowing of the leaf, producing the typical "fern leaf" (Fig. 22.2), and in some cultivars this virus causes chlorotic lines across the middle of the leaf, like the outline of an oakleaf (Lawson, 1966) (Fig. 22.3). In some cases flowers may also show these deformations. Sometimes only parts of the infected plant show symptoms. Symptoms are best visible during spring.

Fig. 22.2. Narrowing of leaf lamina. "Fern leaf" caused by cucumber mosaic virus (CMV).

Tomato spotted wilt (TSWV) and impatiens necrotic spot (INSV) tospoviruses

These viruses have attracted attention because of their extremely wide host range, their unusual physical and chemical properties, their vector relationship (TSWV is transmitted by thrips in a persistent way), and their rapid progression in many ornamental and vegetable crops in the last decade.

First described as "Oakleaf disease" or "Dahlia ring spot" based on symptoms (Brierley, 1933), TSWV has been observed and recorded in many different countries for many years. The recent high incidence of TSWV in dahlia cultures is correlated with the development of the efficient thrips vector *Frankliniella occidentalis* and the presence of many other infected crops grown in greenhouses.

Fig. 22.3. Oakleaf symptoms and chlorotic rings in plants. (From Wang.)

According to the cultivar, different types of "ring spot" can be observed. Another typical symptom is a chlorotic line on the leaves. These lines are more or less pronounced and sometimes become necrotic. In some cultivars it is possible to observe on old leaves the appearance of several irregular and sinuous lines or concentric rings (Fig. 22.4; color plate). Later, often at high temperatures, the chlorotic pattern becomes necrotic. In some cultivars young leaves can express a discrete necrotic line along the main midrib and on the edge of the leaf. We may mention the tendency for symptoms to develop rapidly, particularly in the field, and to become masked.

Tobacco streak ilarvirus (TSV)

TSV is a virus with isometric RNA-containing particles of 27–35 nm in diameter. The single-stranded RNA genome is divided into three segments. It has a wide host range infecting mainly woody plants but is not very common. Readily transmitted by manual inoculation and transmission by thrips, (*Frankliniella* spp., *Thrips tabaci*) has been reported twice (Fulton, 1985). TSV is widespread in dahlias; it seems to have no real economic impact on this plant, but it can be a potential danger for other crops.

Infected dahlias are often symptomless or present inconspicuous leaf symptoms or light mottling. In some cases decoloration and deformation of petals occur.

Identification

Symptoms

Some visible symptoms may be useful for identification:

DaMV — mosaic, vein banding, stunting (Fig. 22.5; color plate)
CMV — narrowing of leaf, filiform leaf, mosaic, oakleaf
TSV — inconspicuous
TSWV, INSV — sinuous lines, ring spot, chlorotic lesions on leaves

Light microscopy

Dahlia mosaic infection may be diagnosed by detection of typical inclusions stained by trypan blue (0.5%). Spherical or ovoid intracellular inclusions can be observed in strips from the lower epidermis along the veins of fully expanded leaves. These inclusion bodies consist of an electron-dense matrix in which virus particles are embedded.

Electron microscopy

Virion particles can be observed in negatively stained leaf dip preparations.

DaMV — isometric particles of 50 nm in diameter
CMV — spherical particles of 28 nm in diameter
TSV — isometric particles of 27, 30 and 35 nm in diameter
Tospovirus — irregular particles 70–90 nm bound by a membrane

Biological assays

Dahlia viruses are readily sap transmissible in a buffer containing nicotine sulfate (2.5%) and antioxidants to reduce the inhibitory effect of tanins and the activity of polyphenol oxidases. Suitable indicator plants are:

DaMV	—	*Verbesina encelioides, Chenopodium quinoa*
CMV	—	*Chenopodium quinoa, Cucumis sativus*
TSV	—	*Chenopodium quinoa, C. amaranticolor, Nicotiana tabacum*
TSWV, INSV	—	*Nicotiana rustica, N. benthamiana, Petunia hybrida*

DaMV induces chlorotic or semi-necrotic lesions on *V. encelioides* within 10 days followed by general chlorosis and distortion. On *C. quinoa* DaMV induces local symptoms and systemic chlorosis after 15 days.

CMV induces orange-colored local lesions on *C. quinoa* after 5–8 days, on *C. sativus* systemic mosaic and stunting.

TSV induces depressive local lesions on *C. quinoa* after 5 days and systemic necrosis after 10 days, on *C. amaranticolor* small local lesions 4–5 days followed by systemic distortion, and on *N. tabacum* local necrotic spots or rings and systemic necrotic lines.

TSWV and INSV induce local necrotic lesions on *N. rustica* and *N. benthamiana* followed by systemic patterns and leaf deformation. On *P. hybrida* local necrotic lesions occur after 2–4 days.

Serology

The four viruses can be routinely tested by conventional double-antibody-sandwich type ELISA with commercial reagents. Detection of viruses in dahlia can be improved by the use of nicotine sulfate (5%) in phosphate buffer to increase the sensitivity and reliability of the test, and for DaMV, addition of urea (4 M) to the extract buffer to provoke the disruption of inclusion bodies and the release of the particles (Wang et al., 1988).

Good results have been obtained in dot immunobinding assays (DIBA) on nitrocellulose membranes (Albouy et al., 1994). It is a simple, rapid, and convenient method for testing large numbers of samples. Samples are prepared in 0.1 M Tris pH 9.0.

Nucleic acid probes

A DNA hybridization technique for detection of DaMV was standardized and developed for practical use, using non-radioactive probes (sulfonated) (Wang et al., 1988). This technique, as sensitive as that using a radioactively labelled probe, was adapted for testing *in vitro* plantlets with low virus titer.

Control

Methods of controlling dahlia viruses consist of growing and multiplying selected healthy stocks. Screening of dahlia crops for virus was done for many years by visual selection or biological indexing and is now done by ELISA tests and tissue culture.

Routine visual inspection of dahlias for symptoms gave no satisfactory results due to the masking of symptoms. With the classical production method, production of virus-free dahlia has failed. Application of ELISA to detect the four different viruses and routinely applied tissue culture enables the growers to propagate virus-free material. The introduction in France of Dahlia meristem-tip culture to produce DaMV virus-free dahlias (Morel and Martin, 1952) led to successful procedures in many crops. Dahlia meristem-tip culture is effective for eradicating dahlia viruses and *in vitro* multiplication conditions allowing rapid increase of pathogen tested plantlets (Albouy et al., 1992).

During the period of greenhouse propagation, isolation, roguing, and the control of aphid or thrips vectors are recommended. Prevention of stylet-borne virus spread is possible by spraying mineral oil and pyrethroid. Tolerant cultivars infected by DaMV should be segregated from the healthy stocks. The presence of reservoirs of DaMV in plants other than dahlias should be avoided; this is more difficult for other viruses such as CMV and TSWV which are widespread in other cultures.

After meristem-tip culture, virus indexation must be carried out several times before a plantlet can be confidently used as starting material for micropropagation by nodal segmentation or during the acclimation stage. CMV and TSV are easily eliminated; DaMV and TSWV have a more delayed resurgence period (3–6 months). Maintenance of pathogen-free (bacteria and viruses) plantlets is also possible by establishing a conservatory of healthy mother plants. The scheme of the propagation including tissue culture has a commercial application. Using these techniques tubers can be ready for commercial distribution within the same year. Differences between virus-infected and virus-free plants are evident with respect to qualitative and quantitative impact (cuttings, tubers).

Selected Literature

Albouy, J., Lemattre, M., Wang, W. and Mevel, A. 1992. Production of pathogen-tested dahlia in France. Acta Hortic. 325:781–786.

Albouy, J., Kusiak, C., Louanchi, M. and Wang, W. 1994. Evaluation of rapid ELISA procedures for detecting viruses in propagation stock of pelargonium, dahlia and orchids. Acta Hortic. 377:189–196.

Brierley, P. 1933. Studies on mosaic and related diseases of dahlia. Boyce Thompson Inst. Contrib. 5:235–288.

Brierley, P. and Smith, F.F. 1950. Some vectors, hosts and properties of dahlia mosaic viruses. Plant Dis. Rep. 35:405–407.

Brunt, A.A. 1971. Some host and properties of dahlia mosaic virus. Ann. Appl. Biol. 67:357–368.

Brunt, A.A. 1971. Dahlia mosaic virus CMI/AAB Description of Plant Viruses, No. 51.

Francki, R.I.B. 1979. Cucumber mosaic virus. CMI/AAB Description of Plant Viruses, No. 213.

Fulton, R.W. 1985. Tobacco streak. CMI/AAB Description of Plant Viruses, No. 307.

Ie, T.S. 1970. Tomato spotted wilt. CMI/AAB Description of Plant Viruses, No. 39.

Lawson, R.H. 1966. Oakleaf chlorosis symptomatic of cucumber mosaic infection in dahlia. Phytopathology 56:343–344.

Morel, G. and Martin, C. 1952. Guérison de dahlias atteints d'une maladie à virus. CR. Acad. Sci. 235:1324–1325.

Wang, W., Tronchet, M., Larroque, N., Dorion, N. and Albouy, J. 1988. Production of virus-free dahlia by meristem culture and virus detection through cDNA probes and ELISA. Acta Hortic. 234:421–428.

23

Freesia

Alan A. Brunt
Horticulture Research International, Littlehampton, Sussex, UK

Introduction

Freesia, named after Fr H. Th. Freese, is a genus originating from South Africa but species of which are now cultivated worldwide. Many cultivars of *F. refracta*, the most important of the seven or so species, are grown for cut flower production although some are increasingly being grown as pot plants in North America.

Freesia is especially important as a glasshouse-grown flower crop in the Netherlands, where 293 ha. were grown in 1992/93; in 1993, the sales of freesias in eight flower auctions totalled 145.2 million Dutch guilders (Anon., 1994). However, there is also significant production in Italy (90 ha.), Japan (62 ha.), and the Channel Islands (16 ha.); but the total production in the US, UK, and other producing countries is thought to be less than 10 ha. (Imanishi, 1993).

The Viruses

The following four viruses and one virus-like agent are known to infect freesias.

Bean yellow mosaic potyvirus (BYMV)

This has a very extensive geographical distribution and natural host range, and is described in greater detail as a pathogen of gladiolus and other ornamental species. BYMV probably occurs in freesias wherever the crop is grown and can severely affect plant growth and the yield and quality of both flowers and corms (Koot et al., 1954; Derks and Vink-van der Abeele, 1987). Infected plants usually have leaves which are chlorotic; chlorotic areas, especially in leaves of intolerant cultivars, often later become necrotic. The most conspicuous feature of infection, however, is the production of flowers which are "broken" and distorted with pistols and stamens protruding beyond the corollas of partially opened flowers (Fig. 23.1).

Leaf yellowing and corm necrosis, found occasionally in field-grown crops of cv. Aurora in the Netherlands, are also induced by BYMV (Derks and Vink-van der Abeele, 1987). BYMV has filamentous particles mostly c. 750 nm long and is a member of the potyvirus genus. The virus is transmitted in a non-persistent manner by many aphid species, but *Macrosiphum euphorbiae* and *Myzus persicae* are its important vectors in freesia crops. The virus is readily identified by ELISA and immunosorbent electron microscopy (ISEM); although c-DNA probes and PCR techniques are available for the rapid detection of BYMV, they have not yet been applied to its detection in freesia.

Freesia mosaic potyvirus (FreMV)

FreMV is known to occur naturally only in freesia (Koot et al., 1954) and alstroemeria (Bellardi et al., 1992). Infection by FreMV is often prevalent in older freesia cultivars wherever they are cultivated, presumably due to past international trade in partially infected stocks. Infected plants of many cultivars remain virtually symptomless, but those of blue, purple, and red cultivars such as cvs Princess Marijke, Rose Marie, Arosa, Aurora, and Romany have flowers which are "broken" or mottled but not distorted (Fig. 23.2); cv. Marion is unusual in that infected plants also produce leaves which are conspicuously chlorotic (Koot et al.,

Fig. 23.1. Broken and distorted flowers in freesia induced by bean yellow mosaic potyvirus.

1954). When infected with a complex of FreMV and the freesia leaf necrosis virus-like agent, plants of many cultivars develop a very damaging disease described as severe leaf necrosis or freesia streak (Dorst, 1973; Brunt, 1974). FreMV, with filamentous particles 750-850 nm long, is a typical member of the potyvirus genus. Like BYMV, it is transmitted efficiently in the non-persistent manner from infected to healthy freesias by *M. euphorbiae* and *M. persicae*, and is readily detected and identified by ELISA and ISEM.

Fig. 23.2. Flower motting in freesia cv. Arosa induced by freesia mosaic potyvirus.

Cucumber mosaic cucumovirus (CMV)

Although CMV has a very extensive geographical distribution and natural host range, infection in freesias has been reported only in Italy (Mondelli and Rana, 1985). Infected plants are stunted and produce chlorotic leaves and "broken" and distorted flowers.

CMV, the type member of the cucumovirus genus, has isometric particles mostly c. 30 nm in diameter. It is transmitted naturally by many aphid species; it is also seed-transmitted in many plant species, although this has not been reported to occur with freesia. It, too, is readily identified by ELISA and, dependent on virus strain, by ISEM.

Tobacco rattle tobravirus (TRV)

TRV is reported to occur in freesias in the Netherlands and Hungary (Asjes, 1989;

Nemethy, 1994). Leaves of infected plants have indented margins, are narrow, and sometimes have brown spots; plants are stunted, flower stems have fewer and smaller flowers, roots are poorly developed, and corm yield is much reduced (I. Bouwen and A.F.L.M. Derks, personal communication). A severe disease affects plants infected with both FreMV and TRV in Hungary (Nemethy, 1994).

TRV, the type member of the tobravirus genus with rod-shaped particles of two modal lengths (c. 190 nm and 45–115 nm depending on the virus strain), has a very extensive natural host range and geographical distribution. It is transmitted naturally by nematodes (*Trichodorus* and *Paratrichodorus* species). It is readily identified by serological methods (ELISA and ISEM) and by molecular hybridization.

Freesia leaf necrosis virus-like agent

A disease since described as freesia leaf necrosis (FLN) has probably occurred in the Netherlands for over 40 years (Koot et al., 1954; Dorst, 1973), although it was not fully described until two decades later (Dorst, 1973). The first few leaves produced by infected corms develop angular chlorotic spots which later coalesce and become necrotic (Fig. 23.3). Leaves produced later are usually symptomless. A very severe disease (Fig. 23.4), previously described as "freesia streak" (Casper and Brunt, 1971) or "severe leaf necrosis" (Dorst, 1973), is now known to be induced by the FLN agent in complex with FreMV (Dorst, 1973; Brunt, 1974). The FLN agent is probably transmitted from infected to healthy freesia plants by zoospores of *Olpidium brassicae* (Dorst, 1975).

It has recently been shown by Bouwen (1994) that the FNL agent is mechanically transmissible with difficulty; it is so transmissible from infected freesia to seedlings of *Chenopodium quinoa*, *Nicotiana occidentalis*, and *N. hesperis*, but only during winter and when inocula are prepared with an extractant containing a reducing agent (sodium bisulphite), a copper-chelating agent (sodium diethyldithiocarbamate) and an adsorbent (activated charcoal). The agent is very unstable *in vitro* and so is difficult to purify. However, preparations contain rod-shaped particles 21–25 nm in diameter and lengths of 160–390 nm; they thus resemble those of tobacco stunt, lettuce big vein, and camellia yellow mottle viruses, all of which are also transmitted by zoospores of *O. brassicae*, but the particles need to be further characterized before they can be recognized as those of a causal virus.

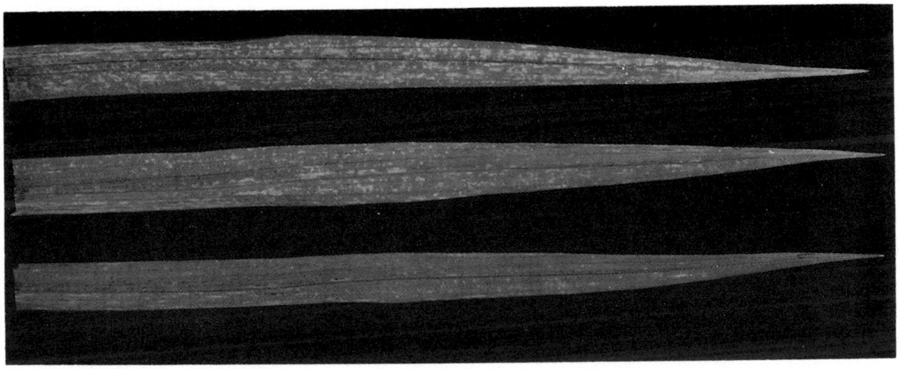

Fig. 23.3. Chlorotic spotting of freesia leaves induced by the freesia leaf necrosis virus-like agent.

Fig. 23.4. Severe freesia leaf necrosis induced by freesia mosaic potyvirus in complex with freesia leaf necrosis virus-like agent.

Infection by the FLN agent can be recognized at present only by the characteristic symptoms it induces in freesia. Better diagnostic procedures are now actively being sought in the Netherlands (Bouwen, 1994).

Control

Virus-induced diseases in freesias are best controlled by planting virus-tested stocks. Such stocks are readily obtained as virus-free plantlets; they are easily produced by meristem-tip culture and then rapidly propagated *in vitro* (Bajaj and Pierik, 1974). The General Netherlands Inspection Service for Ornamental Crops (NAKS) operates a Certification Scheme for freesia in which the health status of crops is monitored for infection with BYMV and FreMV by ELISA and for other viruses and the FLN agent by visual inspection. The European and Mediterranean Plant Protection Organisation has also prepared a Classification Scheme for freesias. To minimize the possibility of virus spread within glasshouse-grown crops, it is necessary to apply insecticides if aphid infestation is likely to occur and to use sterilized tools during crop cultivation and flower harvesting.

Selected Literature

Anon. 1994. Produktsdrop Voor Siergewassen en Produktschap voor Groenten en Fruit Tuinbouwstatistiek 93. The Hague, The Netherlands.

Asjes, C.J. 1989. Tobacco rattle virus and other soil-borne viruses in ornamental bulb culture in The Netherlands. Ann. Appl. Biol. 22:83–86.

Bajaj, Y.P.S. and Pierik, R.L.M. 1974. Vegetative propagation of freesia through callus cultures. Neth. J. Agric. Sci. 22:153–159.

Bellardi, M.G., Bertaccini, A. and Vibio, M. 1992. Natural occurrence of freesia mosaic virus in *Alstroemeria* sp. Plant Dis. 76:643.

Bouwen, I. 1994. Freesia leaf necrosis: some of its mysteries revealed. Acta Hortic. 377: 311–318.

Brunt, A.A. 1974. Freesias. Report of the Glasshouse Crops Research Institute for 1973, p. 117.

Casper, R. and Brunt, A.A. 1971. Freesia streak virus, a new freesia virus in Germany. Nachrichtenblatt des Deutschen Pflanzenschutzdienstes 23:89–90.

Derks, A.F.L.M. and Vink-van der Abeele, J.L. 1987. Leaf yellowing in combination with corm necrosis in freesia caused by bean yellow mosaic virus: factors involved in syndrome development. Neth. J. Plant Pathol. 93:159–166.

Dorst, H.J.M. van. 1973. Two new diseases in freesias. Neth. J. Plant Pathol. 79:130–137.

Dorst, H.J.M. van. 1975. Evidence for a soil-borne nature of freesia leaf necrosis virus. Neth. J. Plant Pathol. 81:45–48.

Imanishi, I. 1993. Freesia. In: The Physiology of Flowerbulbs, A. de Hertogh and M. Le Nard (eds.), Elsevier, Amsterdam.

Koot, Y. van, Slogteren, D.H.M. van, Cremer, M.C. and Camfferman, J. 1954. Virus-verschijnselen in freesias. Tijdschrift over Plantenziekten 60:157–192.

Mondelli, D. and Rana, G.L. 1985. Il virus del mosaico del cetriolo su fresia in Puglia. Informatore Fitopatologica 35:41–42.

Nemethy, I.Z. 1994. Survey on virus diseases of bulbous flowers in Hungary. Acta Hortic. 377:267–274.

24

Gladiolus

A. Stein
Department of Virology, Agricultural Research Organization, The Volcani Center, Bet Dagan 50250, Israel

Introduction

The genus *Gladiolus* belongs to the family Iridaceae. It includes more than 150 species, most of them native to Africa and a few which originated in the Mediterranean area, Asia, and Europe. Cultivars of gladiolus exhibit a great diversity of color, size, shape, flowering time, and bulbing and dormancy behavior. This variability arose from a complex of crosses among several diploid and polyploid botanical species.

The cultivars are classified as spring or summer types. Spring types are short (less than 1 m), cultivated in tropical and subtropical areas, planted in autumn and winter, and they flower in the spring. The most popular spring cultivars arose

Contribution No. 1491-E, 1994 series, from the Agricultural Research Organization, The Volcani Center, Bet Dagan, Israel.

from the diploid *G. nanus* and *G. tubergenii* hybrid groups. Summer types are more vigouros in growth, cultivated mainly in temperate regions, where they are planted in the spring and flower during summer. They originated from complex crosses using a pool of five species from Africa and are referred to as *G.* × *grandiflorus* Hort. Some of the most popular summer cultivars, such as Friendship, Eurovison, Peter Pears, Spic and Span and others, have the ability to produce good quality flowers out of their season during the subtropical winter in Florida and in the Mediterranean region (Cohat, 1993).

Recently, hybrids between species belonging to the two classes have been made and bred for winter flowering in subtropical regions (Cohen and Barzilai, 1991).

Propagation of gladioli is vegetative by cormels and corms. A corm is formed by the swelling of shortened basal internodes of the flowering stalk. A new corm is formed each growing season above the existing corm which shrivels away. The main reproductive organs are cormels which are formed at the tips of branched stolons that develop at the base of the new corm. They are tiny corm structures of 0.5–1 cm in circumference with a hard tunic. A planted cormel gives rise to one shoot that subsequently swells to a small corm and produces several cormels ("cormels from cormels"). The circumference of the corms varies in size from 2–20 cm. After one or two growing seasons, when a corm reaches a certain size, it gives rise to one or more flowering shoots. Corms and cormels experience short or prolonged periods of dormancy after each growing season. Corms can also be produced from seeds which are being used for breeding purposes. However, many cultivars of hybrid origin do not produce viable seeds at all. Propagation can be accelerated by using *in vitro* techniques which include rapid proliferation of shoots and formation of corms (Ziv and Lilien-Kipnis, 1990).

Gladiolus is a popular garden plant and a commercial cut flower crop grown in many countries all over the world. According to the available statistics, during the years 1990–1992, the share of gladiolus in the world cultivated area of flower bulbs was approximately 25% (8500 out of 33,000 hectares). The leading production countries during these years were the Netherlands (2200 ha for corms only) and the US (2500 ha, both corm and flower production) (Cohat, 1993).

The Viruses

Gladioli are affected by a large number of viruses. Bridgmon and Walker (1952) described gladiolus as a virus "reservoir" in the sense of its being a host to four different viruses and a source of infection to other crops. Since then the "reservoir" has increased to no fewer than 14 viruses which have been noted to occur on gladioli, some of them of minor importance. None of the viruses found on gladioli was specific to the crop species or to the Iridaceae. All of them have a wide host range as well as geographic distribution and can be transmitted by external mobile vectors. In several cases specific gladiolus strains were isolated and characterized.

Symptoms and evaluation of prevalence

Symptoms on gladioli vary greatly: flower or leaf color-break, twisted flowers, leaf mosaics, blotching, dwarfing and deformation of the whole plant and corms, or lack of visible symptoms altogether. It is impossible to identify with certainty viruses in gladioli by visual inspection, especially since mixed infections by two or more viruses are the rule. All the commercially grown plants which were tested were found to be infected by bean yellow mosaic virus (BYMV) (Nagel et al., 1983; Bellardi and Pisi, 1985a). It was therefore difficult to attribute specific symptoms to any individual virus other than BYMV which is generally symptomless. The great variety of cultivars of gladioli also contributes to the variability of symptoms. The association of certain viruses with specific symptoms should therefore be approached with caution.

Field data on the prevalence of virus-diseased gladioli in different parts of the world ranged from 20% to approximately 70% when plant symptoms were considered (Heinis, 1954; Beute et al., 1970); however, when tests on symptomless plants were included, 100% prevalence was recorded. Gladiolus is thus a highly susceptible crop that may suffer considerable losses if control measures are not taken.

Main viruses

Two viruses, cucumber mosaic virus (CMV) and BYMV, are the most prevalent in commercial gladioli and present major production problems.

Cucumber mosaic cucumovirus

CMV has a wide host range and geographical distribution. It was first identified as a disease of gladiolus in 1948 in Tasmania (Wade, 1948). Since then CMV has been recognized as the major disease of gladiolus cultivars and presents a serious problem in the production of cut flowers and corms.

Symptoms

Severe color break and deformation of flowers (Fig. 24.1; see color plate) are the most common symptoms associated with CMV infection in gladioli. However, symptoms vary from a few streaks on colored petals to various degrees and forms of distortion of flowers and whole plants (Fry, 1953; Bing, 1972). Chlorophyll break or mosaic is sometimes observed on leaves of young plants but only in a few very susceptible cultivars do they correlate with CMV infections; in general most infections are symptomless until start of flowering. Pitting and discoloration of corms occur in some cultivars.

Primary late infection may remain symptomless for one season or more. When infected corms are planted ("secondary infection"), intensity of symptoms depends on the cultivar and the rate of virus accumulation in the plant.

Identification

CMV in gladiolus can be identified by sap inoculation to indicator plants such as *Cucumis sativus, Vigna unguiculata, Zinnia elegans, Nicotiana* sp., and by reaction with suitable antisera. Flower petal extracts are the most reliable source for sap inoculation (Stein et al., 1979). Different strains of CMV are known to occur on gladioli and probably differ with the region. Thus, from a stock of 100 field-grown gladiolus plants (cv. Peter Pears), four different isolates were characterized on the basis of symptoms on CMV indicators. Strains with low aphid transmissibility were also detected (Kim et al., 1992; Stein, unpublished).

For routine wide-scale detection, ELISA is the most reliable procedure, provided a suitable antiserum is available. Generally, CMV is regarded as a rather poor immunogen so that many of the available antisera may fail to detect it in gladiolus. An antiserum prepared against Price No. 6 strain, according to Loebenstein et al. (1977), enabled reliable detection of CMV in gladioli leaves, corms and cormels (Stein et al., 1979).

Transmission and epidemiology

Many aphid species are capable of transmitting CMV in a non-persitent mode to gladioli even though they do not colonize on the plants (Swenson and Nelson, 1959). Spread of the virus occurs mainly by alate movement between diseased and healthy gladiolus plants in the field. The role of migrating aphids from distant (100–500 m) sources of infection or from neighboring non-gladiolus crops is limited to exceptional situations like abundance of efficient aphid-vectors, appropriate winds, susceptibility of the cultivar, etc. (Swenson and Nelson, 1959; Smith, 1961). Thus, in spite of the wide occurrence of CMV on many crops, infected gladiolus plants are the predominant source for aphid-borne infection of other gladioli in the field. CMV-infected gladioli are also a potential source of infection for other crops. In controlled aphid-confined acquisition and inoculation feeding tests, gladioli plants were found to be a much more efficient donor of CMV than acceptor of the virus from other species (Aly et al., 1986).

Differences in the susceptibility of cultivars to CMV infection were observed in the field and confirmed by confined feeding inoculation with aphids. Two red flowering cultivars, Trader Horn and Eurovision, remained consistently CMV-free under circumstances in which other cultivars such as Peter Pears, Spic and Span, and White Friendship became readily infected (Aly et al., 1986). Most of the commercial cultivars are, however, more or less susceptible to CMV.

Sap inoculation of CMV to a gladiolus cultivar was reported (Sushak, 1977), but no spread of the virus by mechanical contacts in the field could be detected (Stein, unpublished).

The world-wide presence of CMV in gladiolus cultivars is due mainly to the world trade in corms. A few infected plants in the field are sufficient for spread of the disease and may lead to great losses. Thus, with cv. Peter Pears, the presence of 3% infected corms in a planting stock developed under certain conditions in the field to 60% infection within one growing season (Stein, unpublished). It is therefore important to use CMV-tested certified planting material and to take measures to reduce reinfections.

Control

The best way to control CMV is by growing resistant cultivars whenever possible. For the majority of cultivars which are susceptible to CMV, virus-tested planting source material has to be prepared.

CMV-free corms can be selected from regular stocks of most cultivars by thorough testing of individual corms. The selected CMV-free corms are either propagated through cormel production under aphid-free conditions or by micro-propagation followed by *in vitro* corm formation (Logan and Zettler, 1985; Ziv and Lilien-Kipnis, 1990). With both procedures, freedom from CMV has to be redetermined during at least two *ex vitro* growth cycles before corms can be certified as CMV-free nuclear stock. If virus-free source corms are not available, a meristem bud-tip culture has to precede the *in vitro* proliferation steps.

Although young cormels of infected plants were claimed to be CMV-free, this observation could not be confirmed as a general phenomenon.

In the field, reinfection of CMV-free plants can be best controlled by maintaining a distance from CMV-infected gladiolus plants. Partial control was also achieved by using reflective surfaces or coarse net covers combined with a mineral oil spray (Johnson et al., 1967; Aly et al., 1986). These protective measures can be employed on small areas of special quality corms. Escaping the major aphid peaks by early or late planting may be helpful but difficult to achieve in practice.

Bean yellow mosaic potyvirus

BYMV has a wide host range, especially on Leguminosae and Iridaceae species. The virus is transmitted nonpersistently — as are other potyviruses — by many aphid species. BYMV is ubiquitous in gladiolus, and all tested commercial stocks of many cultivars were found almost 100% infected. BYMV causes only mild symptoms in gladioli, and in most plants there are no visual symptoms at all. The effects of BYMV infection on the etiology and symptom expression in the presence of other viruses and non-viral pathogens have not yet been studied, since almost no BYMV-free gladioli are available.

Identification

BYMV in gladiolus leaves can readily be identified and differentiated into subclasses by sap inoculation to seedlings of pea (*Pisum sativum*), bean (*Phaseolus vulgaris*), and broad bean (*Vicia faba*) cultivars. Strains of BYMV that predominate in gladiolus were found to differ from isolates infecting several other species in their host range, and less so in serological properties (Zettler and Abo El Nil, 1977). This separation of pathotypes may limit the range of potential sources of inoculum and leaves infected gladioli to be the main source for reinfecting virus-free propagation material (Nagel et al., 1983). Strains of BYMV in gladioli also seem to differ from area to area. Analysis of gladiolus cultivars with a series of monoclonal antibodies (Jordan and Hammond, 1991) provided evidence for differences between isolates prevalent in each cultivar.

The virus can be reliably detected in leaves and flowers of gladiolus by several methods, irrespective of strain specialization. Polyclonal and monoclonal antisera with wide detection spectra are used in direct double sandwich or in indirect antigen-coated plate ELISA, in immunosorbent electron microscopy (IEM), and in tissue immunoblotting reactions. Dot-blot hybridization and tissue print with cDNA prepared from RNA of a gladiolus isolate of BYMV have also been applied (Hammond and Hammond, 1985). The virus could not be detected in corms of infected plants by any of these methods (in contrast to CMV, which is readily detected in corms). The amounts of BYMV in corms are apparently below the detection limit of these methods. BYMV-RNA can be detected in corms by its amplification through PCR and, if necessary, by further transcription steps (Rosner et al., 1992). If corms are cut-wounded, the virus accumulates in the cut area and can be detected after several weeks by the serology and hybridization methods. The cutting procedure, including storage conditions, has to be adapted to each cultivar (Stein et al., 1986).

Control

As all gladiolus stocks are completely infected with BYMV, elimination of the virus can be accomplished only by meristem or tip culture. The use of an apical corm bud meristem dome with its few subsidiary leaf primordia as explants resulted in elimination of the virus in 60–100% of the surviving offspring. Elimination rates depend greatly on the size of the meristem used or in excision techniques, on the virus strain, and on the gladiolus cultivar. Propagation must be followed by repeated virus testing of individual shoots and cormlets *in vitro* and then, after transplanting, to insect-proof conditions outdoors. Sensitive virus detection methods based on PCR techniques may facilitate the early elimination of infected material in the initial stages of propagation, thus saving labor and expenses (Rosner et al., 1994). Only after this step may virus-free corms, if found, be used as source material for *in vitro* or conventional multiplication.

In the field, BYMV-free plants may rapidly be reinfected if they are not isolated from the regular, BYMV-infected, stocks. Within one growing season, 80% of BYMV-free plants of cv. Peter Pear and 40% of cv Eurovision were found to be reinfected by the virus. Growing the plants under coarse nets reduced reinfection greatly (Stein-Rieger et al., 1990).

Production of BYMV-free gladioli is so far limited to a few commercial stocks in several gladioli-growing centers. Data on the performance of these stocks as compared with regular (entirely infected) ones are not definitive and conclusions have to await additional experimentation.

Other viruses

Most of the viruses listed below are of limited distribution in gladioli. However, under certain circumstances they can spread and cause a severe effect on gladioli and be a source of infection to other crops.

Tobacco ringspot (TobRSV) and tomato ringspot (TomRSV) nepoviruses

These viruses consist of isometric particles with a diameter of about 28 nm. In nature these viruses occur mostly on a wide range of perennial crops. Transmission is by the nematode *Xiphinema americana* and related species. For TobRSV non-nematode vectors such as grasshoppers and aphids have been suggested. Natural spread of the virus in gladioli has been reported in North America where the vectors occur. TomRSV was associated in North Carolina (US) with short flowers and stunted growth (Bozarth and Corbett, 1957). TobRSV may induce milder symptoms or be symptomless (Bing, 1972). Both viruses were probably disseminated from North America to other countries in infected planting material. They were reported from Japan in imported corms (Fukumoto et al., 1982). TobRSV was also reported from Australia (Randles and Francki, 1965), Italy (Bellardi and Marani, 1985), Holland (Asjes, 1970), and Iran (Kaniran and Izadpanah, 1982). TobRSV can be transmitted on tools used during the harvesting of flowers and corms (Brierley and Smith, 1960). A low rate of transmission by gladioli seeds was reported for TobRSV and TomRSV (Sushak, 1977).

Both viruses are readily sap transmitted to diagnostic indicators by which they cannot be distinguished easily from each other. Both can be readily detected and differentiated serologically.

Three other nepoviruses were reported to occur on gladiolus.

Tomato black ring virus (TBRV)

The virus is transmitted by free-living nematodes of the genus *Longidorus*. It has not been reported outside of Europe. TBRV in gladiolus was first reported from Poland (Kaminska, 1978) with symptoms occurring on leaves and flowers. In Italy TBRV was isolated from symptomless plants grown from imported corms (Bellardi and Pisi, 1985b). TBRV can be serologically detected and easily distinguished from other nepos.

Arabis mosaic virus (ArMV) and strawberry latent ringspot virus (SLRV)

These two viruses have been identified in gladioli in a few cases in Italy (Bellardi et al., 1986; Bellardi and Pisi, 1985a).

Tobacco rattle tobravirus (TRV)

This virus has tubular particles of two lengths, 190 and 45–115 nm. TRV is transmitted by nematodes of the Trichodoridae family. Most strains are also sap-transmissible and have a wide host range. Sporadic occurrence of the virus in

Fig. 24.2. Notched leaf margins caused by tobacco rattle virus.

gladiolus was reported from Holland, Israel, Egypt, and Poland. TRV causes notched leaf blade margins (Fig. 24.2), but in general the damage to gladioli is minimal.

Tomato spotted wilt tospovirus (TSWV)

This virus was found on gladioli in South Australia in plants with chlorotic degeneration. A sudden appearance and high incidence of the virus were noted in a single season, although TSWV was present in the area constantly on other crops (Lee et al., 1979). TSWV was also reported to occur on gladioli in the US and in France.

The following viruses have occasionally been found to infect gladioli: Tobacco streak virus, isolated in Italy by Bellardi et al. (1987) from plants with stunted flowers and unique necrotic spots on the leaves (Fig. 24.3; see color plate); tobacco mosaic virus in the Netherlands and Japan; tobacco necrosis virus in Russia; soybean mosaic virus in North Carolina; and narcissus latent virus in the Netherlands.

Mycoplasma-like organisms (MLOs)

Symptoms typical of MLO infections (referred to earlier as "aster yellows") have been reported to occur on gladioli and to be transmitted by the leafhopper *Macrosteles sexnotatus*.

Spiral growth of the flower spike is the most prominent symptom. In sieve tubes and phloem-parenchyma cells of flowers bearing this symptom, MLOs were observed by transmission electron microscope (Bertaccini and Marani, 1980). Infected corms may not emerge or may give poor shoots with a "grassy top" appearance. Epidemic spread of MLO-like symptoms is rare and can be controlled by eradicating the vector.

Concluding Remarks

Viral infections, frequently in combination with fungal diseases (Beute, 1970), affect gladiolus corm and flower production. These infections account in part for the exceptionally high cultivar turnover rate in this crop (Logan and Zettler, 1985). Therefore, the use of virus-tested planting material is beneficial. Specialization in mass production of high quality, pathogen-free corms and cormels is predicted. Mechanized corm production in isolated areas assisted by laboratory and insect-proof facilities should enable the supply of planting material for the gladiolus industry worldwide.

The search for genetic resistance to the major viruses of gladioli is a long-term approach to the problem. Introduction of resistance into commercial or new cultivars can be achieved by sexual crosses or by manipulating the expression of sequences of relevant viral genomes to obtain transgenic resistant plants.

Selected Literature

Aly, R., Stein, A., Levy, S., Raccah, B. and Loebenstein, G. 1986. Spread and control of cucumber mosaic virus in gladiolus. Phytoparasitica 14:205–217.

Asjes, C.J. 1970. Virusziekten in gladiolen. Jversl. Lab. Bloembollenonderzoek, Lisse, 1969–1970, p. 40.

Bellardi, M.G. and Marani, F. 1985. Nepoviruses isolated from gladiolus in Italy. Acta Hortic. 164:297–308.

Bellardi, M.G. and Pisi, A. 1985a. Survey of gladiolus viruses in Italy. Rivista della Ortoflorofrutticoltura Italiana 69:133–144.

Bellardi, M.G. and Pisi, A. 1985b. Identification of three isolates of tomato black ring virus (TBRV) from gladiolus. Rivista della Ortoflorofrutticoltura Italiana, 69:299–309.

Bellardi, M.G., Canova, A. and Gelli, C. 1986. Comparative studies on gladiolus isolates of Arabis mosaic virus (ArMV). Phytopathol. Mediterr. 25:85–91.

Bellardi, M.G., Vicchi, V. and Gelli, C. 1987. Stunting of gladiolus flower spike associated with tobacco streak virus (TSV). Phytopathol. Mediterr. 26:73–80.

Bertaccini, A. and Marani, F. 1980. Mycoplasma-like organisms in *Gladiolus* sp. with malformed and virescent flowers. Phytopathol. Mediterr. 19:121–128.

Beute, M.K. 1970. Effect of virus infection on susceptibility to certain fungus diseases and yield of gladiolus. Phytopathology 60:1809–1813.

Beute, M.K., Milholland, R.D. and Gooding, G.V. 1970. A survey of viruses in field grown gladiolus in North Carolina. Plant Dis. Rep. 54:125–127.

Bing, A. 1972. Virus. In: The World of the Gladiolus, N. Koening and W. Crowley (eds.), Edgewood Press, Edgewood, MD, pp. 182–191.

Bozarth, R.F. and Corbett, M.K. 1957. Tomato ringspot virus associated with stunt or stub disease of gladiolus in Florida. Plant Dis. Rep. 42:217–221.

Bridgmon, G.H. and Walker, J.C. 1952. Gladiolus as a virus reservoir. Phytopathology 42: 65–70.

Brierley, P. and Smith, F.F. 1960. Transmission of tobacco ringspot virus on tools during harvesting flowers and corms of *Gladiolus*. Plant Dis. Rep. 44:463–464.

Cohat, J. 1993. Gladiolus. In: The Physiology of Flower Bulbs, A. De Hetogh and M. Le Nard (eds.), Elsevier, Amsterdam, pp. 297–320.

Cohen, A. and Barzilay, A. 1991. Miniature gladiolus cultivars bred for winter flowering. HortScience 26:216–218.

Fry, P.R. 1953. Two virus diseases of gladiolus. N.Z. J. Sci. Technol. Feb. 1953:460–464.

Fukumoto, F., Ito, Y. and Tochihara, H. 1982. Viruses isolated from *Gladiolus* in Japan. Ann. Phytopathol. Soc. Jpn. 48:68–71.

Hammond, J. and Hammond, R.W. 1985. Nucleic acid probe for detection of bean yellow mosaic virus. Acta Hortic. 164:373–378.

Heinis, J.L. 1954. Gladiolus virus survey in Oregon in 1954. Plant Dis. Rep. 38:733–735.

Johnson, G.V., Bing, A. and Smith, F.F. 1967. Reflective surfaces used to repel dispersing aphids and reduce spread of aphid-borne cucumber mosaic virus in gladiolus plantings. J. Econ. Entomol. 60:16–18.

Jordan, R.L. and Hammond, J. 1991. Comparison and differentiation of potyvirus isolates and identification of strain-, virus-, subgroup-specific and potyvirus group-common epitopes using monocloal antibodies. J. Gen. Virol. 72:25–36.

Kaminska, M. 1978. Some properties of isolates of tomato black ring virus. Zeszyty Problemowe Postepow Nauk Rolniczych 214:109–117.

Kaniran, R. and Izadpanah, K. 1982. Isolation and identification of bean yellow mosaic and tobacco ringspot viruses from gladiolus in Shiraz. Iranian J. Plant Pathol. 17:1–10.

Kim, J.S., Choi, G.S. and Lee, K.H. 1992. An aphid non-transmitting cucumber mosaic virus from gladiolus. Research Reports of the Rural Development Administration, Crop Protection 1992, 34:18–27 (Agr. Sci. Inst. RDA, Suwon Korea Republic).

Lee, T.C., Francki, R.I.B. and Hatta, T. 1979. A serious disease of gladiolus in Australia caused by tomato spotted wild virus. Plant Dis. Rep. 63:343–348.

Loebenstein, G., Cohen, J., Shabtai, S., Coutts, R.H.A. and Wood, K.R. 1977. Distribution of cucumber mosaic virus in systemically infected tobacco leaves. Virology 81:117–125.

Logan, A.E. and Zettler, F.W. 1985. Rapid *in vitro* propagation of virus-indexed *Gladioli*. Acta Hortic. 164:169–180.

Nagel, J., Zettler, F.W. and Hiebert, E. 1983. Strains of bean yellow mosaic virus in relation to gladiolus production in Florida. Phytopathology 73:449–454.

Randles, J.W. and Francki, R.I.B. 1965. Some properties of a tobacco rinspot virus isolated from South Australia. Aust. J. Biol. Sci. 18:979–986.

Rosner, A., Stein, A. and Levy, S. 1992. Transcription amplification of polymerase chain reaction products of bean yellow mosaic virus RNA extracted from gladioli corms. Ann. Appl. Biol. 121:269–276.

Rosner, A., Stein, A., Levy, S. and Lilien-Kipnis, H. 1994. Evaluation of linked PCR-transcription amplification procedure for bean yellow mosaic virus detection in gladioli. J. Virol. Meth. 47:227–236.

Smith, F.F. 1961. Insect vectors of cucumber mosaic virus in gladiolus. Gladio Gram 48: 10–13.

Stein, A., Loebenstein, G. and Koenig, R. 1979. Detection of cucumber mosaic virus and bean yellow mosaic virus in gladioli by enzyme-linked immunosorbent assay (ELISA). Plant Dis. Rep. 63:185–188.

Stein, A., Salomon, R., Cohen, J. and Loebenstein, G. 1986. Detection and characterization of bean yellow mosaic virus in corms of *Gladiolus grandiflorus*. Ann. Appl. Biol. 109: 1467–1154.

Stein-Rieger, A., Levy, S. and Loebenstein, G. 1990. [Propagation of gladioli free from bean yellow mosaic virus (BYMV).] Hassadeh 70:926–927 (in Hebrew).

Sushak, P.M. 1977. Seed transmission of tobacco ringspot and tomato ringspot viruses in gladioli and inoculations of tissue-cultured virus-free gladiolus plants and calli with bean yellow mosaic virus and with cucumber mosaic virus. Gladio Gram 26:2–7.

Swenson, K.G. and Nelson, R.L. 1959. Relation of aphids to the spread of cucumber mosaic virus in *Gladiolus*. J. Econ. Entomol. 52:421–424.

Wade, G.C. 1948. Gladiolus diseases. Tasmania J. Agric. 19:36–40.

Zettler, F.W. and Abo El-Nil, M.M. 1977. Bean yellow mosaic infections of galdiolus in Florida. Plant Dis. Rep. 61:243–247.

Ziv, M. and Lilien-Kipnis, H. 1990. Gladiolus. In: Handbook of Plant Cell Culture, Vol. 5, P.V. Armmirato, D.A. Evans, W.R. Sharp and Y.P.S. Bajaj (eds.), McGraw-Hill, New York, pp. 461–478.

25

Hippeastrum (Amaryllis)

A.F.L.M. Derks
Bulb Research Centre, PO Box 85, 2160 AB Lisse, The Netherlands

Introduction

Hippeastrum, commonly known as *Amaryllis*, belongs to the family of the Amaryllidaceae and consists of 60–70 species, all originating in Central and South America. The true *Amaryllis* (*A. belladonna*), indigenous to the southwestern part of the Cape Province in South Africa, is not widely cultivated (Dahlgren et al., 1985; Okubo, 1993).

Hippeastrum was introduced in Europe in the middle of the 17th century. The first hybrids made were those between *H. vittatum* and *H. reginae* resulting in types with striped flowers. The species *H. leopoldii* and *H. pardinum* were used as breeding parents for many of the large-flowered cultivars grown today. The types with small flowers used for pot plant culture were obtained by backcross breeding of *H.* × *hybridum* with *H. rutilum* (Okubo, 1993). *Hippeastrum* is mainly grown in the Netherlands (110 ha), Israel (30 ha), South Africa (15 ha), and Brasil (11 ha) (J. Doorduin, pers. comm.).

In the Netherlands, 98% of the *Hippeastrum* grown have large flowers. About 10 million bulbs are exported (Okubo, 1993). In 1993 the value of *Hippeastrum* as cut flower at the Dutch auctions was 27.5 million and as pot plants 4.3 million Dutch guilders (VBN, 1994). The main viruses in *Hippeastrum* are given below.

The Viruses

Hippeastrum mosaic potyvirus (HiMV)

The leaves and flower stalks show an irregular light and dark green mosaic pattern (Fig. 25.1; see color plate) (Brunt, 1973; Brölman-Hupkes, 1975).

Hippeastrum mosaic virus is a potyvirus with a particle length of 740–750 nm. It occurs also in *Eucharis*, *Crinum*, *Hymenocallis*, *Ismene*, *Phaedranassa*, and *Urceolina* (Brunt, 1973; Jayasinghe and Dijkstra, 1979). HiMV induces chlorotic local lesions in *Hyoscyamus niger* and vague symptoms in *Nicotiana clevelandii* sometimes followed by a systemic vein-clearing (Brölman-Hupkes, 1975).

All HiMV antisera available are mixed antisera and also react with nerine latent virus. The virus is not detectable with the potyvirus-specific monoclonal antibody PTY1 (Derks, 1992). The virus can be identified in PCR using potyvirus-specific degenerate primers (Langeveld et al., 1991; Langeveld and Derks, unpublished).

HiMV is serologically related to narcissus yellow stripe virus (Jayasinghe and Dijkstra, 1979). The virus has a sedimentation coefficient of 155 S and an A 260/280 ratio of 1:21. Infected cells contain pinwheel and bundle inclusions (Brunt, 1973). HiMV is not transmitted by seed (Brunt, 1973; Brölman-Hupkes, 1975). It is transmitted in a non-persistent manner by the aphids *Myzus persicae* and *Aphis gossypii* (Brunt 1973; Jayasinghe and Dijkstra, 1979).

Nerine latent carlavirus (NeLV)

The virus occurs frequently in *Hippeastrum* × *hybridum* without showing symptoms. NeLV, synonym *Hippeastrum* latent virus, is not seed-borne in *Hippeastrum* (Brölman-Hupkes, 1975). For more details about the virus, see Nerine (Chapter 30).

Cucumber mosaic cucumovirus (CMV)

Infected plants show yellow concentric ring and line patterns on the leaves (Fig. 25.2). Whereas it was reported to occur in the USA more than occasionally

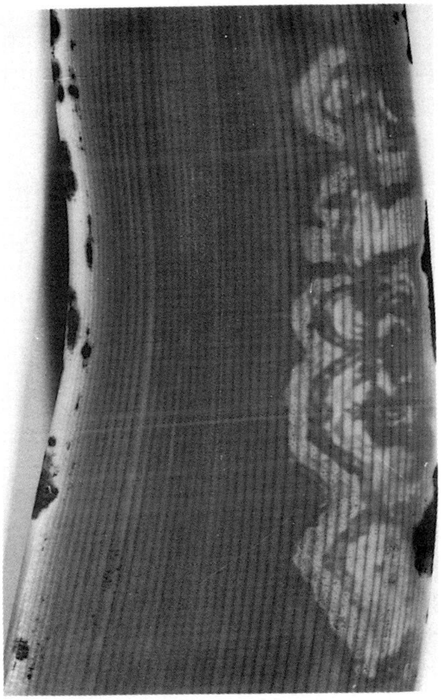

Fig. 25.2. Yellow ring and line patterns on the leaf of a *Hippeastrum* hybrid caused by cucumber mosaic virus (CMV).

Fig. 25.3. Irregular-shaped yellow and dark green spots on the leaves of cv. Scarlet Globe infected with tomato spotted wilt virus (TSWV).

(Kahn and Scott, 1964), it is infrequently found in the Netherlands. The virus can be transmitted by cutting the flowers (Anonymous, 1978). A description of CMV is given in Chapter 5.

Tomato spotted wilt tospovirus (TSWV)

The irregular-shaped yellow, white, and/or dark green spots and line patterns on the leaves (Fig. 25.3) might be confused with those of CMV in some cases. The virus is easily detected with ELISA in symptom-bearing parts of the leaves and often also in bulb scales connected with these leaves. TSWV is occasionally found in the basal plate of infected plants, which might explain why bulbs of infected plants are cured when grown for several years. Moreover, there is only a limited transmission of the virus to the daughter bulbs (Derks, 1995). The virus occurs occasionally. For more details about the virus see Chapter 5.

Control

Plants with symptoms have to be rogued. In glass-house grown stocks control of virus spread is possible by using insecticides. It is important to sterilize the knives regularly while cutting the flowers. Since no transmission of HiMV and NeLV by seed was found, it is adviseable to grow new hybrids isolated from old virus-infected stocks.

Stocks destined for bulb production are visually inspected by the Bulb Inspection Service (BKD) in the Netherlands. The maximum percentages allowed are 0.5% for HiMV and CMV, and 0% for TSWV.

Final Remarks

The average particle length of HiMV mentioned above is more in line with those known from other potyviruses than the length mentioned by Brölman-Hupkes (1975). Her results probably deviated by the different manner of measuring the virus particles. However, she clearly demonstrated the existence of two viruses — HiMV and NeLV — in mosaic diseased plants and their differential test plants.

Other viruses occasionally found in *Hippeastrum* are: tobacco mosaic tobamovirus (De Leeuw, 1972) and *Hippeastrum* streak virus (Van Velzen, 1967). In *Amaryllis belladonna* two viruses have been detected so far: nerine yellow stripe virus and tomato spotted wilt virus (Derks, unpublished).

Selected Literature

Anonymous. 1978. Ziekten en afwijkingen bij bolgewassen. Deel II. Amaryllidaceae, Iridaceae e.a. Ed. Laboratorium voor Bloembollenonderzoek, Lisse, The Netherlands.

Brölman-Hupkes, J.E. 1975. Tentative description of Hippeastrum latent virus in *Hippeastrum hybridum* plants and differentation from Hippeastrum mosaic virus. Neth. J. Plant Pathol. 81:226–236.

Brunt, A.A. 1973. Hippeastrum mosaic virus. CMI/AAB Descriptions of Plant Viruses, No. 117.

Dahlgren, R.M.T., Clifford, H.T. and Yeo, P.F. 1985. The families of the Monocotyledons. Springer-Verlag, Berlin.

De Leeuw, G.T.M. 1972. Tobacco mosaic virus in *Hippeastrum hybridum*. Neth. J. Plant Pathol. 78:69–71.

Derks, A.F.L.M. 1992. Some unusual serological reactions among potyviruses. Arch. Virol. (Suppl 5):77–79.

Derks, A.F.L.M. 1995. Tomato spotted wilt and Impatiens necrotic spot viruses in flowerbulb crops. Acta Bot. Neerlandica, in press.

Jayasinghe, U. and Dijkstra, J. 1979. Hippeastrum mosaic virus and another filamentous virus in *Eucharis grandiflora*. Neth. J. Plant Pathol. 85:47–65.

Kahn, R.P. and Scott, H.A. 1964. Serological relationship of cucumber mosaic virus and certain virus isolates that incite Amaryllis mosaic symptoms. Phytopathology 54: 360–362.

Langeveld, S.A., Dore, J.M., Memelink, J., Derks, A.F.L.M., Van der Vlugt, C.I.M., Asjes, C.J. and Bol, J.F. 1991. Identification of potyviruses using the polymerase chain reaction with degenerate primers. J. Gen. Virol. 72:1531–1541.

Okubo, H. 1993. Hippeastrum (Amaryllis). In: The Physiology of Flower Bulbs, A. De Hertogh and M. Le Nard (eds.), Elsevier, Amsterdam, pp. 321–334.

VBN, 1994. Statistiekboek 1993. Ed. Vereniging van Bloemenveilingen in Nederland, Leiden, The Netherlands.

26

Hyacinth

A.F.L.M. Derks
Bulb Research Centre, PO Box 85, 2160 AB Lisse, The Netherlands

Introduction

The genus *Hyacinthus* consisting of three species belongs to the family of the *Hyacinthaceae*. It is indigenous to south-eastern Europe and Asia Minor (Dahlgren et al., 1985). The first report of hyacinths grown outside the center of origin dates back to 1562. Only the species *H. orientalis* is of horticultural importance (Nowak and Rudnicki, 1993). In 1993 the acreage of hyacinths grown in the Netherlands was about 919 ha, of which 58 ha with scooped and scored bulbs (PVS/BKD, 1994). Other countries where hyacinths are grown are Japan (c. 18 ha) and France. In 1992 at the Dutch auctions, hyacinths were sold as cut flowers with a value of 11.7 million and as pot plants with a value of 13.8 million Dutch guilders (VBN, 1993).

The main viruses in hyacinths are discussed below.

The Viruses

Hyacinth mosaic potyvirus (HyMV)

Infected plants show a mottle on the leaves mainly on the basal parts that range in color from light green to bright yellow. The mottle is stripe-like or consists of many spots and ring patterns (Fig. 26.1). Flower stalks of some cultivars show stripes and ring patterns (Fig. 26.2; see color plate). Flower breaking is possible (Derks and Vink-Van den Abeele, 1980a). HyMV causes a reduction in bulb yield varying from 17% in cv. Pink Pearl to 0% in the tolerant cv. Queen of the Pinks (Vreeburg, 1982).

Hyacinth mosaic virus is a potyvirus with a particle length of 740–750 nm (Atkey et al., 1968; Derks and Vink-Van den Abeele, 1980a). Other natural hosts are *Muscari* spp. (Derks and Vink-Van den Abeele, 1980b, 1981). HyMV is

Fig. 26.1. Light green mottle at the leaf bases in cv. Pink Pearl caused by hyacinth mosaic virus (HyMV).

Fig 26.3. Light green and grey spots and stripes on leaves and flower stalk of cv. Queen of the Pinks; leaves are teared, caused by tobacco rattle virus (TRV).

not transmissible to test plants in leaf extracts of hyacinths because of inhibitory substances (Derks et al., 1983).

HyMV is detectable with a specific antiserum in microprecipitin tests (Derks and Vink-Van den Abeele, 1980a), with the potyvirus group-specific monoclonal antibody PTY1 in an indirect ELISA and with potyvirus-specific degenerate primers in PCR (Langeveld et al., 1991). The virus is most reliably detected in leaf material from flowering up to a month after flowering. Virus detection in bulbs during storage is unreliable. HyMV is serologically closely related to *Ornithogalum* mosaic virus and more distantly to freesia mosaic virus, iris severe mosaic virus, and *Scilla* mosaic virus (Derks and Vink-Van den Abeele, 1980a).

HyMV is not spread from diseased to healthy plants when flowers are removed (for better bulb production) or when bulbs are scooped or scored (for propagation). However, all descendants after vegetative propagation from a diseased bulb will also be HyMV-infected (Derks and Vink-Van den Abeele, 1980b). The virus is transmitted in a non-persistent manner by the aphids *Myzus persicae* and *Macrosiphum euphorbiae* (Derks and Vink-Van den Abeele, 1981; Derks et al., 1983).

Tobacco rattle tobravirus (TRV)

Leaves and flower stalks show light green to yellow spots and stripes. These symptoms may turn to brown or grey necrotic stripes later in the growing season. In some cases, leaves tear and bend along the stripes (Fig. 26.3).

Transparent and in a later stage, yellow or brown spots become visible on the bulb scales (Van Slogteren, 1958). Tobacco rattle virus is described in Chapter 5. The vector, *Trichodorus* spp., occurs in sandy and light loamy soil types where almost all hyacinths are grown. Nevertheless, TRV occurs occasionally in hyacinths themselves (Asjes, 1974).

Control

Symptom-bearing plants have to be rogued as early in the season as possible. This is effective in the control of TRV together with other cultural methods such as late planting of the bulbs (Asjes, 1974). The efficiency of visual selection of HyMV-infected plants varies with the cultivar and growing stage. Plants may be symptomless in case of HyMV tolerant cultivars and of young plants developed on scooped or scored bulbs. Although the cvs. King of the Blues and Queen of the Pinks are tolerant, visual selection is possible during a limited period because of conspicuous symptoms of HyMV on the flower stalks (Derks and Vink-Van den Abeele, 1980a).

Spread of HyMV is considerably reduced by weekly spraying with mineral oils or permethrin (Asjes, 1981). To reduce spread of TRV it is advised to keep soils free from weeds. Soil disinfection can be considered if the damage is expected to be too heavy (Van Slogteren, 1958; Asjes, 1974).

Stocks are field-inspected in the Netherlands by the Bulb Inspection Service (BKD). Stocks destined for propagation by scooping or scoring are not allowed to contain viruses. For other stocks the percentage of HyMV or TRV has to be lower than 1%. Stocks of tolerant cultivars 100% infected with HyMV and grown separate from the normal stocks have to be replaced by virus-free stocks within about 4 years.

Meristem Culture

The tolerant cultivars can be made virus-free by meristem tip culture (Asjes et al., 1974; Blom-Barnhoorn et al., 1986, Derks, 1993). The yield of virus-free plants varies from 1% with culture of vegetative apices of bulbs grown on scooped bulbs (Asjes et al., 1974) to 15–30% by using meristems adventitiously formed on bulb-scale tissue. Virazole does not increase the final percentage of HyMV-free plants (Blom-Barnhoorn et al., 1986). Testing of the plantlets on HyMV by ELISA with the monoclonal antibody PTY1 seems to be more reliable and sensitive than by electron microscopy (Derks, 1993). It is advisable to grow nuclear stocks in gauze houses.

Final Remarks

Viruses occasionally detected in hyacinths are cucumber mosaic cucumovirus (Ainsworth, 1938), arabis mosaic, and tobacco ringspot nepoviruses (Asjes, 1974). A yellows disease in hyacinths caused by a mycoplasma-like organism has occasionally occurred in the Netherlands (Van Slogteren and Muller, 1972; Van Slogteren et al., 1974).

Selected Literature

Ainsworth, G.C. 1938. A note on certain viruses of the cucumber virus 1 type isolated from monocotyledonous plants. Ann. Appl. Biol. 24:867–869.

Asjes, C.J. 1974. Soil-borne virus diseases in ornamental bulbous crops and their control in the Netherlands. Agric. Environ. 1:303–315.

Asjes, C.J. 1981. Control of stylet-borne spread by aphids of tulip breaking virus in lilies and tulips, and hyacinth mosaic virus in hyacinths by pirimicarb and permethrin sprays versus mineral-oil sprays. Meded. Fak. Landbouww. Gent 46 (3):1073–1077.

Asjes, C.J., Bunt, M.H. and Van Slogteren, D.H.M. 1974. Production of hyacinth mosaic virus-free hyacinths and lily symptomless virus-free lilies by meristem tip culture. Acta Hortic. 36:223–228.

Atkey, P.T., Hollings, M., Stone O.M. and Brunt, A.A. 1968. Rep. Glasshouse Crops Res. Inst. 1967:107–108.

Blom-Barnhoorn, G.J., Van Aartrijk, J. and Van der Linde, P.C.G. 1986. Effect of virazole on the production of hyacinth plants free from hyacinth mosaic virus (HMV) by meristem culture. Acta Hortic. 177:571–574.

Dahlgren, R.M.T, Clifford, H.T. and Yeo, P.F. 1985. The families of the Monocotyledons. Springer-Verlag, Berlin.

Derks, A.F.L.M. 1993. Jversl. Lab. Bloembollenonderz., Lisse 1991–1992:28–29.

Derks, A.F.L.M. and Vink-Van den Abeele, J.L. 1980a. Hyacinth mosaic virus: symptoms in hyacinths, serological detection, and relationships with other potyviruses. Acta Hortic. 109:495–502.

Derks, A.F.L.M. and Vink-Van den Abeele, J.L. 1980b. Jversl. Lab. Bloembollenonderz., Lisse, 1979:53.

Derks, A.F.L.M. and Vink-Van den Abeele, J.L. 1981. Jversl. Lab. Bloembollenonderz., Lisse, 1980:49.

Derks, A.F.L.M., Vink-Van den Abeele, J.L. and Hollinger, T.C. 1983. Jversl. Lab. Bloembollenonderz., Lisse, 1982:61.

Langeveld, S.A., Dore, J.M., Memelink, J., Derks, A.F.L.M., Van der Vlugt, C.I.M., Asjes, C.J. and Bol, J.F. 1991. Identification of potyviruses using the polymerase chain reaction with degenerate primers. J. Gen. Virol. 72:1531–1541.

Nowak, J. and Rudnicki, R.M. 1993. *Hyacinthus*. In: The Physiology of Flower Bulbs, A. De Hertogh and M. Le Nard (eds.), Elsevier, Amsterdam, pp. 335–347.

PVS/BKD. 1994. Beplante oppervlakten Bloembollen 1990/91–1993/94: Voorjaarsbloeiers. Ed. Produktschap voor Siergewassen, The Hague, The Netherlands.

Van Slogteren, D.H.M. 1958. Ratelvirus als oorzaak van ziekten in bloembolgewassen en de mogelijkheden de infectie door middel van grondontsmetting te bestrijden. Tijdschr. Pl. Ziekten 64:452–462.

Van Slogteren, D.H.M. and Muller, P.J. 1972. "Lissers", a yellows disease in hyacinths, apparently caused by a mycoplasma. Meded. Fak. Landbouww. Gent 37(2):450–457.

Van Slogteren, D.H.M., Groen, N.P.A. and Muller, P.J. 1974. Yellows disease of gladiolus and hyacinth in the Netherlands. Acta Hortic. 36:303–311a.

VBN. 1993. Statistiekboek 1992. Ed. Vereniging van Bloemenveilingen in Nederland, Leiden, The Netherlands.

Vreeburg, P.J.M. 1982. Teelt van hyacinten: gewasverslag 1981–1982: 31. Ed. Bulb Research Centre, Lisse. The Netherlands.

Bulb and Corm Crops

27

Iris

Cornelia I.M. Van der Vlugt and A.F.L.M. Derks
Bulb Research Centre, PO Box 85, 2160 AB Lisse, The Netherlands

Introduction

Irises have been used for centuries both as ornamentals and as a source of perfumes and medicine. Irises originate in the temperate and subtropical zones of the Northern hemisphere (Köhlein, 1987). More than 200 species of the *Iris* genus are rhizomatous plants which are primarily used in gardens. In the Netherlands about 10 ha of rhizome irises are grown commercially. The most important group for the bulb industry consists of *Xiphium* irises originating from the western Mediterranean, i.e. Spain, Portugal, southern France, Morocco, Algeria, and Tunisia. Crosses between *X. vulgare*, *X. tingitanum*, *X. boissieri*, and *X. filifolium* resulted in the group of Dutch irises: *Iris* × *hollandica* Tub. A large assortment with blue to purple as the dominating flower color has been bred in the last century. Dutch irises are grown commercially in the Netherlands (772 ha), France (103 ha), UK (40 ha), Israel (50 ha), the US (182 ha), and Japan (118 ha)

(De Munk and Schipper, 1993). In 1992 the cut flower production in the Netherlands reached the value of ca. 46 million Dutch guilders (ca. US $24 million) (VBN, 1993).

Dwarf irises belonging to the *Iridodictyum* group are grown on a smaller scale: In the Netherlands, ca. 27 ha were cultivated in 1991 (De Munk and Schipper, 1993). These irises, mainly *I. reticulata* and *I. danfordiae*, originate from the Caucasus, Turkey, and northern Iran (Köhlein, 1987) and are grown for use in gardens and for forcing as pot plants (De Munk and Schipper, 1993).

The Viruses

Iris mild mosaic potyvirus (IMMV)

Almost all commercially available stocks of bulbous irises are infected with IMMV (Asjes, 1979; Brunt et al., 1988). Symptoms tend to develop 1–2 months after emergence (Derks et al., 1985). Leaf symptoms appear as a faint mosaic which may be more conspicuous on the flowerbud sheath (Fig. 27.1; see color plate). Flower symptoms are limited to a few small dark flecks in the fall (Hammond et al., 1985). After reinfection of virus-free irises, first-year symptoms of IMMV can be more severe especially in the inner leaves. Often the top halves of the leaves are light-green to yellow-green with small dark-green stripes (Fig. 27.2). With prolonged propagation, symptoms become less pronounced (Derks and Hollinger, 1989; Derks et al., 1985). IMMV also infects the rhizomatous *I. susiana* (Derks and Hollinger, 1986). The only other natural host known is *Ferraria undulata* (Derks et al., 1985). In the diagnostic host *Chenopodium quinoa*, IMMV causes chlorotic local lesions (Brunt, 1986). IMMV is transmissible in a non-persistent manner by the aphids *Aphis gossypii*, *Macrosiphum euphorbiae*, and *Myzus persicae* (Brunt, 1986). The virus has been partially characterized by Alper et al. (1984), Brunt (1986), Brunt and Phillips (1980), and Derks et al. (1985). Purified virus particles usually sediment as a single component with a sedimentation coefficient ($S^{o}_{20,w}$) of 142–147 S and an A_{260}/A_{280} of 1.18–1.22. IMMV did not show serological relationships in tube precipitin tests to at least six other potyviruses but in indirect ELISA was found to be related to asparagus 1, bean yellow mosaic, pea seed-borne mosaic, potato Y, and tobacco etch viruses (Brunt, 1986). IMMV is readily detected by ELISA in leaves and in bulbs from three weeks after lifting of both primarily and secondarily infected plants (Van Schadewijk et al., 1988).

Fig. 27.2. First year symptoms caused by iris mild mosaic virus (IMMV) in iris cv. Professor Blaauw.

Fig. 27.3. Yellow-green mosaic pattern and necrotic spots in the leaves of iris cv. Professor Blaauw caused by narcissus latent virus (NLV).

Narcissus latent carlavirus (NLV)

NLV in iris (Brunt, 1976, 1977; Brunt and Phillips, 1980; Alper et al., 1984; Derks et al., 1985) has also been described as iris mild yellow mosaic virus (IMYMV, Asjes, 1979).

Several cultivars are infected to a large extent with NLV in addition to IMMV. Leaves of such plants show a yellow-green mosaic pattern, especially at the margins and tips, which is typically more prominent soon after emergence than later in the growing season (Hammond et al., 1985). Necrotic spots appear on the top leaf halves, mostly after flowering (Fig. 27.3). First-year symptoms of NLV are not known (Derks et al., 1985). Although NLV causes circular chlorotic lesions in the diagnostic hosts *Tetragonia expansa* and *Phaseolus vulgaris*, neither is a reliable assay host. Infectivity is usually best assayed in *Nicotiana clevelandii* or *N. megalosiphon*. It is transmissible in a non-persistent manner by the aphids *Acyrthosiphon pisum*, *Aphis gossypii*, and *Myzus persicae* (Brunt, 1976). The virus has been partially characterized by Alper et al. (1984), Brunt (1976, 1977), Brunt and Phillips (1980), and Derks et al., (1985). Purified virus particles sediment as a single component with a $S^{o}_{20,w}$ of 158±4 S and an A_{260}/A_{280} of 1.09. Isolates from narcissus, nerine, and bulbous iris are serologically indistinguishable. No relationship was detected to 14 other carlaviruses (Brunt, 1976). Although the virus is readily detectable in the leaves of secondarily infected iris plants by ELISA, it is difficult or impossible to detect it in the bulbs (Derks et al., 1986).

Iris severe mosaic potyvirus (ISMV)

In combination with IMMV, ISMV causes a mild or severe mosaic in the leaves consisting of pale-green and yellowish-green stripes and wider bands in irregular patterns extending upward from below soil level and contrasting with grey- to blue-green stripes (Fig. 27.4). After leaf emergence, symptoms are apparent and can be confused with those of NLV (Derks et al., 1985). Flowers of bulbous iris show a breaking pattern as stripes or spots (Brunt et al., 1988). The symptoms, especially the flower symptoms, are less pronounced in irises infected with ISMV alone. Bearded and beardless iris mosaic viruses, now known to be ISMV (Derks and Hollinger, 1986; Barnett et al., 1971), infect rhizomatous iris. The only other natural host known is *Crocus*. Its only diagnostic host species is *Belamcanda chinensis*, which is not susceptible to IMMV, BYMV, or NLV. ISMV is transmissible in a non-persistent manner by the aphids *Macrosiphum euphorbiae* and *Myzus persicae* (Brunt et al., 1988). The virus has been partially characterized by Alper et al. (1984), Brunt et al. (1988), Brunt and Phillips (1980), Derks et al. (1985), and Derks and Hollinger (1986). Purified virus particles sediment as a single component with a $S^{o}_{20,w}$ of circa 150 S, some preparations may also contain faster sedimenting virus aggregates. A_{260}/A_{280} is 1.12. ISMV is serologically unrelated to nine other potyviruses (Brunt et al., 1988). The 3′-terminal region of the ISMV genome, including the complete coat protein gene, has recently been sequenced (van der Vlugt et al., 1994; EMBL bank accession number X75939). The virus is readily detectable in leaves of secondarily infected plants by ELISA and hybridization assays (Van der Vlugt et al., unpublished results). In secondarily infected bulbs, detection of ISMV directly after harvesting is unreliable but can be improved by wounding or high-temperature treatment due to multiplication of the virus (Van der Vlugt et al., 1993a, 1993b). After early

Fig. 27.4. Yellowish-green stripes and wider bands extending upward from below soil level in iris cv. Professor Blaauw caused by iris severe mosaic virus (ISMV) in combination with iris mild mosaic virus (IMMV).

primary infection ISMV can be detected in the leaves in varying degree, but after a late infection the virus cannot be detected in the leaves. In contrast, late primary infections can be detected easily in recently lifted bulbs while early primary infections are not detectable at all. In general, application of the wounding treatment improves the detection of ISMV in ELISA for both primarily and secondarily infected iris bulbs (Van der Vlugt, unpublished results).

Bean yellow mosaic potyvirus (BYMV)

BYMV may be associated with flower break in iris and causes a variable foliar mosaic which may be very intense, especially on the central portion of the plant (Hammond et al., 1985; Fig. 27.5). BYMV is of rare occurrence in the Netherlands and in the UK but is more common in the US and Israel (Hammond et al., 1985). For further details see Chapter 24.

Fig. 27.5. Variable mosaic, most intense on the central portion of the plant in iris cv. Professor Blaauw caused by bean yellow mosaic virus (BYMV) in combination with iris mild mosaic virus (IMMV).

Fig. 27.6. Yellow discolorations of the inner leaves with spots and stripes caused by tomato spotted wilt virus (TSWV) in iris cv. Blue Magic.

Tomato spotted wilt tospovirus (TSWV) and impatiens necrotic spot tospovirus (INSV)

These members of the tospovirus group (Derks, 1995) both induce yellowish discoloration of the inner leaves or yellow and sometimes necrotic spots and stripes (Fig. 27.6). In the Netherlands these viruses were incidentally detected in irises grown for flower production in or in the direct neighborhood of glasshouses with thrips infestation. The few bulbs harvested from an infected stock did not show symptoms in the following growing season nor could the viruses then be detected. The viruses were also detected in samples of irises grown in Mediterranean countries. For further details on tospoviruses see Chapter 5.

Additional remarks

IMMV, ISMV, and BYMV are readily distinguished from NLV and from each other serologically, although strains of each of these four viruses occur that may differ in host range and symptom expression. A list of hosts, serology, and the typical symptoms in iris that can be used to differentiate these viruses are reviewed by Hammond et al. (1985). All three potyviruses react with the potyvirus specific PCR primers (Langeveld et al., 1991). Tospoviruses are clearly distinguishable by serology and morphology of the particles.

Other viruses infecting bulbous iris are: turnip mosaic potyvirus (Inouye and Mitsuhata, 1978); tobacco ringspot nepovirus; tobacco mosaic tobamovirus; and tobacco rattle tobravirus — the latter three being found occasionally in the Netherlands (Asjes, 1979).

Other viruses infecting rhizomatous iris are: iris fulva mosaic potyvirus from *I. fulva* × *brevicaulis* (Barnett and Alper, 1977); lilac chlorotic leafspot closterovirus (Lisa, 1980); cucumber mosaic cucumovirus (Brunt, 1975); broad bean wilt fabavirus (Brunt, 1975); tobacco ringspot nepovirus (Brunt, 1975), and a rhabdovirus in *I. germanica* (Lisa, 1980).

Certification Procedures

As commercial iris stocks are still almost totally infected with IMMV, inspection is mainly for ISMV and NLV. Disease control in iris is carried out by visual selection of ISMV and NLV infected plants in the field. Disease control can be hampered by the masking of NLV and ISMV symptoms when IMMV develops simultaneously (Asjes, 1979). For certification procedures samples of these stocks are also grown in greenhouses for inspection. Depending on the certificate, virus percentages may vary from a maximum of 1% (quality class SSE) to 10% (quality class E) for originally virus-free irises. For the standard stocks, which are 100% infected with IMMV, allowable percentages for ISMV and NLV vary from 0.25–0.5% (class I and class II).

Virus-free Bulbous Iris

To obtain virus-free iris stocks, the starting material should contain as few viruses as possible, preferably bulbs with iris mild mosaic virus alone. It is possible nowadays to obtain virus-free iris stocks by meristem culture (Baruch and Quak,

1966; Van der Linde et al., 1988). Several methods are reviewed by De Munk and Schippers (1993). The bulb yield of these virus-free irises can increase up to 50% (Asjes, 1988; Rijnders and Schipper, 1990). In forcing experiments, virus-free plants produced longer flower stems and higher quality cut flowers than the original commercial stocks (Rijnders and Schipper, 1990).

Virus-free stocks were introduced to commercial growers several years ago. Although strongly advised to grow virus-free (virus-tested) material in isolation, in practice this may be very difficult, as neighboring growers may have virus-infected irises, narcissi, croci, gladioli, ixia, and crocosmia growing nearby (Asjes, 1988). Also the occurrence of IMMV, BYMV, and ISMV in rhizomatous irises indicate the potential danger of growing rhizomatous irises in the neighborhood of (virus-free) stocks of bulbous irises (Derks and Hollinger, 1986). To maintain virus-free stocks, they must be sprayed with insecticides such as synthetic pyrethroids or with mineral oil to limit virus spread by aphids (Asjes, 1985). Such mineral oil sprays, however, considerably reduce bulb yields, although the advantages may counteract these losses, as e.g. bulb yield of virus-free plants increases and better quality irises as cut flowers is achieved (Asjes, 1988). Therefore, it is advised at the moment to multiply the virus-free irises initially in gauze houses and in later years to grow this stock apart from the other trade material and to apply weekly sprays of mineral oil and pyrethoids.

Selected Literature

Alper, M., Salomon, R., and Loebenstein, G. 1984. Gel electrophoresis of virus-associated polypeptides for detecting viruses in bulbous irises. Phytopathology 74:960–962.

Asjes, C.J. 1979. Viruses and virus diseases in Dutch bulbous irises (*Iris hollandica*) in the Netherlands. Neth. J. Plant Pathol. 85:269–279.

Asjes, C.J. 1988. Still having to live with viruses. The Grower. 109:28–31.

Asjes, C.J. 1985. Control of field spread of non-persistent viruses in flower-bulb crops by synthetic pyrethroid and pirimicarb insecticides, and mineral oils. Crop Prot. 4:485–493.

Barnett, O.W. and Alper, M. 1977. Characterization of iris fulva mosaic virus. Phytopathology 67:448-454.

Barnett, O.W., de Zoeten, G.A., and Gaard, G. 1971. Bearded iris mosaic virus: transmission, purification, inclusions, and its differentiation from bulbous iris mosaic. Phytopathology 61:926–932.

Baruch, E.R., and Quak, F. 1966. Virus-free plants of iris "Wedgewood" obtained by meristem culture. Neth. J. Plant Pathol. 72:270–273.

Brunt, A.A. 1977. Some hosts and properties of narcissus latent virus, a carlavirus commonly infecting narcissus and bulbous iris. Ann. Appl. Biol. 87:355–364.

Brunt, A.A. 1976. Narcissus latent virus. CMI/AAB Descriptions of Plant Viruses, No. 170.

Brunt, A.A. 1986. Iris mild mosaic virus. CMI/AAB Descriptions of Plant Viruses, No. 324.

Brunt, A.A., Derks, A.F.L.M., and Barnett, O.W. 1988. Iris severe mosaic virus. CMI/AAB Descriptions of Plant Viruses, No. 338.

Brunt, A.A., and Phillips, S. 1980. The detection, separation from naturally occurring complexes, and partial characterisation of four aphid-borne viruses infecting bulbous iris. Acta Hortic. 109:503–508.

Brunt, A.A. 1975. Viruses and virus diseases of irises in Britain. Acta Hortic. 47:45–50.

De Munk, W.J., and Schipper, J. 1993. Iris- Bulbous and rhizomatous, In: The Physiology of Flower Bulbs, A. De Hertogh and M. Le Nard (eds.), Elsevier, Amsterdam, pp. 349–379.

Derks, A.F.L.M. 1995. Tomato spotted wilt and impatiens necrotic spot viruses in flower bulb crops. Acta Botanica Neerlandica, in press.

Derks, A.F.L.M. and Hollinger Th.C. 1986. Similarities of and differences between potyviruses from bulbous and rhizomatous irises. Acta Hortic. 177:555–561.

Derks, A.F.L.M. and Hollinger Th.C. 1989. Ook virusvrije teelt vereist selectie. Virusziekten in iris opnieuw bekeken. Bloembollencultuur 4:40–41.

Derks, A.F.L.M., Hollinger, Th.C. and Lemmers M.E.C. 1986. Ontwikkeling van methoden ten behoeve van de identificatie van virussen en de diagnostiek van virusziekten in bloembolgewassen. Jvers. Lab. Bloembollenonderz., Lisse, 69–70.

Derks, A.F.L.M., Hollinger, Th.C. and Vink-Van den Abeele, J.L. 1985. Identification and symptom expression of four elongated viruses infecting bulbous irises. Acta Hortic. 164: 309–317.

Hammond, J., Brunt, A.A., Derks, A.F.L.M., Inouye, N., Barnett, O.W., Allen, T.C. and Lawson, R.H. 1985. Viruses infecting bulbous iris: a clarification of nomenclature. Acta Hortic. 164:395–397.

Inouye, N. and Mitsuhata K. 1978. Turnip mosaic virus isolated from iris. Nogaku Kenkyu. 57:1–16.

Köhlein, F. 1987. Iris. Timber Press, Portland, Oregon.

Langeveld, S.A., Dore, J.-M., Memelink, J., Derks, A.F.L.M., Van der Vlugt, C.I.M., Asjes C.J. and Bol, J.F. 1991. Identification of potyviruses using the polymerase chain reaction with degenerate primers. J. Gen. Virol. 72:1531–1541.

Lisa, V. 1980. Two viruses from rhizomatous iris. Acta Hortic. 110:39–53.

Rijnders, E. and Schipper, J. 1990. 50% meer opbrengst. Vakwerk. 47:10–11.

Van der Linde, P.C.G., Hol, G.M.G.M., Blom-Barnhoorn, G.J., Van Aartrijk, J. and De Klerk, G.J. 1988. *In vitro* propagation of *Iris holllandica* Tub. cv. Professor Blaauw. I. Regeneration on bulb-scale explants. Acta Hortic. 226:121–128.

Van der Vlugt, C.I.M., Derks, A.F.L.M., Boonekamp, P.M. and Goldbach R.W. 1993a. Improved detection of iris severe mosaic virus in secondarily-infected iris bulbs. Ann. Appl. Biol. 122:279–288.

Van der Vlugt, C.I.M., Lohuis, H., Dijkstra, J., Marquart A., Goldbach R.W. and Boonekamp, P.M. 1993b. Multiplication and distribution of iris severe mosaic virus in secondarily-infected iris bulbs after stress induced by heat and wounding. Ann. Appl. Biol. 123:601–610.

Van der Vlugt, C.I.M., Langeveld, S.A. and Goldbach, R.W. 1994. Molecular cloning and sequence analysis of the 3′-terminal region of iris severe mosaic virus RNA. Arch. Virol. 136:397–406.

Van Schadewijk, A.R., Derks, A.F.L.M., Lemmers, M.E.C. and Hollinger, Th.C. 1988. Detection of iris mild mosaic virus in bulbs and leaves of bulbous iris by ELISA. Acta Hortic. 234:199–206.

VBN. 1993. Statistiekboek 1992. Ed: Vereniging van Bloemenveilingen in Nederland, Leiden, Holland.

28

Lily

A.F.L.M. Derks
Bulb Research Centre, PO Box 85, 2160 AB Lisse, The Netherlands

Introduction

The genus *Lilium* belongs to the family of the *Liliaceae*. Lilies are indigenous to Europe (c. 12 species), Asia (50–60), and North America (c. 24); some are native to the tropics at higher elevations (Beattie and White, 1993). The lilies of horticultural importance can be divided into three groups: *Lilium longiflorum*, Asiatic, and Oriental hybrids. The first commercial success in lily bulb production dates back to the period 1880–1920 when *Lilium longiflorum* was grown in Bermuda. These lilies, native only to Japan, are grown now in the USA (c. 400 ha.) for pot plant production and in Japan (205 ha), the Netherlands (114 ha), and Israel (45 ha) mainly for cut flower production (Miller, 1993; PVS/BKD, 1994; C. Buschman, pers. comm.).

The Asiatic hybrids derived from about 12 species have a large diversity in flower colors and shape. In the Netherlands these hybrids represent the largest group with an acreage of 1437 ha. The Oriental hybrids derived from crosses between *L. auratum*, *L. speciosum* and *L. rubellum* are late flowering and have mostly white and pink flowers. In the Netherlands the acreage of Orientals amounts to 778 ha (Beattie and White, 1993; PVS/BKD, 1994). At the Dutch auctions lilies were sold as cut flowers with a value of 227.8 million and as pot plants of 6.3 million Dutch guilders (VBN, 1993).

The main countries with lily culture are: the Netherlands (2386 ha), USA (475), Japan (400), Poland (140), and Mexico (100), according to the International Flower Bulb Centre (C. Buschman, pers. comm.). The main viruses in lilies are discussed below.

The Viruses

Lily symptomless carlavirus (LSV)

Leaves may show vein-clearing or light green stripes between the veins (Fig. 28.1). After flowering, small light brown spots on the back side of the leaves may pass into yellow, purple, or brown spots on the upper surface of the leaves (Fig. 28.2; see color plate) which leads to premature death of the plants at the endof the growing season (Derks and Hendriks, 1990). Many cultivars remain symptomless when infected with LSV. Plants often show reduced growth, smaller flowers, have a pronounced lower bulb yield, and shorter vase life as cut flowers (Boontjes, 1983). Symptoms develop better at lower temperatures of less than 15°C (Allen, 1972).

LSV is a carlavirus with particles of 640–650 nm long and occurs wherever lilies are grown (Allen, 1972). Other natural hosts are tulip and *Alstroemeria*. No test plants are known (Derks and Asjes, 1975; Phillips and Brunt, 1986).

In leaves LSV is detected most reliably by serological methods using ELISA during the period from flowering until about 3 weeks after flowering (Derks and Vink-Van den Abeele, 1980). However, primary infections can escape from detection in a leaf test (Derks, 1988). Therefore, a bulb test is preferred. After lifting, the bulbs have to be stored for at least 3 weeks at c. 0°C to obtain the best results (Van Schadewijk, 1986). Incubation of bulb extracts with hemicellulase may improve the detection of LSV in some cultivars (Beijersbergen and Van der Hulst, 1980). LSV is detectable in hybridization assays and with LSV specific primers in PCR. A cDNA clone complementary to the 3′-terminal region of the genomic RNA (2613 bp) including the region encoding the coat protein has been sequenced (Memelink et al., 1990).

The virus is serologically related to chrysanthemum virus B, potato virus M, potato virus S, carnation latent virus, and passiflora latent virus.

Fig. 28.1. Light green stripes between the leaf veins of *L. speciosum rubrum* caused by lily symptomless virus (LSV).

Fig. 28.3. Light green stripe-mosaic on the leaves of *L. longiflorum* infected with lily mottle virus (LMoV) and lily symptomless virus (LSV).

LSV has a sedimentation coefficient of 171–172 S and an A 260/280 ratio of 1.20–1.43. In leaf extracts virus particles are often abundant and laterally aggregated. No pinwheels or inclusion bodies have been observed (Allen, 1972; Derks and Vink-Van den Abeele, 1980).

LSV is transmitted in a non-persistent manner by the aphids *Myzus persicae*, *Macrosiphum euphorbiae*, *Aphis gossypii*, *A. fabae*, and *Aulacorthum solani* (Mowat and Stefanac, 1974; Derks and Asjes, 1975). The virus is sap-transmissible and not seed-borne in lilies (Brierley and Smith, 1944a; Allen, 1972).

Lily mottle potyvirus (LMoV)

Leaves show a light to yellowish green mottle or mosaic (Fig. 28.3) and can be twisted or narrowed. Plants are often shorter and die prematurely and have a shorter vase life as a cut flower. Flowers are sometimes malformed or may show a breaking pattern. Bulb scales of some cultivars show brown necrotic spots or rings. The symptoms are intensified (more necrotic stripes on the leaves) when plants are also infected with lily symptomless virus (Brierley and Smith, 1944b; Asjes et al., 1973). When plants become infected shortly after emergence, extreme symptoms such as yellowing and browning of leaves and veins in the stems (Fig. 28.4; see color plate) may occur in some cultivars (Derks, 1993).

Lily mottle virus is a potyvirus with a particle length of 750–770 nm that occurs wherever lilies are grown. The virus is sap-transmissible to tulips. Some isolates are transmissible to test plants, e.g. *Nicotiana benthamiana*, *Tetragonia expansa*, and *Chenopodium* spp. (Asjes et al., 1973; Alper et al., 1982; Dekker et al., 1993).

LMoV is detected most reliably with polyclonal antibodies by DAS-ELISA in leaves during the period from flowering until about 3 weeks after flowering. Bulbs of Asiatic hybrids have to be stored for at least 3 weeks at c. 0°C to obtain the best test results (Van Schadewijk, 1986). With *Lilium longiflorum*-cultivars, bulb scales have to be stored at 20°C with additional lighting prior to testing (Derks, 1993). The detection of LMoV in bulbs of Orientals is unreliable with both storage regimes (Derks, unpublished results).

LMoV is detectable in hybridization assays and with specific primers in PCR (Dekker et al., 1993). The complete coat protein gene has been sequenced after isolation of a cDNA clone encoding the 3′-end of the viral RNA (1586 bp; Langeveld et al., 1991). A comparison with the coat protein gene of tulip breaking virus and other properties of both viruses revealed significant differences. Therefore, they should not be considered identical viruses (Brierley and Smith, 1944b; Dekker et al., 1993, Ohira et al., 1994).

LMoV is serologically related to tulip breaking virus, Rembrandt tulip breaking virus, iris severe mosaic virus, and potato virus Y in ISEM (Dekker et al., 1993). It has a sedimentation coefficient of 155 S (Derks and Vink-Van den Abeele, 1980). In infected lilies cylindrical inclusions such as pinwheels, bundles, and scrolls were observed (Alper et al., 1982).

LMoV is transmitted in a non-persistent manner by the aphids *Aphis gossypii*, *Myzus persicae*, and *Macrosiphum solanifolii*. The virus is sap-transmissible and not seed-borne in lilies (Brierley and Smith, 1944b; Alper et al., 1982).

Cucumber mosaic cucumovirus (CMV)

Leaves show chlorotic or yellow spots or stripes or vein-clearing (Fig. 28.5).

Fig. 28.5. Vein-clearing and leaf-curling in cv. Apeldoorn caused by cucumber mosaic virus (CMV).

Later grey or brown necrotic spots may develop. A coarse breaking pattern in the flowers is possible in some cultivars. Leaves and petals may curl or may be malformed (Fig. 28.5). The virus is not very prevalent in lilies and occurs in a limited number of cultivars. Symptoms in *L. longiflorum* and *L. tigrinum* are very faint when singly infected with cucumber mosaic virus but pronounced when co-infected with lily symptomless virus (Brierley and Smith, 1944a; Mokrá; 1976; anonymous, 1983).

CMV detection in lily leaves and bulbs is possible with DAS-ELISA using the same protocols as described for lily symptomless virus (Hagita, 1989; Derks, unpublished results).

The virus is not seed-borne in lilies. For more details about CMV, see Chapter 5.

Lily X potexvirus (LVX)

Very faint chlorotic spots appear on the leaves and occasionally brown necrotic lesions. In most cultivars there are no symptoms. LVX stimulates the symptom expression of lily symptomless virus (Stone, 1980; Derks and Hendriks, 1990). Lily virus X is a potexvirus with an average particle length of c. 550 nm. The virus occurs occasionally in lilies. No other natural hosts are known. The virus is sap-transmitted to the test plants *Tetragonia expansa*, *Chenopodium murale*, *Gomphrena globosa*, and *Nicotiana clevelandii* (Stone, 1980).

LVX is detectable with DAS-ELISA in leaves and bulbs using the same protocols as for lily symptomless virus. A cDNA clone complementary to the 3′-terminal region of the genomic RNA (2343 bp) has been sequenced (Memelink et al., 1990). The virus is serologically related to Commelina virus X. It does not spread rapidly and is transmitted in a persistent or semi-persistent manner, although the vector is still not known (Stone, 1980; Asjes, 1991).

Control

Plants showing symptoms have to be rogued as early in the season as possible. Growing lilies derived from nuclear stocks or seedlings next to old, virus-infected stocks should be avoided.

The spread of LMoV and LSV is considerably reduced by weekly spraying with mineral oils and/or pyrethroids. Spread of LVX is reduced by pyrethroids and Aldicarb but not significantly by mineral oils. To reduce virus spread further; weed growth has to be minimized, and the plant density of the lilies has to be high (Asjes, 1991). Covering fields with knitted, white, high-density, polyethylene nets leads to a much lower incidence of virus diseases (Hagita et al., 1989).

Since great differences exist between lily cultivars in their resistance against LMoV, selection of new resistant cultivars is possible (Mokrá, 1976). In the Netherlands stocks are field-inspected and samples are tested in ELISA by the Bulb Inspection Service (BKD). Nuclear stocks are not certified unless they are virus-free. For other stocks the maximum percentages of LMoV and LVX vary between 0–1% and for LSV between 1–10%, depending the classification (OEPP/EPPO, 1993).

Meristem Culture

In the 1970s all lilies were infected with LSV. An extensive program was started in the Netherlands to free lilies of this virus by meristem culture. For Asiatic hybrids the methods using adventitiously formed meristems on bulb-scale explants resulted in 50–70% LSV-free plants. With *L. longiflorum* cultivars no virus-free plants were obtained using these meristems. However, the culture of isolated vegetative stem apices of the bulbs can be used to free these lilies from LSV (Asjes et al., 1974; Allen et al., 1980; Blom-Barnhoorn and Van Aartrijk, 1985). The addition of 400 μM virazole to the meristem-inducing medium was effective in raising the percentage of plants without LSV but had no effect on the elimination of LMoV. Detection of LSV with ELISA during culture *in vitro* has a high reliability in contrast to the detection of LMoV (Blom-Barnhoorn and Van Aartrijk, 1985).

Final Remarks

Other viruses incidentally detected in lilies are: arabis mosaic and tobacco ringspot nepoviruses, tobacco rattle tobravirus, citrus tatter leaf virus, narcissus mosaic potexvirus, tobacco mosaic tobamo virus (Mokrá, 1976; Alper et al., 1982; Bellardi et al., 1988) and strawberry latent ringspot virus (Cohen et al., 1995). Furthermore, mycoplasma-like organisms may infect lilies (Bertaccini and Marani, 1982).

Selected Literature

Allen, T.C. 1972. Lily symptomless virus. CMI/AAB Descriptions of Plant Viruses, No. 96.

Allen, T.C., Ballantyne, O., Goodell, J., Anderson, W.C. and Lin, W. 1980. Recent advances in research on lily symptomless virus. Acta Hortic. 109:479–485.

Alper, M., Koenig, R., Lesemann, D.E. and Loebenstein, G. 1982. Mechanical transmission of a strain of tulip breaking virus from *Lilium longiflorum* to *Chenopodium* spp. Phytoparasitica 10:193–199.

Anonymous, 1983. Ziekten en afwijkingen bij bolgewassen. Deel I: Liliaceae. Ed.: Laboratorium voor Bloembollenonderzoek, Lisse, The Netherlands.

Asjes, C.J. 1991. Control of air-borne field spread of tulip breaking virus, lily symptomless virus and lily virus X by mineral oils, synthetic pyrethroids, and a nematicide in the Netherlands. Neth. J. Plant Pathol. 97:129–138.

Asjes, C.J., Bunt, M.H. and Van Slogteren, D.H.M. 1974. Production of hyacinth mosaic virus-free hyacinths and lily symptomless virus-free lilies by meristem tip culture. Acta Hortic. 36:223–228a.

Asjes, C.J., De Vos, N.P. and Van Slogteren, D.H.M. 1973. Brown ring formation and streak mottle, two distinct syndromes in lilies associated with complex infections of lily symptomless virus and tulip breaking virus. Neth. J. Plant Pathol. 79:23–35.

Beattie, D.J. and White, J.W. 1993. *Lilium*-hybrids and species. In: The Physiology of Flower Bulbs, A. De Hertogh and M. Le Nard (eds.), Elsevier, Amsterdam, pp. 423–454.

Beijersbergen, J.C.M. and Van der Hulst, C.Th.C. 1980. Application of enzymes during bulb tissue extraction for detection of lily symptomless virus by ELISA in *Lilium* spp. Neth. J. Plant Pathol. 86:277–283.

Bellardi, M.G., Marani, F. and Bertaccini, A. 1988. Narcissus mosaic virus in lily. Acta Hortic. 234:457–464.

Bertaccini, A. and Marani, F. 1982. Electron microscopy of two viruses and mycoplasma-like organisms in lilies with deformed flowers. Phytopathol. Mediterr. 21:8–14.

Blom-Barnhoorn, G.J. and Van Aartrijk, J. 1985. The regeneration of plants free of LSV and TBV from infected *Lilium* bulb-scale explants in the presence of Virazole. Acta Hortic. 164:163–168.

Boontjes, J. 1983. Virusvrije lelies zijn groter areaal waard. Vakblad voor Bloemisterij 18: 36–39.

Brierley, P. and Smith, F.F. 1944a. Studies on lily virus diseases: the necrotic-fleck complex in *Lilium longiflorum*. Phytopathology 34:529–555.

Brierley, P. and Smith, F.F. 1944b. Studies on lily virus diseases: the mottle group. Phytopathology 34:718–746.

Cohen, J., Gera, A. and Loebenstein, G. 1995. Strawberry latent ringspot virus in lilies. Eur. J. Plant Pathol. 101(2):217–219.

Dekker, E.L., Derks, A.F.L.M., Asjes, C.J., Lemmers, M.C., Bol, J.F. and Langeveld, S.A. 1993. Characterization of potyviruses from tulip and lily which cause flower-breaking. J. Gen. Virol. 74:881–887.

Derks, A.F.L.M. 1988. Virus spread in ornamental crops and its practical consequences. Acta Hortic. 234:529–535.

Derks, A.F.L.M. 1993. Jvers. Lab. Bloembollen Onder., Lisse, 1991–1992:28–31.

Derks, A.F.L.M. and Asjes, C.J. 1975. Lily symptomless virus in tulip. Neth. J. Plant Pathol. 81:14–21.

Derks, A.F.L.M. and Hendriks, C.H.M. 1990. Symptomen en schade door lelievirus X in lelies. Vakblad voor Bloemisterij 16:30–31.

Derks, A.F.L.M. and Vink-Van den Abeele, J.L., 1980. Purification of lily symptomless virus. Use and value of antisera against intact and pyrrolidine-degraded virus for testing lilies and tulips. Neth. J. Plant Pathol. 86:239–250.

Hagita, T. 1989. Detection of cucumber mosaic virus and lily symptomless virus from bulb scales of Maximowicz's lily by enzyme-linked immunosorbent assay (ELISA). Ann. Phytopathol. Soc. Jpn. 55:344–348.

Hagita, T., Kodama, F. and Akai, J. 1989. The virus diseases of lily in Hokkaido. Ann. Phytopathol. Soc. Jpn. 55:1–8.

Langeveld, S.A., Dore, J.M., Memelink, J., Derks, A.F.L.M., Van der Vlugt, C.I.M., Asjes, C.J. and Bol, J.F. 1991. Identification of potyviruses using the polymerase chain reaction with degenerate primers. J. Gen. Virol. 72:1531–1541.

Memelink, J., Van der Vlugt, C.I.M., Linthorst, H.J.M., Derks, A.F.L.M., Asjes, C.J. and Bol, J.F. 1990. Homologies between the genomes of a carlavirus (lily symptomless virus) and a potexvirus (lily virus X) from lily plants. J. Gen. Virol. 71:917–924.

Miller, W.B. 1993. *Lilium longiflorum*. In: The Physiology of Flower Bulbs, A. De Hertogh and M. Le Nard (eds.), Elsevier, Amsterdam, pp. 391–422.

Mokrá, V. 1976. Virus diseases of lilies. The lily yearbook of the North American Lily Society 29:20–23.

Mowat, W.P. and Stefanac, Z. 1974. Aphid-transmitted viruses from lilies in Britain. Ann. Appl. Biol. 76:281-287.

OEPP/EPPO, 1993. Certification scheme. Bull. OEPP/EPPO Bull. 23:215–224.

Ohira, K., Namba, S., Miyagawa, M., Kusumi, T. and Tsuchizaki, T. 1994. Nucleotide sequence of the coat protein coding region of tulip breaking virus. Virus Genes 8:165–167.

Phillips, S. and Brunt, A.A. 1986. Four viruses of *Alstroemeria* in Britain. Acta Hortic. 177: 227–233.

PVS/BKD. 1994. Beplante oppervlakten Bloembollen 1992 en 1993: Lelie. Ed. Produktschap voor Siergewassen, The Hague, The Netherlands.

Stone, O.M. 1980. Two new potexviruses from monocotyledons. Acta Hortic. 110:59–63.

Van Schadewijk, A.R. 1986. Detection of tulip-breaking virus and lily symptomless virus in lily bulbs by means of ELISA. Acta Hortic. 177:121–128.

VBN. 1993. Statistiekboek 1992. Ed. Vereniging van Bloemenveilingen in Nederland, Leiden, The Netherlands.

29

Narcissus

Alan A. Brunt
Horticulture Research International, Littlehampton, Sussex, UK

Introduction

Narcissus, a genus of the Amaryllidaceae, contains c. 63 species; these are mostly of Mediterranean origin, but some are now cultivated in temperate countries worldwide. Some species are now commercially very important flowerbulb crops in many such countries, bulbs mainly being grown and sold for garden and amenity planting, for forced cut flower production, and for use as pot plants. *Narcissus pseudonarcissus* (trumpet daffodil) is commercially the most important species, although *N. tazetta* (Chinese sacred lily, or paperwhite) is also of special importance in some countries including Israel, Japan, and China. Other commercially less important species are *N. triandrus* (Angel's tears), *N. viridiflorus* (Autumn daffodil), *N.* × *odorus* (Campernelle jonquil), *N. bulbocodium* (Hoop-Petticoat), and *N. poeticus* (poet's narcissus, pheasant's eye).

It is estimated that world production of narcissus in the late 1980s was c. 7200 ha./annum (Hanks, 1993). The UK (3972 ha.) and the Netherlands (1639 ha.) were the largest producers, although there was significant production also in the US (467 ha.), Channel Islands (300 ha.), Scotland (255 ha.), Poland (200 ha.), and Israel (144 ha.) and, to a lesser extent, Italy (80 ha.), Denmark (36 ha.), Japan (28 ha.), Germany (29 ha.), Ireland (26 ha.), France (21 ha.), and South Africa (12 ha).

The Viruses

Twenty-one viruses have been reported to occur naturally in narcissus (Table 29.1). Of these, nine are transmitted by aphids, one by thrips, one by a fungus, and seven by nematodes; the remaining three viruses have no known vectors. Of these 21 viruses, eight (BBWV, CarLV, NWSV, OYDV, TSWV, Tom RSV, RRSV, and NVQ) either occur in narcissus in restricted geographical locations or are generally considered to be of only minor, if any, economic importance. Thus, BBWV and TomRSV have been reported to occur in narcissus only in Japan (Iwaki and Komuro, 1971, 1972), RRSV in Scotland (Mowat, 1968), TSWV in Australia (G. Guy, personal communication), CarLV in Israel (Brunt, 1977a), TNV in the Netherlands (Asjes, 1974), and NVQ in northern Europe (S. Phillips, W.P. Mowat and A.A. Brunt, unpublished information); the reported occurrence of OYDV in *N. tazetta* and *N. jonquilla* in the US (Brierley and Smith, 1946) has yet to be confirmed. Moreover, although NWSV has been reported to be a distinct virus (Blanton, 1939; Haasis, 1939b; McWhorter, 1957; Mowat et al., 1988), it has yet to be characterized and might well be a synonym of NLSYV (C.J. Asjes, personal communication).

Seven of the viruses (NDV, NLV, NYSV, NLSYV, NMV, NTNV, and NVQ) are generally considered to be specific pathogens of narcissus, although NLV and NMV are also found occasionally infecting other related species. NDV is known only to occur in *N. tazetta* (Brunt, 1980), but the other viruses also infect other species and some such general pathogens have very extensive natural host ranges.

The viruses often occur in complexes in narcissus; up to five viruses have been isolated from older cultivars such as King Alfred, Golden Harvest, and Grand Soleil d'Or (Brunt, 1966a, 1966b, 1977a, 1977b).

The thirteen viruses of economic importance in narcissus are each considered in further detail.

Table 29.1. Viruses reported to occur naturally in narcissus

1. Transmitted by aphids	
Broadbean wilt (BBWV)	Iwaki and Komuro, 1972
Carnation latent (CarLV)	Brunt, 1977a
Cucumber mosaic (CMV)	Brunt, 1966a; Iwaki and Komuro, 1972
Narcissus degeneration (NDV)	Brunt, 1966a, 1970, 1971b, 1980
Narcissus latent (NLV)	Brunt and Atkey, 1967; Brunt, 1977b; Mowat et al., 1991
Narcissus late season yellows (NLSYV)	Brunt, 1977a, 1980; Mowat et al., 1988
Narcissus white streak (NWSV)	Haasis, 1938b; Blanton, 1939
Narcissus yellow stripe (NYSV)	Dod, 1984; Haasis, 1939a; Blanton and Hassis, 1939, 1940, 1942; Brunt, 1971a
Onion yellow dwarf (OYDV)	Brierley and Smith, 1946
2. Transmitted by thrips	
Tomato spotted wilt (TSWV)	G. Guy (personal communication)
3. Transmitted by nematodes	
Arabis mosaic (AMV)	Broadbent et al., 1962; Brunt, 1966a; Asjes, 1971, 1972a
Raspberry ringspot (RRSV)	Mowat, 1968
Strawberry latent ringspot (SLRV)	Brunt, 1966a
Tobacco ringspot (TobRSV)	Asjes, 1968, 1971, 1972a
Tomato black ringspot (TBRSV)	Broadbent et al., 1962; Brunt, 1966b
Tomato ringspot (TomRSV)	Iwaki and Komuro, 1971
Tobacco rattle (TRV)	Slogteren, 1958; Brunt, 1966a; Asjes, 1971, 1972b
4. Transmitted by a fungus	
Tobacco necrosis (TNV)	Asjes, 1974
5. No known vector	
Narcissus mosaic (NMV)	Slogteren and Ouboter, 1946; Slogteren, 1955; Brunt, 1966b
Narcissus tip necrosis (NTNV)	Mowat et al., 1977
Narcissus Q (NVQ)	W.P. Mowat, personal communication, S. Phillips, personal communication

Aphid-borne viruses

Cucumber mosaic cucumovirus (CMV)

This occurs commonly in *N. tazetta* in Britain and Japan, but only rarely in

Fig. 29.1. Leaf symptoms induced in *Narcissus tazetta* cv. Grand Soleil d'Or by narcissus degeneration virus.

trumpet daffodils; it invariably occurs together with other viruses, so its symptomatology when alone is unknown (Brunt, 1966a; Iwaki and Komuro, 1972).

Narcissus degeneration potyvirus (NDV)

This is probably the most important of the five viruses which occur, often in complexes, in *N. tazetta* cv. Grand Soleil d'Or in S.W. Britain (Brunt, 1966a). The virus, which induces chlorotic leaf striping in *N. tazetta* plants (Fig. 29.1), is apparently a typical but distinct member of the potyvirus genus (Brunt, 1969; Mowat et al., 1988).

Narcissus latent macluravirus (NLV)

This virus was first detected in narcissus in Britain (Brunt and Atkey, 1967), but is now known to occur also in bulbous iris, gladiolus, and *Acidanthera* sp. (Brunt, 1977b; unpublished information). It often occurs together with other viruses in daffodils, but alone induces symptomless infection or mild leaf chlorosis especially at the leaf tips of some cultivars; however, it probably exacerbates the symptoms of other viruses with which it commonly occurs in narcissus.

NLV has long been considered to be a distinct carlavirus (Brunt, 1977b). However, it is now known to induce pinwheel inclusions in infected plants (Mowat et al., 1991; Brunt et al., 1994); it is thus now best considered to be a member of the Potyviridae and, moreover, probably to be included with the similar maclura mosaic virus in a new taxon (Mowat et al., 1991) tentatively designated the macluravirus genus (Shukla et al., 1994).

Narcissus late season yellows potyvirus (NLSYV)

Daffodils infected with this virus remain symptomless throughout much of the growing season but characteristically develop chlorotic striping and extensive leaf chlorosis (Fig. 29.2) late in the season (Brunt, 1977a). NLSYV is now considered to be a distinct potyvirus (Mowat et al., 1988). Silver or white streak leaf symptoms (Fig. 29.3), previously reported to be induced by narcissus white streak virus (Blanton, 1939; Haasis, 1939a, 1939b; McWhorter, 1957; Mowat et al., 1988), might also be attributable to NLSYV (C.J. Asjes, personal communication).

Narcissus yellow stripe potyvirus (NYSV)

Symptoms attributable to this virus have been known to occur in narcissus in Britain and the Netherlands for over 100 years (Dod, 1894). The virus probably causes the most important virus-induced disease of narcissus, and now possibly occurs wherever daffodils are grown.

Infected plants are often stunted and have leaves, sometimes also distorted, with chlorotic streaks particularly in their upper parts (Fig. 29.4); the flowers of such plants are also often "broken". Leaf symptoms appear early in the season and can often be seen soon after leaf emergence. The yield of infected plants is severely reduced (Haasis, 1939a).

The suggestion by Mowat et al. (1991) that yellow stripe symptoms in narcissus are possibly attributable to some isolates of NLV needs to be further investigated.

Nematode-borne viruses

Tobacco rattle tobravirus (TRV)

This was first found infecting narcissus in the Netherlands (Slogteren, 1958), but was later also found infecting stocks with leaf mottling-mosaic symptoms in Britain (Brunt, 1966a). The virus occurs commonly in some cultivars such as Golden Harvest (Fig. 29.5) and Carlton, but only infrequently or not at all in many others.

TRV can occur symptomlessly in some susceptible cultivars (Asjes, 1974). However, in others it induces yellow or chlorotic streaks especially in leaf bases, leaf distortion, and sometimes "mild breaking" of flowers. Leaf symptoms of TRV, like those of NYSV, appear early in the season, and those of the two viruses can be easily confused.

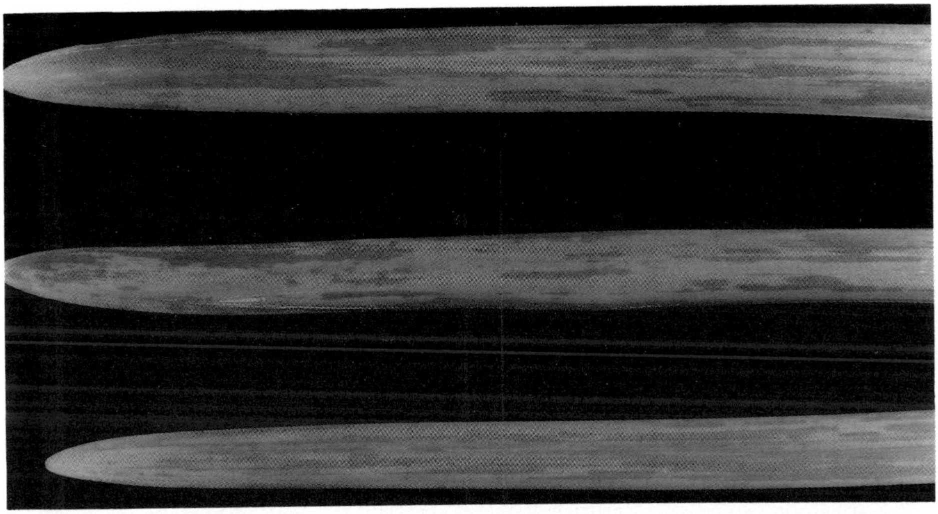

Fig. 29.2. Leaf symptoms induced by narcissus late season yellows virus in *N. pseudonarcissus* cv. Golden Harvest.

Fig. 29.3. Silver streak leaf symptoms in *N. pseudonarcissus* cv. Golden Harvest.

Soil-borne ringspot viruses (nepoviruses)

Although TomRSV has been isolated only from daffodils with mild mosaic leaf symptoms in Japan (Iwaki and Komuro, 1971) and RRSV has been found

Fig. 29.4. Leaf symptoms induced by narcissus yellow stripe virus in *N. pseudonarcissus* cv. Golden Harvest.

infecting daffodils only in Scotland (Mount, 1968), four other nepoviruses occur in trumpet daffodils and *N. tazetta*, perhaps wherever the crops are grown.

Arabis mosaic nepovirus (ArMV) and tomato black ring nepovirus (TBRV)

These viruses were first detected in narcissus in Scotland and S.W. England (Broadbent et al., 1962), but these and strawberry latent ringspot nepovirus (SLRV) are now known to occur commonly in narcissus in N.W. Europe. Tobacco ringspot nepovirus (TobRSV) was first detected in narcissus as well as tulips, irises, and hyacinths in the Netherlands by Asjes (1969). In narcissus, it induces twisting and pale chlorosis of leaf centres (Asjes, 1974).

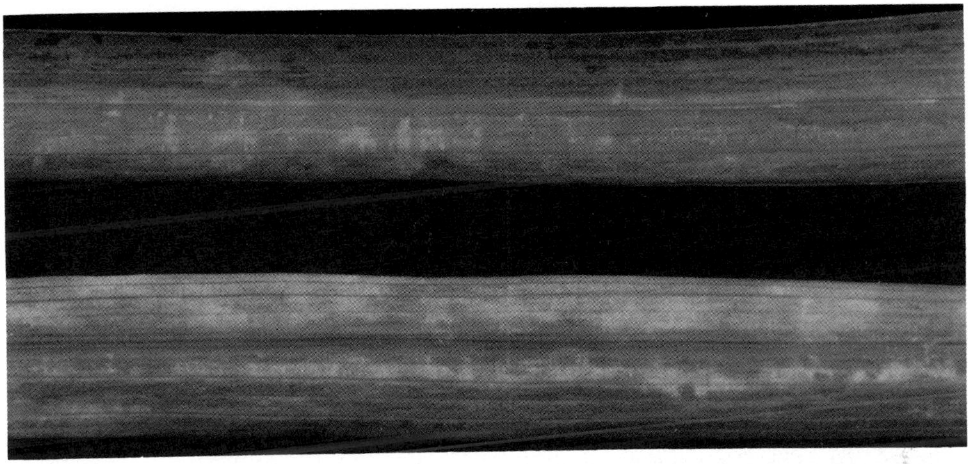

Fig. 29.5. Leaf symptoms inducd by tobacco rattle virus in *N. pseudonarcissus* cv. Kilmorack.

Daffodil seedlings infected with ArMV, SLRV, or TBRV by mechanical inoculation developed no conspicuous symptoms, observations suggesting that these viruses in daffodils are much less damaging than those transmitted by aphids (Brunt, 1966a). However, when occurring in complexes, they possibly exacerbate the symptoms induced by other viruses and, as nepoviruses have extensive natural host ranges, infected daffodils are possibly more important as foci of infection for other more susceptible crops.

Viruses with no known invertebrate vectors

Narcissus mosaic potexvirus (NMV)

This virus was first identified by Slogteren and Ouboter (1946) and later shown to differ morphologically (Cremer and Veken, 1964) and serologically (Slogteren, 1955) from NYSV. The virus, a typical but distinct potexvirus, alone induces mild mosaic leaf symptoms or dark brown oval spots mainly on the top halves of the leaves in infected daffodils (Brunt, 1966b; C.J. Asjes, personal communication). However, it frequently occurs in complexes with other viruses in more severely affected plants.

Narcissus tip necrosis carmovirus (NTNV)

A virus subsequently designated NTNV was independently detected in daffodils in Scotland, the Netherlands and England (Mowat et al., 1977).

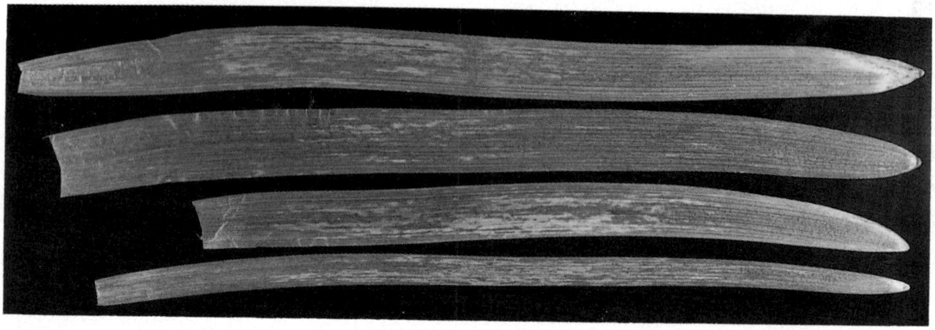

Fig. 29.6. Leaf symptoms induced by narcissus tip necrosis virus in *N. pseudonarcissus* cv. Sempre Avanti. Photograph courtesy of the Scottish Crop Research Institute.

The virus is widespread in commercially important daffodil cultivars; in some it induces no conspicuous symptoms but in others it is associated with the development during the late season of leaf tip chlorosis; such areas later become necrotic, and the necrosis continues basipetally, ultimately resulting in premature leaf senescence. Symptoms, however, are dependent upon environmental conditions, milder symptoms only developing after prolonged periods of dry, sunny weather (Mowat et al., 1977). The reaction of cv. Sempre Avanti to infection differs from that of other cultivars; infected plants develop chlorotic and necrotic streaks mainly in the mid-region of abaxial leaf surfaces, and such leaves senesce prematurely (Fig. 29.6).

Transmission

The aphid-borne viruses infecting narcissus are all transmitted in the non-persistent manner by one or more aphid species including *Aphis fabae*, *Acrythosiphon pisum*, *Macrosiphum euphorbiae*, and *Myzus persicae*. Aphids rarely colonize narcissus, although some species do so under exceptionally favorable climatic conditions (Blanton and Haasis, 1942; Slogteren and Ouboter, 1946; Broadbent et al., 1962). Spread of the viruses is thus largely attributable to migratory alatae which briefly settle and feed on daffodils. The specific narcissus viruses can be acquired only from infected daffodils, so their spread is only from foci within — or closely adjacent to — daffodil crops; the general pathogens, by contrast, can also be acquired by aphids from unrelated infected plant species.

Aphids are rarely detected in daffodil crops before mid-May in S.E. England and probably at comparable times in other temperate countries. As the daffodil season extends only for another 6–8 weeks, exposure to vector aphids is restricted to the latter part of the growing season. Nevertheless, the extent and rate of spread is closely correlated with the size of aphid populations and the number of primary foci of infection (Broadbent et al., 1962; Haasis, 1939a).

The nematode-borne viruses occur in a wide range of weed and cultivated plant species, and so can be acquired by nematodes from daffodils and/or other species. TRV is transmitted by *Trichodorus* and *Paratrichodorus* species, and the other viruses by *Longidorus* and *Xiphinema* species. The nematode-borne ringspot viruses are also seed-borne in daffodils, a mode of transmission perhaps of importance only to breeders.

The mode of transmission of three viruses (NTNV, NMV, NVQ) is unknown.

Identification and Control

All the important viruses infecting daffodils can be rapidly detected and/or identified by serological procedures such as ELISA and immunosorbent electron microscopy and some by cDNA hybridization (e.g. Harrison et al., 1983; Brunt et al., 1984; Mowat et al., 1988, 1989). The availability of general primers for potyviruses also permits the detection of such viruses in daffodils by PCR (Langeveld et al., 1991). Digestion of such amplified fragments with restriction enzymes permits three potyviruses to be distinguished (S.A. Langeveld and A.F.L.M. Derks, unpublished results). These potyviruses can also be detected, but not specifically identified, in diagnostic tests using the general potyvirus monoclonal antibody supplied by Agdia Inc. (Elkhart, Indiana, USA).

Viruses are prevalent in stocks of older commercial cultivars. The best strategy to minimize virus-induced crop losses, therefore, is to select or obtain virus-tested (putative virus-free) plants and, by their rapid multiplication using twin-scaling, chipping, or *in vitro* methods, obtain elite stocks which thereafter should be protected as far as possible from reinfection. By repeated careful testing, it is sometimes possible to select apparently virus-free plants from a stock of bulbs, especially those of newer cultivars. However, it is often necessary to produce such plants by meristem-tip culture; culture media and growing conditions have been defined (e.g. Stone et al., 1978; Brunt et al., 1984) and, as described above, effective procedures for virus testing and rapid bulb multiplication are now available.

A pilot scheme with *N. tazetta* cv. Grand Soleil d'Or showed that virus-tested stocks produced greatly increased bulb and flower yields and flowers of better quality (Stone et al., 1978). Virus-tested stocks of another 78 cultivars have since been produced in the UK at Horticulture Research International, Littlehampton and The Scottish Crop Research Institute, Dundee (Brunt et al., 1984; Mowat et al., 1986) and the productivity of some tested in field trials.

Virus-tested stocks grown in Scotland and England are usually superior to those of normal commercial stocks (Stone et al., 1978; Brunt et al., 1984; Mowat et al., 1986). For example, bulb yields of cvs Fortune and Carlton grown in S.W. England were 50% and 15–25% higher, and those of cvs Ice Follies and White Lion were also higher; the quality and yield of flowers from virus-tested stocks, however, did not differ significantly from those of commercial stocks. The performance of virus-tested stocks in eastern England did not always produce such large benefits, probably because virus-tested stocks were more suited to the earlier and shorter growing season and/or their ability to resist leaf diseases which are more prevalent in S.W. England. Similar trials are planned for elite stocks of cvs Gold Harvest, St. Keverne, Dutch Master, and Tête-a-Tête, and virus-free bulbs of another 30 cultivars are being propagated for possible future use. The wider use of improved bulb stocks is being encouraged as their cultivation should lead to increased quality and yield of bulbs and also to a reduction in the number of stocks rejected for export because of virus infection.

A voluntary Plant Health Propagation Scheme for narcissus is in operation in England (Ebbels, 1986) and a Narcissus Certification Scheme in Scotland (Mowat et al., 1986). Certification and Classification Schemes have been prepared for wider use by the European and Mediterranean Plant Protection Organisation.

Selected Literature

Asjes, C.J. 1968. Virusziekten in narcissen. Jaarverslag Laboratorium Bloembollenonderzoek, Lisse 1967–1968, pp. 37–38.

Asjes, C.J. 1971. Virus diseases in narcissus in The Netherlands. Daffodil Journal 8:3–11.

Asjes, C.J. 1972a. Virus diseases in narcissus. Gardeners Chronicle 172:28–29.

Asjes, C.J. 1972b. Virus diseases in narcissus, tobacco rattle virus. Gardeners Chronicle 172:32–33.

Asjes, C.J. 1974. Soil-borne virus diseases in ornamental bulb crops and their control in The Netherlands Agric. Environ. 1:303–315.

Blanton, F.S. 1939. Aphid transmission of the virus causing white streak of narcissus. J. Econ. Entomol. 32:726–727.

Blanton, F.S. and Haasis, F.A. 1939. Transmission of narcissus mosaic by aphids. J. Econ. Entomol. 32:469–470.

Blanton, F.S. and Haasis, F.A. 1940. Three additional species of aphids transmitting narcissus mosaic. J. Econ. Entomol. 33:942.

Blanton, F.S. and Haasis, F.A. 1942. Insect transmission of the virus causing narcissus mosaic. J. Agric. Res. 65:413–419.

Brierley, P. and Smith, F.F. 1946. Reaction of onion varieties to yellow dwarf virus and to three similar viruses isolated from shallot, garlic and Narcissus. Phytopathology 38: 292–296.

Broadbent, L., Green, D.E. and Walker, P. 1962. Narcissus virus diseases. Daffodil and Tulip Yearbook 28:1–7.

Brunt, A.A. 1966a. The occurrence of cucumber mosaic and four nematode transmitted viruses in British narcissus crops. Plant Pathol. 15:157–160.

Brunt, A.A. 1966b. Narcissus mosaic virus. Ann. Appl. Biol. 58:13–23.

Brunt, A.A. 1969. Narcissus viruses. Report of the Glasshouse Crops Research Institute 1968, pp. 105–106.

Brunt, A.A. 1970. Virus diseases of narcissus. Daffodil and Tulip Yearbook 37:18–37.

Brunt, A.A. 1971a. Narcissus yellow stripe virus. CMI/AAB Descriptions of Plant Viruses, No. 76.

Brunt, A.A. 1971b. Occurrence and importance of viruses infecting narcissus in Britain. Acta Hortic. 23:292–299.

Brunt, A.A. 1977a. Bulb crops; narcissus. Report of the Glasshouse Crops Research Institute 1976, pp. 122–123.

Brunt, A.A. 1977b. Some hosts and properties of narcissus latent virus, a carlavirus commonly infecting narcissus and bulbous iris. Ann. Appl. Biol. 87:355–364.

Brunt, A.A. 1980. A review of problems and progress in research on viruses and virus diseases of narcissus in Britain. Acta Hortic. 110:23–30.

Brunt, A.A. and Atkey, P.T. 1967. Rapid detection of narcissus yellow stripe virus and two other filamentous viruses in crude negatively stained narcissus sap. Report of the Glasshouse Crops Research Institute 1966, pp. 155–169.

Brunt, A.A., Stone, O.M. and Phillips, S. 1984. Recent progress in the production, propagation and distribution of virus-free flower bulbs. Report of the Glasshouse Crops Research Institute 1983, pp. 109–116.

Brunt, A.A., Atkey, P.T., Frost, C. and Phillips, S. 1994. Cytoplasmic inclusions induced by narcissus latent virus. Acta Hortic. 377:275–280.

Cremer, M.C. and Veken, J.A. van der 1964. Localisation of viral antigen in narcissus leaves infected with yellow stripe virus, determined by means of a fluorescein conjugated antiserum. Neth. J. Plant Pathol. 72:105–113.

Dod, C.W. 1984. Variegation in Narcissus. J. Royal Soc. 17:22.

Ebbels, D.L. 1986. Certification of varieties and varieties of certification. In: Healthy Planting Material, D. Rudd-Jones and F.A. Langton (eds.), British Crop Protection Council, Thornton Heath, UK, pp. 35–37.

Haasis, F.A. 1939a. Studies on narcissus mosaic. Memoirs of Cornell University Agricultural Experiment Station 224.

Haasis, F.A. 1939b. White streak of narcissus. Phytopathology 29:890–895.

Hanks, G.R. 1993. Narcissus. In: The Physiology of Flowerbulbs, A. de Hertogh and M. Le Nard (eds.), Elsevier, Amsterdam.

Harrison, B.D., Robinson, D.J., Mowat, W.P. and Duncan, G.H. 1983. Comparison of nucleic acid hybridisation and other tests for detecting tobacco rattle virus in narcissus plants and potato tubers. Ann. Appl. Biol. 102:331–338.

Iwaki, M. and Komuro, Y. 1971. Viruses isolated from narcissus Narcissus spp. in Japan. II. Tomato ringspot virus and its transmission by *Xiphinema americanum*. Ann. Phytopathol. Soc. Jpn. 37:108–116.

Iwaki, M. and Komuro, Y. 1972. Viruses isolated from narcissus (*Narcissus* spp.) in Japan. III. Cucumber mosaic virus, tobacco rattle virus and broad bean wilt virus. Ann. Phytopathol. Soc. Jpn. 38:137–145.

Langeveld, S.A., Dore, J-M., Memelink, J., Derks, A.F.L.M., van der Vlugt, C.I.M., Asjes, C.J. and Bol, J.F. 1991. Identification of potyviruses using the polymerase chain reaction with degenerate primers. J. Gen. Virol. 72:1531–1541.

McWhorter, F.P. 1957. Bulb Growing and Forcing, Northwest Bulb Growers Association Handbook, pp. 127–130.

Mowat, W.P. 1968. NEPO viruses in narcissus and tulip Report of the Scottish Horticultural Research Institute 1967, p. 57.

Mowat, W.P., Dixon, G.R. and Fordyce, W. 1986. The production of virus-tested narcissus in Scotland. In: Healthy Planting Material, D. Rudd-Jones and A.F. Langton (eds.), British Crop Protection Council, Thornton Heath, UK, pp. 193–198.

Mowat, W.P., Asjes, C.J. and Brunt, A.A. 1977. Occurrence, purification and properties of narcissus tip necrosis virus. Ann. Appl. Biol. 86:189–198.

Mowat, W.P., Duncan, G.H. and Dawson, S. 1988. Narcissus late season yellows potyvirus: symptoms, properties and serological detection. Ann. Appl. Biol. 113:531–544.

Mowat, W.P., Dawson, S. and Duncan, G.H. 1989. Production of antiserum to a non-structural potyviral protein and its use to detect narcissus yellow stripe and other potyviruses. J. Virol. Meth. 25:199–210.

Mowat, W.P., Dawson, S., Duncan, G.H. and Robinson, D.J. 1991. Narcissus latent, a virus with filamentous particles and a novel combination of properties. Ann. Appl. Biol. 119:31–46.

Shukla, D., Ward, C.M. and Brunt, A.A. 1994. The Potyviridae. CAB International, Wallingford, pp. 4–6.

Slogteren, E. van 1955. Serological diagnosis of plant diseases. Ann. Appl. Biol. 42:122–128.

Slogteren, E. van and Ouboter, M.P. de Bruyn 1946. Investigations on virus diseases of narcissus. Daffodil Tulip Yearbook 12:3–20.

Slogteren, D.H.M. van 1958. Ratelvirus als oorzaak van ziekten in bloembolgewassen en de mogelijkheden de infectie door middel van grondontsmetting te bestrijden. Tijdschrift voor Plantenziekten 64:452–462.

Stone, O.M., Brunt, A.A. and Hollings, M. 1978. Methods, logistics and problems in the production, distribution and use of virus-free clones of *Narcissus tazetta* cv. Grand Soleil d'Or. Report of the Glasshouse Crops Research Institute 1977, pp. 149–167.

30

Nerine

A.F.L.M. Derks[1] and D.Z. Maat[2]
[1]*Bulb Research Centre, PO Box 85, 2160 AB Lisse, The Netherlands*
[2]*DLO Research Institute for Plant Protection, PO Box 9060, 6700 GW Wageningen, The Netherlands*

Introduction

The genus *Nerine* belongs to the family of the *Amaryllidaceae* and includes about 30 species. It has a pronounced center of origin in the winter rainfall coastal area of South Africa (Dahlgren et al., 1985). In the middle of the 17th century Dutch sailors were responsible for the spread of bulbs of *Nerine sarniensis* to the isle of Guernsey (old name: Sarnia), where from that time on this species has been grown in gardens (Koomen, 1990). The most popular species grown today is *N. bowdenii*, responsible for about 95% of all nerines grown in the Netherlands. Other species grown are *N. sarniensis* (3%), *N. undulata* (= *N. crispa*), *N.* × "Mansellii", and *N. flexuosa*. The acreage of nerine grown in the Netherlands amounts to about 75 ha. In 1993 about 22.7 million stems were sold at the Dutch auctions. Other countries where substantial amounts of nerine are grown are: South Africa, New Zealand, Japan, and Columbia (N. Groen, C. Buschman; pers. comm.).

The Viruses

Nerine yellow stripe potyvirus (NeYSV)

Infected plants often show chlorotic stripes on the leaves and flower stalks (Fig. 30.1). Other types of leaf symptoms are a faint mosaic or dark green oval spots (Fig. 30.2). Flowers may show a breaking pattern and malformation. Depending on the cultivar, symptoms may appear or disappear later in the growing season. Symptomless infections are also possible (Derks and Maat, unpublished). Nerine yellow stripe virus is a potyvirus with particles of 770–800 nm long. It occurs frequently in cultivars of *N. bowdenii* and in *N. sarniensis*. Other hosts are *Lycoris* and *Amaryllis belladonna* (Derks, unpublished). Test plants are not known.

An antiserum prepared by Maat is useful for the detection of NeYSV in DAS-ELISA. The virus can be identified by PCR with potyvirus specific, degenerate primers followed by RFLP analysis (Langeveld et al., 1991).

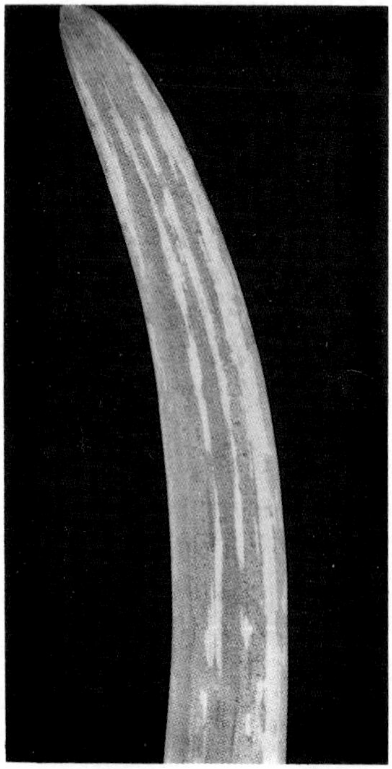

Fig. 30.1. Chlorotic stripes on a leaf of a *Nerine* hybrid caused by nerine yellow stripe virus (NeYSV).

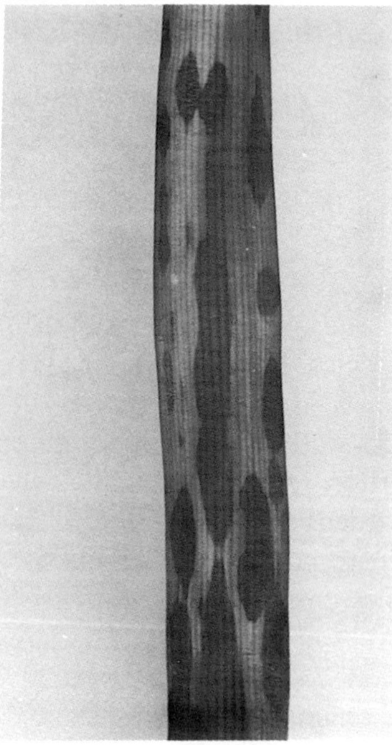

Fig. 30.2. Dark green oval spots on a leaf of a *Nerine* hybrid caused by nerine yellow stripe virus (NeYSV).

NeYSV is not transmitted by seed (Hakkaart et al., 1975). Vectors are unknown.

Nerine latent carlavirus (NeLV)

The virus occurs frequently in cultivated *Nerine* species and hybrids without showing symptoms (Brunt et al., 1970; Maat et al., 1978). Other hosts are *Hippeastrum* (Brölman-Hupkes, 1975) and *Eucharis* (Jayasinghe and Dijkstra, 1979).

Nerine latent virus is a carlavirus with a particle length of 660–665 nm. The virus induces local lesions in the test plants *Chenopodium quinoa*, *C. amaranticolor*, and *Gomphrena globosa* and is transmissible to *Nicotiana clevelandii* producing a local symptomless infection. Nerine latent virus is serologically closely related to carnation latent virus and more distantly to potato virus S, potato virus M, chrysanthemum virus B, pea streak virus, and lily symptomless virus. The sedimentation coefficient of NeLV is 155 S (Maat et al., 1978) and the A 260/280 ratio 1.19 (Brölman-Hupkes, 1975). The virus is readily detected with an antiserum in DAS-ELISA (Maat et al., 1978).

NeLV is not transmitted by seed (Hakkaart et al., 1975). Vectors are unknown.

Nerine X potexvirus (NVX)

NVX produces only a faint mosaic on the leaves; in most cases symptoms are absent. The virus occurs in *N. sarniensis*, *N. fothergillii*, and *N. bowdenii* (Maat, 1976; Brunt 1977a; Balasingam et al., 1988). The same or a closely related virus occurs in *Agapanthus* (Phillips and Brunt, 1980; Balasingam et al., 1988).

Nerine virus X is a potexvirus with an average particle length of c. 540 nm. It is transmissible with difficulty to *Chenopodium quinoa*. The virus is detectable with an antiserum useful for DAS-ELISA (Maat, 1976; Phillips and Brunt, 1980) and with a 1.8 kb cDNA in hybridization assays (Balasingam et al. 1988). NVX is serologically distantly related to narcissus mosaic virus and clover yellow mosaic virus (Maat, 1976). No data are available about virus transmission through seed or by vectors.

Tobacco rattle tobravirus (TRV)

Leaves of infected plants show white to yellowish green spots and stripes, which may turn into grey necrotic stripes later in the season. In some cases the leaves tear and bend along the stripes (Fig. 30.3).

Fig. 30.3. Yellowish green and white stripes on leaves of cv. Van Roon caused by tobacco rattle virus (TRV).

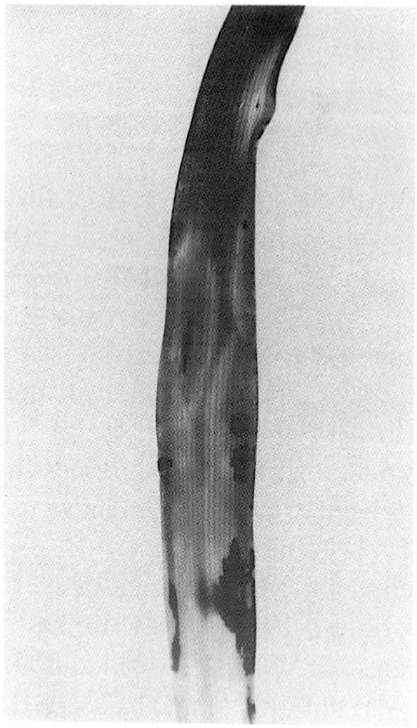

Fig. 30.4. Irregular-shaped yellow and dark green spots on a leaf of cv. Favourite infected with tomato spotted wilt virus (TSWV).

Tobacco rattle virus may occur in *N. bowdenii* grown on sandy and light loamy soil types infested with nematodes of *Trichodorus* spp. The virus is described in Chapter 5.

Frequently, virus detection is difficult using ELISA, electronmicroscopy, or test plants because of low virus concentrations. PCR using TRV-specific primers is a more sensitive detection method from both leaf and bulb extracts in comparison with ELISA and EM (Derks, Lemmers and Langeveld, unpublished).

Cucumber mosaic cucumovirus (CMV)

Field-grown infected plants may show yellow to light green spots on the leaves. It is not certain whether these symptoms are caused by CMV alone or by the combined infection of CMV and NeYSV. The symptoms can be confused with those caused by tomato spotted wilt virus (Derks, unpublished). However, CMV-infected plants often remain symptomless. The virus occurs in *N. bowdenii* and *N. sarniensis* (Hakkaart et al., 1975). For more details about this virus see Chapter 5.

Tomato spotted wilt tospovirus (TSWV)

Infected plants show irregular-shaped, yellow, white, and/or dark green spots (Fig. 30.4). TSWV occurs occasionally in nerine. It is easily detected in glasshouse- and field-grown plants by its pronounced symptoms. Antisera are available for serological confirmation; the virus is easily detected in symptom-bearing parts of leaves. When TSWV infected bulbs are grown for several years, the virus percentage diminishes gradually (Derks, 1995). For further details see Chapter 5.

Control

Stocks destined for bulb production are visually inspected in the Netherlands by the NAKB (in 1993; about 52 ha from the 75 ha grown). Plants with severe symptoms are rogued. From the descriptions above it is clear that this includes the selection for the absence of tobacco rattle virus and tomato spotted wilt virus, and, only partially, of nerine yellow stripe virus.

Virus-free stocks obtained by meristem tip culture (Hakkaart et al., 1975) are grown on limited scale in glasshouses. Since no viruses were detected in seedlings (Hakkaart et al., 1975), it is advised to grow new hybrids in glasshouses separated from old virus-infected stocks.

Final Remarks

It is not certain whether the nerine yellow stripe virus described here corresponds with the NeYSV from A. Brunt with a particle length of c. 750 nm or with nerine virus Y with a particle length of c. 800 nm. The latter virus is easily detectable with a 1.5 kb cDNA probe that does not hybridize with NeYSV from A. Brunt (Balasingam et al., 1988).

A second potyvirus occasionally occurring in *N. sarniensis* in the Netherlands has particles with an average length of c. 740 nm. The symptoms observed with this virus are chocolate brown stripes and oval ring patterns especially at the top halves of the leaves, sometimes accompanied with a stripe-like mosaic. The virus is mechanically transmissible to *Hyoscyamus niger* and is detectable with a specific antiserum in ELISA (Maat and Derks, unpublished).

Other viruses occasionally detected in nerine are narcissus yellow stripe potyvirus (Brunt, 1971), narcissus mosaic potexvirus (Maat, 1976; Koenig et al., 1973), narcissus latent carlavirus (Brunt, 1977b), and arabis mosaic nepovirus (Brunt, 1977a).

Selected Literature

Balasingam, G., Milne, K.S. and Forster, R.L.S. 1988. Sensitive and specific detection of two filamentous viruses from nerines using cloned cDNA probes. Acta Hortic. 234: 267–274.

Brölman-Hupkes, J.E. 1975. Tentative description of Hippeastrum latent virus in *Hippeastrum hybridum* plants and differentation from Hippeastrum mosaic virus. Neth. J. Plant Pathol. 81:226–236.

Brunt, A.A. 1971. Narcissus yellow stripe virus. CMI/AAB Descriptions of Plant Viruses, No. 76.

Brunt, A.A. 1977a. Rep. Glasshouse Crops Res. Inst. 1976:123.

Brunt, A.A. 1977b. Some hosts and properties of narcissus latent virus, a carlavirus commonly infecting narcissus and bulbous iris. Ann. Appl. Biol. 87:355–364.

Brunt A.A., Hollings, M. and Stone, O.M. 1970. Rep. Glasshouse Crops Res. Inst. 1969: 138.

Dahlgren, R.M.T., Clifford, H.T. and Yeo, P.F. 1985. The families of the Monocotyledons. Springer-Verlag, Berlin.

Derks, A.F.L.M. 1995. Tomato spotted wilt and impatiens necrotic spot viruses in flower bulb crops. Acta Bot. Neerlandica, in press.

Hakkaart, F.A., Maat, D.Z. and Quak, F. 1975. Viruses and meristem culture of Nerine. Acta Hortic. 47:51–53.

Jayasinghe, U. and Dijkstra, J. 1979. Hippeastrum mosaic virus and another filamentous virus in *Eucharis grandiflora*. Neth. J. Plant Pathol. 85:47–65.

Koenig, K., Lesemann, D., Brunt, A.A. and Kühne, H. 1973. Narcissus mosaic virus found in *Nerine bowdenii*. Identification aided by anomalies in SDS PAGE. Intervirology 1: 348–353.

Koomen, N.J. 1990. Gestrand VOC-schip zorgde voor mythevorming. Bloembollencultuur 1990 (19):11.

Langeveld, S.A., Dore, J.M., Memelink, J., Derks, A.F.L.M., Van der Vlugt, C.I.M., Asjes, C.J. and Bol, J.F. 1991. Identification of potyviruses using the polymerase chain reaction with degenerate primers. J. Gen. Virol. 72:1531–1541.

Maat, D.Z. 1976. Two potexviruses in Nerine. Neth. J. Plant Pathol. 82:95–102.

Maat, D.Z., Huttinga, H. and Hakkaart, F.A. 1978. Nerine latent virus: some properties and serological detectability in *Nerine bowdenii*. Neth. J. Plant Pathol. 84:47–59.

Phillips, S. and Brunt, A.A. 1980. Some hosts and properties of an isolate of Nerine virus X from *Agapanthus praecox* subsp. *orientalis*. Acta Hortic. 110:65–70.

31

Ornithogalum

J.T. Burger[1] and M.B. Von Wechmar[2]
[1]*Vegetable and Ornamental Plant Institute, P/Bag X293, Pretoria 0001, South Africa*
[2]*Department of Microbiology, University of Cape Town, P/Bag, Rondebosch 7700, South Africa*

Introduction

The aesthetic value of *Ornithogalum* spp. has been appreciated for a long time: the oldest description of a South African *Ornithogalum* species is that of Clusius in 1611 (Obermeyer, 1978). The genus *Ornithogalum* belongs to the family Hyacinthaceae of the order Asparagales (Dahlgren and Clifford, 1982) and consists of approximately 245 species in Africa, Europe, and western Asia. In South Africa (SA) where ornithogalum is commonly known as the chinkerinchee, 120 species are indigenous (Ferreira and Hancke, 1986). The genus includes several horticulturally important species and a few that are extremely poisonous to livestock (Watt and Breyer-Brandwijk, 1962).

Ornithogalum has excellent potential as an ornamental cut flower crop: it is resistant to extreme temperatures; longevity of flowers after cutting is exceptionally high (both in cold storage before distribution, and in the vase); minimal pigment change occurs in flowers after cutting; flower stalks are long and sturdy; flowering periods can be manipulated, and plants are easily and rapidly vegetatively propagated.

A breeding program with *Ornithogalum* spp. was started in South Africa during the 1960s and is being continued at the Vegetable and Ornamental Plant Institute (VOPI), using mainly *O. thyrsoides* Jacq., *O. conicum* Jacq., *O. dubium* Houtt., *O. maculatum Jacq.*, and *O. pruinosum* Leighton. To date, the VOPI has released seven cultivars for commercial cultivation. In SA several commercial ornithogalum growers are producing cut flowers for the local and overseas markets. A few of the larger growers have breeding programs and develop their own cultivars that are not commercially available. Due to protected interests, the exact value and size of the market are not known.

The greatest natural concentration of *Ornithogalum* species is found in the south-western Cape region of SA where predominantly *O. thyrsoides* grow densely in large patches (Fig. 31.1). Growing areas represent great extremes in climate varying from arid rocky areas to sandy flats, marshes, river banks, and lower mountain slopes.

Fig. 31.1. a. Natural stand of *Ornithogalum thyrsoides* in a marsh. b. Flower detail.

The Viruses

The first record of what became known as ornithogalum mosaic virus (OMV) was written by Nance in 1940 (according to Smith and Brierley, 1944). Until 1986, no real attempts were made to isolate and characterize the virus(es) which was becoming a serious threat to the success of the breeding programs. Some of the ornithogalum cultivars which have been commercially released are currently being re-evaluated as a result of the high virus incidence in the bulbs. The first isolation and characterization of OMV, the cause of mosaic disease in ornithogalum, was reported by Burger and Von Wechmar (1989). OMV was shown to be a distinct but typical potyvirus (see Chapter 5) (Burger and Von Wechmar, 1989; Burger et al., 1990). A second, icosahedral virus was isolated from OMV-infected ornithogalum but was not fully characterized (Burger, 1991) and was later identified as tobacco necrosis virus (TNV) (unpublished results).

Symptoms

OMV produce very distinct symptoms in ornithogalum (Burger, 1991). Infected leaves show typical mosaic patterns of light- and dark-green mottling (Fig. 31.2a). Symptoms on flower stalks are generally more pronounced and consist of larger light- and dark-green blotches, with the lighter green areas often sunk into surrounding tissue. In cases of severe infection usually associated with cultivation, flower stunting and deformation occur (Fig. 31.2b). TNV causes necrosis of tissue with mosaic patterns.

Transmission

OMV has an extremely narrow host range (Burger, 1991). The virus can be transmitted mechanically to virus-free ornithogalum and lachenalia (the only other natural host) and by aphid in a non-persistent manner (see Chapter 10). Typical systemic mosaic symptoms appear 2–3 weeks post-inoculation. Virus transmission to none of the common propagation hosts — *Chenopodium quinoa*, *C. giganteum*, *C. album*, *C. schaderanum*, *C. capitatum*, *Nicotiana tabacum*, *N. clevelandii*, *N. benthamiana*, *Tetragonia tetragonioides*, *Phaseolus vulgaris*, *Pisum sativum*, *Petunia hybrida*, *Tropaeolum majus* — could be demonstrated. Similarly, OMV was not transmissible to a variety of other commercial bulbous ornamentals including amarine, nerine, Dutch iris, brodiaea, hyacinth, narcissus, sparaxis, freesia, tritonia, ixia, and gladiolus. In reciprocal experiments, viruses naturally infecting some of these plants were not transmissible to ornithogalum plants (Burger, 1991).

Fig. 31.2. Ornithogalum mosaic virus symptoms on (a) ornithogalum leaves and (b) flower stunting and deformation in ornithogalum caused by severe OMV infection.

Diagnostic assay

In evaluating different virus diagnostic tests, the following considerations are important: the earliest possible detection; the tissue type to be tested; reliability; simplicity, both in terms of materials and technical skill required; and the cost of such a test. Considering all these criteria, an OMV diagnostic test that combined a simple, single-vial sap extraction method with DAS-ELISA was developed. OMV could be reliably detected in as little as 0.1 gram leaf tissue from sprouting bulbs (Burger and Von Wechmar, 1988). The reliability, simplicity, and relatively low cost of this non-destructive assay make it suitable for large-scale routine testing, even in less sophisticated laboratories such as those of larger bulb growers.

Control

Ornithogalum mosaic virus control focuses on preventative measures. Good glasshouse sanitation and control of natural vectors (aphids) are the most practical methods of preventing OMV spread. Aphids shelter undetected for prolonged periods in young leaf whorls and emerge as alatae adults. A regular spray program is essential to control these aphids. *In vitro* disease elimination is another method being used to obtain virus-free plant material, but this is a labor-intensive and expensive practice, and re-infection can occur. In spite of extensive breeding programs, no OMV-resistant lines have yet emerged. The use of coat protein-mediated resistance holds promise as an OMV control measure and is in progress (Burger and De Villiers, unpublished results).

Selected Literature

Burger, J.T. and Von Wechmar, M.B. 1988. Rapid diagnosis of ornithogalum and lachenalia viruses in propagation stock. Acta Hortic. 234:31–38.

Burger, J.T. and Von Wechmar, M.B. 1989. Purification and some properties of South African isolates of Ornithogalum mosaic virus. Phytopathology 79:385–391.

Burger, J.T., Brand, R.J. and Rybicki, E.P. 1990. The molecular cloning and nucleotide sequencing of the 3'-terminal region of Ornithogalum mosaic virus. J. Gen. Virol. 71: 2527–2534.

Burger, J.T. 1991. The Characterisation of Ornithogalum Mosaic Virus. PhD Thesis, University of Cape Town.

Dahlgren, R.M.T. and Clifford, H.T. 1982. The Monocotyledons: A Comparative Study. Academic Press, London.

Ferreira, D.I. and Hancke, F.L. 1986. Indigenous flower bulbs of South Africa: a source of new genera and species for ornamental bulb cultivation. Acta Hortic. 177:405–410.

Obermeyer, A.A. 1978. Ornithogalum: a revision of the southern African species. Bothalia 12:323-376.

Smith, F.F. and Brierley, P. 1944. Ornithogalum mosaic. Phytopathology 34:497–503.

Watt, J.M. and Breyer-Brandwijk, M.G. 1962. Medicinal and Poisonous Plants of Southern and Eastern Africa, 2nd ed. E and S Livingstone, Edinburgh and London.

32

Ranunculus

Assunta Bertaccini and Maria Grazia Bellardi
Istituto di Patologia Vegetale, Bologna, Italy

Introduction

Ranunculus or Persian buttercup (*Ranunculus asiaticus* L.; Ranunculaceae) is a popular ornamental plant grown for cut flowers, bedding, and potted plants. There are up to 250 species of this perennial tuberous plant, but the most widely cultivated are divided into four groups: "Tecolote" or Californian hybrids, INRA or French hybrids, "Barbaroux", and "Grazia". It is cultivated in many areas of the world, in particular in the United States (southern California), Europe (France and Italy), Israel, Japan, and South Africa.

The Viruses

At least ten viruses, not all identified, have been isolated. The most widespread are cucumber mosaic virus (CMV), tobacco necrosis virus (TNV), tobacco rattle virus (TRV), tomato spotted wilt virus (TSWV), impatiens necrotic spot virus

(INSV), and some potyviruses. Mixed infections are often found in the same ranunculus plant so it is very difficult to attribute a disease to a specific virus. Symptoms are mottling, chlorosis, deformation, stunting, vein-yellowing in the leaves, and color breaking in the flowers.

Cucumber mosaic cucumovirus

CMV is the most widespread virus in the Mediterranean; generally it induces color breaking in flowers and crinkle and deformation in leaves (Fig. 32.1; see color plate), and infected plants appear dwarfed (Devergne et al., 1969; Avgelis, 1973; Bellardi et al., 1989).

Tobacco necrosis necrovirus and tobacco rattle tobravirus

Two other dangerous viruses in the same Mediterranean areas are the soilborne TNV and TRV. The first induces necrotic leaf spots, vein and/or general yellowing (Fig. 32.2; see color plate), and stunting. Flowers are smaller and do not bloom; some necrosis at root level, and latent infections are also described (Devergne et al., 1969; Ragozzino, 1972; Bellardi et al., 1989). After the first report in Italy (Lovisolo, 1963), TRV was found in France in plants cv Barbaroux with leaf mosaic and deformations (Devergne et al., 1969).

Potyviruses

A potyvirus was repeatedly isolated from crops growing in Liguria (northern Italy) with fern-like leaf symptoms. This virus was often found in asymptomatic (Bellardi et al., 1988) plants or in mixed infection with varying degrees of leaf chlorosis and deformation, depending on the virus combinations (CMV and TNV). The virus (750 nm) appeared to be an isolate of PVY and was called PVY-R. Pinwheels, scrolls, tubes, laminated aggregates, and bundles were observed in infected cells (Fig. 32.3). PVY-R was mechanically transmitted to six plant species belonging to the Chenopodiaceae and Solanaceae families. On the basis of its properties, PVY-R was attributed to the group Y_0, the most widespread PVY group; this is a rare case of PVY presence in an ornamental plant. More recently, severe outbreaks caused by a potyvirus named RV (707 nm) were reported in southern California and Japan (Elliott et al., 1988) in plants with severe systemic leaf mottle, deformation, and stunting. RV was mechanically and aphid-transmitted to *Consolida ambigua* and to *R. asiaticus* seedlings. RV induced pinwheels and

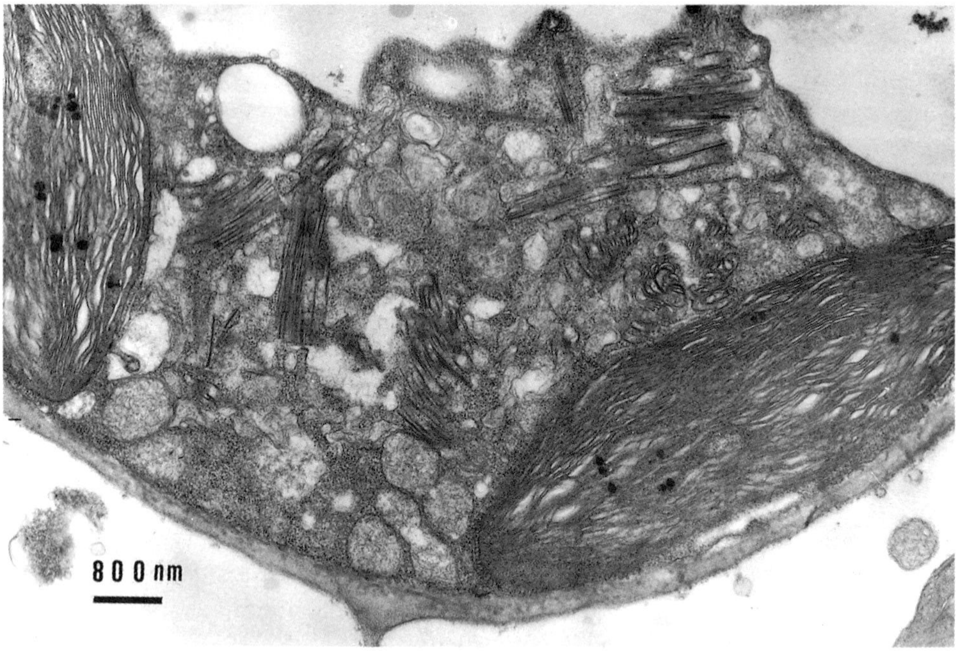

Fig. 32.3. Micrograph of ranunculus parenchymatic cell infected with PVY-R showing some of the different types of cytoplasmic inclusions induced by this virus such as tubes, pin-wheels, and scrolls.

laminated aggregates and reacted with serum to watermelon mosaic virus-2 (WMV-2) but not with the sera to 11 other potyviruses including potato virus Y (PVY). This virus may be synonymous with the ranunculus mottle virus found in southern California (Laird and Dickson, 1967). Other potyviruses were reported in India and in the UK (Elliott et al., 1988).

Tospoviruses

TSWV was first found in wild buttercup in Louisiana (US); more recently, in Liguria (Italy), severe diseases in *R. asiaticus* were caused by tospoviruses (TSWV or INSV), in mixed infections with TNV and/or TRV. The plants showed necrotic spots on the leaves (Fig. 32.4; see color plate) and in some cases midrib necrosis, stem stripes and tip necrosis; in some plants PVY-R was also detected

(Lisa et al., 1990; Vaira et al., 1992). An occasional report of TMV in association with TNV (Bellardi et al., 1989) confirmed the finding in 1955 of a 300 nm virus in the UK.

Mycoplasma-associated diseases

The presence of mycoplasma-like organisms (MLO)-associated diseases is reported in ranunculus mainly in cultivations of southern France (Devergne and Lovisolo, 1969) and northwestern Italy (Liguria). Recently outbreaks of this disease were reported in both Italy and France (Bertaccini et al., 1988; Bertaccini et al., 1994). Infected plants showed flower virescence, marked phyllody with rosette-like structures, and stunting (Fig. 32.5; see color plate). The first symptoms appear in November and remain for the entire growing season. Two different types of symptoms were observed: flower virescence in plants of normal size, and marked phyllody with "rosette-like" structures in stunted plants; leaf yellowing was also present.

Identification and Assay Methods

Pathogen identification is sometimes possible on the basis of symptoms exhibited by ranunculus plants. Mechanical inoculations and leaf dip preparations can be useful for partial virus identification. Immunological assays (double diffusion in agar and IEM) must be used to confirm the virus identity. In some cases, ultrastructural features such as the presence of viral particles for tospoviruses or cylindrical and cytoplasmic inclusions in epidermal leaf strips (under light microscope) for potyviruses can also help virus identification (Bellardi et al., 1988; Elliott et al., 1988).

MLO presence in the sieve tubes can also be detected by electron microscopy of embedded tissues. Up to now ELISA tests on ranunculus have only been successfully applied for tospoviruses (Vaira et al., 1992) using commercial kits. DAPI and DNA probes are more suitable techniques for MLO diagnosis, in particular DAPI is easy and quick to perform and not very expensive. In addition, DNA probes can be used in dot hybridizations to examine numerous samples at the same time. PCR and RFLP analyses of 16S rDNA indicated that the MLOs infecting ranunculus in Italy and France belong to two aster yellows subgroups, I-C and I-B (Lee et al., 1993; Bertaccini et al., 1994).

Transmission

Tuberous roots are involved in virus and MLO transmission; there are no reports on seed infection. During cultivation, CMV, PVY, and RV can be transmitted by aphids in a non-persistent manner; TSWV and INSV are also transmitted by thrips (*Frankliniella occidentalis*). TRV and TNV are soilborne; the first is transmitted by nematodes (*Paratrichodorus* and *Tricodorus*), the second by *Olpidium* sp., especially common in greenhouses. TMV can be easily transmitted by plant and root contacts and also by common agronomical practices. There are no reports on the MLO vector for ranunculus crops: in view of the areas where MLO infections were reported and their genetic group, it is possible that leafhoppers belonging to genus *Euscelis*, *Euscelidius*, and *Macrosteles* are involved.

Control

Since ranunculus is infected by viruses transmitted by many vectors (insects, fungi, and farmers!), all available measures should be adopted to achieve the best flower quality and avoid serious losses. It is necessary to eliminate all the infected plants and keep the cultivations free from weeds (perennial and annual) that can be dangerous virus reservoirs. Recently, the use of "reflective" sprays and detergents were suggested as alternative controls for non-persistent virus spreading. It is difficult to control both the active forms (larvae and adults) and quiescent forms (prepupae and pupae) of thrip infestations. The soil should be checked for nematodes and fungi and treated to avoid dangerous infections by soilborne viruses, especially in the open field. Due to the persistence of *Olpidium* zoospores in soil, crop rotation is an important control measure for TNV infections.

MLO infections can only be avoided by using healthy plants, especially for tuberous roots that are the commercial propagation material. It is possible to recover symptomatic plants with oxytetracycline-hydrochloride (OTC) (Bertaccini et al., 1988), but this kind of therapy is expensive and can be dangerous for the environment; like all antibiotic treatments, OTC is forbidden in the European Union. The only control measure after the disease has appeared is to eliminate all the symptomatic material, the potential vectors, and the weed reservoirs.

Selected Literature

Avgelis, A. 1973. Virosi del ranuncolo in Puglia. Inf. tore Fitopatol. 3:5–8.

Bellardi, M.G., Bertaccini, A. and Marani, F. 1988. A PVY isolate infecting *Ranunculus*. Phytopathol. Mediterr. 27:157–162.

Bellardi, M.G., Marani, F. and Pizzi, L. 1989. Indagine sulle malattie da virus del ranuncolo in Italia. Inf. tore Fitopatol. 2:73–78.

Bertaccini, A., Marani, F. and Rapetti, S. 1988. Phyllody and virescence in ranunculus hybrids. Acta Hortic. 234:123–128.

Bertaccini, A., Vibio, M., Lee, I.-M. and Davis, R.E. 1994. Genetic characterization of mycoplasmalike organisms (MLOs) infecting flower and vegetable crops in Italian and French Riviera. 24th Int. Hortic. Congr., Kyoto, Jpn., p. 102.

Devergne, J.C. and Lovisolo, O. 1969. Les viroses des plantes ornamentales dans le bassin mediterraneen. Ann. Phytopathol. HS 1:285–301.

Devergne, J.C., Cardin, L. and Marais, A. 1969. Isolement de quelques virus infectant le Renoncule cultivée, *Ranunculus asiaticus* L., dans le sud-est de la France. Annals Phytopathol., HS 1:321–328.

Elliott, M.S., Zettler, F.W., Gallegati, M. and Cho, N.-J. 1988. A potyvirus infecting *Ranunculus asiaticus*. Acta Hortic. 234:39–44.

Laird, E.F., Jr. and Dickson, R.C. 1967. Ranunculus mottle virus: field incidence and insect vector relationships at Ponto, California. J. Econ. Entomol. 60:10–13.

Lee, I.-M., Hammond, R.W., Davis, R.E. and Gundersen, D.E. 1993. Universal amplification and analysis of pathogen 16S rDNA for classification and identification of mycoplasmalike organisms. Phytopathology 83:834–842.

Lisa, V., Vaira, A.M., Milne, R.G., Luisoni, E. and Rapetti, S. 1990. Tomato spotted wilt virus in cinque specie coltivate in Liguria. Inf. tore Fitopatol. 12:34–41.

Lovisolo, O. 1963. The tobacco rattle in Italy. Proc. 2nd Conf. Europ. Ass. Potato Res., Pisa, p. 209.

Ragozzino, A. 1972. Due malattie da virus del ranuncolo in Campania. Riv. Ortoflorofrutt. Ital. 56:157–161.

Vaira, A.M., Lisa, V. and Dellavalle, G. 1992. Distribution of tospovirus in Liguria, Italy. 5th Int. Plant Virus Epidemiology Symposium, "Viruses, vectors and the environment", Valenzano, Bari, pp. 23–24.

33

Tulip

W.P. Mowat
Formerly of Scottish Crop Research Institute, Invergowrie, Dundee DD2 5DA, UK

Introduction

Of the several genera of Liliaceae which comprise some of the choicest ornamental plants and a large part of the florists bulbs, *Tulipa* is outstanding in its historic interest and commercial importance. A genus of over 100 species with hardy bulbs, the cultivated tulip was already a substantially developed and highly prized ornamental plant in Turkey before its introduction to Europe in the second half of the 16th century. How these introductions from Turkey relate to native species is somewhat conjectural as they had already been subjected to breeding and selection. Wild species occur from the Mediterranean region, including North Africa, across Asia to Japan with Central Asia as the likely primary gene center (Hoog, 1972). Today there are several thousand named cultivars resulting from hybridization, and a few hundred cultivars have been selected from sports (Dix, 1963; Anon., 1981). The nomenclature and classification of tulip cultivars is governed by the Royal General Bulbgrowers' Society of Hillegom in the Netherlands, which provides a register of tulip names with 15 divisions (Anon., 1981). Tulips are marketed as cut flowers, potted plants, and bulbs for garden and amenity planting as well as propagation stock. The principal types grown are Triumph, Darwin Hybrid, Single Early, Double Early, and Single Late which, in the Netherlands, comprise about 80% of the tulip crop. The world production is estimated at about 9,000 hectares, of which the Netherlands, with just under 8,000 hectares, is by far the largest producer. In 1990 about 900 million cut flowers valued at about £75 million ($130 million) were produced in the Netherlands. Other bulb and cut flower producers include several European countries, Japan, Australia, Canada, and the US.

The Viruses

The 22 named viruses reported to occur in tulip are taxonomically diverse: four are insufficiently characterized to classify and the remainder represent 11 genera. Transmission is either by aphid, fungus, or nematode, but the natural mode of spread of eight is unknown. Infection of tulip by manual inoculation has often proved difficult. The role of 11 of the viruses in the etiology of their associated diseases has still to be established by experimental infection. Table 33.1 gives the salient symptoms in tulip of the viruses dealt with in detail in this chapter. Some of the viruses which have been reported only briefly in the literature or received preliminary study are mentioned when describing a virus with which it has some affinity. Others not noted in this way are pseudo-Augusta, which has particles 700 nm long (Asjes, 1994b); tulip severe mosaic, a possible closterovirus (Nagao et al., 1988; Asjes, 1994b); and a possible rhabdovirus (van Schadewijk and Eggink, 1982). The consensus is that the viruses which most commonly infect tulip are tulip breaking, tobacco necrosis, and tobacco rattle. The relative occurrence of the remainder is less clear. Viruses, such as carnation ringspot and potato X (Asjes, 1994b), reported only from individual plants, may indeed represent rare, chance infections. However, it is possible that too much reliance on symptom diagnosis may have resulted in underestimating the occurrence of some of the others. Thus the similar symptoms expressed by several viruses (Table 33.1), in particular those of leaf necrosis and flower break, do not always allow reliable diagnosis. Therefore, with these types of symptoms, at the very least it is imperative to confirm the identity of the virus by a diagnostic host test or a serological test. For many of the viruses detection and identification is straightforward. However, readily applied tests have yet to be developed to distinguish

Table 33.1. Viruses[a] infecting tulip and their salient symptoms

Genus	Virus	Disease	Symptom in	
			Leaf	Tepal
Potexvirus	Tulip X	Tulip X	Chlorosis and necrosis	Pigmented streaks
Carlavirus	Lily symptomless	Unnamed	None	Pigmented streaks
Potyvirus	Tulip breaking	Tulip breaking	Chlorosis and silvering	Color break
	Tulip chlorotic blotch	Tulip chlorotic blotch	Chlorosis and silvering	Color break
Tobamovirus	Tobacco mosaic	Unnamed	Necrosis	Necrotic blotches
Tobravirus	Tobacco rattle	Unnamed	Chlorosis	Pigmented streaks
Cucumovirus	Cucumber mosaic	Tulip corky fleck	Necrosis and/or chlorosis	Color break
Necrovirus	Tobacco necrosis	Augusta	Necrosis	Necrotic streaks
Nepovirus	Arabis mosaic	Unnamed	Chlorosis and necrosis	Increased pigmentation
	Tobacco ringspot	Tulip veinal streak	Chlorosis	Chlorotic streaks
	Tomato black ring	Unnamed	Chlorosis and necrosis	Not reported
Tombusvirus	Tomato bushy stunt	Unnamed	Necrosis	Unpigmented streaks and necrosis
Unassigned	Tulip halo necrosis	Tulip halo necrosis	Necrosis	None

[a]Further details given in text; viruses described in the same order as in the table.

between some of the proposed different potyvirus infections obtained from plants with flower break, although differences in the break pattern have been described for some (Asjes, 1994b). Also, special procedures are needed to detect tobacco ringspot virus, the role of which in the etiology of tulip veinal streak has still to be substantiated; and the putative tulip halo necrosis virus, the precise nature of which is unknown.

Propagation of the commercially grown tulip is by natural production of daughter bulbs, and the tendency for many of the viruses to pass into the progeny bulbs presents difficulties in maintaining the health of stocks. The main means of controlling the spread of viruses in tulip are by the removal of virus sources inside and outside the crop and evading vectors or reducing their numbers. For the aphid-borne tulip breaking potyvirus, transmission can be reduced by chemical sprays. Sources of host resistance to infection by tulip breaking virus have been recognized, and observations suggest that these may also exsist for tobacco rattle. Meristem-tip culture and micropropagation, which have an important role in the production of healthy stocks of other ornamentals, have proved intractable in their development and are not yet available routinely.

A comprehensive historical review of the subject is outside the scope of this chapter; therefore, some of the early and notable works on tulip virus diseases by such as P. Bradley and F.P. McWhorter in the US, D.M. Cayley and A.W. Mckenney-Hughes in Britain, and M.P. de Bruyn Ouboter and E. van Slogteren in the Netherlands, have not received their due mention or emphasis.

Tulip X (tulip X potexvirus)

Outbreaks of this disease were first reported in 1980 from bulb farms in Britain and shown to be caused by a previously undescribed potexvirus, named tulip virus X (TVX) (Mowat, 1982). The virus has since been reported from Japan (Fujiwara et al., 1994). Because the leaf symptoms are similar to those induced by several other viruses, it is possible that TVX is more widespread.

Symptoms

The virus causes chlorotic and necrotic light brown to grey elongate streaks or elliptical markings in leaves (Fig. 33.1). Tepals develop streaks of intensified pigmentation, mainly at the margins (Fig. 33.2.); some lack pigment or are necrotic. The effects on flowers of white or yellow cultivars have not been studied. The pigmented streaks in flowers are intermediate in width between those caused by lily symptomless virus (narrower) and tobacco rattle virus (wider).

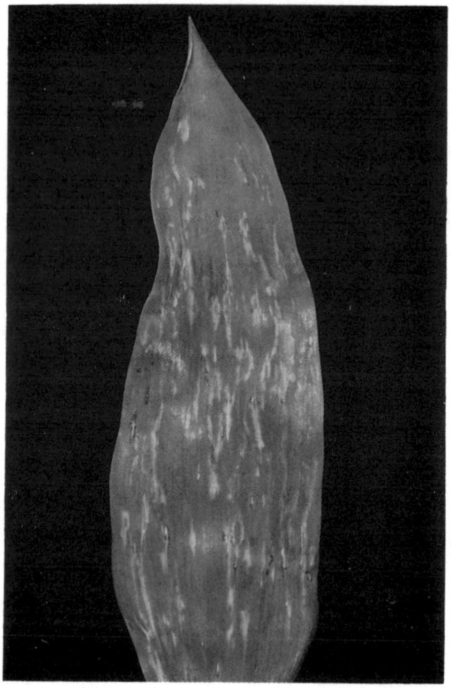

Fig. 33.1. Tulip cv. Smiling Queen naturally infected with tulip virus X showing chlorotic and necrotic lesions in leaf.

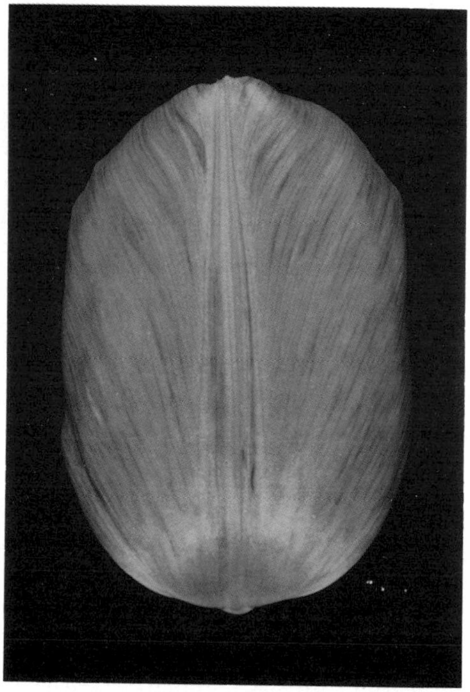

Fig. 33.2. Tulip cv. Smiling Queen naturally infected with tulip virus X showing streaks of intensified pigmentation in tepal.

Where streaks are necrotic these may be indistinguishable from those shown by Augusta disease (tobacco necrosis virus). Although TVX infection does not distort or dwarf plants, as does tobacco necrosis virus, it nevertheless renders flowers unsaleable.

Causal agent

TVX is a definitive potexvirus with filamentous particles c. 495 nm$\times$23 nm. It has a wide experimental host range, infecting species of ten plant families, but most are infected only erratically. The virus is best propagated in *Chenopodium quinoa* and assayed in *C. amaranticolor*; these become regularly infected. Virus particles occur in high concentration in leaf tissue (500 μg/g *C. quinoa* leaf), are readily purified and provide high titer antisera. For further information on the properties of the virus and methods of purifying particles, see Mowat (1982, 1984).

Transmission

Like other potexviruses, the natural mode of transmission is unknown. Although the virus is transmissible by inoculation of sap, infection of tulip in this way is only moderately efficient (Mowat, 1982), and therefore the virus may not be readily spread in the field by contact between plants. The virus is perpetuated in progeny bulbs, but the extent to which this occurs has not been assessed in detail.

Detection and identification

A careful assessment of the symptoms in tepals and recognition of the elliptical markings in leaves should lead to correct diagnosis. However, symptom diagnosis may not be reliable when other viruses are present. TVX is readily distinguished from other viruses infecting tulip by the characteristic line patterns induced in systemically infected leaves of *C. amaranticolor*, by not infecting *Nicotiana clevelandii* and in particle morphology. Serodiagnosis by ELISA provides a rapid and reliable means of identification (Mowat and Dawson, 1987).

Control

Control measures have not been studied; spread in the field by contact between plants should not be assumed and requires investigation. The removal of affected plants to avoid perpetuation through the bulb would seem an obvious precaution.

Remarks

Other viruses with particles c. 500 nm long which have been reported from tulip are potato X (Mowat, 1981; Asjes and Segers, 1988) and tulip mild mosaic (Asjes, 1994b). Therefore partical morphology alone is not diagnostically dependable.

Lily symptomless carlavirus

Although this virus is prevalent in lily, there are few reports of it infecting tulip. This contrasts with the ready infection of tulips and lilies by potyviruses, also transmitted by aphids in the non-persistent manner. Naturally infected plants have been found only in the Netherlands (Derks and Asjes, 1975) and the US (Allen, 1972). The incidence of infection within plantings is usually less than 1%; outbreaks of high incidence (Derks et al., 1982) seem exceptional.

Symptoms

Symptoms are confined to the tepals. Those of pink flowered cultivars such as Rose Copland, Peerless Pink, and Azida develop fine veinal streaks of intensified pigment (Fig. 33.3; see color plate) (Allen, 1972; Derks and Asjes, 1975). Similar streaks develop in the cultivar Preludium by the loss of pigment (Derks and Asjes, 1975). Symptom expression is often delayed until the year following infection. The virus is transmitted to most of the progeny bulbs. Although infection does not affect the number of progeny bulbs formed, bulb yield can be reduced by c. 7% (Derks and Asjes, 1975).

Properties

LSV has particles c. 640 nm long and infects only its natural hosts — lily, tulip (Allen, 1972), and alstroemeria. It is transmissible by mechanical inoculation of sap. Isolates differing in host range or virulence have not been reported. For additional information, see Chapter 28.

Transmission

Aphids shown to transmit the virus to tulip are *Macrosiphum euphorbiae*, *Myzus persicae*, and *Aphis gossypii*. Transmission is in the non-persistent manner.

Detection and identification

The virus particles are readily detected by electron microscope examination of sap extracts. The particle concentration in tepals is considerably higher than in leaves. Identification requires serodiagnosis for which microprecipitin and single drop immunodiffusion tests are of similar sensitivity (Derks and Vink-van den Abeele, 1980). DAS ELISA has superseded these methods for the large-scale indexing of tulip stocks (Derks et al., 1982).

Control

The effects of the virus are so slight that it is not economically important. Control measures for tulip have not been studied. The main obstacles to providing control are the inconspicuous symptom, the likelihood of latent infection, and the readiness with which the virus passes to progeny bulbs. Unrecognized infections in tulip stocks would constitute a potential hazard to neighboring stocks of lilies.

Remarks

Another virus with particles c. 650 nm long has been associated with streaks and bands of necrotic tissue in leaves and narrow dark streaks in tepals of plants of the cv. Blenda. This virus, yet to be studied in detail, is named tulip necrotic streak (Asjes, 1994b).

Tulip breaking (tulip breaking potyvirus and variants)

One of the earliest recorded plant diseases, tulip breaking today persists as a prevalent and damaging disease of tulips. It has been reported from many countries in both hemispheres and is likely to appear wherever tulips are grown. It became notorius during the "Tulip mania" of 17th century Europe when, through speculative trading, specimens of broken tulips attained preposterous values. The causal virus, tulip breaking potyvirus (TBV) (van Slogteren, 1971), is also responsible for a serious disease of lilies, but not all genotypes may be susceptible to infection (Derks et al., 1994).

Symptoms

Symptoms are usually present throughout the growing season and are expressed in all above ground parts of the plant. The main types of leaf symptoms are an inconspicuous mottle, a distinct chlorotic mottle or streaks or broad bandings of chlorotic tissue, and silver or necrotic elliptical rings or line patterns (Figs. 33.6 and 33.7 show the same effects caused by tulip chlorotic blotch virus). In cultivars with flowers containing anthocyanins, break patterns are formed in tepals by the intensification and/or the removal of color in sectors (Fig. 33.4; see color plate). In some cultivars the break is expressed only by intensification of pigment, and when this occurs in flowers which are normally deeply colored, e.g. cv. Renown, the break symptom may not be readily noticed. The three most widely recognized types of break are full break (deficient in pigment), self-break (color intensified), and average break (full and self-break in the same tepals). Suggestions that these are caused by different viruses or strains have not been substantiated. A more acceptable explanation is that the type of break produced depends on the cultivar and the stage of plant development at which infection occurs (Yamaguchi, 1963). There are no pigmentation effects in white or yellow flowered cultivars although flower shape may be changed. Infection also alters the color of the stigma (Fig. 33.5; see color plate), and for the non-pigmented cultivars this is a useful diagnostic symptom. For a comprehensive description of symptoms, see Asjes and Elbertsen (1982).

The development of current year symptoms depends on infection occurring at an early stage of growth (Yamaguchi and Hirai, 1967). Consequently, in the Netherlands where plant development is well advanced when vectors first appear in the field, current season symptoms tend to occur mainly in late flowering cultivars (van Slogteren and Asjes, 1970). Differences in the susceptibility of cultivars to infection have been shown (Romanow et al., 1990), and susceptibility to infection seems to alter during plant growth (Asjes, 1994b). Infection reduces bulb yield, often severely, and cut flowers are smaller.

Causal agent

The virus has filamentous particles c. 750 nm long and 12 nm wide and is manually transmissible by inoculation of sap. Traditionally its host range was considered to be restricted to the natural hosts tulip and lily. However, isolates with experimental dicotyledonous hosts have been reported by Alper et al. (1982), Mowat (1985), and Hammond and Chastagner (1988), as have isolates from tulip which do not infect lily (Mowat, 1985; Dekker et al., 1993). The particles of TBV are readily purified from fresh but not from frozen leaves of either lily or tulip, and antisera suitable for use in ELISA can be prepared (Derks et al., 1982). Studies on particle serology and genome RNA sequence of some isolates associated with tulip breaking disease reveal a wide range of variation. Some isolates, including tulip chlorotic blotch and tulip top breaking virus, show in some respects a relationship with turnip mosaic or bean yellow mosaic viruses (Hammond and Chastagner, 1989; Dekker et al., 1993). Others have been proposed as distinct potyviruses: Rembrandt-tulip breaking, tulip band breaking, and lily mottle (Dekker et al., 1993; Derks at al., 1994; Ohira et al., 1994).

Transmission

Natural transmission is by several species of aphid in the non-persistent manner. As with other potyviruses, the efficiency of transmission can differ with the species of vector. Those shown to transmit are *Myzus persicae, Aphis gossypii, Macrosiphon euphorbiae*, *Neomyzus circumflexus*, and *Doralis fabae*. Less well known vectors that may be overlooked are the cereal aphids *Metopolophium dirhodum*, *Rhopalosiphum padi* (Thomsen, 1980a), and *Sitobion avenae*. Transmission by *Anuraphis tulipae* during bulb storage is also reported (Hughes, 1934). Progeny bulbs usually become infected and develop symptoms and, in the absence of rogueing, the virus can accumulate rapidly in a stock. Limited transmission can result from beheading tulips and, more likely, from harvesting flowers (van Slogteren and Asjes, 1970).

Detection and identification

ELISA, with either polyclonal antisera or monoclonal antibodies, is effective for indexing a large number of samples. DAS-ELISA is recommended for use with polyclonal antisera (van Schadewijk and Eggink, 1984) and a plate trapped antigen form of ELISA for monoclonal antibodies (Hammond and Chastagner, 1989). Selection of antisera by screening them against a number of field isolates is advisable as they may differ in efficacy of detection (Derks et al., 1982; Derks et al., 1994). However, surveys in Britain, the Netherlands, and the US revealed few plants with tulip break that failed to react in ELISA with antiserum to TBV. The identification and taxonomy of potyvirus isolates from tulips with flower break symptoms is not straightforward (see above). There are no procedures suitable for large-scale indexing which will identify serological variants or pathovars, although screening with antisera to additional potyviruses, e.g. bean yellow mosaic, tulip chlorotic blotch, and turnip mosaic, would be helpful in selecting potential variants for further examination. Hybridization tests using c-DNA probes have been used to distinguish between lily isolates (lily mottle virus) and TBV (Derks et al., 1994).

Control

Effective control relies on the removal of infected plants within the crop and those in adjacent plantings of tulip and lilies. Rogueing needs to start before aphid activity and to be repeated regularly during the growing season. White and yellow flowered cultivars can be major sources of infection because symptoms are difficult to detect. In Australia a separation of 275 m gave effective protection from external sources (Sutton and Garret, 1978). A combination of the mineral oil Luxan oil H and the synthetic pyrethroid deltamethrin, applied at weekly intervals, can significantly reduce virus spread but lowers bulb yield (Asjes, 1990, 1994b). However, it is a valuable treatment in the build up of stocks of high health status by limiting spread from symptomless plants arising from current year infections or from diseased plants which have escaped rogueing.

Remarks

Because of the relationship of some of the "tulip breaking" isolates to turnip mosaic or bean yellow mosaic viruses, tulip needs to be considered as a potential host for these viruses. Due regard to this possibility is required where tulip is introduced into new cropping systems in which these viruses may already be established.

Tulip chlorotic blotch (tulip chlorotic blotch potyvirus)

Found in Australia in 1977, the symptoms of this disease are the same as those caused by tulip breaking virus but are induced by another potyvirus, named tulip chlorotic blotch virus (TCBV) (Mowat, 1985). In limited surveys the disease was not found in Britain (Mowat, 1985) and only rarely in the US (Hammond and Chastagner, 1988).

Symptoms

Like tulip breaking virus, TCBV causes self-break in cvs Apeldoorn and Prominence and average break in cvs Halcro, Kees Nelis, Orange Marvel (Fig. 33.4; see color plate), Preludium, Renown, and Rose Copland. Leaf symptoms (Figs. 33.6 and 33.7) show a similar range of variation to that expressed by tulip breaking virus. While the color break will adversely affect the marketing of flowers, the effect of the virus on bulb yield is not known.

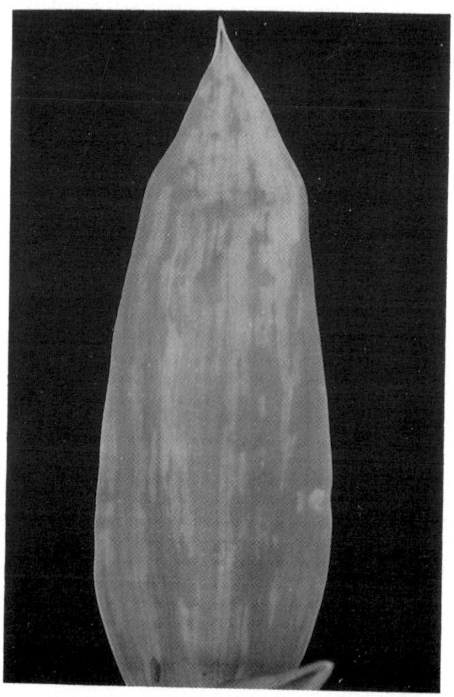

Fig. 33.6. Leaf symptoms in tulip cv. Merry Widow experimentally infected with tulip chlorotic blotch virus.

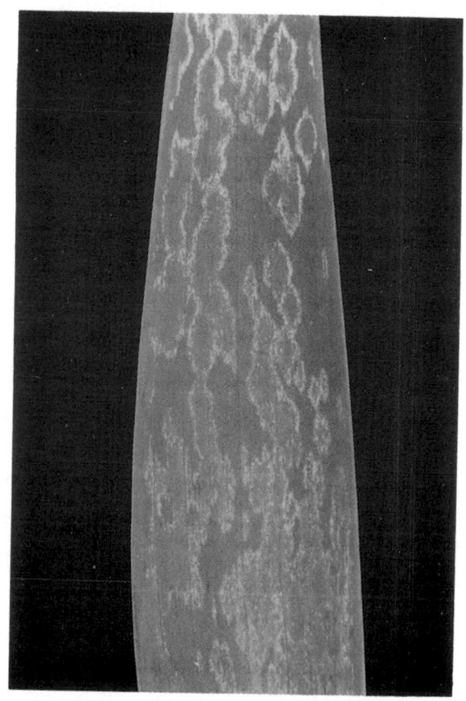

Fig. 33.7. Leaf symptoms in tulip cv. Paul Richter experimentally infected with tulip chlorotic blotch virus.

Causal agent

The virus particles are c. 720 nm long and 12 nm wide. It has a moderately wide host range including *Nicotiana clevelandii* (a suitable propagation host) and *Chenopodium amaranticolor* (a local lesion assay host). Non-hosts are turnip (*Brassica rapa*) and *Lilium formosanum*. The virus particles are readily purified (c. 1 mg/100 g *N. clevelandii* leaf) and are good antigens. Serologically they are not related to those of tulip breaking virus but distantly so to those of bean yellow mosaic virus (Mowat, 1985) and more closely to those of turnip mosaic virus (Hammond and Chastagner, 1988). Nucleic acid hybridization tests also indicate homology with turnip mosaic virus (Hammond and Chastagner, 1988). For further details of properties and method of purifying particles, see Mowat (1985).

Transmission

Myzus persicae is an efficient vector following short acquisition feeds (Mowat, 1985). The virus is transmitted to progeny bulbs but no details are available.

Detection and identification

The virus is readily detected in leaf or tepal extracts of symptom bearing plants by inoculation of *C. amaranticolor* (local lesions in c. 7 days) or by ELISA (Mowat and Dawson, 1987). Serological identification of potyvirus isolates may not be clear cut, and results will be affected by the type of serological test and the particular antiserum used. While a panel of antisera representing bean yellow mosaic, tulip breaking, tulip chlorotic blotch, and turnip mosaic viruses (see section on tulip breaking virus for additional viruses/variants) may distinguish between isolates, the serological relationship indicated may not indicate host range differences.

Control

The epidemiological significance of the potentially wider host range of this virus has not been studied. However, it would seem possible that the control of spread of TCBV and other similar viruses could be significantly affected by their extended natural host range compared with that of tulip breaking virus. This emphasises the need for vigilance where tulips are introduced to new crop environments or when unexpected breakdown of previously effective control regimes occur.

Remarks

Although serologically distinct from tulip breaking virus, TCBV is related to other potyviruses including turnip mosaic virus (Hammond and Chastagner, 1988). As mentioned in the section on tulip breaking, the identification and taxonomy of potyvirus isolates from tulips with flower break is not straightforward, and readily applied discriminatory tests are not available.

Tobacco mosaic tobamovirus

Reported from Czechoslovakia and the Netherlands, its effects can be confused with the several other viruses that induce necrotic symptoms in tulip.

Symptoms

Necrotic irregular blotches or elongate lesions develop in leaves and sometimes in tepals. Leaves are often distorted and plants can be stunted. The disease has been found in many cultivars representing Double Early, Triumph, Darwin Hybrid, Single Late, Lily-flowered, and Fosteriana groups and in *Tulipa eichleiri* cv. Regel.

Properties

The virus has particles 300 nm × 18 nm and is readily sap transmissible. It has a wide host range and a worldwide distribution. Many serological and symptom variants are known. Particles occur in high concentration in several hosts, but others support only low levels of replication. Particle concentration in tulip is generally c. 20–50 μg/ml of sap, but it can be higher (Mokra et al., 1973). Preparations of purified particles and high-titered antisera to particles are readily obtained (Zaitlin and Israel, 1975).

Transmission

Natural specific vectors are not known. It is possible that inadverent mechanical inoculation occurs through feeding by some insects. Virus particles and fragments of infected tissue in the soil may be sources of infection of roots. Field outbreaks in Czechoslovakia have been attributed to soilborne infection (Mokra et al., 1973).

Detection and identification

Nicotiana glutinosa and *N. tabacum* cv. Samsun NN are useful detection and diagnostic hosts producing necrotic local lesions at temperatures below 28°C. The estimated particle concentration in tulip leaves would be adequate for the application of ELISA.

Control

Discarding bulbs of a circumference less than 10 cm would seem a useful approach as diseased plants from progeny bulbs were restricted to those of smaller grades (Mokra et al., 1973).

Tobacco rattle tobravirus

Tobacco rattle (TRV), tobacco necrosis, and tulip breaking viruses are the most common virus infections in tulip. TRV has a very wide host range of wild and cultivated plants and is transmitted by soil inhabiting nematodes which favor light structured soils. It occurs in many temperate regions (Martelli and Taylor, 1989) including Europe, Japan, New Zealand, North America, and the Russian Federation and is likely to infect tulips grown in these regions. Other ornamental crops infected by TRV include Allium, anemone, aster, crocus, freesia, gladiolus, hyacinth, narcissus and Ornithogalum.

Symptoms

Leaves of infected plants develop chlorotic flecks, streaks, or oval lesions (Fig. 33.8) which appear early in the growing season. Tepals of red flowered cultivars have streaks of intensified pigment (Fig. 33.8), whereas in those with white or yellow flowers the streaks are translucent. Infected plants are usually shorter and in some cultivars flowers are sometimes misshapen. Infection may cause only slight reductions in bulb yield (Asjes, 1994a).

Properties

TRV has tubular particles of two predominant lengths: 190 nm and 50–115 nm (depending on the isolate). The two lengths of particle each contain a different species of RNA, termed RNA-1 and RNA-2, respectively. Many strains have been described but are not reliably distinguished by symptoms in test plants. There are also many serotypes, and usually members of different serotypes are serologically distantly related. The particles of some strains are poorly immunogenic, but others can provide antisera with moderately high titers. Isolates differ in the ease with which particles can be purifed but some can yield 10–100 mg of particles per kg of leaf; for methods and additional information, see Robinson and Harrison (1989). Isolates which have only RNA-1 (called NM-types) do not produce particles and are transmitted with difficulty by inoculation with sap. This type of isolate is relatively common in potato but has not been reported from tulip.

Fig. 33.8. Symptoms of tobacco rattle virus in naturally infected tulip cv. Apeldoorn showing color streaks in tepals and chlorotic streaks in leaves.

Transmission

Natural transmission is by several species of ectoparasitic, root feeding nematodes in *Paratrichodorus* and *Trichodorus*. A high degree of specificity occurs between virus serotype and vector species (Ploeg et al., 1991). Surveys in Britain, Denmark, and the Netherlands show that sandy loam and loamy sand soils are commonly infested with the vectors (Alphey and Boag, 1976; Thomsen, 1980b; Ploeg et al., 1991). Restriction of the disease to patches in a field is usually indicative of primary infection. Generally, in cultivated soils, a high proportion of vector populations are viruliferous. The virus can be retained for several months by non-feeding nematodes and is thought not to be transmitted through nematode eggs or molts. Contributing to the persistence of the virus in soils is its transmission through seed to a small proportion of progeny plants of several weed species (Cooper and Harrison, 1973). In tulip it passes to some of the progeny bulbs, and the disease incidence in a stock may diminish only gradually in subsequent years (van Hoof and Silver, 1976; Asjes, 1994a).

Detection and identification

Detection by serological methods is not completely reliable because of the extreme antigenic diversity of isolates. No one antiserum or combination of antisera will ensure detection of all variants. The most effective methods of serological detection will be those forms of ELISA providing broad serological reactivity, e.g. $F(ab')_2$, plate trapped antigen, and dot blotting on nitrocellulose membrane. The virus is more reliably detected by the inoculation of sap to an indicator host (Asjes et al., 1992). The concentration of virus particles in samples of flower, stem, and leaf is usually adequate for both ELISA and infectivity tests. However, its detection in the scale tissue and shoot of dormant bulbs is not reliable (van der Vlugt et al., 1988; Asjes et al., 1992). Diagnostic hosts include *Chenopodium amaranticolor* (necrotic local lesions in 3–5 days; usually not systemic); *Nicotiana tabacum* cv Samsun NN (necrotic spots or rings in inoculated leaves and sporadic systemic distortion and/or necrosis); and *Phaseolus vulgaris* cv The Prince (pin-point necrotic local lesions in 2–4 days; not systemic). Conclusive identification of all isolates, including NM forms, is by nucleic acid hybridization tests with RNA-1 probes (Robinson, 1989). An equally reliable method, and one which avoids the use of radioactive probes, is a test based on polymerase chain reaction (Robinson, 1992). For infectivity assays of soil, seedlings of *Cucumis sativus* or *Petunia hybrida* are suitable bait plants. After c. 4 weeks growth in a soil sample, roots of the bait plants are tested for infectivity by inoculation to indicator hosts. Testing roots of some weed species, e.g. *Stellaria media*, taken directly from field sites, is an effective alternative, particularly when the plants have overwintered.

Control

Cultural methods include (1) pre-selection of sites based on infectivity assays of the soil; (2) choice of previous crop to favor those associated with the suppression of disease, e.g. *Hordeum vulgare*, and to avoid crops such as *Lolium multiflorum* which enhance disease incidence (Asjes, 1994a); (3) delayed planting which, as with Augusta disease, reduces the incidence of TRV infection (Asjes, 1974). Populations of the vector can be decreased to low levels by treating field soils with the fumigants dichloropropene-dichloropropane or dazomet. Although the treatment is never fully effective, infection may be significantly curtailed during growth of the tulip crop.

Remarks

The distribution of infected tulip bulbs may be a means of establishing the virus in soils already containing vector nematodes.

Tulip corky fleck (cucumber mosaic cucumovirus)

This disease name describes the symptoms produced in the stored bulbs. Although the causal virus, cucumber mosaic (CMV), has a wide host range and is widespread in temperate regions, there have been few reports of this disease in tulip. Even so, severe outbreaks of up to 70% have been noted in commercial stocks (van Slogteren, 1966), but these are not common (Asjes, 1994b). Possibly the disease is more of a problem in gardens than in commercial bulb fields where reservoirs of the virus are less likely to occur.

Symptoms

Leaf symptoms can vary from chlorotic to necrotic markings in the form of a mottle, spots, and streaks (Fig. 33.9). Necrosis can vary from grey to brown and may emulate that of Augusta disease (Mowat, 1970). The pigment break which develops in tepals closely resembles that caused by tulip breaking virus but is more likely to be confined to the margin. Bulb scales develop grey-brown sunken

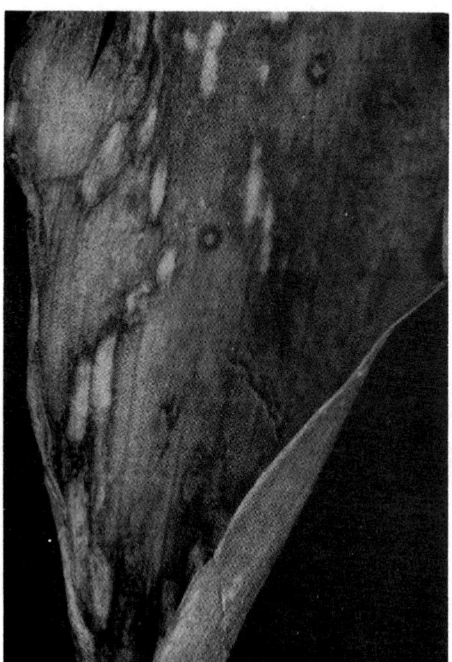

Fig. 33.9. Necrotic lesions in leaf of tulip cv. Rose Copland naturally infected with cucumber mosaic virus.

necrotic lesions in the form of spots, arcs, and rings. Typically necrotic symptoms in bulbs develop during storage and the necrosis may spread and kill the bulb. Although the virus is perpetuated in progeny bulbs from primary infected plants, the second generation bulbs are generally free from the the virus (van Slogteren, 1966).

Causal agent

The virus has isometric particles c. 28 nm in diameter. It has a wide natural and experimental host range in monocotyledons and dicotyledons, including many ornamental species, and is readily transmissible by mechanical inoculation of sap. Isolates of the virus are relatively unstable, and the addition of stabilizing agents to the leaf extraction medium is advisable (e.g. 0.1% thioglycollic acid or 0.01 M sodium diethyl dithiocarbamate). Many variants differing in host reactions have been described, and several serotypes are recognized. Some isolates contain a satellite RNA species which potentially modifies symptoms in some hosts.

Transmission

The virus is transmissible by many aphid species in the non-persistent manner.

Detection and identification

Detection is by inoculation of sap to indicator hosts such as *Chenopodium amaranticolor, C. quinoa* (necrotic local lesions in a few days), and *Nicotiana clevelandii* (systemic mottle). Serological identification of isolates of CMV by gel double diffusion tests is not always reliable even in the presence of a stabilizing agent such as 0.001 M ethylene diamine tetra-acetic acid. A form of ELISA with a broad antigenic spectrum is a more reliable test (Mowat and Dawson, 1987).

Control

During dry bulb storage, periodic inspections are required to detect and remove bulbs which develop necrotic lesions on the fleshy scales. Diseased plants should be removed from field plantings as early in the season as possible to prevent transmission of the virus to tulips and weed hosts.

Augusta disease (tobacco necrosis necrovirus)

This disease, which for a time was also named tulip necrosis, was first recognized in 1928 and shown later to be caused by tobacco necrosis virus (TNV) (de Bruyn Ouboter and van Slogteren, 1949; Kassanis, 1949). Disease outbreaks are sporadic and usually fewer than 5% of plants are affected, but entire stocks may collapse. After tulip breaking it is probably the most common disease in tulip and is found worldwide.

Symptoms

Typical symptoms consist of streaks and elliptical or angular spots of necrosis on stem, leaves, and petals and the plant is often stunted and distorted (Fig. 33.10). Sunken necrotic lesions on bulb scales may also occur. Field and glasshouse grown plants show similar symptoms. Although several other viruses induce comparable leaf symptoms, the fine necrotic lines in tepals are diagnostic of TNV (Fig. 33.11). Outbreaks of the disease can be caused either by current year infection or by bulb-borne virus. However, following natural root infection, tulip plants mostly do not develop symptoms whether field or glasshouse grown, either in the year of infection or in the one following (Mowat, 1970; Lange, 1976). In such plants the virus is usually confined to the roots, and when detected in leaves it is in low concentration. Symptom development seems to be influenced by soil type. The disease develops earlier but affects fewer plants when they are grown on sandy soils as compared with heavy loams (Asjes, 1993). It has been recorded in more than fifty cultivars of the Single Early, Mendel, Triumph, Darwin, Darwin Hybrid, Cottage, and Parrot groups of tulip.

Fig. 33.10. Necrotic leaf symptoms and stunt in tulip cv. Korneforos experimentally infected with a D strain of tobacco necrosis virus.

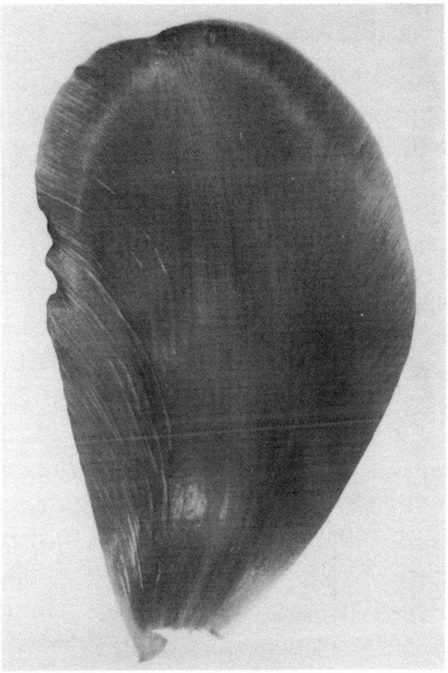

Fig. 33.11. Fine necrotic lines in tepals of tulip cv. Princess Beatrix experimentally infected with an A strain of tobacco necrosis virus.

Fig. 33.12. Tulip cv. London naturally infected with tomato black ring virus showing necrotic and chlorotic leaf symptoms.

Causal agent

The virus is the type member of the necrovirus group. It is readily transmitted by manual inoculation to a wide range of host plants but usually does not infect them systemically. Several strains, which comprise two serological groups, have been distinguished (Babos and Kassanis, 1963). Particles of most strains are stable, are readily purified, and act as good immunogens. The virus can be associated with a satellite virus which is serologically distinct from TNV but is dependent on it for replication. Symptoms in tulips seems to be unaffected by the presence of the satellite virus and by the strain of TNV (Mowat, 1970).

Transmission

The virus is transmitted by zoospores of the root infecting, chytrid fungus *Olpidium brassicae*. Transmission depends on a specificty between the *Olpidium* strain, the virus strain and the host plant species (Kassanis and Macfarlane, 1965).

Vector strains of *O. brassicae* colonise lettuce but not turnip whereas the reverse applies to non-vector strains. The virus does not survive in the resting spore of the fungus. *O. brassicae* can colonize the roots of many plant species but usually does not complete its life cycle on tulip (van Slogteren and Visscher, 1967; Mowat, 1970).

Detection and identification

Most strains of TNV are reliably detected on indicator hosts, such as *Chenopodium quinoa* and *Phaseolus vulgaris* cv. The Prince, by inoculation of sap extracted in 0.07 M phosphate buffer; local necrotic lesions appear in 3–4 days. Serodiagnostic tests require antisera representing the two serotypes, A and D. Double diffusion tests in gel or agarose are effective with stable strains. The concentration of virus particle antigen in diseased leaf tissue should be adequate for ELISA. However, detection in bulb scales by ELISA was not reliable because of the uneven distribution of virus antigen (Asjes, 1994b). Lettuce (*Lactuca sativa*) is a suitable bait plant for testing the infectivity of soil. After 1–2 weeks growth in well watered soil, the roots are tested for infectivity on an indicator host. Mung bean (*Vigna aureus*) roots develop large, brown, necrotic lesions often 1–2 days after infection and have been used as a diagnostic host for testing soil (Teakle, 1962).

Control

Prediction of disease incidence in progeny bulbs from that expressed in the parent stock is unreliable (Asjes, 1994b). Delaying planting reduces incidence but may be impractical for mechanical planting in heavy soils (Asjes, 1974, 1993). Soil, steam sterilized or treated with dazomet (van Slogteren and Visscher, 1967), can control disease outbreaks from current year infection in glasshouse culture but is not considered practical for field plantings. In other crops methyl bromide fumigation has successfully controlled TNV infection (Thomas, 1973).

Nepoviruses: arabis mosaic and tomato black ring

In addition to tobacco ringspot virus (see tulip veinal streak), two other nepoviruses have been implicated in diseases of tulip reported in Britain and the Netherlands but seem to occur rarely.

Symptoms

Different types of symptoms have been associated with arabis mosaic virus

(ArMV) infection: yellow streaks in leaves, malformation of flowers, and severe stunt (Brunt and Jenkins, 1970); mosaic with grey necrotic ringspots in leaves and intensification of pigment in flowers (Asjes, 1976). Symptomless infection may also occur (Asjes, 1976). The effects attributed to tomato black ring virus (TBRV) infection are necrotic leaf symptoms similar to those of Augusta disease and severe stunt (Moore, 1979). Chlorotic leaf lesions have also been associated with this disease (Fig. 33.12).

Properties

Both viruses have a wide natural host range and also infect many cultivated plants. They have isometric particles of similar appearance, c. 28 nm in diameter, which are readily purified and produce high-titered antisera. The viruses are readily transmitted by inoculation of sap (Murant, 1970a, 1970b).

Transmission

ArMV and TBRV are transmitted specifically by certain species of ectoparasitic and soil-inhabiting nematodes as *Xiphinema* and *Longidorus*, respectively. Although the vectors are widespread in Europe and have been reported in North America and Australasia, they are localized in distribution and the incidence of infective populations is low (Cotten, 1979). Infection of tulips is therefore likely to be restricted and have only local importance. As with tobacco rattle virus, the presence of patches of disease in field plantings indicate primary infections.

Detection and identification

Infection is readily detected by inoculation of sap to a suitable indicator like *Chenopodium quinoa* and isolates identified by serological tests such as double diffusion in gels. Serological tests are essential as not all nepoviruses can be distinguished by host reaction tests. Also TBRV may be accompanied with raspberry ringspot virus and ArMV with strawberry latent ringspot virus as these pairs of viruses share the same vectors.

Control

Control measures specific for tulip have not been investigated as these infections are not considered to be important. However, the principles evolved from investigations on other crops would apply viz. pre-selection of sites based on infectivity tests of the soil, the perpetuation of the virus in alternative weeds and cultivated hosts, seed transmission in weeds, persistence of the virus in the vectors, and the vertical distribution of vectors in the soil (Harrison, 1977).

Tulip veinal streak (tobacco ringspot nepovirus?)

This disease has long been recognized by growers as a serious problem in the Netherlands and Britain. The etiology of the disease has still to be established. Its inclusion as a virus disease depends solely on the detection, at near subliminal concentrations, of tobacco ringspot virus (TobRSV) as a general infection in tulips (Asjes, 1972).

Symptoms

Leaf symptoms consist of translucent chlorotic streaks of varying length which may develop a central necrotic streak (Fig. 33.13). Symptoms may develop on all leaves and often are most pronounced on the basal part of the lower leaves. Similar symptoms develop on tepals, and flowers may be arrested in development. Plants may or may not be malformed. Bulbs show no symptoms. The development of the disease in glasshouse grown tulips is affected by temperature and water regimes (Asjes and Muller, 1973). It occurs in many cultivars of Mendel, Triumph, Darwin, and Darwin Hybrid tulips. The cultivars Bing Crosby, Korneforos, Merry Widow, Paul Richter, and William Copland seem especially prone

Fig. 33.13. Tulip cv. Orange Korneforos with symptoms of veinal streak viewed by transmitted light.

to the disease, and it may have been a factor in the demise of the Copland cultivars. Field grown and glasshouse grown stocks totally affected by the disease have been reported in the Netherlands and Britain. Affected plants are unsalable.

Causal agent

The detection of TobRSV, a possible cause of this disease, has been achieved in only one laboratory (Asjes, 1972). The virus was regularly detected in diseased and symptomless plants and in those showing other disease symptoms. The virus has three types of isometric particles c. 28 nm in diameter sedimenting at 51, 91, and 126 S and containing 0, 35%, and 43% RNA, respectively. The two RNA species, mol. wt. 1.4×10^6 and 2.4×10^6, are both required for infection (Stace-Smith, 1985; Chapter 5). Isolates from hosts other than tulip are readily obtained by manual transmission to a range of herbaceous plants. The isolates obtained from tulip are difficult to establish in culture, and particle concentration in tulip is very low (Asjes, 1972).

Transmission

Studies in other crops show that natural transmission is primarily by the nematode vector *Xiphinema americanum*, but other types of vectors have been implicated (Stace-Smith, 1985). TobRSV is commonly seed transmitted in several hosts. Seed transmission in tulip, following infection of the original breeding material, is the postulated explanation of the widespread occurence of the virus (Asjes, 1972). Vector transmission of the virus to tulip has not been studied.

Detection and identification

In contrast to other natural hosts of the virus, inoculum prepared from tulip by extraction in 0.07 M phosphate buffer, pH 7.2, is non-infective to species such as *Chenopodium quinoa, Nicotiana rustica*, and *N. tabacum* cv. White Burley. Isolates of the virus are only obtained from tulip when partially purified preparations of virus particles are used as inoculum (Asjes, 1972). Inoculated plants of *N. tabacum* cv White Burley produce only few lesions after c. 2 weeks, and one or more passages are necessary to obtain systemic infection which occurs in c. 3–4 weeks. Microprecipitin and gel double diffusion tests are suitable for identification. The use of ELISA for the detection or diagnosis of TobRSV in tulip has not been reported.

Control

The removal of diseased plants from field grown stocks is of doubtful use as progeny plants from totally affected stocks may remain symptomless or show a low incidence of mild symptoms. Disease expression under glasshouse conditions is suppressed by using light soil for planting and covering the bulbs in standing out grounds, shallow planting, and avoidance of overwatering (Asjes and Muller, 1973).

Tomato bushy stunt tombusvirus

Originally reported from a single specimen of tulip in Britain, tomato bushy stunt virus (TBSV) has since been found in tulip in Sweden. It is readily confused with tobacco necrosis virus in the symptoms it induces in tulip and in some experimental host species.

Symptoms

Leaf symptoms are identical to those of Augusta disease (tobacco necrosis virus) (Figs. 33.14 and 33.15), but those in flowers are different (Mowat, 1972). TBSV induces broad streaks of colorless tissue in contrast to the narrow necrotic lines in tepals which are characteristic of Augusta disease; however, sometimes necrotic streaks may also develop. TBSV seems unlikely to present a potential problem at its present level of occurrence although as severe in its effects as Augusta disease.

Properties

TBSV is the type member of the tombusvirus genus with isometric particles c. 30 nm in diameter which generally are present in high concentration in leaf tissues. It has a wide, experimental host range, and in most species infection does not develop systemically. The virus also naturally infects several dicotyledons and has been reported from some European countries, and North and South America. Natural infections are usually sporadic. It is readily transmissible to experimental hosts by manual inoculation of sap, and its transmission to tulip appears to be improved by the use of 300 mesh (cf 600 mesh) carborundum powder as an abrasive. Strains or variants of the virus have been distinguished serologically and by the symptoms they cause in some experimental hosts. Serologically the original isolate from tulip was placed in the pelargonium leaf curl group (Hollings and Stone, 1975). This isolate and the type isolate induce identical symptoms in tulip. Virus particles are readily purified and provide high titered antisera. For a detailed description of the virus and methods of purification, see Martelli et al. (1971).

Fig. 33.14. Symptoms in tulip cv. Princess Beatrix experimentally infected with tomato bushy stunt virus.

Fig. 33.15. Symptoms in tulip cv. Princess Beatrix experimentally infected with tomato bushy stunt virus.

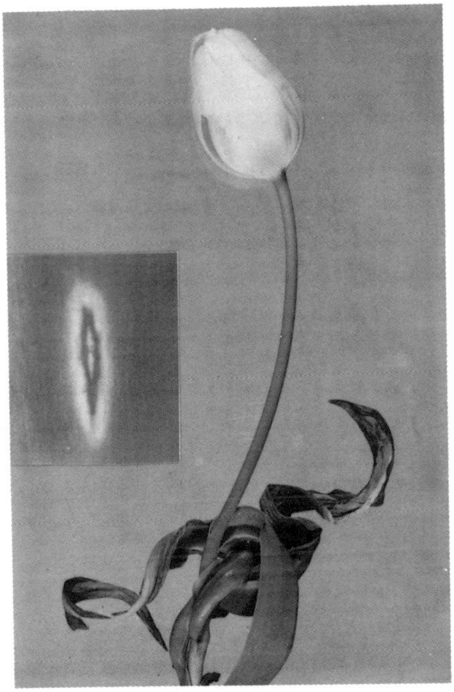

Fig. 33.16. Naturally infected tulip cv. Andes with tulip halo necrosis: flower typically symptomless. Insert shows detail of leaf lesion with halo of translucent tissue.

Transmission

No natural vector is known. The virus is transmitted to progeny bulbs which develop the same symptoms as in primary infections.

Detection and identification

The broad colorless streaks in tepals are a useful diagnostic feature. The occurrence of necrotic streaks, however, can be confused with those caused by tobacco necrosis virus. *Phaseolus vulgaris*, often used as a diagnostic host for tobacco necrosis virus, is unreliable when used by itself as both viruses cause similar symptoms. *Nicotiana clevelandii*, a useful additional diagnostic host as TBSV, but not tobacco necrosis virus, causes systemic symptoms. The concentration of particle antigen in leaf sap for both these viruses is sufficient to allow the use of double diffusion gel-tests and ELISA for detection and identification.

Control

The removal of diseased plants to reduce the perpetuation of infection through progeny bulbs would seem prudent.

Tulip halo necrosis (tulip halo necrosis virus)

Reported only from Britain affecting six cultivars in three outbreaks (Mowat, 1971), this disease can be confused with other necrotic virus diseases of tulip. Special requirements are necessary to detect and culture the associated virus, tulip halo necrosis (THNV).

Symptoms

Characteristically the necrotic lesions of tulip halo necrosis are surrounded by a border of translucent tissue, or halo, and some lesions are not necrotic (Fig. 33.16). Stem lesions, when present, are not sunken in parts as are those in plants affected by Augusta disease. Flowers of plants infected with the virus usually appear normal. This is in contrast to the other viruses inducing necrotic symptoms in tulip which usually also have effects on the flowers.

Causal agent

No particle has been found, but the distribution of infectivity after centrifugation of sap strongly suggests a virus pathogen. The virus is very labile in crude sap but can be consistently transmitted by manual inoculation to *Nicotiana clevelandii* plants using freshly prepared extracts made in 0.07 M phosphate buffer at pH 8, containing a stabilizing agent such as 0.2 M 2-mercaptoethanol or 0.01 M dithiothreitol and using Carborundum (600 mesh) as the abrasive. Stabilized extracts of leaves (2 ml/g of leaf) of tulip or *N. clevelandii* are usually infective at a dilution of 1/16 but rarely at 1/32. Cultures of the virus can be maintained in *N. clevelandii* kept at 14°C or 18°C but not at 21°C (Mowat, 1971).

Transmission

No vector is known, and the virus was not transmitted by *Myzus persicae* following short or long acquisition feeds (Mowat, 1971).

Detection and identification

Although the halo necrotic lesions on tulip leaves are useful indicators, a confirmatory infectivity test on an indicator host is essential. Suitable indicator plants are: *Chenopodium quinoa* (necrotic local lesions, sometimes fragmented, in 6–7 days; not systemic); *Nicotiana clevelandii* (necrotic and/or chlorotic local lesions in 7–8 days and systemic infection in 12–14 days; vein banding followed by mottle at 18°C but a more necrotic reaction at 14°C); *Phaseolus vulgaris* cv. The Prince (brown local lesions, usually ring shaped). However, in mixed infection with tobacco necrosis virus (which causes local lesions only in *N. clevelandii*), the systemic infection of *N. clevelandii* by THNV is not reliable. It is not possible, therefore, to show that tulips with halo necrotic lesions and infected with tobacco necrosis virus also contain THNV (Mowat, 1971).

Control

No control measures have been devised, but as the virus passes to progeny bulbs rogueing diseased plants would be appropriate.

Remarks

For the successful detection of THNV in mixed outbreaks with tobacco necrosis virus, it is essential to test a sufficiently large number of plants to obtain one containing only the unstable virus.

Acknowledgement

The illustrations were supplied courtesy of the Scottish Crop Research Institute.

Selected Literature

Anon. 1981. Classified List and International Register of Tulip Names. The Royal General Bulbgrowers' Association, Parklaan 5, Hillegom, Holland.

Allen, T.C. 1972. Lily symptomless virus. CMI/AAB Descriptions of Plant Viruses, No. 96.

Alper, M., Koenig, R., Lesemann, D.E. and Loebenstein, G. 1982. Mechanical transmission of a strain of tulip breaking virus from *Lilium longiflorum* to *Chenopodium* spp. Phytoparisitica 10:193–199.

Alphey, T.J.W. and Boag, B. 1976. Distribution of trichodorid nematodes in Great Britain. Ann. Appl. Biol. 84:371–381.

Asjes, C.J. 1972. Tulip veinal streak, a disorder probably caused by tobacco ringspot virus. Neth. J. Plant Pathol. 78:19–28.

Asjes, C.J. 1974. Soil-borne virus diseases in ornamental bulbous crops and their control in The Netherlands. Agric. Environ. 1:303–315.

Asjes, C.J. 1976. Arabis mosaic virus in tulips. Neth. J. Plant Pathol. 82:187–189.

Asjes, C.J. 1990. Production for the virus freedom of some principal bulbous crops in the Netherlands. Acta Hortic. 266:517–529.

Asjes, C.J. 1993. Occurrence and recurrence of symptoms of Augusta disease caused by tobacco necrosis virus in tulip. In: Proc. of the Second Symposium of the International Working Group on plant viruses with fungal vectors, Ch. Hiruk (ed.). McGill University, Montreal, pp. 141–144.

Asjes, C.J. 1994a. Occurrence of tobacco rattle virus in ornamental bulbous crops in The Netherlands. Acta Hortic. 377:349–356.

Asjes, C.J. 1994b. Viruses in tulip in the Netherlands. Acta Hortic. 377:289–300.

Asjes, C.J. and Muller, P.J. 1973. Tulip veinal streak: a viral disorder of forced tulips in the glasshouse under certain environmental conditions. Phytophathol. Z. 76:328–342.

Asjes, C.J. and Elbertsen, M. 1982. Tulpemozaiekvirus in tulpen-de symptomen en het zeikzoeken. Ed. Ministerievoor Landbouw en Visserij, Consulentschap in Algemene Dienst, Laboratorium voor Bloembollenonderzoek en Plantenziektenkundige Dienst.

Asjes, C.J. and Segers, L.Ch. 1988. Diagnostiek van virusziekten en identificatie van virussen in bloembolgewassen. Jvrs. 1987 Lab. Bloembollenonderzoek, Lisse, pp. 52–53.

Asjes, C.J., van Schadewijk, A.R. and Segers, L.Ch. 1992. Detection of tobacco rattle virus by ELISA and test plants in main sprouts of tulip during storage at different temperatures. Neth. J. Plant Pathol. 98:213–219.

Babos, P. and Kassanis B. 1963. Serological relationships and some properties of tobacco necrosis virus strains. J. Gen. Microbiol. 32:135–144.

Brunt, A.A. and Jenkins, J.E.E. 1970. Arabis mosaic virus infecting tulips. Plant Pathol. 19:100.

Bruyn Ouboter, M.P. de and van Slogteren, E. 1949. Het Augustaziek der tulpen, een virusziekte van het tabaks-necrose type. Tijdschr. Plantenziekten 55:262–267.

Cooper, J.I. and Harrison, B.D. 1973. The role of weed hosts and the distribution and activity of vector nematodes in the ecology of tobacco rattle virus. Ann. Appl. Biol. 73: 53–66.

Cotten, J. 1979. The effectiveness of soil sampling for virus-vector nematodes in MAFF certification schemes for fruit and hops. Plant Pathol. 28:40–44.

Dekker, E.L., Derks, A.F.L.M., Asjes, C.J., Lemmers, M.E.C., Boll, J.F. and Langeveld, S.A. 1993. Characterization of potyviruses from tulip and lily which cause flower breaking. J. Gen. Virol. 74:881–887.

Derks, A.F.L.M. and Asjes, C.J. 1975. Lily symptomless virus in tulip. Neth. J. Plant Pathol. 81:14–21.

Derks, A.F.L.M. and Vink-van den Abeele, J.L. 1980. Purification of lily symptomless virus. Use and value of antisera against intact and pyrrolidone-degraded virus for testing lilies and tulips. Neth. J. Plant Pathol. 86:239–250.

Derks, A.F.L.M., Vink-van den Abeele, J.L. and van Schadewijk, A.R. 1982. Purification of tulip breaking virus and production of antisera for use in ELISA. Neth. J. Plant Pathol. 88:87–98.

Derks, A.F.L.M., Lemmers, M.E.C. and van Gemen, B.A. 1994. Lilly mottle virus in lilies: characterization, strains and its differentiation from tulip breaking virus in tulips. Acta Hortic. 377:281–288.

Dix, J.F.Ch. 1963. Vegetative sports of tulips. The Daffodil and Tulip Year Book. 1964, pp. 109–114.

Fujiwara, Y., Saito, N. and Kimura, S. 1994. Tulip virus X isolated from tulip imported from the Netherlands under post-entry quarantine. Res. Bull. Pro. Jpn. 30:99–103.

Hammond, J. and Chastagner, G.A. 1988. Natural infection of tulips with turnip mosaic virus and another potyvirus isolate distinct from tulip breaking virus in U.S.A. Acta Hortic. 234:235–242.

Hammond, J. and Chastagner, G.A. 1989. Field transmission of tulip breaking virus and serologically related potyviruses in tulip. Plant Dis. 73:331–336.

Harrison, B.D. 1977. Ecology and control of viruses with soil-inhabiting vectors. Annu. Rev. Phytopathol. 15:331–360.

Hollings, M. and Stone, O.M. 1975. Serological and immunoelectrophoretic relationships among viruses in the tombusvirus group. Ann. Appl. Biol. 80:37–48.

Hoog, M.H. 1972. On the origin of Tulipa. Lilies, 1973, and other Liliaceae. The Royal Horticultural Society, London, pp. 47–64.

Hughes, A.W.Mck. 1934. Aphids as vectors of "breaking" in tulips. Ann. Appl. Biol. 21: 112–119.

Kassanis, B. 1949. A necrotic disease of forced tulips caused by tobacco necrosis viruses. Ann. Appl. Biol. 36:14–17.

Kassanis, B. and Macfarlane, I. 1965. Interaction of virus strain, fungus isolate, and host species in the transmission of tobacco necrosis virus. Virology 26:603–612.

Lange, L. 1976. Augusta disease in tulips. Field experiments concerning tobacco necrosis virus and its vector *Olpidium brassicae*. Tidschrift Planteavl 80:153–169.

Martelli, G.P. and Taylor, C.E. 1989. Distribution of viruses and their nematode vectors. Adv. Dis. Vector Res. 6:151–189.

Martelli, G.P., Quacquarelli, A. and Russo, M. 1971. Tomato bushy stunt virus. CMI/AAB Descriptions of Plant Viruses, No. 69.

Mokrá, V., Čech, M., Pozděna, J. and Brčák, J. 1973. Tulip necrosis caused by tobacco mosaic virus. Phytopathol. Z. 76:46–56.

Moore, W.C. 1979. Diseases of Bulbs. Ministry of Agriculture, Fisheries and Food. Reference Book HPD1. Her Majesty's Stationary Office, London, p. 31.

Mowat, W.P. 1970. Augusta disease in tulip — a reassessment. Ann. Appl. Biol. 66:17–28.

Mowat, W.P. 1971. Stablization, culture and some properties of tulip halo necrosis virus. Ann. Appl. Biol. 69:147–155.

Mowat, W.P. 1972. A necrotic disease of tulip caused by tomato bushy stunt virus. Plant Pathol. 21:171–174.

Mowat, W.P. 1981. Scottish Horticultural Research Institute Ann. Rep. 1980, p.97.

Mowat, W.P. 1982. Pathology and properties of tulip virus X, a new potexvirus. Ann. Appl. Biol. 101:51–63.

Mowat, W.P. 1984. Tulip virus X. CMI/AAB Descriptions of Plant Viruses No. 276.

Mowat, W.P. 1985. Tulip chlorotic blotch virus, a second potyvirus causing tulip flower break. Ann. Appl. Biol. 106:65–73.

Mowat, W.P. and Dawson, S. 1987. Detection and identification of plant viruses by ELISA using crude sap extracts and unfractionated antisera. J. Virol. Meth. 15:233–247.

Murant, A.F. 1970a. Arabis mosaic virus. CMI/AAB Descriptions of Plant Viruses No. 16.

Murant, A.F. 1970b. Tomato black ring virus. CMI/AAB Descriptions of Plant Viruses No. 38.

Nagao, N., Nishio, T. and. Kobayashi, T. 1988. Fine structure of tulip leaves infected with tulip severe mosaic virus, occurring in The Netherlands. Res. Bull. Plant Prot. Jpn. 24:57–61.

Ohira, K., Namba, S., Miyagawa, M., Kusumi, T. and Tsuchizaki, T. 1994. Nucleotide sequence of the coat protein coding region of tulip breaking virus. Virus Genes 8: 165–167.

Ploeg, A.T., Asjes, C.J. and Brown, D.J.F. 1991. Tobacco rattle virus serotypes and associated nematode vector species of Trichodoridae in the bulb-growing areas in the Netherlands. Neth. J. Plant Pathol. 97:311–319.

Robinson, D.J. 1989. Tobacco rattle tobravirus: variation among strains and detection by cDNA probes. EPPO Bull. 19:619–623.

Robinson, D.J. 1992. Detection of tobacco rattle virus by reverse transciption and polymerase chain reaction. J. Virol. Meth. 40:57–66.

Robinson, D.J. and Harrison, B.D. 1989. Tobacco rattle virus. AAB Descriptions of Plant Viruses, No. 346.

Romanow, L.R., van Eijk, J.P., Eikelboom, W., van Schadewijk, A.R. and Peters, D. 1990. Determining levels of resistance to tulip breaking virus (TBV) in tulip (*Tulipa* L.) cultivars. Euphytica 51:273–280.

Stace-Smith, R. 1985. Tobacco ringspot virus. AAB Descriptions of Plant Viruses, No. 309.

Sutton, J. and Garrett, R.G. 1978. The epidemiology and control of tulip breaking virus in Victoria. Aust. J. Agric. Res. 29:555–563.

Teakle, D.S. 1962. Necrotic symptoms of tobacco necrosis virus in roots. Phytopathology 52:1037–1040.

Thomas, W. 1973. Control of *Olpidium brassicae*, the vector of cucumber systemic necrosis and bean stipple streak virus diseases. New Z. J. Exp. Agric. 1:92–96.

Thomsen, A. 1980a. Infection trials with tulip mosaic virus. Acta Hortic. 109:91–92.

Thomsen, A. 1980b. Tulip bulbs grown on rattle virus infected soil. Acta Hortic. 109: 475–477.

van der Vlugt, C.I.M., Linthorst, H.J.M., Asjes, C.J., van Schadewijk, A.R. and Bol, J.F. 1988. Detection of tobacco rattle virus in different parts of tulip by ELISA and cDNA hybridization assays. Neth. J. Plant Pathol. 94:149–160.

van Hoof, H.A. and Silver, C.N. 1976. Natural elimination of tobacco rattle virus in tulip "Apeldoorn". Neth. J. Plant Pathol. 82:252–256.

van Schadewijk, A.R. and Eggink, J. 1982. A new virus in tulips as one of the possible explanations for serological distinction between antisera against tulip breaking virus (TBV). Acta Bot. Neerlandica 31:242–243.

van Schadewijk, A.R. and Eggink, J. 1984. Detection of tulip breaking virus in tulips by means of ELISA. Acta Bot. Neerlandica 33:238.

van Slogteren, D.H.M. 1966. Necrosis in the bulb scales of sensitive tulip varieties caused by cucumber mosaic virus. Meded. Rijksfac. LandbWetensch. Gent 31:986–994.

van Slogteren, D.H.M. 1971. Tulip breaking virus. CMI/AAB Descriptions of Plant Viruses, No. 71.

van Slogteren, D.H.M. and Asjes, C.J. 1970. Virus diseases in tulip. Daffodil and Tulip Year Book 35:85–97.

van Slogteren, D.H.M. and Visscher, H.R. 1967. Transmission of tobacco necrosis virus, causing Augusta disease, to the roots of tulip by zoospores of the fungus *Olpidium brassicae* (Wor.) Dang. Meded. Rijksfaculteit Landbouw-wetenschappen Gent, 32:927–938.

Yamaguchi, A. 1963. Current season symptoms in tulip flowers by the tulip breaking virus. Phytopathol. Z. 49:355–361.

Yamaguchi, A. and Hirai, T. 1967. Symptom expression and virus multiplication in tulip petals. Phytopathology 57:91–92.

Zaitlin, M. and Israel, H.W. 1975. Tobacco mosaic virus (Type strain). CMI/AAB Descriptions of Plant Viruses, No. 151.

CUT FLOWER CROPS (VEGETATIVELY PROPAGATED)

34

Carnation

Vittoria Lisa
Istituto di Fitovirologia applicata CNR, Strada delle Cacce 73, Torino, Italy

Carnation (*Dianthus caryophyllus*), family Caryophyllaceae, is a major flower crop grown worldwide; its cultivation and trade involve large economic interests. Carnations have been grown since ancient times, and several varieties were known in Europe before the seventeenth century. Breeding that led to the recurrent flowering hybrids, the ancestors of today's cultivars, started in France in 1840, and currently several types of carnation are grown. These can be divided into three main groups: the standard, or single flower carnation, that can be further split into the Mediterranean and the Sim (or American) ecotypes; the spray or miniature, characterized by a single stem that bears several flowers; and the mini-carnation. This last group comprises hybrid lines, often interspecific, and may be further divided into different subgroups. The demand for new cultivars, with unusual combinations of colors and suitability to different climatic conditions and new cropping practices, is high, and breeding work is continuous. The main production areas are in Western Europe, Latin America, Japan, Eastern Europe, and

Australia. Other important production areas are South-East Asia, the US, Central America, and Israel. An estimate based on commercial data and including all types of carnations suggests that in the years 1993–94 production (in millions of plants) was about 155 in Western Europe, 90 in Israel (for local plantings and exports), 80 in Latin America, 70 in Japan, 40 in Africa, 30 in Eastern Europe, 15 in Australia, 8 in South-East Asia, 4 in the US, and 3 in Central America.

Carnations can be infected by several viruses, that cause significant losses (see review by Lovisolo and Lisa, 1978). All types of carnations are susceptible to virus infection, although to different degrees. Carnation viruses can be divided into those that infect only carnations in the field or a few other species in the genus *Dianthus* (and therefore in this text will be indicated as specific to carnation), and those with a wide range of natural hosts. Most of the major viruses, i.e. those widespread in the crop and causing economically important diseases, are specific to carnations. The plants are commonly propagated vegetatively, and this practice allows the viruses, once established in a cultivar, to be perpetuated. Wide international trade in cuttings has then spread the viruses wherever carnations are grown. Up to about 20 years ago, before massive clean-up programs were started, it was quite easy to find cultivars totally infected by one or more viruses.

The Viruses

Carnation mottle carmovirus (CarMV)

CarMV (Hollings and Stone, 1970) is the most common virus that infects carnations. It is highly infectious, infects a moderately wide number of test plants, has no known vector, and is easily transmitted by contact and by cropping operations. Although CarMV is considered specific to carnations, natural infection of other ornamental plants has been reported: *Begonia elatior* and *Begonia* × *Cheimantha* in Denmark (Paludan and Begtrup, 1985) and two species of *Daphne* in New Zealand (Morris-Krsinich and Milne, 1977). In begonias CarMV was associated with disease, while in the other species infections were symptomless.

Symptoms on carnation are generally indistinct, consisting of a mild mottling of the leaves, apparent only in soft growth; the wax on mature leaves on plants grown "hard" masks the symptoms so that the flowering plant may appear symptomless. Despite the apparent mildness of the disease induced, CarMV is important because it diminishes the number, size, and quality of flowers, and renders plants more susceptible to fungal diseases. Furthermore, CarMV causes a strong synergistic effect in mixed infections with other viruses. In France, CarMV was reported to be more frequent in Sim than in Mediterranean carnations, and different degrees of susceptibility were found within a single variety (Poupet et al., 1973). CarMV generally occurs in high concentration in infected plants. However, an "attenuated" form in which virus particles are present at extremely low concentration can occur in clones freed from viruses by thermotherapy and meristem-tip culture, and even in plants not subjected to any treatment (Hollings and Stone, 1970; Poupet et al., 1972; Luisoni et al., 1984). This form escapes routine serological indexing and has been detected only by a series of sap inoculations to *Chenopodium* sp.; occasionally it may revert to the normal form during the propagation of virus-tested material, thus becoming a source of infection. Strains of CarMV, differing serologically or in symptoms on test plants, have been described (Hollings and Stone, 1970; Rana and Castellano, 1984).

Carnation vein mottle potyvirus (CVMV)

On carnations CVMV (Hollings and Stone, 1977) causes chlorotic mottle or green vein mottle on leaves, flower breaking in particularly susceptible cultivars, and marked decreases in flower yield. Symptoms are much worse in mixed infections with CarMV (Fig. 34.1; see color plate). CVMV is sap and aphid transmissible and has a restricted experimental host range. In the field, apart from carnations, it has been found only in *Dianthus barbatus*. The main vector is considered to be *Myzus persicae*. Serologically CVMV is related to a few other potyviruses but is clearly distinct from them. A few strains of CVMV differentiated by the symptoms caused on test plants have been reported.

Carnation etched ring caulimovirus (CERV)

Natural infections of CERV (Lawson et al., 1977) have been reported only on carnations. Symptoms consist of necrotic flecks, rings, and line patterns "etched" on leaves. In many cultivars symptoms are visible only if CarMV is also present. In the cv Peace River symptom severity was temperature-dependent: infected plants kept at 21°C showed symptoms while at 26°C they were symptomless (Lawson et al., 1977). Single infection by CERV on the cv Joker caused a one-week delay in flowering and reduced production of first quality cut flowers (Hakkaart, 1977). CERV is non-persistently transmitted by aphids, in particular by *Myzus persicae*, and is sap-transmissible only to carnations and some other Caryophyllaceae such as *Silene armeria* and *Saponaria vaccaria* (Hakkaart, 1968, 1972). *S. vaccaria* is a suitable host for virus purification (Hsu and Lawson, 1985).

Carnation necrotic fleck closterovirus (CNFV)

CNFV (Inouye 1974) — synonyms *carnation yellow fleck* and *carnation streak* (Poncet and Devergne, 1988) — is specific to carnations on which it causes whitish or necrotic streaks and reddish-purple flecks and spots on leaves and stems (Fig. 34.2; see color plate) and complete necrosis on older leaves. In observations made in California, infected plants grown at low temperature had mild or no symptoms, and disease severity increased with temperature (Mayhew, 1979). The virus is sap transmissible, with difficulty, only to carnations and a few other Caryophyllaceae such as *Dianthus barbatus* and *D. chinensis*, and is easily transmitted by aphids, in particular by *Myzus persicae*, in a semi-persistent manner. Carnation, *D. barbatus*, *D. chinensis*, and *Saponaria vaccaria* are suitable hosts for virus purification (Smookler and Loebenstein, 1974; Poncet and Devergne, 1988).

Carnation latent carlavirus (CLV)

CLV (Wetter, 1971) is less widespread in carnation crops than the viruses mentioned above. It does not induce specific symptoms on carnations, but infected plants are less vigorous and produce fewer good flowers. CLV is sap transmissible, has a limited experimental host range that includes Caryophyllaceae, some Chenopodiaceae and Solanaceae, and is efficiently transmitted by *Myzus persicae*.

Carnation ringspot dianthovirus (CRSV)

CRSV (Tremaine and Dodds, 1985) used to be one of the most severe and widespread viruses of carnations, but today, although still listed among the main carnation viruses, it has almost disappeared from crops due to roguing of infected plants and sanitation programs. However, cases of infection still occur (Magid and Robb, 1990), indicating that CRSV may yet be a threat for carnations. Infected plants show rings, line patterns, and mottling on leaves, which also show lateral bending and buckling. Plants are stunted, flowers are smaller, and bleaching, color breaking, and petal crinkle may also occur. CRSV has been found to naturally infect several species besides carnations, mostly fruit trees and the weed *Stellaria media*. It is easily sap transmissible, has a wide experimental host range, and can be transmitted by contact. The virus is believed to be soil-borne; nematode transmission has been reported, but the data have not been confirmed.

Viruses of Minor Importance

For the viruses listed in this paragraph, references up until 1977 are cited in Lovisolo and Lisa (1978).

Carnation bacilliform, possible rhabdovirus

This virus, uncharacterized, has bacilliform particles sized about 260×50 nm and was found in New Zealand.

Carnation cryptic cryptovirus (CCV)

Although CCV (Lisa et al., 1986) and possibly a second cryptic virus are very common in carnations and probably present in most carnation growing areas worldwide, they are considered to be of minor importance because, like all cryptoviruses, they cause no symptoms, have no synergistic effect in mixed infections, are not transmissible by sap or grafting, and occur in low concentration in the plant. CCV is transmitted with high frequency in seeds of Mediterranean and hybrid carnations.

Carnation Italian ringspot tombusvirus (CIRSV)

CIRSV has been isolated from carnations in the UK, in plants coming from Italy and the US, and from a creek in a forested area in Germany (Buttner et al., 1987). Infected carnations developed chlorotic spots and oval rings on the younger leaves.

Carnation yellow stripe necrovirus (CYSV)

This virus was isolated in Italy from plants with bright yellow stripes on the leaves (Gallitelli et al., 1979).

Arabis mosaic nepovirus (ArMV)

Infection of carnations with ArMV has been described in Holland and New Zealand. The disease was characterized by shortening of the internodes and excessive lateral sprouting.

Cucumber mosaic cucumovirus (CMV), carnation strain

CMV has been isolated from carnation in Italy. Infected plants showed green mosaic.

Four other viruses, namely *alfalfa mosaic alfamovirus*, *sowbane mosaic sobemovirus*, *tobacco ringspot nepovirus*, and *turnip crinkle carmovirus* have been sporadically isolated in England. These viruses were generally found in mixed infections with CarMV or some of the other main carnation viruses, and no indication of field disease associated with them is available.

Identification and Assay Methods

Visual inspections do not provide reliable indication for infection of all carnation viruses. For example, CarMV or CERV as single infections may easily be almost symptomless and be overlooked. Sap inoculation to *Chenopodium amaranticolor* and *C. quinoa* will detect the main viruses of carnation except CERV and CNFV and most of the minor viruses; inoculation to *Saponaria vaccaria* or *Silene armeria* will also detect CERV infections. All the viruses can be easily identified by serological methods. CRSV or "normal-type" CarMV, both present at high concentration in plant tissues, can be detected in crude sap by simple methods such as immunodiffusion tests. However, ELISA is by far the best assay method for the main viruses of carnations since it can also reliably detect low concentrations of viruses and can easily be adopted by commercial growers. Specific antisera have been produced to all the main viruses of carnations (Devergne et al., 1982; Lommel et al., 1983; Luisoni et al., 1984; 1988; Hsu and Lawson, 1985; Moran et al., 1985; Poncet and Devergne, 1988) and are commercially available. In studies made in France it was shown that the content of CVMV or CNFV in carnations is not constant during the vegetative period but may undergo fluctuations; February–March and September–October are the periods of highest virus concentration and therefore most suitable for indexing (Poupet et al., 1981). "Attenuated" CarMV, as already pointed out, can only be detected by a series of inoculations to chenopodiums. Detailed indications for detection of the main viruses of carnations particularly suitable for routine indexing in certification schemes have been issued by the EPPO (European and Mediterranean Plant Protection Organization) (EPPO Bull., 1991). Identification of CCV is possible by electron microscopy, serology, and detection of dsRNA (Lisa et al., 1986). The minor viruses can be identified through conventional techniques of plant virology such as electron microscopy, indicator plants, or serology.

Control

Control of the main viruses affecting carnations is feasible through production of virus-free plants by therapy, generally consisting of heat treatment combined with meristem-tip culture, careful diagnosis for early detection and elimination of infected plantlets, and propagation of cured plants under conditions that avoid reinfection. Such programs have been implemented for at least two decades now, and carnation crops have improved dramatically with respect to viral diseases. Carnations are particularly easy to keep free of viruses by phytosanitary measures because the main viruses are specific (or almost) to the genus *Dianthus*, and so sources of infection are limited to other carnation crops.

Clean-up programs should start with careful selection of parent clones for desirable characteristics and with diagnosis of viruses present in them. Heat treatment of parent plants (4–8 weeks at 33–40°C), although not strictly necessary, is advisable since it reduces the amount of virus potentially present in the young shoots and is particularly useful for eradication of CERV (Paludan, 1970). High light intensity, daily alternation of low and high temperatures, and increased levels of CO_2 during the light period were found to be advantageous to carnation thermotherapy (Cohen, 1977). Eradication of viruses by tissue culture involves the excision of the meristem dome plus one or more leaf primordia (larger portions can be excised if the plants have been subjected to preventive heat treatment), aseptic culture of the explant in proper media, and finally the establishment of the plantlet in soil. Techniques of thermotherapy and tissue culture for virus eradication have been reviewed, among others, by Quak (1977) and Walkey (1991), and are described with respect to ornamentals (see Chapter 17). Elimination of viruses from carnations using these techniques has been described by several authors, for example, Stone (1968), Hakkaart and Jordanova (1968), Hollings et al., (1977), and Paludan (1970). Callus cultures from petals or ovule cultures or a combination of meristem-tip culture with treatment with the antiviral agent ribavirin have also been used to eradicate CarMV (Wang et al., 1990).

Great care has to be taken to avoid reinfection of virus-free material, especially by highly contagious viruses such as CarMV. The wide international trade in propagation material associated with the carnation industry can easily bring infected material close to virus-free stocks. Cleaning of contaminated hands, tools, implements, and growing containers with efficient horticultural disinfectants should be carefully applied. Protocols for production of regenerated nuclear stocks, their maintenance, multiplication of the material, and production of certified stock under the certification scheme for production of pathogen-tested material of carnations considering the six main viruses of carnations as listed in this chapter, and hygiene guidelines for certification schemes have been proposed by EPPO (EPPO Bull., 1991, 1993).

Carnation Stunting Syndrome

This severe disease of carnations has been reported in different Mediterranean countries, Portugal, the US, and Japan (Lommel and Morris, 1983; Lisa et al., 1985; Boccardo et al., 1988; Kanematsu et al., 1989). The symptoms are stunting, abnormal shoot proliferation giving a grassy appearance to the plant, poor flower setting, and abnormal flower budding with malformation of the calyx and suppression of petals (Figs. 34.3 and 34.4; see color plate). Plants that show symptoms at an early stage may set no flowers at all. Symptoms generally appear on young plants two to three months after planting and are more severe under conditions of high temperature and light intensity. At lower light intensities and temperature, plants may set new healthy looking shoots and flowers, but in some cases a return of the disease has been observed when the previous conditions were restored.

Symptoms have been reported in Mediterranean, Sim, and spray carnations. Viruses, MLOs, and environmental conditions have been excluded as agents of the disease. A viroid-like RNA is consistently associated with the diseased plants (Lommel and Morris, 1983; Lisa et al., 1984; Kanematsu et al., 1989) and is transmissible from diseased plants or from polyacrylamide gels after purification, to healthy carnations in which the RNA replicates but has not reproduced the field symptoms. The viroid-like RNA, named *carnation stunting associated viroid* (CarSAVd), was also found in symptomless plantlets just planted out and in some adult healthy looking plants (Boccardo et al., 1988). Further studies have shown that CarSAVd has little sequence homology with most of the better known viroids and that it exists in two circular forms of different electrophoretic mobility. Cloned DNA or RNA acid probes can be used for diagnosis (Boccardo et al., 1988; Palukaitis et al., 1992).

A different approach to define the etiology of the disease has been attempted in Israel where a similar or identical disease is present, characterized by stunting of plants and fasciation and deformation of flowers. From symptomatic plants a gram-positive non-motile bacterium, identified as *Rhodococcus* (*Corynebacterium*) *fascians*, was consistently isolated. A high correlation between presence of the bacterium and diseased plants was found (Zutra et al., 1994). However, attempts to reproduce the field symptoms were again not successful (G. Loebenstein, pers. comm.).

MLO Associated Diseases

There are no recent report of diseases of carnations associated with MLOs. In 1953 in the US a flower malformation was described, occurring only at high temperatures, that was assumed to be caused by aster yellows (Thomas, 1953); malformation and virescence of carnation flowers has also been reported in central Europe (Schmelzer and Schmidt, 1960).

Selected Literature

Boccardo, G., d'Aquilio M., Palukaitis, P., Marzachì, C. and Lisa V. 1988. Infectivity of the viroid-like RNA associated with the carnation stunting syndrome. Acta Hortic. 234: 109–114.

Buttner, C., Jacobi, V. and Koenig, R. 1987. Isolation of carnation Italian ringspot virus from a creek in a forested area South West of Bonn. J. Phytopathol. 118:131–134.

Cohen, D. 1977. Thermotherapy and meristem-tip culture of some ornamental plants. Acta Hortic. 78:381–387.

Devergne, J.C., Cardin, L. and Bontemps J. 1982. Indexage biologique et immunoenzymatique (ELISA) pour la production d'oeillet indemnes de virus de la marbrure (CarMV). Agronomie 2: 655–666.

EPPO Bull. 1991. Certification scheme. Pathogen-tested material of carnation. EPPO Bull. 21:279–290.

EPPO Bull. 1993. Certification scheme. Nursery requirements-recommended requirements for establishments participating in certification of fruit or ornamental crops. EPPO Bull. 23:249–252.

Gallitelli D., Castellano, M.A., Di Franco A. and Rana, G.L. 1979. Properties of carnation yellow stripe virus, a member of the tobacco necrosis virus group. Phytopathol. Mediterr. 17:31–40.

Hakkaart, F.A. 1968. Silene armeria, a test plant for carnation etched ring virus. Neth. J. Plant Pathol. 74:150–158.

Hakkaart, F.A. 1972. Detection of carnation viruses with the test plant *Saponaria vaccaria* "Pink Beauty". Acta Hortic. 36:35–46.

Hakkaart, F.A. 1977. The effect of carnation etched ring virus on the productivity of the carnation cultivar "Joker". Acta Hortic. 71:191–193.

Hakkaart, F.A. and Jordanova, J. 1968. Heat treatment experiments with carnations for the elimination of carnation mottle and etched ring viruses. Neth. J. Plant Pathol. 74: 146–149.

Hollings, M. and Stone, O.M. 1970. Carnation mottle virus. CMI/AAB Descriptions of Plant Viruses, No. 7.

Hollings, M., Stone. O.M., Atkey, P.T. and Barton, R.J. 1977. Investigations on carnation viruses. IV. Carnation vein mottle virus. Ann. Appl. Biol. 85:59–70.

Hsu, H.T. and Lawson, R.H. 1985. Detection of carnation etched ring virus using mouse monoclonal antibodies and polyclonal chicken and rabbit antisera. Acta Hortic. 164: 199–205.

Inouye, T. 1974. Carnation necrotic fleck virus. CMI/AAB Descriptions of Plant Viruses, No. 136.

Kanematsu, S., Uematsu, S., Namba, S. Yamashita, S. Tuchizaki, T. and Kameya-Iwaki, M. 1989. Viroid-like RNA detected from carnation in Japan. Ann. Phytopathol. Soc. Jpn. 55:545 (Abstr.).

Lawson, R.H., Hearon, S.S. and Civerolo, E.L. 1977. Carnation etched ring virus. CMI/AAB Descriptions of Plant Viruses, No. 182.

Lisa, V., D'Agostino, G. and Boccardo, G. 1985. A new disease causing stunting and shoot proliferation in carnation. Acta Hortic. 164:63–70.

Lisa, V., Luisoni, L. and Milne, R.G. 1986. Carnation cryptic virus. AAB Descriptions of Plant Viruses, No. 315.

Lommel, S.A. and Morris, T.J. 1983. A viroid-like RNA associated with a stunting syndrome in carnations. Phytopathology 73:791 (Abstr.).

Lommel, S.A., McCain, A.H., Mayhew, D.E. and Morris, T.J. 1983. Survey of commercial carnation cultivars for four viruses in California by indirect enzyme-linked immunosorbent assay. Plant Dis. 67:53–56.

Lovisolo, O. and Lisa, V., 1978. Virus diseases of carnation. Fitopatologia Brasileira 3: 219–233.

Luisoni, E., Lisa, V. and Roggero, P. 1984. Carnation mottle virus in Italia: rilevamento e prevenzione delle forme normale e attenuata. Atti. Giorn. Fitopatol. 3:427–436.

Luisoni, E., Vecchiati, M. and Roggero, P. 1988. ELISA detection of carnation vein mottle and carnation latent viruses. Acta Hortic. 234:221–226.

Magid, A.M.A. and Robb, S.M. 1990. Investigations on viruses of pinks (*Dianthus* sp.) in the United Kingdom. Arab J. Plant Prot. 8:121–126.

Mayhew, D.E. 1979. Carnation necrotic fleck virus in California. Plant Dis. Rep. 63: 978–980.

Moran, J.R., Wilson, J.M., Garrett, R.G. and Smith. P.R. 1985. ELISA indexing of commercial carnations for carnation mottle virus using a urease-antibody conjugate. Plant Pathol. 34:467–471.

Morris-Krsinich B.A. and Milne, K.S. 1977. Natural infection of daphne by carnation mottle virus. Plant Dis. Rep. 61:675–678.

Paludan, N. 1970. Carnation etched ring virus. Survey, infection trials, thermotherapy and meristem culture. Tidsskr. PlAvl 74:75–86.

Paludan, N. and Begtrup, J. 1985. Carnation mottle virus demonstrated in *Begonia eliator* and *Begonia x Cheimantha* showing vein clearing, leaf curl and flower break. Acta Hortic. 164:33–39.

Palukaitis, P.F., Kurath, G. and Boccardo, G. 1992. Viroids — purification, characterization, diagnosis, and detection of sequence variation: application to the carnation stunt associated viroid RNAs. In: Viroids and Satellites: Molecular Parasites at the Frontiers of Life, K. Maramorosh (ed.), CRC Press, Boca Raton, pp. 59–77.

Poncet, C. and Devergne, J.C. 1988. Immunological detection of carnation yellow fleck virus (CYFV) in infected carnations. Acta Hortic. 234: 209–220.

Poupet, A., Beck, D. and Maia, E. 1972. Relation entre le virus de la marbrure de l'oeillet (carnation mottle virus) et différent plantes-hotes. I. Mise en evidence et propriétés (biologiques et serologiques) d'une "souche atténuée" du virus de la marbrure de l'oeillet. Ann. Phytopathol. 4:325–335.

Poupet, A., Marais, A. and Beck, D. 1973. Relation entre le virus de la marbrure de l'oeillet (carnation mottle virus) et different plantes-hotes. II. Fluctuations de la teneur en antigène viral dans l'oeillet méditerranéens au cours d'un cycle de culture. Ann. Phytopathol. 5: 273-280.

Poupet, A., Marais, A., Beck, D. and Bettachini, B. 1981. Comportement de deux virus filamentaux (carnation vein mottle virus, carnation streak virus) dans deux espèces du genre *Dianthus*. Agronomie 1:231–234.

Quak, F. 1977. Meristem culture and virus-free plants. In: Applied and Fundamental Aspects of Plant Cell, Tissue, and Organ Culture, In: J. Reinert and Y.P.S. Bajaj (eds.), Springer-Verlag, Berlin, pp. 598–615.

Rana, G.L. and Castellano, M.A., 1984. Una variante naturale del virus della maculatura del garofano in Puglia. Inf. Fitopatol. 34 (11):31–33.

Schmelzer, K. and Schmidt, H.O. 1960. Blutenvergrunungen an Zierpflanzen und Unkrautern. Dt. Gartenbau 7:214–217.

Smookler, M. and Loebenstein, G. 1974. Carnation yellow fleck virus. Phytopathology 64: 979–984.

Stone, O.M. 1968. The elimination of four viruses from carnation and sweet william by meristem tip-culture. Ann. Appl. Biol. 62:119–122.

Tremaine, J.H. and Dodds, J.A. 1985. Carnation ringspot virus. AAB Descriptions of Plant Viruses, No. 308.

Thomas, W.D. 1953. Malformation of carnation blossom caused by aster yellows virus. Plant Dis. Rep. 37:284.

Walkey, D.G.A. 1991. Production of virus-free plants. In: Applied Plant Virology, 2nd ed., Chapman and Hall, London, pp. 270–292.

Wang, B., Lu, M.K., Yu, S.Q. and Zhang, R.P. 1990. Elimination of carnation mottle virus from carnation plants. Chin. J. Virol. 341–346.

Wetter, C. 1971. Carnation latent virus. CMI/AAB Descriptions of Plant Viruses, No. 61.

Zutra, D., Cohen, J., Gera, A. and Loebenstein, G. 1992. Association of *Rhodococcus (Corynebacterium) fascians* with stunting-fasciation syndrome in Israel. Abstracts, 8th Int. Symp. Virus Diseases Ornamental Plants, Prague, p. 45.

Cut Flower Crops (vegetatively propagated)

35

Chrysanthemum

Inge Bouwen[1] and Annemarie van Zaayen[2]

[1]*DLO Research Institute for Plant Protection (IPO-DLO), PO Box 9060, 6700 GW Wageningen, The Netherlands*

[2]*Inspection Service for Floriculture and Arboriculture (NAKB), Johan de Wittlaan 12, 2517 JR Den Haag, The Netherlands*

Introduction

Chrysanthemum is an important cut-flower. Worldwide trade is estimated at 4.5 billion flower-stems in 1993. The leading country is Japan with a trade of approximately 2 billion flower-stems. Turnover of chrysanthemums at the Dutch auctions in 1993 was about 1.3 billion flower-stems produced in the Netherlands on a total cropping area under glass of 782 ha. Sales totaled 607.5 million Dutch guilders, ranking this crop second after roses and far ahead of tulips (Produktschap Voor Siergewassen en Produktschap voor Groenten en Fruit, 1994).

Chrysanthemum is propagated by cuttings. Reproduction through seed is used mainly in breeding and for the production of certain varieties of potted chrysanthemums. In the Netherlands approximately 1.4 billion cuttings are traded per year.

All chrysanthemum cultivars, either cut-flowers or pot plants, belong to the *Chrysanthemum* Indicum group (family: Compositae). The name of this cultivar group is based on the Latin name *Chrysanthemum indicum*, a botanical species from Japan, which probably played an important role in the development of the current assortment. The name *Chrysanthemum indicum* was given by Linnaeus, who accumulated a wide variety of species in the genus *Chrysanthemum*. In 1855 Des Moulins moved a number of species from *Chrysanthemum* to a genus newly described by himself: *Dendranthema*. One of these species was *Chrysanthemum indicum*, which was renamed as *Dendranthema indicum*. This view of Des Moulins has been disregarded for a long time, but botanists have been studying the genus *Chrysanthemum* since the mid-seventies and did accept it. Consequently, most cultivars now belong to the *Dendranthema* Indicum group (family: Asteraceae, tribe: Anthemideae, subtribe: Artemisiinae). Other known species that earlier belonged to *Chrysanthemum* have also been moved to other genera, such as *Ajania*, *Argyranthemum*, *Coleostephus*, *Hymenostemma*, *Leucanthemum*, *Tanacetum*, and *Tripleurospermum* (Bremer and Humphries, 1993; Hetterscheid, 1993).

Because the crop has been propagated vegetatively for many years, viruses and viroids have also multiplied. Several viruses and viroids have been described. Since the early descriptions were based on symptomatology only, enormous confusion was created. The most important viruses and viroids that have been found in chrysanthemum for many years are chrysanthemum virus B, tomato aspermy virus, and chrysanthemum stunt viroid, which is an EPPO A2 quarantine pest. The importance of tomato spotted wilt virus has increased significantly after the introduction in European glasshouses of its vector, the western flower thrips, *Frankliniella occidentalis*.

The Viruses

Chrysanthemum B carlavirus (CVB)

Symptoms

Most cultivars are infected without showing visible symptoms. In some cultivars a slight loss of flower quality, a mild leaf mottling (Fig. 35.1), vein clearing, or a combination of these is found (Hollings and Stone, 1972). Mosaic, malformation, and slight to severe necrosis can also be found (Hakkaart and

Fig. 35.1. Mild mottle on a chrysanthemum leaf infected with chrysanthemum virus B. Photograph: Research Station for Floriculture, Aalsmeer.

Matt, 1974). The symptoms in chrysanthemums can be divergent, which is illustrated by the fact that many different viruses have been described in the past based on various observed symptoms. Examples are the five "viruses" of Brierley and Smith (1958) and the members of the leaf mottling complex of Hollings and Stone (1967). These were considered to be strains of CVB by Hakkaart and Maat (1974) like the six symptomatologically different entities they distinguished.

Description of the virus

CVB was first described by Noordam (1952) and belongs to the carlavirus group (see also Chapter 5, carlavirus).

The virus particles are slightly flexuous, rod-shaped, approximately 685 nm long, and 12 nm wide. They have a sedimentation coefficient of about 168 S, an A 260/280 of 1.55 (Hollings and Stone, 1972), and consist of a single-stranded RNA genome and a single coat protein (Koenig, 1982).

The virus is a good immunogen and is serologically related to carnation latent, elderberry carla, helenium S, lilac mottle, lily symptomless, nerine latent, passiflora latent, potato M, potato S, and red clover vein mosaic viruses (Koenig, 1982).

Geographical distribution and damage caused

CVB is widespread wherever chrysanthemums are grown. An infection with CVB alone will cause little damage, although the keeping quality of the cut-flowers may be negatively influenced. Symptoms will be more severe in mixed infections with tomato aspermy virus or chrysanthemum stunt viroid.

Methods of assay

Indexing for CVB can be done by inoculation on test plants, electron microscopy or ELISA. Most isolates of CVB induce local yellow spots, but some isolates cause necrotic or green spots or yellow rings after 2–5 weeks on leaves of *Petunia hybrida*. Symptomless systemic infections rarely occur (Hollings and Stone, 1972). The infectivity test requires much time and glasshouse space and can be unreliable in summer and also in the case of a co-infection with tomato aspermy virus (Hakkaart, 1969; Koenig et al., 1979).

Electron microscopy and ELISA are reliable throughout the year (Koenig et al., 1979). However, for large-scale routine testing, ELISA is generally used. A double antibody sandwich (DAS)-ELISA with a polyclonal antiserum (Maat, 1981) has been used in the Netherlands for many years. Reliable results are obtained, where one fully expanded leaf is picked per plant, and two leaves are combined in ELISA.

Natural and experimental transmission

CVB is transmitted in the non-persistent manner by several aphid species. It is also readily mechanically transmissible (Hollings and Stone, 1972). This means that CVB is easily transmitted from one mother-plant to the other when cuttings are taken without disinfection. Seed transmission has not been demonstrated.

Control methods

Chrysanthemum plants can be freed easily from CVB by meristem culture preceded by heat treatment. Depending on cultivar and duration of heat treatment, 100% plants free from CVB can be obtained (Hakkaart and Quak, 1964).

Once the virus has been eliminated, measures must be taken to prevent re-infection. Cultivation should start with certified material, and this should be done in aphid-proof glasshouses (OEPP/EPPO, 1993b). Infected plants should be eliminated. Since CVB often occurs symptomless, visual examination is not reliable. In the Netherlands plants of nuclear, propagation, or certified stock are regularly indexed for the presence of viruses (Van Ruiten, 1988, 1993). The NAKB-laboratory tests approximately 20,000 chrysanthemum samples each year.

The European and Mediterranean Plant Protection Organization (EPPO) has presented a certification scheme in which nuclear stock is checked once by mechanical inoculation to *Petunia* sp. or by ELISA, and propagation/certified stock is tested by applying ELISA to randomly selected plants (OEPP/EPPO, 1993a).

Tomato aspermy cucumovirus (TAV)

Symptoms

TAV causes severe breaking, dwarfing, and distortion of the flowers (Fig. 35.2) and sometimes chlorotic flecks or rings on the leaves, but most infected cultivars show no leaf symptoms and no loss of vegetative vigor (Hollings and Stone, 1971). Synonyms of TAV are chrysanthemum aspermy virus, chrysanthemum mosaic virus (Hollings and Stone, 1971), and probably chrysanthemum mild mottle virus described by Tochihara (1973).

Description of the virus

TAV belongs to the cucumovirus group (see also Chapter 5, cucumovirus). The virus particles are isometric, 29 nm in diameter, and built from 180 identical polypeptides of M_r 26.1×10^3 with $T=3$ symmetry. The particles encapsidate four species of single-stranded RNA of M_r 1.3 to 0.43×10^6 packaged so that the particles sediment at about 99 S (Francki, 1985). The A 260/280 is between 1.73 and 1.77 (Hollings and Stone, 1971).

Fig. 35.2. Right: flower distortion caused by tomato aspermy virus. Left: flower from a virus-free plant. Photograph: Research Station for Floriculture, Aalsmeer.

TAV is a good immunogen. It is serologically related to cucumber mosaic virus (CMV) and peanut stunt virus. No relationship was detectable in ELISA using CMV-specific antibodies. In indirect ELISA, however, all available strains of TAV were detected (Rybicki and Von Wechmar, 1985).

Geographical distribution and damage caused

TAV is widespread and common in cultivated chrysanthemum but uncommon in tomatoes (Hollings and Stone, 1971). Infection with only TAV will cause little damage, but symptoms will be more severe in mixed infections with CVB.

Methods of assay

Plants can be indexed for TAV by an inoculation on test plants, or by ELISA. *Chenopodium quinoa*, *Nicotiana glutinosa*, *N. tabacum* and *Petunia* sp. are suitable test plants. TAV induces numerous chlorotic or necrotic local lesions in *C. quinoa* after 3–7 days. Inoculated leaves of *N. glutinosa* may develop chlorotic or semi-necrotic spot or ring lesions, later followed by systemic vein-clearing, and severe light and dark green blister-mottle with distortion and dwarfing. *N. tabacum* develops severe systemic mottle, dwarfing, and distortion (Hollings and Stone, 1971). On *Petunia*, TAV produces local grey spots after 6–10 days, sometimes rings with a green center, systemic green mosaic, and severe leaf malformation (OEPP/EPPO, 1993a).

For large-scale routine testing, ELISA is generally used. In the Netherlands a DAS-ELISA with a polyclonal antiserum (Maat, 1981) is generally used, as for CVB.

Natural and experimental transmission

Although TAV can be transmitted through all the known modes of transmission, its natural transmission is mainly by several aphid species in the non-persistent manner. Experimental transmission is by mechanical inoculation of test plants. TAV can also be transmitted through seed, e.g. in *Stellaria media*, and some *Cuscuta* spp. have been reported to play a modest role (Hollings and Stone, 1971).

Control methods

Plants can be freed effectively from TAV by meristem culture combined with heat treatment. Control methods for TAV are in general the same as for CVB.

Tomato spotted wilt tospovirus (TSWV)

Symptoms

Symptoms caused by TSWV in chrysanthemum may differ greatly from one cultivar to the other. A variety of symptoms can be found, including chlorotic spots, ring patterns, bronzing or necrosis on leaves, and stem distortion and/or stem cankers and stunting. Of all these symptoms, chlorotic spots and foliar necrosis predominate. A characteristic necrosis starts in the leaf and extends through the petiole to the stem, forming longitudinal streaks (Fig. 35.3; see color plate). In the Palisade, Icecap, and Polaris groups, stem canker and stunting are prevalent. In contrast, these symptoms are not noted in the Marble and Super groups. TSWV can also influence the flower quality by reduction in fresh weight, reduction in petal number, and reduction in flower number per stem, especially for spray-type cultivars (Matteoni and Allen, 1989). The plant may outgrow the symptoms, and symptomless infections occur frequently. The highest incidence of latent infections tends to occur in the less susceptible cultivars, and when plants are grown in a cool regime (21/16°C day/night) (Allen et al., 1990).

Description of the virus

See Chapter 5, tospovirus.

Geographical distribution and damage caused

TSWV is common in temperate and subtropical regions throughout the world, wherever the vector, the western flower thrips, *Frankliniella occidentalis* occurs (Ie, 1970).

TSWV can cause substantial economic damage since it affects the quality of the flower. Foliar necrosis, stunting, and smaller flowers will make the product unmarketable. Usually, when TSWV is reported, extensive roguing of infected and adjacent plants should be done, resulting in crop loss.

Methods of assay

TSWV is mechanically transmissible to several test plants, such as *Petunia hybrida*. It induces local necrotic lesions after 2–4 days in *Petunia*. The infection does not become systemic.

TSWV belongs to the only plant virus group with large, spherical, enveloped particles and should therefore be easily recognizable in electron microscopy.

However, the virus concentration is generally low, and the particles are not very stable. Consequently, particles are often hard to find.

For large-scale routine testing the method of choice currently is ELISA with polyclonal antibodies or with monoclonal antibodies specific to capsid proteins and glycoproteins. The results of ELISA should be interpreted with care, as different biological variants of TSWV exist, and new strains might be generated (German et al., 1992). For testing, preferably leaves with symptoms should be sampled, although TSWV generally cannot be detected in plant parts with necrotic symptoms. From asymptomatic plants, the older leaves with high levels of feeding damage due to thrips are the best candidates for virus detection (Allen et al., 1990). In the Netherlands, TSWV is detected with a polyclonal antiserum prepared by D. Peters (De Avila et al., 1990).

Natural and experimental transmission

TSWV can be transmitted in a persistent manner by several thrips species. The western flower thrips, *Frankliniella occidentalis*, is the most important vector, even when other vector species are present, as it has polyphagous feeding habits, high fecundity, and dispersal capacity (German et al., 1992). The virus is acquired by the larvae when feeding on virus-infected plants, but it is transmitted by both larval and adult thrips which remain infectious for life. Experimental transmission is usually by mechanical inoculation.

Control methods

Cultivation should start with certified material, and precautions should be taken to prevent infestation with thrips. Visual inspection of the crop for symptoms as well as for the thrips vector should be combined with random testing by ELISA, inoculation to test plants, or both. If an infection is noticed, thrips should be adequately controlled, and then infected and adjacent plants should be destroyed. Weeds inside and outside the glasshouse can be host plants for TSWV and should be destroyed.

Preferably cultivars should be used that are unattractive to thrips and resistant to TSWV. However, few cultivars meet both requirements. The best alternative seems to be the use of moderately to highly resistant cultivars, even though some may be strongly preferred by thrips (Allen et al., 1990).

The Main Viroid

Chrysanthemum stunt viroid (CSVd)

Symptoms

The symptoms which develop after infection with CSVd are dependent on the cultivar. A typical reaction in many infected cultivars is a combination of stunted growth resulting in a reduction of one-half to two-thirds the normal plant size (Fig. 35.4; see color plate), with an acceleration of flowering from a few days to three weeks. Flowers and leaves are smaller and paler than normal. The flower color of yellow and white cultivars can change to pink. CSVd generally has a long incubation period: a few weeks to several months depending on the circumstances (Teyssier and Dunez, 1971; Hollings and Stone, 1973; Lawson, 1987).

Description of the viroid

For general information on viroids, see Chapter 6. CSVd consists of a covalently closed circular RNA of 356 nucleotides with 70% of base pairing (Haseloff and Symons, 1981).

Geographical distribution and damage caused

CSVd was first recognized in the US in the 1940s and spread rapidly, first to Canada and in the 1950s to all areas of the world where chrysanthemums are grown, probably through the international transport of cuttings (Lawson, 1987). CSVd seriously affects the plants and flowers of several cultivars. In some cultivars the plants are severely stunted, yielding an unsalable product.

Methods of assay

Grafting on an indicator cultivar is a reliable method, but it is time-consuming and labor-intensive.

Slab gel electrophoresis (Monsion et al., 1980), bi-directional electrophoresis, and molecular hybridization methods (Huttinga et al., 1987) can also be used. In the Netherlands the NAKB-laboratory applies bi-directional electrophoresis (Huttinga et al., 1987), which is as sensitive as the molecular hybridization method, but has proved to be more practical and does not need hazardous chemicals. The presence of viroid RNA can readily be detected in symptomless infected cultivars and long before the appearance of symptoms in the susceptible cultivars. The technique is, however, rather laborious: only 32 samples are handled at a time. As one sample may represent 15 leaves from five plants (three leaves per plant, from top, middle, and bottom), in principle 160 plants can be indexed at a time. Annually, approximately 1500 samples are tested in the Netherlands.

Viroids cannot be detected by classical immunological methods such as ELISA, since they are not encapsidated and do not encode any proteins.

Natural and experimental transmission

CSVd is transmitted by sap: by foliar contact, handling during cultivation (e.g. disbudding), and by cutting knives (e.g. taking cuttings).

Experimental transmission by inoculation of sap can be very erratic, and 100% success is uncommon. Better results are obtained by grafting (Hollings and Stone, 1973).

Transmission by insects and dodder has not been demonstrated (Hollings and Stone, 1973), and the situation of seed transmission is unclear. Hollings and Stone (1973) could not prove it, whereas Monsion et al. (1973) reported a seed transmission of 11% in chrysanthemums.

Control methods

Since no natural resistance or tolerance to CSVd has been identified, control measures primarily consist of efforts to prevent transmission by hands, tools, and plant-to-plant contact (Lawson, 1987), although several disinfectants did not seem to have an inhibitory effect on transmission (Hollings and Stone, 1973).

Meristem(-tip) culture can be used to free plants from CSVd, but only a low percentage of CSVd-free plants will be obtained (Lawson, 1987). Heat treatment is not effective, as CSVd is very heat tolerant: it withstands 10 minutes at 98–100°C. The best way to eliminate CSVd is by a cold treatment of plants at 5°C for at least 4 months, followed by meristem culture. Paludan (1985) applied this combined treatment and obtained 67% CSVd-free plants.

In the Netherlands CSVd-free plants are selected by visual inspection of the flowering crop. The reliability of this visual inspection can reach 90–95%, depending on the expertise of the inspector. Suspected plants will be tested by bi-directional electrophoresis. According to the certification scheme presented by EPPO, every plant of the nuclear stock is once tested by grafting, electrophoresis, or probe. Propagation and certified stock are randomly tested as for nuclear stock (OEPP/ EPPO, 1993a).

Final Remarks

Chrysanthemum may harbor other viruses and viroids of minor importance, such as chrysanthemum chlorotic mottle viroid, cucumber mosaic virus, and soybean mosaic virus.

1. Chrysanthemum chlorotic mottle viroid (CCMVd) causes a striking mottling on leaves with acute symptoms followed by complete chlorosis on leaves with chronic symptoms. Additional symptoms include dwarfing of leaves, flowers, and the entire plant and a delay in flower development. CCMVd has a narrow host range, and is less contagious as compared to other viroids (Horst, 1987). CCMVd is only associated with leaf mottle in a limited number of susceptible varieties (Monsion et al., 1980). Control methods are similar to those for CSVd.

2. Cucumber mosaic cucumovirus is only found in the Netherlands when chrysanthemums are cultivated outside the glasshouse. It is also reported to naturally infect chrysanthemums in India, where it causes diffuse chlorosis, chlorotic dots near the veins, and stunting of the plants (Srivastava et al., 1992).

3. Soybean mosaic potyvirus has been reported to infect *Argyranthemum frutescens* (formerly *Chrysanthemum frutescens*) in Italy (Bellardi et al., 1993). It caused marked chlorotic spots on the leaves.

Acknowledgements

The authors wish to thank J. Schuuring, NAKB, for sharing his extensive experience on the large-scale routine testing of chrysanthemums for several viruses and viroids.

Selected Literature

Allen, W.R., Matteoni, J.A. and Broadbent, A.B. 1990. Susceptibility of cultivars of florist's chrysanthemum to tomato spotted wilt virus. Can. J. Plant Pathol. 12:417–423.

Bellardi, M.G., Marani, F. and Bertaccini, A. 1993. Detection of soybean mosaic virus (SMV) in *Chrysanthemum frutescens*. Phytopathol. Mediterr. 32:156–158.

Bremer, K. and Humphries, C.J. 1993. Generic monograph of the Asteraceae-Anthemideae. Bull. Nat. Hist. Museum London (Bot.) 23(2):71–177.

Brierley, P. and Smith, F.F. 1958. Some characteristics of eight mosaic and two rosette viruses of chrysanthemum. Plant Dis. Rep. 42:752–763.

De Avila, A.C., Huguenot, C., Resende, R. de O., Kitajima, E.W., Goldbach, R.W. and Peters, D. 1990. Serological differentiation of 20 isolates of tomato spotted wilt virus. J. Gen. Virol. 71:2801–2807.

Francki, R.I.B. 1985. The viruses and their taxonomy. In: The Plant Viruses, Vol. 1. Polyhedral Virions with Tripartite Genomes, R.I.B. Francki (ed.), Plenum Press, New York and London, pp. 1–18.

German, T.L., Ullman, D.E. and Moyer, J.W. 1992. Tospoviruses: Diagnosis, molecular biology, phylogeny, and vector relationships. Annu. Rev. Phytopathol. 30:315–348.

Hakkaart, F.A. 1969. A comparison of electron microscopy and serology with infectivity tests for the detection of chrysanthemum virus B. Neth. J. Plant Pathol. 75:355–359.

Hakkaart, F.A. and Maat, D.Z. 1974. Variation of chrysanthemum virus B. Neth. J. Plant Pathol. 80:97–103.

Hakkaart, F.A. and Quak, F. 1964. Effect of heat treatment of young plants on freeing chrysanthemums from virus B by means of meristem culture. Neth. J. Plant Pathol. 70: 154–157.

Haseloff, J. and Symons, R.H. 1981. Chrysanthemum stunt viroid: Primary sequence and secondary structure. Nucleic Acids Res. 9:2741–2751.

Hetterscheid, W.L.A. 1993. Drastische verandering naamgeving *Chrysanthemum* en *Dracaena*. Vakblad voor de Bloemisterij 48(46):28–29.

Hollings, M. and Stone, O.M. 1967. Chrysanthemum viruses. Ann. Rep. Glasshouse Crops Res. Inst. 1966:93.

Hollings, M. and Stone, O.M. 1971. Tomato aspermy virus. CMI/AAB Descriptions of Plant Viruses, No. 79.

Hollings, M. and Stone, O.M. 1972. Chrysanthemum virus B. CMI/AAB Descriptions of Plant, Viruses No. 110.

Hollings, M. and Stone, O.M. 1973. Some properties of chrysanthemum stunt, a virus with the characteristics of an uncoated ribonucleic acid. Ann. Appl. Biol. 74:333–348.

Horst, R.K. 1987. Chrysanthemum chlorotic mottle. In: The Viroids, T.O. Diener (ed.), Plenum Press, New York and London, pp. 291–295.

Huttinga, H., Mosch, W.H.M. and Treur, A. 1987. Comparison of bi-directional electrophoresis and molecular hybridization methods to detect potato spindle tuber viroid and chrysanthemum stunt viroid. Bull. OEPP/EPPO Bull. 17:37–43.

Ie, T.S. 1970. Tomato spotted wilt virus. CMI/AAB Descriptions of Plant Viruses, No. 39.

Koenig, R. 1982. Carlavirus group. CMI/AAB Descriptions of Plant Viruses, No. 259.

Koenig, R., Dalchow, J., Lesemann, D.E. and Preil, W. 1979. ELISA as a reliable tool for the detection of chrysanthemum virus B in chrysanthemums throughout the year. Plant Dis. Rep. 63:301–303.

Lawson, R.H. 1987. Chrysanthemum stunt. In: The Viroids, T.O. Diener (ed.), Plenum Press, New York and London, pp. 247–259.

Maat, D.Z. 1981. Serologie van virussen als ziekteverwekkers in land en tuinbouwgewassen. Jversl. Inst. Plantenziektenkundig Onderz. 1980: 81-82.

Matteoni, J.A. and Allen, W.R. 1989. Symptomatology of tomato spotted wilt virus infection in florist's chrysanthemum. Can. J. Plant Pathol. 11:373–380.

Monsion, M., Bachelier, J.-C., and Dunez, J. 1973. Quelques propriétés d'un viroïde: Le rabougrissement du Chrysanthème. Annales de Phytopathologie 5:467–469.

Monsion, M., Macquaire, G., Bachelier, J.-C., Faydi, C. and Dunez, J. 1980. Detection of chrysanthemum stunt and chlorotic mottle viroids by slab gel electrophoresis. Acta Hortic. 110:321–328.

Noordam, D. 1952. Virusziekten bij chrysant in Nederland. With a summary: virus diseases of *Chrysanthemum indicum* in the Netherlands. Tijdschrift over Plantenziekten 58: 121–190.

OEPP/EPPO, 1993a. Certification scheme. No. 6. Pathogen-tested material of chrysanthemum. Bull. OEPP/EPPO Bull. 23:239–247.

OEPP/EPPO, 1993b. Certification scheme. No. 7. Nursery requirements-recommended requirements for establishments participating in certification of fruit or ornamental crops. Bull. OEPP/EPPO Bull. 23:249–252.

Paludan, N. 1985. Inactivation of viroids in chrysanthemum by low-temperature treatment and meristem-tip culture. Acta Hortic. 164:181–186.

Produktschap Voor Siergewassen en Produktschap voor Groenten en Fruit, 1994. Tuinbouwstatistiek 93. The Hague, The Netherlands.

Rybicki, E.P. and Von Wechmar, M.B. 1985. Serology and immunochemistry. In: The Plant Viruses, Vol. 1. Polyhedral Virions with Tripartite Genomes, R.I.B. Francki (ed.), Plenum Press, New York and London, pp. 221–222.

Srivastava, K.M., Raj, S.K. and Singh, B.P. 1992. Properties of a cucumber mosaic virus strain naturally infecting chrysanthemum in India. Plant Dis. 76:474–477.

Teyssier, D. and Dunez, J. 1971. Le rabougrissement du chrysanthème: symptômes et détection. Annales de Phytopathologie 3:65–73.

Tochihara, H. 1973. Serological relationship among chrysanthemum mild mottle virus, tomato aspermy virus and cucumber mosaic virus. Ann. Phytopathol. Soc. Jpn. 39: 272–274.

Van Ruiten, J.E.M. 1988. Production and trade of healthy propagating material: A need for a quality control system. Acta Hortic. 234:523–528.

Van Ruiten, J.E.M. 1993. Quality inspection and certification schemes for ornamental plants and fruit crops in the Netherlands. In: BCPC Monograph No 54: Plant Health and the European Single Market, D. Ebbels (ed.), BCPC/AAB/BSPP, UK, pp. 83–88.

Cut Flower Crops (vegetatively propagated)

36

Orchid

Roger H. Lawson and Hei-ti Hsu
US Department of Agriculture, Agricultural Research Service, US National Arboretum, Floral and Nursery Plants Research Unit, Beltsville, MD 20705, USA

Introduction

Orchids are the largest family of flowering plants with more than 800 genera and over 25,000 species. Orchids grow naturally in a wide range of habitats ranging from temperate through tropical climates in many parts of the world. Grown commercially in many different countries, orchids have perhaps the highest unit value of any ornamental pot plant. The procedure of mericloning developed in the 1960s greatly reduced the cost of orchid plants. As a result, culture expanded rapidly, and the hobby of orchid growing became popular in many different countries.

The market for orchid plants and cut flowers has expanded as new hybrids and cultivars have been introduced. Cut flower production has been concentrated in Asia and Hawaii in the US. The principal genera of cut flowers include *Dendro-*

bium, *Oncidium*, and intergeneric *Vanda* hybrids. In 1990 more than 2150 kilograms of flowers were produced in Asia.

Until recently, most mass market potted orchids were produced in Europe. In 1987 more than one million potted *Cymbidium*, *Paphiopedilum*, *Phalaenopsis*, and intergeneric *Odontoglossum* hybrids were sold in Europe. Production of *Cymbidium* and *Paphiopedilum* remained fairly steady from 1990 to 1992, but *Phalaenopsis* sales in the Netherlands expanded very significantly (Table 36.1). In 1993 a single company in Taiwan produced over one million *Phalaenopsis* with 80% of the plants exported to the US.

Because of the intensive nature of orchid culture, spread of viruses in the crop remains a serious threat to production. This chapter summarizes information on the principal viruses, symptoms, transmission characteristics, and assay methods used in virus detection. Control procedures involving the production of virus-tested propagation stock are also presented.

Table 36.1. Quantity and price (x1,000/million of US dollars) of pot plant orchids sold at auction in the Netherlands 1990–1993. Data from J.H. Post, Floricultura, The Netherlands.

	1990		1991		1992		1993	
	Quantity	Price	Quantity	Price	Quantity	Price	Quantity	Price
Cymbidium	892	5.51	849	5.40	985	6.10	963	6.00
Paphiopedilum	245	1.03	249	1.04	209	1.05	336	1.50
Phalaenopsis	1.599	6.40	1.771	7.50	2.54	9.92	3.15	12.17

The Viruses

Cymbidium mosaic potexvirus

Cymbidium mosaic virus (CyMV) is the most widespread orchid virus infecting a wide range of genera (Lawson, 1990). Foliar symptoms include chlorotic streaks in many *Cymbidium* cultivars and black necrotic spots and necrotic line patterns in *Phalaenopsis* species. Necrotic symptoms also occur in *Cattleya*, *Dendrobium* and many other genera. Although foliar symptoms are common, infected plants may also be asymptomatic.

Flower necrosis has been observed in *Cattleya* species for many years (Fig. 36.1; see color plate) and was once regarded as a symptom of senescence. This was particularly true in cultivars that showed flower necrosis only several days after the flowers were fully open. It was shown, however, that these symptoms were induced by CyMV (Lawson, 1970) which also occurs in *Laelia*, *Cymbidium*, *Phalaenopsis*, *Epidendrum*, and *Vanda*.

CyMV has been purified using a polyethylene glycol precipitation method and one cycle of differential centrifugation followed by separation on a 10–40% sucrose-gradient (Hsu et al., 1992). The virus was further purified by equilibrium density gradient centrifugation through a 35% cesium chloride gradient. Purified virus has been used as an immunogen to prepare polyclonal (Lawson, 1970) and monoclonal antibodies (Hsu et al., 1992).

Tobacco mosaic tobamovirus

A strain of tobacco mosaic virus (TMV) first isolated from *Odontoglossum grande* was described as odontoglossum ringspot virus (ORSV). This virus has been referred to as the orchid strain of TMV or TMV-O (Corbett, 1967). TMV-O induces a wide variety of symptoms in a range of orchid genera (Fig. 36.2; see color plate). Foliage symptoms include chlorotic and necrotic spots, streaks, and rings. In lavender- flowered cattleya cultivars the foliage patterns are often associated with pronounced red or purple pigmentation. Spotting and chevron-line patterns of chlorotic or red-pigmented tissues are common. *Oncidium* leaves may show a combination of sunken chlorotic and necrotic spots while *Epidendrum* shows only chlorotic spotting.

Flower breaking is a prominent symptom of TMV-O infection in lavender-flowered cattleyas. Streaks and dark pigmentation may be observed on the tight bud. Color breaking symptoms are often irregular, and some flowers in a cluster may show symptoms while others are symptomless (Lawson, 1990).

Recently it has been shown that orchids are frequently infected with two tobamoviruses, ORSV and orchid strains of TMV (Van Regenmortel et al., 1988). The orchid strains of TMV differ from type TMV by only a few amino acid exchanges in the coat protein and are usually not distinguishable with monoclonal antibodies to TMV. In contrast, ORSV differs greatly from TMV in amino acid sequence and is only distantly related. None of the five monoclonal antibodies prepared to ORSV was able to bind to orchid strains of TMV (Van Regenmortel et al., 1988).

Other viruses

Several viruses, in addition to CyMV and TMV-O, have been reported and the effects of these viruses on plant growth and flowering have been described

Table 36.2. Viruses reported in orchids.

Virus	Particle morphology	Dimensions (nm)	Symptoms on flowers	Means of transmission	Local lesion hosts	Geographical distribution
Rod-shaped						
Cymbidium mosaic virus (CyMV)	Flexuous	415-475 x 13-18	Necrosis of white colored cattleya flowers (11): Color break in *Dendrobium superbum*; White cell necrosis in lavender-flowered cattleyas	Sap, nursery tools, pots, and irrigation water	*Cassia* species *Chenopodium amaranticolor* *C. quinoa* *Datura stramonium* *Tetragonia expansa*	Worldwide
Orchid tobacco mosaic virus strains including ORSV	Rigid	280-320 x 15-24	Dark color breaks	Sap, nursery tools, pots irrigation water and plant debris	*Chenopodium* species *Gomphrena globosa*, *Tetragonia expansa* and some tobacco species	Worldwide
Bean yellow mosaic virus	Flexuous	750 x 13	None reported	Sap and aphids	*C. quinoa* *Vicia faba*	Japan United States Germany
Dendrobium mosaic virus	Flexuous	750 x 13	No symptoms	Sap and aphids	*C. quinoa* *C. amaranticolor*	Japan
Clover yellow vein virus	Flexuous	750 x 12	Necrosis	Sap and aphids	*Vicia faba*, *Pisum sativum*	Japan
Dendrobium vein necrosis virus	Flexuous	1865	Veinal necrosis (23)	None reported	None reported	Germany United States
Cypripedium filamentous virus	Flexuous	772	None reported	Mechanical transmission unsuccessful	None	Germany
Turnip mosaic virus	Flexuous	745	None reported	Sap and aphids	*Nicotiana clevelandii* *C. quinoa*	Germany
Tobacco rattle virus	Rigid	65 and 193	Misshaped flowers	Sap	*Nicotiana tabacum* cv Xanthi	Germany
Vanilla mosaic virus	Flexuous	767	None reported	Aphids	None reported	French Polynesia
Isometric						
Cymbidium ringspot virus	Isometric	30	None reported	Sap, soilborne	*Chenopodium amaranticolor* *Emilia sagittata* *Nicotiana clevelandii*, *Phaseolus vulgaris* cv. The Prince	England
Cucumber mosaic virus	Isometric	30	Mild color break	Sap and aphid	*Chenopodium amaranticolor* *Citrullus vulgaris*; *D. stramonium*, and *Sesamum*	Japan
Cymbidium mild mosaic virus	Isometric	28	None reported	Sap	*C. amaranticolor*	Korea
Trichopilia isometric virus	Isometric	28	None reported	Not reported	Not reported	Germany
Masdevallia isometric virus	Isometric	Not reported	Not reported	Mechanical transmission not successful	Not reported	Colombia

Table 36.2. (continued).

Virus	Particle morphology	Dimensions (nm)	Symptoms on flowers	Means of transmission	Local lesion hosts	Geographical distribution
Tomato ringspot virus	Isometric	30	Not reported	Sap (to indicator species but not back to orchid)	*Nicotiana tabacum* var. Kentucky 35, *Vigna unguiculata*	United States
Bacilliform and Bullet-shaped Viruses (Small Rhabdovirus-type Viruses)						
Short orchid rhabdovirus	Bullet-shaped bacilliform	47 x 105	None reported	Sap	*C. quinoa* *C. amaranticolor* *N. clevelandii*	Denmark Germany
Dendrobium virus	Bullet-shaped bacilliform	50 x 120	None reported	None reported	None reported	*D. inoglossum* from New Guinea, Germany
Orchid fleck virus	Bullet-shaped bacilliform	40 x 150	None reported	Sap, at temperatures 30 C	*C. amaranticolor* *C. quinoa*, *N. tabacum*, *N. glutinosa*	Japan
Unnamed rhabdovirus	"Rod-like"	40 x 100-120	None reported	None reported	None reported	Brazil
Grammatophyllum bacilliform virus	Bullet-shaped, bacilliform	40-42 wide particles with variable length	None reported	None reported	None reported	Maryland (U.S.)
(Large Rhabdovirus-type Viruses)						
Dendrobium rhabdovirus	Bullet-shaped bacilliform	180, 320 x 85	White streak but CyMV also present and effect of bacilliform virus alone unknown	None reported	None reported	Hawaii (U.S.)
Long orchid rhabdovirus	Bullet-shaped bacilliform	176 x 83	None reported	None reported	None reported	Germany
Laelia red leafspot virus	Bullet-shaped bacilliform	190-220 x 80	None reported	None reported	None reported	Germany
(Other)						
Unnamed bacilliform virus from Phalaenopsis	Bacilliform	119 x 29	None reported	None reported	None reported	Germany
Tospoviruses						
Tomato spotted wilt virus	Quasispherical	85 nm	None reported	Sap and thrips	*Petunia hybrida* *N. benthamiana*	Hawaii (U.S.)
Impatiens necrotic spot virus	Quasispherical	85 nm	None reported	Sap and thrips	*Petunia hybrida* *N. benthamiana*	California (U.S.)

Information adapted from Lawson, 1990.

(Table 36.2). They include several aphid transmitted viruses as well as soil-borne viruses (Table 36.2). The aphid transmitted potyvirus, bean yellow mosaic virus, was reported from *Calanthe* in Japan and from *Masdevallia* in the US and Germany (Lesemann and Koenig, 1985). In Germany the virus was found in many species of *Masdevallia* from only one nursery. Since it was not present in this genus from other nursery sources, the authors postulated that the virus was efficiently spread in the one nursery (Lesemann and Koenig, 1985). This example emphasizes the importance of accurate diagnosis and early detection of specific viral agents. This is particularly important where an efficient vector may be present that results in the rapid movement of a virus from plant-to-plant in the same genus or where transmission may occur in genera not previously recognized as susceptible hosts.

The aphid transmitted potyvirus, turnip mosaic (TuMV), was isolated from five different orchid genera growing in close proximity to each other (Lesemann and Vetten, 1985). The isolate from *Orchis militaris* was identified as a separate strain of TuMV. The authors of this report concluded that the TuMV isolate from *Orchis* was very efficiently spread, although the vector was not identified (Lesemann and Vetten, 1985). It is important, however, to carefully control aphids since the aphid transmitted viruses may cause severe disease in orchids.

Clover yellow vein, vanilla mosaic, dendrobium mosaic, and cucumber mosaic viruses are also aphid transmitted. Dendrobium vein necrosis is probably aphid transmitted (Table 36.2). There is no evidence, however, that CyMV, TMV-O, cymbidium ringspot or cymbidium mild mosaic viruses are spread by aphids.

Some viruses listed in Table 36.2 have not been reported since they were first described. The source of these viruses and their transmission to other orchids are not known. It is important, however, to take adequate precautions in attempts to limit vector spread. Tests are available for detection of several of these viruses, and bioassay hosts have been reported (Table 36.2). These methods should be rigorously applied in an effort to maintain a virus-free collection.

Mechanical Transmission

CyMV and TMV-O are transmitted in plant sap from infected to healthy plants when contaminated cutting tools are used to divide plants or harvest flowers. Bean yellow mosaic virus (BYMV), the short orchid rhabdovirus, tomato ringspot, dendrobium mosaic, cymbidium ringspot, cucumber mosaic, and tobacco rattle viruses may all be mechanically transmitted with differing efficiency (Table 36.2). Because there are no precise data on transmission efficiency, good sanitary practice and sterilization of cutting tools is prudent.

Identification and Assay Methods

Bioassay, serology, and electron microscopy are all useful procedures for detection of orchid viruses. Selection of a sample is the most important step in detection and diagnosis. Reliability of results depends on selection of tissue that is most likely to contain the highest concentration of virus. Samples should be removed from two or three leaves in different stages of growth. These should include new growth as well as tissue formed 2 or 3 years earlier. Although leaf tips are commonly sampled, removal of small leaf pieces from notches made at the base of the leaf is recommended since TMV-O and CyMV may be detected first in samples taken from this location in plants that have been recently infected.

Bioassay is the test that is most easily performed. This method has been widely used for detection of CyMV using *Cassia occidentalis* and *Datura stramonium* as local lesion hosts. TMV-O can be detected in a local lesion reaction on *Chenopodium amaranticolor* or *Gomphrena globosa*. Growing conditions and plant age may greatly influence host susceptibility and/or symptom expression. For example, *G. globosa* grown at 32/29°C day/night temperature may show only small necrotic lesions (Fig. 36.3). In contrast, plants grown at 21/18°C day/night temperature show an abundance of large red lesions with necrotic centers (Fig. 36.3). In addition, the short orchid rhabdovirus, orchid fleck, tomato ringspot, cymbidium ringspot, cucumber mosaic, and dendrobium mosaic can all be detected in bioassay. Tissue extracts can be prepared in phosphate buffer as described in the chapter on virus transmission.

Serology has been commonly used to detect CyMV and TMV-O. ELISA tests can be reliably performed for both viruses and antisera are commercially available. Recently a tissue-blotting technique for detection of viruses was developed (Hsu and Lawson, 1991). The method uses direct blotting of a freshly cut tissue surface onto a nitrocellulose membrane followed by incubation with pathogen-specific primary antibodies and enzyme-labelled immunogobulin-specific secondary antibodies. Detection in a test sample by dot-blot immunoassay is very much more sensitive than ELISA. A larger sample volume can be applied to nitrocellulose than in ELISA, and the dilution of the sample is not a

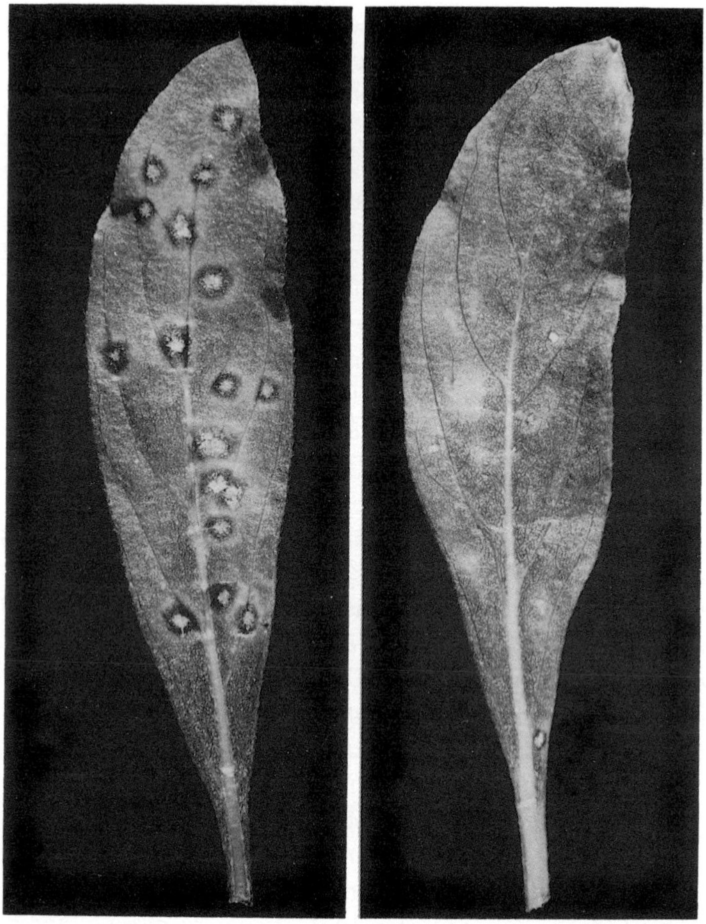

Fig. 36.3. Leaves of *Gomphrena globosa* two weeks after inoculation with orchid tobacco mosaic virus. Left: Leaf showing large red spots with necrotic centers grown at 21/18°C day/night temperatures. Right: Leaf showing small necrotic lesions from a plant grown at 32/29°C day/night temperatures. Plants were grown with 1,500 foot candles of illumination.

limiting factor as it is in ELISA. Further testing should be performed to compare the tissue blot procedure with other methods of serological detection where antisera are available.

CyMV can be detected by polyclonal (Lawson, 1990) or monoclonal antibodies (Hsu et al., 1992) in double antibody sandwich ELISA. Forty-five monoclonal antibodies reacted with the homologous antigen, which was trapped first

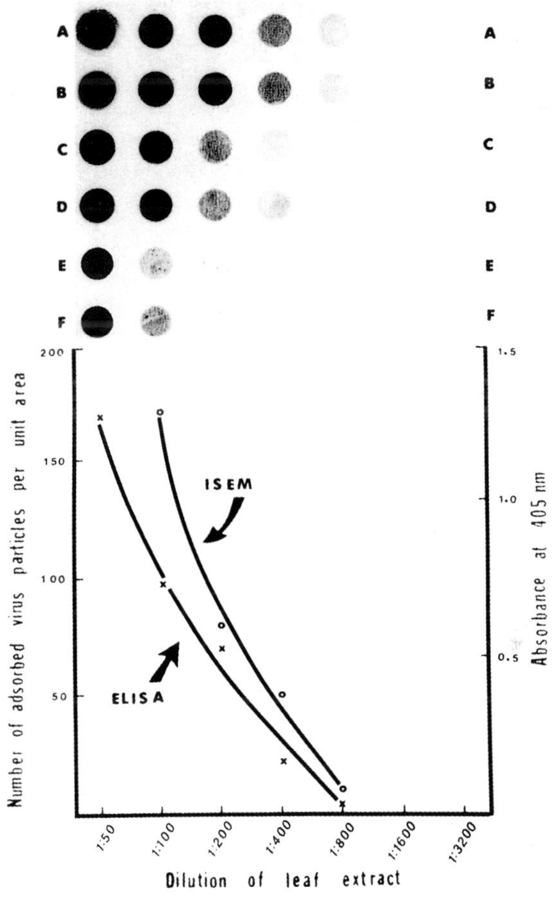

Fig. 36.4. Titrations of cymbidium mosaic virus (CyMV) antigens in leaf sap in an infected *Cattleya* orchid by dot-blot immunoassay (DBIA), enzyme-linked immunoassay (ELISA), and immunosorbent electron microscopy (ISEM). Five microliters per well of serial dilutions of (A and B) CyMV-infected *Cattleya* leaf extract. (C and D) CyMV-infected *Phalaenopsis* leaf extract. (E and F) Healthy *Phalaenopsis* leaf extract. Extracts were applied to a nitrocellulose membrane. CyMV titers were determined by using mouse monoclonal antibodies followed by alkaline phosphatase-labelled goat anti-mouse immunoglobulin antibodies. The apparent positive spots on the membrane with control samples (E and F) were due to green pigment in the preparation. The sample dilutions from infected *Cattleya* were incubated on mouse monoclonal antibody-treated carbon-coated Formvar grids for ISEM analysis or on rabbit polyclonal antibody-treated PVC plates followed by mouse monoclonal antibodies and alkaline phosphatase-labelled goat anti-mouse immunoglobulin antibodies in ELISA analysis.

by CyMV rabbit antiserum coated microtiter plates (Hsu et al., 1992). Twenty-nine of the 45 monoclonal antibodies also reacted with CyMV when the antigen was coated directly on microtiter plates. The sensitivity of the dot blot immunoassay was about eight times that of ELISA (Fig. 36.4). Orchid leaf samples that gave negative or marginal positive results in ELISA may produce a positive reaction in a tissue-blot immunoassay for CyMV (Hsu et al., 1992).

Electron microscopy can detect CyMV and TMV-O at dilutions of plant sap that reveal neither virus in bioassay or serological tests. The leaf dip method is the most common one used to prepare samples for electron microscopy. Visualization of virus particles can be enhanced by first coating the grids with specific antibody before the virus preparation is applied. This immunosorbent procedure significantly increases the number of particles attached to the grid surface. Detection of CyMV in crude sap from infected orchid leaves by immunosorbent electron microscopy was about twice as sensitive as by ELISA (Fig. 36.4). In addition to CyMV and TMV-O, this technique can be used to detect other rod-shaped viruses including bean yellow mosaic virus, dendrobium mosaic, dendrobium vein necrosis, cypripedium filamentous virus, turnip mosaic, tobacco rattle, and vanilla mosaic viruses. Isometric viruses and bacilliform and bullet-shaped virus particles are much more difficult to detect because they cannot always be distinguished from normal host plant constituents.

Nucleic acid hybridization utilizing complementary nucleic acids (cDNAs) to viral nucleic acids can detect specific viruses and differentiate closely related viruses. Although this method is not routinely used for detection of orchid viruses, the procedure offers great promise. A complementary copy of the viral RNA is first produced. A practical method of detection would extract sap from infected orchid plants and apply it to a nitrocellulose membrane. Labelled cDNA would next be applied. After washing the membrane would be treated with a reagent that develops a specific color reaction if the virus being tested is present.

Control — How to Produce a Virus-free Collection

Developing a virus-free collection of orchids is easy for the beginning collector. Plants should be purchased only from growers who test plants for the presence of virus. The buyer should know the history of the plant and how frequently it has been tested. Plants that tested negatively several years earlier may have become reinfected. As a matter of practice, it is a good idea to isolate newly acquired orchids from the rest of a collection until they have been evaluated. If there is any indication of symptom development, plants should be tested for the most common viruses.

Orchid growers at one time thought that tissue culture propagation was a method to produce virus-free plants. Although cymbidiums have been produced free of CyMV in tissue culture of small shoot tips, this method has generally not been successful in producing virus-free cattleyas and other genera. Some success has been achieved, however, in obtaining plants free of CyMV and TMV-O by treating small excised pieces of infected tissue with virus-specific antiserum at the time the tissue explant is removed and placed in the tissue culture medium (Inouye, 1983). This procedure may prove to be important in freeing orchids of CyMV and TMV-O.

Infection of orchids grown from seed is probably rare, but has been reported. In the green pod culture methods of seedling production, tissue from the female parent is transferred to a culture medium where the seedlings develop. In one report, one out of 123 *Dendrobium* seedlings from green pod culture was infected with CyMV when the female parent was infected (Yuen et al., 1979). This report has not been confirmed, however.

Development of a healthy nuclear stock collection of orchids requires careful attention to good sanitary practices combined with a virus testing program.

Selected Literature

Ali, S., Lawson, R.H. and Ishii, M. 1974. A bacilliform virus in white-streaked *Dendrobium phalaenopsis* flowers. Am. Orchid Soc. Bull. 43:529–533.

Chang, M.U., Arai, K., Doi, Y. and Yora, K. 1976. Morphology and intracellular appearance of orchid fleck virus. Ann. Phytopathol. Soc. Japan 42:156–157.

Corbett, M.K. 1967. Some distinguishing characteristics of the orchid strain of tobacco mosaic virus. Phytopathology 57:164–172.

Corbett, M.K. 1974. Intranuclear bacilliform virus-like particles in *Grammatophyllum scriptum*. Phytopathology (Abstr.). Potomac Div. Meeting of Am. Phytopathol. Soc., March 28–29, 1974.

Doi, Y., Chang, M.U. and Yora, K. 1977. Orchid fleck virus. CMI/AAB. Description of Plant Viruses, No. 183.

Doi, Y. and Yora, K. 1978. Cymbidium mild mosaic virus, isolation of cymbidium mild mosaic virus. Korean J. Plant Prot. 17:131–138.

Francki, R.I.B. 1970. Cymbidium mosaic virus. CMI/AAB. Description of Plant Viruses, No. 27.

Goff, L.M. and Corbett, M.K. 1977. Association of tomato ringspot virus with a chlorotic leaf streak of *Cymbidium* orchids. Phytopathology 67:1096–1100.

Hollings, M. and Stone, O.M. 1977. Cymbidium ringspot virus. CMI/AAB. Description of Plant Viruses, No. 178.

Hsu, H.T. and Lawson, R.H. 1991. Direct tissue blotting for detection of tomato spotted wilt virus in Impatiens. Plant Dis. 75:292–295.

Hsu, H.T., Vongsasitorn, D. and Lawson, R.H. 1992. An improved method for serological detection of cymbidium mosaic potexvirus infection in orchids. Phytopathology 82: 491–495.

Hu, J.S., Wang, M., Ferreira, S. and Ogata, D. 1992. Tomato spotted wilt virus on *Oncidium* orchids in Hawaii. Plant Dis. 76:426.

Inouye, N. 1969. Cucumber mosaic virus isolated from *Dendrobium*. Agric. Res. (Kurashiki) 53:49–60.

Inouye, N. 1971. Virus diseases of orchids. V. Symptoms and properties of viruses in *Dendrobium*. Jpn. Orchid Soc. Bull. 17:3–7.

Inouye, N. 1973. A new virus isolated from *Dendrobium*. Ann. Phytopathol. Soc. Jpn. 39: 367–368.

Inouye, N. 1983. Effect of antiserum treatment on the production of virus-free *Cymbidium* by means of meristem culture. Nogaku Kenkyu 60:123–133.

Inouye, N., Maeda, T. and Mitsuhata, K. 1988. A strain of clover yellow vein virus isolated from *Calanthe* sp. Acta Hortic. 234:61–68.

Kitajima, E.W., Blumenschein, A. and Costa, A.S. 1974. Rod-like particles associated with ringspot symptoms in several orchid species in Brazil. Phytopathol. Z. 81:280–286.

Lawson, R.H. 1970. Etiology of flower necrosis in Cattleya orchids. Phytopathology 60: 36–40.

Lawson, R.H. 1990. Orchid viruses and their control. In: Handbook on Orchid Pests and Diseases, A.M. Pridgeon and L.L. Tillman (eds.), Am. Orchid Soc,, West Palm Beach, FL, pp. 66–101.

Lesemann, D.-E. 1977. Long, filamentous virus-like particles associated with vein necrosis of *Dendrobium phalaenopsis*. Phytopathol. Z. 71:257–269.

Lesemann, D.-E. and Doraiswamy, S. 1975. Bullet-shaped virus-like particles in chlorotic and necrotic leaf lesions of orchids. Phytopathol. Z. 83:27–39.

Lesemann, D.-E. and Koenig, R. 1985. Identification of bean yellow mosaic virus in *Masdevallia*. Acta Hortic. 164:347–354.

Lesemann, D.-E. and Vetten, H.J. 1985. The occurrence of tobacco rattle and turnip mosaic viruses in *Orchis* sp., and of an unidentified potyvirus in *Cypripedium calceolus*. Acta Hortic. 164:45–54.

Mayhew, D.E., Cooke, A.L. and Raabe, R.D. 1992. Special report. A new virus is reported for *Phalaenopsis*. Am. Orchid Soc. Bull. 61:574–577.

Petzold, H. 1971. Der elektronenmikroskopische nachweis eines bazilliforen virus an blattfleckenkranken dendrobien. Phytopathol. Z. 70:43–52.

Van Regenmortel, M.H.V., Dekker, E.L., Dore, I., Porta, C., Weiss, E. and Burckard, J. 1988. Recent advances in serodiagnosis of plant virus diseases. Acta Hortic. 234:175–183.

Wisler, G.C., Zettler, F.W. and Mu, L. 1987. Virus infections of *Vanilla* and other orchids in French Polynesia. Plast Dis. 71:1125–1129.

Yuen, C.K.K.H., Kamemoto, H. and Ishii, M. 1979. Transmission of cymbidium mosaic virus through seed propagation in *Dendrobium*. Am. Orchid Soc. Bull. 48:1245–1247.

Cut Flower Crops (Vegetatively Propagated)

37

Phlox

A. Gera and J. Cohen
Department of Virology, Agricultural Research Organization, The Volcani Center, Bet Dagan 50250, Israel

Introduction

The genus *Phlox* is a member of the Polemoniaceae family, which is found from the tropics to high latitudes in both hemispheres. Most species occur in North America, Europe, and north Asia. Phlox is a perennial herb which has been recently cultivated. The important horticultural genus consists of 70 species (Everett, 1985). In addition there are numerous variants and a few hybrids. The genus name comes from the Greek word "phlox" or flame. This is an illusion to the brilliant colors of the leaves. Phlox includes annual, deciduous and evergreen herbaceous perennials, and subshrubs. *P. paniculata*, *P. maculata*, *P. subulata*, and *P. bifida* are the most useful ornamental garden perennials, with white, pink

Contribution from the Agricultural Research Organization, The Volcani Center, Bet Dagan, Israel, No. 1356-E, 1994 series.

and rose-purple flowers. Perennial flowering phlox are familiar to gardeners as named horticultural varieties, listed and often beautifully illustrated in color in dealers catalogs. They are very numerous, and new hybrids are introduced yearly. Flower color ranges from purest white to deepest crimson and includes many variants of pink and red, lavender and purple. In addition to decorating gardens they are useful as cut flowers and pot plants. *P. paniculata* and *P. maculata* are widely grown in Holland and recently have been introduced into Israel for cut flower production.

The Viruses

Cucumber mosaic cucumovirus (CMV)

The first report on CMV in *Phlox drummondii* was by Wellman (1935). CMV has also been found in phlox in India (Rangaraju and Chenulu, 1968) and Hungary (Horvath, 1980). Recently the virus has been detected in *P. paniculata* and *P. maculata* in Israel (Gera and Cohen, 1990). Naturally infected plants show mosaic, necrotic spotting, and ring patterns (Fig. 37.1). The virus is easily sap-transmissible.In aphid transmission tests conducted in India and Israel, *Aphis gossypii* efficiently transmitted the virus from CMV-infected to healthy phlox and other host plants.

Tobacco rattle tobravirus (TRV)

This virus was detected in Holland, both in greenhouse- and open-field-growing phlox (Hakkaart and Versluijs, 1985). Symptoms include irregular stem necrosis, growth reduction, leaf curl, malformation, and necrotic rings. Symptoms are more severe in greenhouse-grown phlox than in the field where the disease often appears as visible patches. The virus has a wide host range and is transmitted by soil-inhabiting nematodes (*Trichodorus* spp.).

Strawberry latent ringspot (SLRV) and raspberry ringspot (RRSV) nepoviruses

These viruses were detected concomitantly in *P. subulata* in Finland (Bremer, 1985). Infected plants are stunted with small and crinkled leaves. Infected leaves show light yellow flecks, and often turn yellow and wither. All tested samples were found to have mixed infection of SLRV and RRSV. Symptoms of the

Fig. 37.1. Shoot of *Phlox paniculata* naturally infected with cucumber mosaic virus (CMV) showing mild mosaic symptoms.

individual virus were not described because of the mixed infection. SLRV and RRSV are readily sap-transmissible and infect the seed of several host plants. In nature, both viruses are transmitted by the nematodes *Xiphinema* and *Logidorus* spp., respectively.

Tomato black ring nepovirus (TBRV)

The virus was detected in *P. paniculata* at two different sites in Sweden (Ryden, 1965). Symptoms in naturally infected plants are highly variable; they include ringspots, light mosaic, and leaf deformation (Fig. 37.2). TBRV occurs mainly in Europe. The virus is easily sap-transmissible and in nature is transmitted by soil-inhabiting nematodes (*Longidorous* spp.).

Fig. 37.2. Leaves of phlox naturally infected with tomato black ring virus (TBRV) showing ringspots and light mosaic symptoms. (Courtesy of K. Ryden, Swedish University of Agriculture, Sweden.)

Identification and Assay Methods

All known viruses infecting phlox are readily sap-transmissible to herbaceous indicator plants where they cause characteristic symptoms. *Chenopodium amaranticolor* and *C. quinoa* are the most reliable local lesion hosts for all known viruses infecting phlox. Antisera against these viruses are commercially available. Serological identification is generally easy and is a reliable means for early detection.

Control

Phlox is propagated vegetatively by cuttings or through seeds. Careful choice of virus-free mother plants is necessary for obtaining virus-free planting material. CMV occurs naturally in many species in wild and cultivated crops, providing a reservoir of virus. The virus is non-persistently transmitted by aphids. Care should be taken to avoid introduction of aphids into glass- or plastic-houses.

Control of the soilborne SLRSV, RRSV, and TRV can be achieved by soil sterilization prior to planting. Treatment with nematocides has been shown to reduce the incidence of TRV in phlox (Hakkaart and Versluijs, 1985).

Selected Literature

Bremer, K. 1985. Strawberry latent ring spot in ornamental plants in Finland. Ann. Agric. Fenniae 24:97–98.

Everett, T.H. (ed). 1985. The New York Botanical Garden Illustrated Encyclopedia of Horticulture. Vol. 8. Garland Publishing, New York and London, pp. 2559–2605.

Gera, A. and Cohen, J. 1990. Occurrence of cucumber mosaic virus in phlox in Israel. Plant Pathol. 39:558–560.

Hakkaart, F.A. and Versluijs, J.M.A. 1985. Tabaksratel-virus in *Phlox paniculata* "Rembrandt". Vakbl. Bloemisterij 49:49–50.

Horvath, J. 1980. Viruses of lettuce II. Host ranges of lettuce mosaic virus and cucumber mosaic virus. Acta Agric. Aca. Scien. Hung. 29:337–352.

Rangaraju, R. and Chenulu, V.V. 1968. Studies on the transmission of phlox mosaic by *Aphis gossypii*. Indian Phytopathol. 21:269–274.

Ryden, K. 1965. Phlox-ringflack -en svar virussjukdom orsakad av tomat-svartringvirus. Sartryck ur Vaxtskyddsnotiser 6:1–4.

Wellman, F.L. 1935. The host range of the southern celery-mosaic virus. Phytopathology 25:377–404.

38

Rose

G. Loebenstein
Department of Virology, Agricultural Research Organization, The Volcani Center, Bet Dagan 50250, Israel

Introduction

Cultivation of the rose is as old as any branch of horticulture. Roses are known to have been grown in northern Africa and Mesopotamia 4000–5000 years ago. Sargon the First (2630–2648 BC) mentioned that he "sent rose bushes from the Taurus mountains to his country". Homer (1000 BC) alluded to the rose in the *Iliad* and the *Odyssey*. Sappho, the Greek poetess (ca. 600 BC), designated the rose as the Queen of Flowers in the following lines: "Would you appoint some

Contribution No. 1365-E, 1994 series, from the Agricultural Research Organization, The Volcani Center, Bet Dagan, Israel.

flower to reign; In matchless beauty on the plain; The Rose (Mankind will all agree); The Rose the Queen of Flowers should be". The "Rose of Sharon" and the "Rose of Jericho" in the Bible probably refer to other flowers (Lilium). However, in the Book of Ben-Sira (24,14) (200 BC), we find a clear mention of the rose.

The rose is the most popular garden plant in the world, as well as one of the most important commercial cut flowers grown under cover. The main commercial growing areas are outlined in Table 38.1.

In 1992, worldwide rose sales were valued at least US $1.1 billion, including $153 million of imports into European countries; $90 million US imports, and $175 million US production; and $3 million Japanese imports and $180 million Japanese production.

Many rose cultivars are known, and their classification has become increasingly complicated as hybridization has proceeded. The major groups of roses are: Hybrid Tea (mostly *Rosa dilecta*), with one or few terminal flowers; Polyantha (a cross between *R. multiflora* and *R. chinensis*), with clusters of many small flowers; Floribunda, hybrids between Polyantha and Hybrid Tea; and Grandiflora, hybrids between Floribunda and Hybrid Tea. The latter two have an intermediate number of flowers in the inflorescence. There are also climbing roses with intermediate growth and lateral flowering; this group includes cultivars with both recurrent and nonrecurrent flowering.

The main understock cultivars used are *R. manetti*, generally the choice in the US; *R. canina*, often used in Europe; and *R. indica* "Major" used in Europe and in the Mediterranean region, including Israel.

The Viruses

Rose mosaic disease

Rose mosaic disease is widely distributed throughout the world and occurs wherever roses are grown. It has been reported in Australia, the former Czechoslovakia, Denmark, France, Germany, India, Israel, Italy, New Zealand, the Netherlands, Russia, South Africa, Spain, the US, and the Ukraine.

Symptoms

Rose mosaic is not a well defined disease but rather a "catch-all" name for a range of symptoms. The symptoms are highly variable and include ringspots

Table 38.1. Main commercial rose production areas

Area/main countries	Hectares			% Worldwide	Year
	Under cover	Open field	Total		
Western Europe			3492	53.4	1989–93
Italy	779	200	979	15.0	1989
Netherlands	894		894	13.6	1992
Spain & Canary Islands			541	8.3	1989
Germany	241	218	459	7.0	1992
France			420	6.4	1990
South America			1307	20.0	1992
Colombia			1000	15.2	1992
Ecuador			173	2.6	1992
North America			526	8.0	1992
US			426	6.5	1992
Mexico			100	1.5	1992
Asia			800	12.2	1991, 93
Japan			468	7.2	1991
Israel	200		200	3.1	1993
Taiwan			126	1.9	1991
Africa			384	5.8	
Zimbabwe			150	2.3	1993
Total, including minor production areas			6545		1989–93

Sources: Personal communication from A. Malter, based on: *Yearbook of the International Horticultural Statistics*, Vol. 41, Institut für Gartenbauokonomie der Universität Hannover, Germany, 1993; Israel Flower Board and Agrexco, *Onat Y'tsu 1992/93*, Tel-Aviv, 1993 (in Hebrew); JETRO (Japan External Trade Resources Organization), The Japanese market for cut flowers, *Tradescope*, April 1992, pp. 7–18; USDA, *Foreign Agricultural Trade of the US: Calendar Year 1992 Supplement*, Economic Research Service, Washington, D.C., June 1993; USDA, *Floriculture Crops 1992*, National Agricultural Statistics Service, Agricultural Statistics Board, Washington, DC, April 1993; and VBN (Association of Dutch Flower Auctions), *Statistiekboek 1992*, Leiden, Holland, January 1993.

Fig. 38.1a. Symptoms of rose mosaic on rose leaves: ringspots.

Fig. 38.1b. Symptoms of rose mosaic on rose leaves: line pattern.

(Fig. 38.1a), line patterns (Fig. 38.1b), chlorotic banding, oak-leaf patterns, and mosaic. At temperatures of 10–15°C and high light intensities, leaf symptoms become anthocyanin-reddish on *R. indica*. Thomas (1981) observed that chlorotic line patterns, ringspots, or mottles are generally induced by the cherry serotype of prunus necrotic ringspot virus (PNRSV), while an apple serotype (apple mosaic virus, ApMV) causes stunted, distorted, and puckered leaflets (Thomas, 1984) (Fig. 38.2).

Symptoms in roses are not always conspicuous and tend to disappear, and generally there is little adverse effect on growth. Virus-tested plants of some varieties gave an increase in flower production or quality when compared with infected ones (Moran et al., 1988). However, a marked decrease in graft unions was observed when infected scions were grafted on healthy rootstocks compared with graft-takes with healthy scions (Stein et al., 1988).

Causal Agents

Rose mosaic disease has generally been associated with isolates of PNRSV or ApMV or mixtures of the two. Both belong to subgroup III of the Ilarviruses.

Fig. 38.2. Distorted and puckered rose leaflets induced by apple mosaic virus (from Thomas, 1984).

Particles of PNRSV are quasi-isometric with a diameter of ca. 23 nm, some of the particles being bacilliform. The particles are unstable in neutral phosphotungstate (PTA) unless previously fixed, but remain intact in uranyl acetate. The genome is tripartite with four molecules of positive sense, single-stranded RNA (RNAs 1 to 4) of Mr 1.3, 0.89, 0.69 and 0.31×10^6. The three larger RNAs carry all the genetic information of the virus. The nucleotide sequence specifying the coat protein, contained in part of RNA 3, is also carried on RNA 4, the subgenomic RNA. The three larger RNAs are packaged separately, and two or more of the small RNAs are packaged together. RNAs 1–3 plus coat protein or RNA 4 are required for infectivity. The molecular weight of the coat protein is 25×10^6.

Particles of ApMV are quasi-isometric 29 and 25 nm in diameter, some of the particles being bacilliform.

Arabis mosaic nepovirus (ArMV), either singly or in mixed infections with PNRSV, has also been associated with rose mosaic symptoms.

Transmission

There is no evidence for vector transmission of PNRSV and ApMV in roses. Although ArMV is known to be transmitted by free-living nematodes (especially *Xiphinema diversicaudatum* and *X. coxi*), no such transmissions have been reported in roses, probably because in many greenhouses the soil is treated against soil-borne organisms before planting. PNRSV is pollen-transmitted in fruit trees. It was also reported that several thrips species enhanced PNRSV transmission by pollen from fruit trees by creating wounds during their feeding (Greber et al., 1992). This may occur also in roses. Spread of rose mosaic is, therefore, attributable almost only to dissemination of propagation material, when infected buds, scions or rootstocks are grafted to healthy plants.

PNRSV from roses can be transmitted with difficulty to *Cucumis sativus*, *Chenopodium quinoa* (necrotic local lesions, followed by systemic infection causing severe distortion), *Momordica balsamina* (necrotic lesions), and *Cyamopsis tetragonoloba* (large dark local lesions); ApMV to *C. sativus*, *Vigna unguiculata*, and *Catharanthus roseus*; and ArMV to *Chenopodium amaranticolor*, *C. quinoa*, *C. sativus*, and *Petunia hybrida*. It is sometimes advantageous to graft infected rose buds to peach seedlings and to transmit the virus mechanically from them to herbaceous species. For maintaining PNRSV *C. quinoa* is a commonly used host.

Identification and assay methods

PNRSV in roses can be detected by enzyme-linked immunosorbent assay (ELISA). It is advantageous to absorb the enzyme-conjugated γ-globulin with healthy rose or peach leaf extract, thereby greatly reducing non-specific reactions (Thomas, 1980; Stein et al., 1987). Recently, using polyclonal antibody-coated ELISA plates and a monoclonal antibody for the diagnostic reaction, reliable detection was obtained without an absorption step (A. Stein, pers. comm.).

ApMV is detectable in rose leaf samples by immuno-electron microscopy (IEM) (Thomas, 1981) or by ELISA (A. Stein, pers. comm.). Petals or young leaf tissue gave the best results for both PNRSV and ApMV, but sampling should not be done during the warm season.

For indexing of nuclear stock plants it may be advantageous to graft buds from the plant to be tested onto flowering cherry *(Prunus serrulata* var. "Shirofugen"). Shirofugen reacts to grafts of PNRSV and ApMV-infected buds with a distinctive local hypersensitive reaction of necrosis and gumming at the graft site about 30–50 days after budding (Fig. 38.3). As the reaction is only local, a number of buds can be grafted to each branch of Shirofugen. Shirofugen, however, requires a cold period to break its dormancy. Indexing can therefore also be done on *Prunus tomentosa*, in which rose mosaic causes a systemic dieback reaction. However, a large greenhouse area is required as each *P. tomentosa* plant can be grafted with buds from only one candidate rose plant.

At least ten buds should be sampled from the basal part of the candidate's branches and indexing should be repeated in the following year, due to the uneven distribution of the virus (Fleisher et al., 1971). Similar sampling should be used for ELISA.

Purification

For propagation, good results are obtained with *C. quinoa* (Mink, 1980) and

Fig. 38.3. Necrosis and gumming at the graft site of Prunus serrulata "Shirofugen" grafted with prunus necrotic ringspot virus-infected budwood (left) compared with healthy buds (right).

Fig. 38.4. Yellow flecking on rose leaf infected with strawberry latent ringspot virus.

Nicotiana clevelandii (Thomas, 1981). For purification, methods developed by Mink (1980) or Thomas (1981) or variations based on these methods give satisfactory results.

Strawberry latent ringspot nepovirus

Strawberry latent ringspot virus (SLRV) (Murant, 1974) occurs naturally in many wild and cultivated plant species. SLRV was isolated from roses in England showing distinct small angular yellow flecks in young leaves (Fig. 38.4) and a reduction in size of the leaflets. Infected plants are often stunted. Symptoms vary in severity and some rose cultivars show no symptoms. On plants infected with SLRV, there is frequent scion bud death.

Particles of SLRV (a putative nepovirus) are isometric and ca 30 nm in diameter. Sap from infected tobacco or cucumber plants, when negatively stained, showed the particles to occur frequently in tubular sheaths.

SLRV in roses can be detected by ELISA and IEM, the latter being twice as sensitive (Thomas, 1980). SLRV can be mechanically transmitted from young rose leaves to *C. amaranticolor*, *C. quinoa*, and cucumbers. Adding insoluble polyvinyl pyrrolidone (PVP) gave a marked increase in transmission efficiency (Ikin and Frost, 1976). SRLV is transmitted by the nematode *X. diversicaudatum*, but nematode transmission in rose plantings is of little or no significance.

Rose rosette disease

Rose rosette disease occurs in wild roses (*Rosa multiflora*) and in commercial roses in California and the midwestern states of the US. Symptoms include bright red misshapen or stunted leaves developing from auxiliary buds and a witches' broom phyllody or rosette appearance of the stems. Affected stems are often thorny and slow to mature. The plant usually dies as symptoms spread.

Electron microscopy revealed spherical virus-like particles (120–150 nm), singly or grouped in the cytoplasm of affected plants (Gergerich and Kim, 1983). The particles are similar to those associated with wheat spot mosaic and fig mosaic diseases.

The disease can be transmitted by grafting or by the eriophyid mite *Phyllocoptes fructiphilus*.

Attempts are being made to use rose rosette as a biocontrol agent against multiflora bushes.

Rose streak

Rose streak disease occurs in Europe, the US, and South Africa. Symptoms on young leaves include irregular pale green or chlorotic streaks along the veins. As the leaves mature, the lines become brown while the tissue encircled by the necrotic lines changes from pale green to yellow. On old leaves necrotic rings or line pattern are seen on both the lower and upper side of the leaves (Fig. 38.5). Leaves drop prematurely.

The disease is graft-transmissible. Several rose cultivars such as "Butterfly" and "Ophelia" can serve as indicators. Necrosis and blackening develop around infected buds grafted to these indicators within 2 months after union between bud and stock.

Rose wilt

Rose wilt disease has been reported from New Zealand, Australia and California.

Fig. 38.5. Ring and line pattern caused by rose streak disease (from Klesser, 1967).

The causal agent is not known. Symptoms may vary. Newly infected plants exhibit downward curling of leaves and leaf epinasty. Loss of apical dominance occurs resulting in a rosetted growth with small recurved leaves — causing "balling" of leaves (Fig. 38.6).

The disease can be transmitted by double-budding. Rose wilt may be confused with wilt caused by *Verticillium alboatrum*. However, the latter causes wilting of leaves at the tips of young stems and yellowing of the lower ones; and *Verticillium* has not been isolated from plants with rose wilt symptoms.

Rose ring pattern

Rose ring pattern has been reported to occur in California and Oregon. Symptoms that are similar to those caused by rose mosaic include rings, fine line patterns, and chlorotic flecks (Fig. 38.7) in leaves and sometimes color-break rings in petals.

Rosa multiflora "Burr" is a reliable indicator. Symptoms on this rootstock include severe stunting and deformity and mottling on the leaflets. Graft inoculation to "Shirofugen" does not cause necrosis, a reaction differing from that induced by rose mosaic. The causal agent has not been identified.

Other virus and virus-like diseases

Tobacco ringspot, tobacco streak, and a tobamovirus have been isolated from roses. In addition, other virus-like diseases have been described. These include

Fig. 38.6. Shoot of rose cv. "Queen Elisabeth" infected with rose wilt disease, showing epinastic balling (from Fry and Hammett, 1971).

Fig. 38.7. Chlorotic patches caused by rose ring pattern in hybrid tea rose (from Secor and Nyland, 1978).

rose spring dwarf, rose flower break, and rose flower proliferation. No causal agent has so far been identified for any of them.

Control

The best way to control virus and virus-like diseases is to use certified propagation material of budwood and stocks. Indexing on indicator plants and serology are used for rose mosaic and some of the other identified viruses, but it is also necessary to have periodic visual inspections, especially for those diseases for which the causal agent is not known. Samples for indexing should be taken from several branches of the rose bush, and it is of advantage to re-index the certified plant every 3–4 years.

Preparation of certified rootstocks is the practice in many countries. However, certification of variety budwood is more complex due to changes in cultivars, especially when breeders release new cultivars and the growers are interested in receiving them without delay.

Heat treatment and meristemming have been used in certain cases to obtain virus-free budwood. Heat treatment can be used to obtain buds free of rose mosaic. Holding rose plants at 38 °C for 4 weeks results in freeing more than 50% of the surviving buds from PNRSV (Bjarnason et al., 1985). Likewise, lateral "buds" propagated from plants infected with rose ring pattern and held at 38 °C for 3 weeks or longer become free of infection. To control SLRV it is also advisable to sterilize the soil to reduce nematode populations.

Selected Literature

Bjarnason, E.N., Hanger, B.C., Moran, J.R. and Cooper, J.A. 1985. Production of prunus necrotic virus-free roses by heat treatment and tissue culture. N.Z. J. Agric. Res. 28: 151–156.

Fleisher, Z., Drori, T. and Loebenstein, G. 1971. Evaluation of Shirofugen as a reliable indicator for rose mosaic virus. Plant Dis. Rep. 55:431–433.

Fry, P.R. and Hammett, K.R.W. 1971. Rose wilt virus in New Zealand. N.Z. J. Agric. Res. 14:735–743.

Gergerich, R.C. and Kim, K.S. 1983. A description of the causal agent of rose rosette disease. Arkansas Farm Res. 32:3–7.

Greber, R.S., Teakle, D.S. and Mink, G.I. 1992. Thrips-facilitated transmisison of prune dwarf and prunus necrotic ringspot viruses from cherry pollen to cucumber. Plant Dis. 76:1039–1041.

Horst, R.K. 1983. Compendium of Rose Diseases. Amer. Phytopathological Society, St. Paul, MN.

Ikin, R. and Frost, R.R. 1976. Virus diseases of roses. II Strawberry latent ringspot virus. Phytopathol. Z. 87:205–223.

Klesser, P.J. 1967. Occurrence of rose streak in South Africa. S. Afr. J. Agric. Sci. 10: 849–856.

Mink, G.I. 1980. Identification of rugose mosaic-diseased cherry trees by enzyme-linked immunosorbent assay. Plant Dis. 64:691–694.

Moran, J.R., Faragher, J.D. and Baker, D.M. 1988. The effects of prunus necrotic ringspot virus on production and quality of rose flowers. Acta Hortic. 234:429–434.

Murant, A.F. 1974. Strawberry latent ringspot virus. CMA/AAB Descriptions of Plant Viruses, No. 126.

Secor, G.A., Kong, M. and Nyland, G. 1977. Rose virus and virus-like diseases. Calif. Agric. 31:4–7.

Secor, G.A. and Nyland, G. 1978. Rose ring pattern: a component of the rose-mosaic complex. Phytopathology 68:1005–1010.

Stein, A., Levy, S. and Loebenstein, G. 1987. Detection of prunus necrotic ringspot virus in several rosaceous hosts by enzyme-linked immunosorbent assay. Plant Pathol. 36:1–4.

Stein, A., Levy, S. and Loebenstein, G. 1988. Effects of prunus necrotic ringspot virus on graft union and flower quality of glasshouse grown roses in Israel. Acta Hortic. 234:435.

Thomas, B.J. 1980. The detection by serological methods of viruses infecting the rose. Ann. Appl. Biol. 94:91–101.

Thomas, B.J. 1981. Studies on rose mosaic disease in field-grown roses produced in the United Kingdom. Ann. Appl. Biol. 98:419–429.

Thomas, B.J. 1984. Rose mosaic disease: symptoms induced in roses by graft inoculation with both prunus necrotic ringspot and apple mosaic viruses. Plant Pathol. 33:155–160.

CUT FLOWER CROPS (SEED PROPAGATED)

39

Asclepias

A. Gera and J. Cohen
Department of Virology, Agricultural Research Organization, The Volcani Center, Bet Dagan 50250, Israel

Introduction

The genus *Asclepias* is a member of the family Asclepiadaceae that comprises approximately 220 genera and 2000 species. Members of this family are distributed throughout the world, primarily in tropical and subtropical regions (Lyons, 1985). The ornamental plant industry has developed the potential of many genera within this family, including *Stephanotis* and *Hoya*. Although *Asclepias* is known for its weedy characteristics, one member of the genus, *A. tuberosa*, is gaining interest as a garden ornamental and recently as a cut–flower crop in Europe and

Contribution from the Agricultural Research Organization, The Volcani Center, Bet Dagan, Israel, No. 1357-E, 1994 series.

Israel (Lyons, 1985). *A. tuberosa* is a common wild flower in habitats across eastern North America, eastern Canada, and Mexico. It consists of a stout taproot from which numerous annual stems are produced in the spring. Plants flower from May to July, but occasional individuals may be found in flower as late as September. *A. tuberosa* is distinguished by bright orange flowers, and occasionally red or yellow. The plant has been introduced in Israel as a cut–flower where it is usually grown in plastic houses.

The Viruses

Asclepias rhabdovirus

This virus is specific for *Asclepias*. The virus particle is bacilliform, measuring 300 nm in length and 50 nm in diameter. The virus occurs sometimes together with a flagellate protozoan (*Phytomonas* spp.). Symptoms appear as diffuse leaf chlorosis or leaf mosaic in *Asclepias curassavica* in Ecuador (Dollet and Gargani, 1989). The virus has not been characterized.

Cucumber mosaic cucumovirus (CMV)

The virus was detected in *A. syriaca* growing near commercial lettuce fields in New York State (Rist and Lorbeer, 1989) and in Hungary and Yugoslavia (Salamon et al., 1989). It was found recently in *A. tuberosa* in commercial fields at various sites in Israel (Gera, 1994). Naturally infected plants show severe mosaic, vein clearing, line pattern, and distortion on leaves (Fig. 39.1).

Tomato spotted wilt tospovirus (TSWV)

On *A. tuberosa* TSWV causes yellowing, mild mosaic, and necrosis on leaves (Fig. 39.2). The virus was detected at two different sites in Israel (Gera, unpublished).

Identification and Assay Methods

Asclepias rhabdovirus was detected by electron microscopy in sections from infected leaves. CMV and TSWV are readily sap–transmissible to indicator plants. *Chenopodium amaranticolor* and *Petunia hybrida* are reliable local–lesion hosts for CMV and TSWV, respectively. Poly– and monoclonal antibodies are commercially available for detection of both viruses by ELISA and tissue blotting.

Fig. 39.1. Leaves of *Asclepias tuberosa* naturally infected with cucumber mosaic virus showing mild mosaic symptoms.

Fig. 39.2. Shoot of *Asclepias tuberosa* naturally infected with tomato spotted wilt virus showing mild mosaic and necrosis.

Control

Cultivated *Asclepias* is propagated sexually by seeds (Borland, 1987) or vegetatively by rooted cuttings (Albrecht and Lehmann, 1989). Recently *A. tuberosa* was successfully propagated from short root segments (Ecker and Barzilay, 1993). Control of CMV and TSWV is difficult and uncertain. Fine mesh covers and chemical control of insects and thrips should be used to minimize the risk of spread of the viruses within the crop.

Selected Literature

Albrecht, M.L. and Lehmann, J.T. 1991. Daylength, cold storage, and plat production method influence growth and flowering of *Asclepias tuberosa*. HortScience 26:120–121.

Borland, J. 1987. Clues to Butterfly milkweed germination emerge from a literature search. Am. Nurseryman 165:91–96.

Dollet, M. and Gargani, D. 1989. Mixed infection of *Asclepias curassavica L.* in Ecuadur: Rhabdovirus-like particles and *Phytomonas* sp. J. Phytopathol. 125:269–273.

Ecker, R. and Barzilay, A. 1993. Propagation of *Asclepias tuberosa* from shoot segments. Sci. Hortic. (Amst) 56:171–174.

Gera, A. 1994. The natural occurrence of cucumber mosaic virus in ornamentals in Israel. Acta Hortic. 377:99–106.

Lyons, R.E. 1985. *Asclepias tuberosa*. In A.H. Halevy (ed.), Handbook of Flowering, Vol. 6. CRC Press, Boca Raton, FL, pp. 22–28.

Rist, D.L. and Lorbeer, J.W. 1989. Occurrence and overwintering of cucumber mosaic virus and broad bean wilt virus in weeds growing near commercial lettuce fields in New York. Phytopathology 79:65–69.

Salamon, P., Horvath, J., Mamula, D., Juretic, N. and Hunyadi, K. 1989. *Asclepias syriaca L.* (common milkweed), a new natural host of cucumber mosaic virus in Hungary and Yugoslavia. Acta Phytopathol. Entomol. Hung. 24:363–374.

CUT FLOWER CROPS (SEED PROPAGATED)

40

Lisianthus

Vittoria Lisa[1] and A. Gera[2]
[1]*Istituto di Fitovirologia applicata CNR, Strada delle Cacce 73, Torino, Italy*
[2]*Department of Virology, Agricultural Research Organization, Volcani Center, Bet Dagan 50250, Israel*

Introduction

Lisianthus (*Eustoma grandiflorum*, syn. *E. russellianum* or *Lisianthus russellianum*), a member of the family Gentianaceae, also known as "prairie gentian", is endemic to the prairies of the south midwestern US. The genus is composed of three species of herbaceous, annual, biennial, or perennial (Balley, 1976). Lisianthus forms tap roots, leafy stems, and a rosette of elegant flowers, somewhat tulip-shaped, in the axis of the upper leaves. The plant was introduced into Japan some decades ago where dwarf forms with a range of flower colors from white to purple were produced. Lisianthus is a popular cut flower in Japan, and in the last ten years of production, either for cut flowers or as pot plants, has increased worldwide. It is normally propagated from seeds. Rooting of cuttings and tissue culture has also been applied with success (Roh and Lawson, 1984; Semeniuk and Griesbach, 1987; Ruffoni et al., 1990). Production is usually under glass.

Lisianthus is highly susceptible to virus diseases; about 10 viruses have been reported in this crop. Two of these are specific to lisianthus, while the others have wide natural host ranges. Detailed information on these last viruses is given in Chapter 5.

The Viruses

Lisianthus necrosis necrovirus (LNV)

This virus (Iwaki et al., 1987) is specific to lisianthus. Field-infected plants show necrotic spots and rings on leaves and stems, top necrosis and white stripes on flowers (Fig. 40.1; see color plate). The virus is sap-transmissible and experimentally infects several species, mostly with local lesions only. LNV has isometric particles about 30 nm in diameter and properties similar to those of necroviruses, while the RNA and coat protein differ somewhat in mobility from those reported for known necroviruses. In agarose gels the RNA of LNV forms a single band of estimated M_r 1.67 MDa, while co-migrating RNA of tobacco necrosis necrovirus (TNV) gave a value of 1.5 MDa. LNV coat protein migrated in PAGE as a single component of 35 kDa, although in similar conditions TNV coat protein gave a band of 27 kDa. LNV is soil-borne, and infectivity is associated with the presence of *Olpidium* sp. It is not seed-transmitted in lisianthus. LNV has so far been found only in Japan where it was first found sporadically in Chiba Prefecture in 1984, and it was again detected in 1993 in Hyogo and Yamaguchi Prefectures (M.K. Iwaki, pers. comm.).

Lisianthus line pattern ilarvirus (LLPV)

LLPV (Lisa et al., 1994) is also a virus specific to lisianthus, on which it causes growth reduction, necrotic spots, and line patterns on leaves (Fig. 40.2; see color plate). Flowers are apparently unaffected. The virus is easily sap-transmissible and has a wide experimental host range like most ilarviruses. Virus particles are quasi-isometric, about 30 nm in diameter, and contain one coat protein of M_r 24.7 kDa, as estimated by PAGE. The RNA forms five bands of estimated M_{rs} 1.03, 0.85, 0.64, 0.30, and 0.25 MDa after electrophoresis in agarose. LLPV is serologically related to, but distinct from, parietaria mottle ilarvirus and more distantly related to tobacco streak ilarvirus. It was first reported in northern Italy in 1988 and has been isolated again in the following years; it may also be present in Israel (Gera, unpublished). No information is available on the natural way of transmission of

LLPV in lisianthus or on its real incidence in crops. Pollen and seed-transmission of LLPV in lisianthus have not been investigated.

Bean yellow mosaic potyvirus (BYMV)

On lisianthus BYMV causes pronounced stunting, leaf curling, mosaic, chlorotic spotting, and color break on flowers. The disease was first reported in northern Italy in 1984 where it severely damaged one of the first lisianthus crops in Italy (Lisa and Dellavalle, 1987) and was later found in Israel (Gera and Cohen, 1990).

Broad bean wilt fabavirus (BBWV)

This virus has been repeatedly reported in Japan. Naturally infected plants show chlorotic line patterns on the upper leaves (Iwaki et al., 1987; Yamashita, 1990).

Cucumber mosaic cucumovirus (CMV)

CMV appears to be the commonest virus in lisianthus, reported in Japan, Israel, Italy, and in the US (Provvidenti, 1985; Iwaki et al., 1987; Gera and Cohen, 1990; Lisa and Dellavalle, unpublished). The diseases induced by CMV in Israel and in the US were severe with mosaic and distortion on leaves, stunting, color break, and malformation of flowers. Symptoms associated with the CMV isolate found in Italy were milder: slight reduction in growth and chlorotic mottle on leaves. In aphid transmission tests conducted in Israel, *Aphis gossypii* efficiently transmitted CMV from lisianthus to cucumber (80%) and from lisianthus to lisianthus at 100%.

Tomato spotted wilt tospovirus (TSWV)

On lisianthus TSWV causes yellowing, mosaic and necrosis on leaves, color breaking on flowers, and poor growth. Infections of TSWV have been reported in the US, Israel, and Italy (Lisa et al., 1990; R. Lawson, pers. commun.; A. Gera, unpublished). In experimental tests lisianthus seedlings were also susceptible to impatiens necrotic spot tospovirus, with symptoms similar to those of TSWV.

Lettuce mosaic potyvirus (LMV)

LMV infections on lisianthus were observed in 1989 in Italy. Field-infected plants showed stunting, severe chlorotic mosaic on leaves, color breaking, and malformation of flowers (Lisa and Dellavalle, unpublished).

Tobacco mosaic (TMV) and tomato mosaic (ToMV) tobamoviruses

TMV has been reported in Israel from plants that showed mosaic that persisted as the leaves aged (Fig. 40.3; see color plate). ToMV was found in Italy; field-infected plants had retarded growth, mosaic, and chlorotic ring patterns on leaves and whitish stripes on flowers (Lisa and Dellavalle, 1989; Gera and Cohen, 1990).

Tomato yellow leaf curl geminivirus (TYLCV)

TYLCV infections were observed in lisianthus in 1992 in Israel. The symptoms include distortion of the plant growing apex, appearance of "cup-shaped" leaves, and typical enations on the lower surface of the leaves. Symptoms are shown in Fig. 40.4 (see color plate) (Cohen et al., 1993). TYLCV is a member of the geminivirus group of plant viruses. The group is characterized by geminate morphology, 18–20 nm isometric particles. The genome consists of a single stranded DNA molecule of 2766 nucleotides. The virus is well known as the causal agent of devastating disease in tomato crops in the Middle East and other tropical and subtropical regions in the world. The virus is transmitted persistently by the whitefly *Bemisia tabaci*.

Unidentified viruses

Some viruses not yet identified have been reported in different countries. A virus with spherical particles about 28 nm in diameter, not related to CMV, was isolated in the US from lisianthus showing stunting, yellow mosaic, and flower break (Mayhew and Sorrell, 1986). Two viruses — one with isometric particles about 30 nm in diameter, differing from CMV and associated with symptoms of mosaic and chlorotic-necrotic spots; and a second with elongated particles 750 nm long, not related to BYMV — were found in Israel (Gera and Cohen, 1990). A virus with similar elongated particles, serologically distinct from BYMV and LMV, was found in Italy in plants with green mosaic (Lisa and Dellavalle, unpublished). Further tests will be necessary to identify these viruses.

Identification and Assay Methods

The known viruses infecting lisianthus, with the exception of TYLCV, are easily sap-transmissible to common indicator plants, causing characteristic symptoms.

Chenopodium amaranticolor and *C. quinoa* are good assay hosts since they give local lesions with all the sap-transmissible viruses and react systemically to LLPV and LMV. Antisera to these viruses are available, and serological identification is generally easy. Immunoelectron microscopy can be applied in laboratory diagnosis, and ELISA can be used for large-scale testing. TYLCV can be detected using a DNA molecular probe, whitefly transmission, and PCR.

Control

Control of the soil-borne LNV can be achieved through soil sterilization prior to planting. No information is available on the natural transmission of LLPV, and the possibility that it may be seed-transmitted, similar to other ilarviruses (Francki et al., 1991), has to be considered. Seed should be collected from virus-free plants and nurseries surveyed for early detection and roguing of diseased seedlings. The possible role of thrips in transmission, as shown for other ilarviruses (Sdoodee and Teakle, 1987) should also be investigated.

Control of BBWV, BYMV, CMV, and LMV — all non-persistently transmitted by aphids (see Chapter 10) — is difficult and uncertain. These viruses have wide natural host ranges, in both wild and crop plants, that may serve as reservoirs of infection. Care should be taken to avoid introduction of aphids into glasshouses, possibly by using nets to cover openings. Aphid populations should be controlled by chemicals, and early roguing of infected plants can minimize the risk of spread of the viruses within the crop. Similar precautions should be adopted for the control of TSWV, transmitted by thrips, particularly by *Frankliniella occidentalis*, keeping in mind that control of thrips by insecticides may be extremely difficult and that nets which prevent aphids are ineffective for thrips.

Infections of TMV and ToMV, transmitted by contact, can be controlled by strict hygiene. These viruses may be transmitted to lisianthus by contaminated hands, clothing, and tools from other susceptible crops such as pepper and tomato, and are spread by man during cultivation. Hygiene measures include washing of hands before working on lisianthus crops, treatment of implements, destruction of infected plant remains, and soil sterilization between cultural cycles. Since the viruses can be transmitted by handling, removal of infected plantlets before "pinching" would avoid contamination.

No information is available on freeing lisianthus clones from virus infection. Possibly this approach has not yet been considered since lisianthus is mainly propagated by seed. However, propagation is possible by cuttings (Griesbach et al., 1988) and by *in vitro* culture of explants from shoot tips, internodal stem sections, petioles, or leaves (Semeniuk and Griesbach, 1987; Ruffoni et al., 1990).

Agamic multiplication of selected clones would allow large-scale and uniform propagation of individuals with good characteristics, especially desirable for production of pot plants. In this context a careful choice of virus-free mother plants would be necessary, and obtaining virus-free plants of particularly valuable clones might be important.

Selected Literature

Balley, L.H. 1976. Hortus Third: A concise dictionary of plants cultivated in the United States and Canada: McMillan, New York.

Cohen, J., Gera, A., Gokkes, M., Aker, R. and Antignus, Y. 1993. Tomato yellow leaf curl virus (TYLCV) is the causal agent of a new disease in lisianthus (*Eustoma russellianum*). Hassadeh 74:169–171 (in Hebrew).

Francki, R.I.B., Fauquet, C.M., Knudson, D.L. and Brown, F., eds. 1991. Classification and nomenclature of viruses. Fifth Report of the Int. Comm. on Taxonomy of Viruses, Arch. Virol. Supplementum 2.

Gera, A. and Cohen, J. 1990. The natural occurrence of bean yellow mosaic, cucumber mosaic and tobacco mosaic viruses in lisianthus in Israel. Plant Pathol. 39:561–564.

Griesbach, R.J., Semeniuk, P., Roh, M. and Lawson, R.H. 1988. Tissue culture in the improvement of *Eustoma grandiflorum*. HortScience 23(4):658.

Iwaki, M., Hanada, K., Maria, E.R.A. and Onogi, S. 1987. Lisianthus necrosis virus, a new necrovirus from *Eustoma russellianum*. Phytopathology 77:867–870.

Lisa, V. and Dellavalle, G. 1987. Bean yellow mosaic virus in *Lisianthus russellianum*. Plant Pathol. 36:214–215.

Lisa, V. and Dellavalle, G. 1989. Malattie da virus del lisianthus. Giorn. Studio sul Lisianthus, Sanremo, June 1989.

Lisa, V., Vaira, A.M., Milne, R.G., Luisoni, E. and Rapetti, S. 1990. Tomato spotted wilt in cinque specie coltivate in Liguria. Inf. Fitopatol. 40 (12):34–41.

Lisa, V., Vaira, A.M., d'Aquilio, M., Dellavalle, G., Masenga, V. Milne, R.G. and Boccardo, G. 1994. Characterization of an ilarvirus from lisianthus (*Eustoma grandiflorum*). Acta Hortic. 377:81–89.

Mayhew, D.E. and Sorrell, M.E. 1986. A new virus disease of Eustoma (Lisianthus). Phytopathology 76:1126 (Abstr).

Provvidenti, R. 1985. Two newly recognized hosts of cucumber mosaic virus: *Eustoma grandiflorum* and *Peristrophe angustifolia*. Plant Dis. 69:542.

Roh, M.S.M. and Lawson, R.H. 1984. The lure of lisianthus. Greenhouse Manager 2(11):103–121.

Ruffoni, B., Damiano, C., Massabò, F. and Esposito, P. 1990. Organogenesis and embryogenesis in *Lisianthus russellianum* Hook. Acta Hortic. 280:83–88.

Sdoodee, R. and Teakle, D.S. 1987. Transmission of tobacco streak virus by *Thrips tabaci*: a new method of plant virus transmission. Plant Pathol. 36:377–380.

Semeniuk, P. and Griesbach, R.J. 1987. *In vitro* propagation of praire gentian. Plant Cell, Tissue and Organ Culture 8:249–253.

Yamashita, K. 1990. Mosaic diseases occurred on gentian and praire gentian in Aomori Prefecture. Ann. Rep. Soc. Plant Prot. North Jpn. 41:80–82.

Pot and Bedding Plants

41

Begonia

Josette Albouy
Station de Pathologie Végétale, INRA, Versailles, France

Introduction

Many cultivars of the genus *Begonia* are grown in northern Europe (Belgium, France, Germany, Denmark, etc.) and in the US. Two main types are grown outdoors: *B. tuberhybrida* cv. *grandiflora* types that are multiplied by seed and *B. tuberhybrida* cv. *multiflora* that are vegetatively propagated by cuttings derived from corms. Greenhouse cultivation of *Begonia* × *hiemalis* race Rieger or *B.* × *elatior* race Rieger has rapidly increased in recent years and is important in potted plant production. They are also vegetatively propagated by cuttings. Mainly because of this vegetative propagation, begonias can be infected by viruses. Knowledge of specific viruses of begonia is relatively limited compared to information available on viruses infecting other ornamental crops. The degree of economic impact seems, nevertheless, to vary from region to region.

Many different viruses have been found and reported in the literature in different begonias. The incidence of these viruses varies from overall infection [cucumber mosaic virus (CMV) in *B. multiflora*] to viruses frequently found [tospoviruses: tomato spotted wilt virus (TSWV), impatiens necrotic spot virus (INSV), tobacco necrosis, and nepoviruses] to those observed only occasionally [carnation mottle (CarMV) in *B. elatior* (Paludan, 1985) and broad bean wilt (BBWV) in *B. semperflorens* (Lockhart, 1982)].

Nepoviruses reported are: tobacco ringspot (TobRSV) in *B. semperflorens* (Lockhart, 1979) and arabis mosaic (ArMV) in *B. multiflora* (Welvaert, 1978). Virus-like symptoms have been described in some countries where begonia is cultivated. They are observed especially in winter in *B. elatior*: leaf curl, vein clearing, poor keeping quality of the plants. Other virus-like foliar symptoms such as leaf mottling and ring spot associated to color break are also reported.

Among the different viruses occurring in *Begonia* spp., those that cause the most severe damage in cultures are the thrips transmitted TSWV and INSV. These tospoviruses have become in recent years in different countries the most prevalent and damaging viral diseases, especially in greenhouse cultivated begonias. For begonias grown outdoors, the most frequently isolated virus is the aphid-transmitted CMV. Nevertheless, the occurrence of multiple infections makes it difficult to assess the symptoms due to each individual virus.

The Viruses

Tomato spotted wilt (TSWV) and impatiens necrotic spot (INSV) tospoviruses

TSWV has been reported from all five continents (Ie, 1970) and is responsible for serious losses in different crops. Since 1987, the appearance of TSWV and INSV in Europe has been reported in begonia as in other ornamental fruit and vegetable crops in correlation with the widespread occurrence and efficiency of the main vector *Frankliniella occidentalis* and the trade of infected plant material (German, 1992). These viruses (see Chapter 5) cause severe diseases and important losses mainly in greenhouses. TSWV appears more ubiquitous than INSV; it has been found in more than 400 plant species, including weed species. INSV, previously considered to be a strain of TSWV, mainly infects ornamental crops. Begonia can be infected by either TSWV or INSV and expresses similar symptoms. Due to infection rates commonly reaching 50–90%, damage to crops can be extremely high.

Symptoms

Symptoms induced by TSWV and INSV in *Begonia* spp. are variable according to the stage of vegetation at the moment of infection, environmental conditions, and time of the year. A variety of symptoms can be observed: distortion, chlorosis (see Fig. 41.1) or bronzing, and concentric rings with leaf necrosis are predominant (see Fig. 41.2; color plate). The quality of the flowers is also affected. When plants are infected at an early age, they can die. Visual inspection may provide some indications of infection, but viruses should be detected in affected plants by means of bioassay and ELISA tests to confirm the identification and to build up clean nursery stocks.

Bioassays: inoculation to indicator plants

Plants can be tested by mechanical inoculation to many indicator plants such as *Nicotiana rustica*, *N. benthamiana*, and *Petunia hybrida* that give typical symptoms. The symptoms expressed are local necrotic lesions, then a systemic necrosis, one week after inoculation in *Nicotiana*, and local necrotic lesions within 2 or 3 days on *P. hybrida*.

Serology

TSWV and INSV can reliably be detected by different ELISA procedures using high-quality polyclonal or monoclonal antibodies, available commercially. Routine tests for serological indexation of begonia are possible only when using a high molarity extract buffer (phosphate buffer saline 0.4 M) or a Tris buffer (0.1 M, pH 9), as begonia sap has a very low pH (pH 2). Due to the great variation in concentration of TSWV in begonia and to avoid false-negative results with

Fig. 41.1. Chlorotic rings on leaf of plant infected with impatiens necrotic spot virus (INSV). (From GRISP, Antibes.)

virus-infected plants, it is necessary to perform several tests and to sample different parts of the whole plant.

Some problems occur associated with serologically distinct isolates. Detection using test plants or ELISA is not always reliable due to variation in virus concentration and uneven distribution of the virus in the plant. Molecular techniques are being developed.

Electron microscopy

Tospoviruses are readily identified by leaf dip preparations or thin sections of infected plants. The particles are 80–100 nm, spherical and membrane bound.

Nuclear acid probes

Riboprobes and cDNA probes are available for diagnostic purposes, but until now their use requires special equipment.

Cucumber mosaic cucumovirus (CMV)

CMV (see Chapter 5) is characterized by having three pieces of single-stranded RNA packaged in three classes of icosahedral particles about 28 nm in diameter. It has an extremely wide host range and a worldwide geographic distribution. CMV is transmitted by numerous species of aphids in a non-persistent manner and is also sap transmissible.

CMV can cause epidemic attacks, and the incidence of infection is higher for outdoor begonias than for the others. Almost 100% of cultivars of *Begonia tuberhydrida multiflora* were infected by CMV (Samyn et al., 1987) before virus-free plants became available.

Symptoms

CMV causes discrete and variable leaf symptoms: mottling and chlorosis. Internodes can be shortened, and a general decrease of vigor is observed.

Bioassay

Erratic distribution of CMV in the plant and the acidity of the sap render mechanical inoculation rather difficult. For *B. multiflora* reliable determination by indexing on the test plant *Nicotiana megalosiphon* is proposed (Welvaert et al., 1987). This test plant allows the detection of all known viruses that infect begonias (ArMV, TomRSV, TSWV). After 5 days on inoculated leaves, CMV provokes necrotic brown spots, then irregular necrotic spots on systemically infected leaves.

The best indexation period is the early spring, but indexation is practically useless during part of the year.

Serology

Serological diagnosis of CMV in begonia is possible using commercial reagents for DAS-ELISA and can be done throughout the year.

Control

Recent methods consist of controlling the potential vectors and the quality of plant material exchanged. During recent years virus-free plants of the vegetatively propagated cultivars of *Begonia* spp. became available, produced by thermotherapy (*B. tuberhybrida*) (Welvaert et al., 1980) and by meristem-tip culture, in combination with a rapid *in vitro* multiplication. Improvement of the quality of virus-free plantlets propagated from tissue culture is evident: production of bigger tubers, more cuttings and better flowers.

Some European countries (Denmark, France, etc.) have adopted a scheme for the production of certified pathogen-tested material of begonia vegetatively propagated by cuttings. This fits into the general scheme of certification covering various other pests such as *Xanthomonas campestris* pv. *begoniae* and *Verticillium* sp. Propagative material can be divided into three certification categories: nuclear stock, propagation stock, and certified stock. Nuclear stock is absolutely healthy propagative material derived from individually selected and tested plants. Individual plants are indexed throughout the year by visual inspection for symptoms; test methods and control measures are taken against the vectors. Propagation stock is directly derived from nuclear stock. In certified stock derived from the propagation stock, plants are randomly checked for tospoviruses. Cuttings taken from the final propagation stocks are the certified (stock) cuttings for sale.

Selected Literature

German, T.L. 1992. Tospoviruses: diagnosis, molecular biology, phylogeny, and vector relationships. Ann. Rev. Phytopathol. 30:315–348.

Lockhart, B.E. and Betzold, J.A. 1979. Begonia yellow spot: a disease caused b tobacco ring spot virus infection. Plant Dis. Rep. 63:1046–1047.

Lockhart, B.E. 1992. Broad bean wilt in begonia in Minnesota. Plant Dis. Rep. 66:72–73.

Ie, T.S. 1970. Tomato spotted wilt. CMI/AAB Description of Plant Viruses, No. 39.

Paludan, N. and Begtrup, J. 1985. Carnation mottle virus demonstrated in *Begonia elatior* and *Begonia* × *Cheimantha* showing vein clearing, leaf curl and flower break. Acta Hortic. 164:33–39.

Samyn, G. and Welvaert, W. 1987. Improvement of quality and profit-earning capacity with virus-free culture of *Begonia* × *tuberhybrida multiflora* group. Med. Fac. Landbouw., RUG 52:1059–1064.

Welvaert, W., Samyn, G. and Van Wijmersch, E. 1980. On the production of virus-free *Begonia tuberhybrida* cv. *multiflora* varieties. Acta Hortic. 110:253–257.

42

Cyclamen

Assunta Bertaccini and Maria Grazia Bellardi
Istituto di Patalogia Vegetale, Bologna, Italy

Introduction

Cyclamen (*Cyclamen persicum* Mill.; Primulaceae) is one of the most common flowering pot plants in the world. In Italy more than 9 million pots are produced annually.

The Viruses

Only a few viruses are known to infect cyclamen worldwide, but the diseases induced by some of them are severe and cause important crop losses.

Seven viruses reportedly infect cyclamen: cucumber mosaic virus (CMV), tobacco mosaic virus (TMV), and potato virus X (PVX) have been isolated sporadically in the Netherlands and Denmark; tomato aspermy virus (TAV) and tobacco rattle virus (TRV) have been mechanically transmitted to cyclamen but are not known to occur naturally in this plant (Allen and Matteoni, 1988). The most widespread are tospoviruses tomato spotted wilt virus (TSWV) and impatiens necrotic spot virus (INSV) that induce diseases mainly consisting of leaf necrosis or deformation. In some cases these are present in mixed infection with CMV, especially when the plants are greenhouse grown. The presence of specific vectors such as *Frankliniella occidentalis* Pergande and *Thrips tabaci* Lindeman can increase the spread of these viruses.

The first report of TSWV infecting cyclamen was in Canada where a severe outbreak of disease occurred in 1985 and 1986 in six commercial greenhouses; losses of up to 95%, including over 5,500 plants, were reported (Allen and Matteoni, 1988). The same virus has been reported to infect cyclamen in North Carolina (US) in greenhouses and open fields with damage ranging from minimal to severe (Jones and Baker, 1990).

Fig. 42.2. Cyclamen infected by MLOs: the flowers are virescent and some of the leaves are stunted.

Recently, a single infection of TSWV was also found in Italy in plants showing concentric necrotic rings on the leaves (Fig. 42.1; see color plate). The symptoms were more evident at the blooming stage. Flowers were dwarfed and malformed and petals were twisted (Vicchi et al., 1992). The same virus in mixed infection with CMV induces necrotic rings, spots, and interveinal lines (Vaira et al., 1992).

INSV was detected in Pennsylvania (US) on plants exhibiting different leaf symptoms such as distortion, stunting, mottle, mosaic, necrosis, ring spots, and chlorosis (Hausbeck et al., 1992). In Italy INSV has always been found in mixed infections with CMV in cyclamen plants with leaves showing thin concentric chloro-necrotic rings (Vaira et al., 1992).

Mycoplasma-associated diseases

Mycoplasma-like organisms (MLO) were detected for the first time on cyclamen hybrid plants in a nursery in northern Italy (Bertaccini, 1990). During the first year, the infected plants showed severe phyllody and virescence on the flowers and continued to grow at the end of the flowering period (Fig. 42.2). During the second year, the plants stopped producing flowers and the new leaves were dwarfed, resembling those of the wild cyclamen. For this reason the disease was called cyclamen little leaf (CyLL).

Identification and Assay Methods

Pathogen identification is possible on the basis of symptoms in the infected leaves or flowers. In fact, the concentric necrotic rings on the leaves are typical of tospovirus infections, and the virescent flowers indicate the presence of MLO. Mechanical inoculation can be useful in some cases to identify TSWV: the test plants are *C. quinoa*, *N. benthamiana*, *N. tabacum* "Samsum" and "White Burley", *P. hybrida*, and *N. glutinosa*. Leaf graft has been used to verify that TSWV is responsible for ringspot disease in Canada (Allen and Matteoni, 1988). In general, tospoviruses can be identified using ELISA tests with polyclonal antisera to the different viruses. Routine tests of cyclamen plants often combine ELISA with mechanical inoculation tests (Hausbeck et al., 1992; Vaira et al., 1992). Leaf dip preparations, ISEM, and ultrastructure observations of embedded tissues can be helpful in detecting tospoviruses (Allen and Matteoni, 1988).

MLO detection has been only achieved by electron microscopy of embedded tissues (Fig. 42.3); preliminary trials with MLO-DNA probes specific for aster yellows MLOs were negative (Bertaccini, 1990).

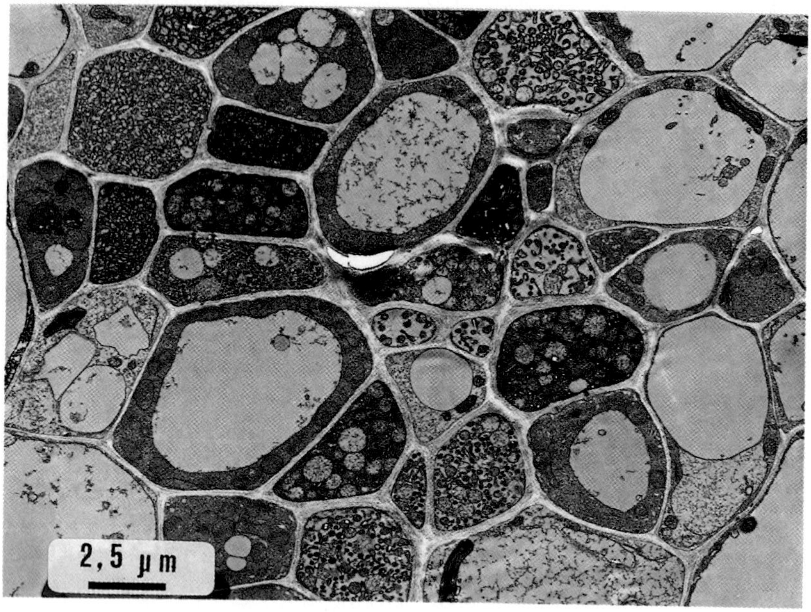

Fig. 42.3. Electron micrograph of cross section of an infected cyclamen phloematic area: all the sieve tubes are full of MLOs with different morphological shapes.

Transmission

Thrips such as *Frankliniella occidentalis* Pergande and *Thrips tabaci* Lindeman are mainly responsible for tospovirus transmission in the open field and greenhouses. The transmission of MLO in cyclamen has not been studied since the incidence of CyLL disease is very low and it was detected in greenhouses free of leafhoppers. Attempts to transmit the disease to periwinkle through several species of dodder have been unsuccessful (Bertaccini and Carraro, unpublished). Since cyclamen is mainly propagated through seed, the transmission of CyLL MLO can only be due to vectors.

Control

It is important to prevent the spread of tospovirus infections on cyclamen by using control measures against thrips which are responsible for their transmission. It is also important to control weeds or other ornamentals, in some cases asymptomatic, which are possible sources of viral infections (Hausbeck et al., 1992). Latent or mild infections can be reservoirs for spreading TSWV, especially in some seasons of the year with low temperatures. Furthermore the stress on plants during transportation can increase their susceptibility to viral infections. It is important to avoid mixing bedding plants with asexually propagated plants and with unsold perennial material (Hausbeck et al., 1992). Allen and Matteoni (1988) report that the ringspot disease of cyclamen due to TWSV is a particularly costly disease. This is because of the long growing period of the plants from seed and the latent period of the disease — 4 months or longer. When symptoms appear, most of the plants in the greenhouse may already be infected.

MLO infections can easily be detected by visual inspection of the plants at flowering. Infected plants show typical symptoms, are unsuitable for the market, and usually are eliminated by the growers.

Selected Literature

Allen, W.R. and Matteoni, J.A. 1988. Cyclamen ringspot: epidemics in Ontario greenhouses caused by the tomato spotted wilt virus. Can. J. Plant Pathol. 10:41–46.

Bertaccini, A. 1990. Cyclamen: a new host of micoplasma-like organisms. Phytopathol. Mediterr. 29:213–214.

Hausbeck, M.K., Welliver, R.A. and Derr, M.A. 1992. Tomato spotted wilt virus survey among greenhouse ornamentals in Pennsylvania. Plant Dis. 76:795–800.

Jones, R.K. and Baker, R. 1990. TSWV: symptoms, host range and spread. Proceedings of a USDA Workshop, "Virus-thrips-plant interactions of tomato spotted wilt virus", Beltsville, Maryland, pp. 89–93.

Vaira, A.M., Gallo, S. and Lisa, V. 1992. Nuove infezioni da due Tospovirus (tomato spotted wilt e impatiens necrotic spot) in Liguria. In. tore Fitopatol. 10:37–42.

Vicchi, V., Bellardi, M.G. and Bertaccini, A. 1992. Le infezioni da TSWV dilagano in Italia: nuovi casi in altre tre regioni. L'inf. tore Agrario 23:53–56.

43

Dianthus barbatus

Vittoria Lisa
Istituto di Fitovirologia applicata CNR, Strada delle Cacce 73, Torino, Italy

Introduction

Little information is available on virus diseases of *Dianthus barbatus*, family Caryophyllaceae, commonly called sweet william. *D. barbatus* is a popular bedding plant, grown everywhere in temperate climates. It is also grown, though to a lesser extent, as a pot plant or for production of cut flowers. The plant is mainly propagated by seed, although selected clones are vegetatively multiplied. The viruses of *D. barbatus* are essentially those infecting carnations (see Chapter 34).

The Viruses

Carnation vein mottle potyvirus (CVMV)

This is the most common virus in *D. barbatus*. Stone (1968) and Hollings et al. (1977) reported that infections of CVMV on *D. barbatus* were common in gardens in Britain. Recently the virus was reported in *D. barbatus* grown for cut flower production in glasshouses in the Czech Republic (Mokra and Gotzova, 1992) and is possibly present in other countries. Early symptoms consist of vein clearing, followed by conspicuous light and green leaf mottle, flecks, rings and spots on the lateral veins, and dwarfing and color break in red or pink flowered varieties (Kassanis, 1955; Hollings and Stone, 1971).

Carnation ringspot dianthovirus (CRSV)

This virus causes on *D. barbatus* a severe disease consisting of leaf mottle, ringspots, stunting and distortion (Hollings and Stone, 1970). CRSV is possibly present in *D. barbatus* in the geographical areas where the virus has been found in carnations, but precise references are not available.

Carnation mottle carmovirus (CarMV)

Infection of CarMV on *D. barbatus* is generally symptomless, and the only alteration reported is flower break, occurring as narrow pink lines on red-flowered clones (Hollings and Stone, 1964). The same authors found a certain degree of resistance to the virus in this plant, for only part of the healthy clones used for infection tests could be infected. Plants derived from meristem-tip culture contained the "attenuated" CarMV (Stone, 1968).

Carnation etched ring caulimovirus (CERV)

Infections of CERV associated with CarMV were found in Denmark on *D. barbatus* × *D. carthusianorum* hybrids. Symptoms were generally confined to younger leaves; *D. barbatus* plants experimentally infected by the two viruses developed yellow mottling not present in plants infected with CarMV alone (Paludan, 1970).

Carnation latent carlavirus (CLV)

This virus does not cause symptoms in *D. barbatus*. No certain reference to field infections of CLV on *D. barbatus* is available. Stone (1968) states, referring to this species: "it is possible that CLV may also occur in this host, for alone it is symptomless".

Cucumber mosaic cucumovirus (CMV)

Natural infection of CMV on *D. barbatus* has been reported in India, and infected plants showed stunting and leaf crinkle. The virus occurred in low concentration in field-infected *D. barbatus* and could be detected in crude sap by Western immunoblotting but not by agargel immunodiffusion (Raj et al., 1993).

MLO-Associated Diseases

A severe disease of *D. barbatus*, characterized by phyllody, virescence, sometimes by generalized yellowing and reddening of leaves (Fig. 43.1) and associated with MLOs, has been reported in Italy and other European countries and in Taiwan (Schmelzer and Schmidt, 1960; Yang, 1975; Ragozzino et al., 1977; Savio and Conti, 1983). Generally the disease occurs sporadically. In experimental conditions *Euscelis obsoletus* transmitted the disease from *D. barbatus* to *Trifolium pratense*, which may act as source of inoculum for crops (Savio and Conti, 1983).

Fig. 43.1. *D. barbatus* flower with virescence, associated with MLOs; field infection. Photo Istituto Fitovirologia Applicata CNR, Torino.

Diagnosis and Control

Diagnosis of the viruses that infect *D. barbatus* is feasible following the indications given for carnation viruses. Large-scale diagnosis can be made by ELISA using antisera to the appropriate carnation viruses or to CMV.

Symptoms on field-infected plants are a clear indication of the MLO-associated disease in *D. barbatus*. Diagnosis should be confirmed by electron microscopy of thin sections for the presence of MLOs in the phloem or by other techniques as specified in Part One, Chatper 7.

Control of CarMV, CVMV, CRSV, CLV and CERV is feasible following the clean-up programs used for carnations. *D. barbatus* can also be freed of viruses by meristem-tip culture (Stone, 1968), combined with thermotherapy for eradicating CERV (Paludan, 1970). Control of CMV or of the MLO-associated disease may be more troublesome, since for both pathogens the sources of infection can be numerous.

Selected Literature

Hollings, M. and Stone, O.M. 1964. Investigation on carnation viruses. I. Carnation mottle. Ann. Appl. Biol. 53:103–118.

Hollings, M. and Stone, O.M. 1970. Carnation ringspot virus. CMI/AAB Descriptions of Plant Viruses, No. 21.

Hollings, M. and Stone, O.M. 1971. Carnation vein mottle virus. CMI/AAB Descriptions of Plant Viruses, No. 78.

Hollings, M., Stone, O.M., Atkey, P.T. and Barton, R.J. 1977. Investigations on carnation viruses. IV. Carnation vein mottle virus. Ann. Appl. Biol. 85:59–70.

Kassanis, B. 1955. Some properties of four viruses isolated from carnation plants. Ann. Appl. Biol. 43:103–113.

Mokra, V. and Gotzova, B. 1992. Detection of carnation vein mottle virus in *Dianthus barbatus*. Abstracts, 8th Int. Symp. Virus Diseases Ornamental Plants, Prague, August 24–28, p. 33.

Paludan, N. 1970. Carnation etched ring virus. Survey, infection trials, thermotherapy and meristem culture. Tidsskrift for Planteavl 74:75–86.

Ragozzino, A., Angelaccio, C. and Stefanis, D. 1977. Su alcune malattie da micoplasmi rinvenute in Campania. Ann. Fac. Sci. Agric., Napoli, Ser. IV, 11:1–10.

Raj, S.K., Aminuddin, Srivastava, K.M. and Singh, B.P. 1993. Natural infection of cucumber mosaic virus on *Dianthus barbatus* in India. Plant Patholol. 42:811–813.

Savio, C. and Conti, M. 1983. Epidemiologia e trasmissione di micoplasmi dei vegetali. Atti 13th Congr. Naz. It. Ent., Sestriere, Torino, pp. 407–414.

Schmelzer, K. and Schmidt, H.O. 1960. Blutenvergrunungen an Zierpflanzen und Unkrautern. Dt. Gartenbau 7:214–217.

Stone, O.M. 1968. The elimination of four viruses from carnation and sweet william by meristem-tip culture. Ann. Appl. Biol. 62:119–122.

Yang, I.L. 1975. Taiwan Agric. Q. 11:93–108; quoted at page 579 in: The mycoplasmas, Vol. 5. Spiroplasmas, acholeplasmas, and mycoplasmas of plants and arthropods. R.F. Whitcomb and J.G. Tully, eds., Academic Press, 1989.

44

Dieffenbachia, Caladium, and Zantedeschia

F.W. Zettler[1] and R.D. Hartman[2]
[1]*Plant Pathology Department, 1453 Fifield Hall, University of Florida, Gainesville, FL 32611-0680, USA*
[2]*Twyford Plant Laboratories, 11850 Twitty Road, Sebring, FL 33870, USA*

Introduction

Dieffenbachias (*Dieffenbachia* spp.) and caladium (*Caladium hortulanum)* are vegetatively propagated, neotropical members of the Araceae widely used as foliage ornamentals. Dieffenbachias have oblong, generally green and cream-colored glabrous leaves borne on thick cane-like stems, whereas caladiums have arrow-shaped green leaves uniquely variegated in pink, red, or white that emerge from corms. Calla lilies (*Zantedeschia* spp.), indigenous to Africa, likewise produce corms, but are grown primarily for their attractive pink, white, or yellow colored inflorescences.

Although more than 30 species and 100 distinct types of dieffenbachia are known, only a few have been grown extensively, including *D. amoena* "Tropic Snow", *D. maculata* "Exotica", *D. maculata* "Rudolph Roehrs", *D.* × bausei (*D. maculata* × *D. weiri*), and *D.* × memoria-Corsii (*D. maculata* × *D. wal-*

lisii). Sports derived from *D. maculata* "Exotica" include "Camille", "Compacta", and "Perfection". "Camille" and "Compacta" have become particularly popular in recent years and currently account for well over three-fourths of the US dieffenbachia market. Over 25 million dieffenbachias are produced annually in North America, mostly in Florida. Of these, about 3.0 million are propagated from tissue culture. Dieffenbachia cultivation in Europe is similar to that in the United States, except that a smaller proportion is derived from tissue culture. Other members of the Araceae commonly cultivated together in greenhouses or shadehouses with dieffenbachias as foliage ornamentals are *Aglaonema*, *Anthurium*, *Epipremnum*, *Philodendron*, *Spathiphyllum*, and *Syngonium*. Unlike dieffenbachia, caladiums and calla lilies are typically field-grown. Florida produces over 90% of the world's caladiums, consisting of an annual production of 40–50 million corms grown on about 1000 acres in Highlands County. Although many dozens of varieties have been described, only a few are commercially available, including "Candidum", "Candidum Jr.", "Carolyn Whorton", "Fannie Munson", "Frieda Hemple", "Rosebud", and "White Christmas". Several species and hybrids of calla lilies are grown commercially in California, Europe, and New Zealand, including *Z. aethiopica*, *Z. albomaculata*, *Z. elliottiana*, and *Z. rehmannii*.

The Viruses

The most frequently encountered virus of dieffenbachia, caladium, and calla lily is the potyvirus, dasheen mosaic (DsMV), which was discovered in 1969. DsMV occurs wherever aroids are grown commercially and may be ubiquitous in plantings of caladium, taro (*Colocasia esculenta*), and *Xanthosoma* spp. (Zettler and Hartman, 1987). Dieffenbachias and calla lilies infected with DsMV have been reported in many parts of the world including Australia, Europe, South Africa, South America, and the US. DsMV is transmitted in a stylet-borne fashion by certain aphids including *Myzus persicae* and *Aphis gossypii*, but apparently not by *Pentalonia nigronervosa*. Mean particle lengths of 750–754 nm have been reported for this virus (Zettler and Hartman, 1987; Zettler et al., 1978, 1990).

Unlike caladium or dieffenbachia, calla lilies also may be infected with cucumber mosaic cucumovirus (CMV) or tomato spotted wilt tospovirus (TSWV). CMV, like DsMV, is transmitted in a stylet-borne manner by various aphid species, whereas TSWV is transmitted by certain species of thrips, including *Frankliniella fusca*, *F. occidentalis*, *F. schultzei*, and *Thrips tabaci*.

The discovery of DsMV coincided with the "foliage plant boom" of the 1970s, during which the wholesale market value of foliage ornamentals increased exponentially, expanding from US $28 million to $300 million dollars in Florida alone. At the same time, substantial advances were being made in the technology and commercialization of micropropagation. By 1989, over 75 million foliage plants, including almost 4 million dieffenbachias, were being produced annually in the US through micropropagation (Zettler et al., 1991). Since it was now possible to generate over 70,000 dieffenbachia microcuttings in one year from a single shoot tip, new cultivars free of DsMV rapidly began to infiltrate the market (Taylor and Knauss, 1978; Zettler and Hartman, 1987). Inevitably, the incidence of DsMV in dieffenbachia began to decline. The ease by which DsMV and other pathogens could be eliminated from dieffenbachia by micropropagation, the proliferation of commercial tissue culture laboratories, the ability to detect infected plants on the basis of symptoms, and the market demand for a high quality product were all important factors in controlling this virus. Moreover, secondary spread by aphid vectors of DsMV was minimized since dieffenbachias are usually maintained in greenhouses with strict insect control programs in effect. Based on surveys conducted in 1969–1971, we postulated that DsMV infections were ubiquitous in commercial plantings of the "Exotica" and "Perfection" cultivars of *D. maculata* (Fig. 44.1; see color plate). In the early 1980s, DsMV was still being found in as many as 60% of the "Perfection" plants surveyed in Florida (Chase and Zettler, 1982). Now that the cultivars "Exotica" and "Perfection" have been supplanted by their micropropagated sports "Camille" and "Compacta", DsMV incidence has dropped precipitously.

A similar phenomenon has occurred for calla lily in that tissue culture and the introduction of new hybrids have played a significant role in reducing virus incidence in commercial plantings. However, as field-grown aroids, transmission by alate aphids of DsMV to calla lily cannot be prevented. In contrast to dieffenbachia and calla lily, DsMV incidence in caladium remains extremely high, despite the ease by which it can be propagated *in vitro* (Hartman, 1974). Micropropagation of caladium as a tenable commercial venture is impeded in large part because of this crop's seasonal production cycle, limited acreage, and relatively low cash value. Caladium sales have not increased appreciably since the 1960s and, at US $5,000/acre, caladium has a much lower cash value than dieffenbachia, which is worth US $100,000/acre. Moreover, there is no significant market demand for new caladium cultivars (Zettler and Hartman, 1987; Zettler et al., 1991). In addition, maintenance of virus-free stock under field conditions in

Highlands County, where most of the world's caladiums are grown, is compromised because of its proximity to sources of DsMV inoculum.

Symptoms

The severity of DsMV symptoms varies considerably among dieffenbachia cultivars (Zettler et al., 1980; Chase and Zettler, 1982). In *D. amoena*, symptoms are inconspicuous, causing little or no appreciable damage. In contrast, they can be much more severe in most *D. maculata* cultivars, ranging from slight mosaic to severe leaf distortion. In *D. maculata* "Rudolph Roehrs", DsMV induces prominent ringspot symptoms (Fig. 44.2; see color plate). Acute symptoms developing immediately after inoculation are much more severe than the chronic symptons developing subsequently (Zettler et al., 1980; Chase and Zettler, 1982). Acute symptoms of *D. maculata* "Perfection" plants include epinasty and a horizontal orientation and necrosis of the newest unfurling leaf. Whereas the new leaves produced by these plants about 20 days after inculation are distorted, those forming 60 days there-after may be asymptomatic (Chase and Zettler, 1982). *D.* × *bausei* and *D.* × *memoria-Corsii* are hypersensitive to this virus; manually inoculated plants develop large, spreading, necrotic local lesions and die 25–45 days after inoculation (Chase and Zettler, 1982). DsMV symptom expression in dieffenbachia may vary according to season (Zettler et al., 1980). In a Florida study, over 80% of the leaves produced by *D. maculata* "Perfection" plants during the spring and autumn months had symptoms, whereas during the summer or winter months, slightly less than half (47%) did.

Foliar symptoms induced by DsMV in caladium are usually subtle and not readily apparent (Fig. 44.3; see color plate). DsMV symptoms are much more conspicuous in calla lily, however, consisting of prominent mosaic and pronounced distortion (Rana et al., 1983; Zettler and Hartman, 1987) (Fig. 44.4; see color plate).

Whereas calla lily leaves infected with CMV have mild chlorotic mottle and veinal yellowing symptoms, those with TSWV suffer much more serious damage, including chlorotic or brown lesions, streaks, and ringspotting. Leaves with TSWV can also be distorted, crinkled, or twisted, and the inflorescences are often of poor quality and irregularly spotted.

Substantial quantitative and qualitative losses due to DsMV have been described for *D. maculata* "Perfection" plants (Zettler et al., 1980). In controlled greenhouse experiments, an average of only 10.1 cuttings per plant were harvested from each of the DsMV-infected mother plants compared to 18.8 from the healthy ones. Similarly, almost all (98.4%) of the 564 cuttings harvested from 30 healthy plants were visually assessed as being salable compared to less than half (47.3%) for the 606 cuttings harvested from 60 infected plants. Fresh weights of cuttings, however, did not differ appreciably (31.4 and 29.3 g/cutting, respectively). In field trials total corm weights of virus-free "Candidum" caladium plants derived from tissue culture exceeded those of their commercial counterparts by 22%.

Virus Identification and Assay Methods

Philodendron selloum seedlings are commonly used for biological assays to detect DsMV. *Philodendron verrucosum* is reportedly a local lesion host, but is not readily available (Zettler et al., 1978, 1990; Shimoyama et al., 1992). *Chenopodium amaranticolor*, *C. ambrosioides*, *C. quinoa*, *Nicotiana benthamiana*, *Saponaria vaccaria*, and *Tetragonia expansa* have also been reported as local lesion hosts of DsMV in Europe or Japan, but these results have not been confirmed elsewhere (Zettler et al., 1978, 1990; Rana et al., 1983; Shimoyama et al., 1992).

DsMV is a good immunogen (Zettler et al., 1978, 1990). Serological methods include double radial immunodiffusion tests conducted in agar gels containing sodium dodecyl sulfate, western blots, and various ELISA (enzyme-linked immunosorbent assay) protocols using either DsMV polyclonal antiserum or monoclonal antiserum against a shared epitope of potyviruses. To overcome nonspecific interference in their tests by *D. picta* leaf extracts, Rodoni and Moran (1988) used an extraction buffer containing 10% polyvinyl pyrrolidone and 5% mannose.

Antiserum of different CMV and TSWV strains are readily available from a variety of sources. Those wishing to test plants for either CMV or TSWV have the option of contacting commercial laboratories equipped to index samples by ELISA or other means.

Light microscopy methods using immunogold labelling (Liang et al., 1994), or azure A or calcomine orange and "Luxol" brilliant green stains have been used to detect intracellular inclusions of DsMV, CMV, and TSWV. Likewise, DsMV virions can readily be detected either in negatively stained leaf extracts or on grids treated with DsMV antiserum (Zettler et al., 1978, 1990). DsMV particles are much more likely to be found in extracts of symptomatic dieffenbachia leaf tissue than in tissues without symptoms. In one experiment, 494 virions were counted in symptomatic tissues but only seven in asymptomatic tissues from the same leaves (Chase and Zettler, 1982). In another trial (Zettler et al., 1980), 136 particles were counted in symptomatic tissues compared to none in the symptomless counterparts.

Control

Virus-free plants of caladium, dieffenbachia, and other aroids can be readily obtained through micropropagation of 1–3 mm^3 lateral buds and shoot tips (Hartman, 1974; Knauss, 1976; Taylor and Knauss, 1978). DsMV-free dieffenbachia plants can also be obtained from diseased plants through conventional propagation, but only when cuttings devoid of symptoms are selected (Zettler et al., 1980).

Seed propagation is another means of obtaining healthy dieffenbachia, caladium, and calla lily plants, but this approach cannot be used commercially because of the phenotypic variability of their progeny (e.g. Hartman et al., 1972; Zettler and Abo El-Nil, 1979).

Caladiums can be grown free of DsMV under field conditions in Florida for several growing seasons provided they are sufficiently isolated from commercial plantings (Zettler and Abo El-Nil, 1979; Zettler and Hartman, 1987). Such a population was maintained for three years in Lee County, Florida, on a six-acre site located about 40 miles from commercial fields in Highlands County.

Field transmission of CMV and TSWV is much more likely in some geographical regions than others, depending upon the source of primary inoculum and the relative prevalence of particular vector species. In southern Florida, for example, transmission of CMV in crops such as celery (*Apium graveolens*) and Easter lilies (*Lilium longiflorum*) has been correlated with high populations of *Commelina* spp. growing in muck soils along adjacent ditchbanks. High incidences of TSWV in peanut (*Arachis hypogaea*) and tomato (*Lycopersicon esculentum*) plantings in northern Florida in recent years are correlated with unusually high numbers of the western flower thrips *F. occidentalis*.

Selected Literature

Chase, A.R. and Zettler, F.W. 1982. Dasheen mosaic virus infection of dieffenbachia cultivars. Plant Dis. 66:891–893.

Hartman, R.D. 1974. Dasheen mosaic virus and other phytopathogens eliminated from caladium, taro, and cocoyam by culture of shoot tips. Phytopathology 64:237–240.

Hartman, R.D., Zettler, F.W., Knauss, J.F. and Hawkins, E.M. 1972. Seed propagation of caladium and dieffenbachia. Proc. Fla. State Hortic. Soc. 85:404–409.

Knauss, J.F. 1976. A tissue culture method for producing *Dieffenbachia picta* cv. Perfection free of fungi and bacteria. Proc. Fla. State Hortic. Soc. 89:293–296.

Liang, Y.G., Yang, T.C. and Chen, C.T. 1994. Identification of dasheen mosaic virus in Taiwan by immunolabellings. Rep. Taiwan Sugar Res. Inst. 143:41–50.

Rana, G.L., Vovlas, C. and Zettler, F.W. 1983. Manual transmission of dasheen mosaic virus from Richardia to nonaraceous hosts. Plant Dis. 67:1121–1122.

Rodoni, B.C. and Moran, J.R. 1988. The detection of dasheen mosaic virus using the enzyme-linked immunosorbent assay method (ELISA). Acta Hortic. 234:281–285.

Shimoyama, J., Kameya-Iwaki, M., Hanada, K., and Gunji, T. 1992. Konjak mosaic virus, a new potyvirus infecting konjak, *Amorphophallus konjac*. Ann. Phytopathol. Soc. Jpn. 58:706–712.

Taylor, M.E. and Knauss, J.F. 1978. Tissue culture multiplication and subsequent handling of known pathogen-free *Dieffenbachia maculata* cv. "Perfection." Proc. Fla. State Hortic. Soc. 91:233–235.

Zettler, F.W. and Abo El-Nil, M.M. 1979. Mode of inheritance of foliage color in caladium. J. Heredity 70:433–435.

Zettler, F.W., Abo El-Nil, M.M. and Hartman, R.D. 1978. Dasheen mosaic virus. Descriptions of Plant Viruses, No. 191.

Zettler, F.W., and Hartman, R.D. 1987. Dasheen mosaic virus as a pathogen of cultivated aroids and control of the virus by tissue culture. Plant Dis. 71:958–963.

Zettler, F.W., Hartman, R.D., Knauss, J.F., Taylor-Knauss, M.E. and Chase, A.R. 1980. Evaluation of *Dieffenbachia maculata* "Perfection" plants. Acta Hortic. 110:259–263.

Zettler, F.W., Hartman, R.D. and Logan, A.E. 1991. Feasibility of producing pathogen-free aroid root crops commercially by micropropagation. In: Twenty Sixth Annual Meeting of the Caribbean Food Crops Soc., 29 July–4 August, 1990, Mayaguez, Puerto Rico, pp. 80–85.

Zettler, F.W., Hiebert, E. and Rodoni, B.C. 1990. Dasheen mosaic potyvirus. In: Viruses of Tropical Plants, A.A. Brunt et al. (eds.). CAB International, Wallingford, Oxon, UK, pp. 237–240.

Pot and Bedding Plants

45

Dimorphotheca sinuata

Vittoria Lisa
Istituto Fitovirologia applicata CNR, Strada delle Cacce 73, Torino, Italy

Introduction

Dimorphotheca sinuata, family Compositae, commonly called "African daisy" or "Star of the Veldt", is an attractive bedding plant native of South Africa that forms small shrubs bearing a large number of daisy-shaped flowers with colors ranging from white to yellow to purple. It is also grown as a pot plant. Propagation is mainly by cuttings and also through seed. Plantlets are grown in pots under glass until planting in the open or delivery to the market as flowering pots. Although considered a "minor" crop, *D. sinuata* has a consistent market, is produced in all temperate areas, and is of economic interest to many countries. Reports on virus or virus-like diseases of *D. sinuata* are rather recent. Most of the cases have occurred in nurseries with consequent economic losses. Mixed infections of two or more viruses may easily occur.

The Viruses

Alfalfa mosaic alfamovirus (AMV)

This virus has been reported in Germany, without information on the symptoms observed (Kleinhempel, 1991).

Cucumber mosaic cucumovirus (CMV)

This has been reported in Italy where the disease appeared widespread. Affected plants are stunted and have distorted leaves with green mosaic; they yield fewer flowers (Fig. 45.1). Experimentally infected *D. sinuata* seedlings reproduced the field symptoms, confirming the severity of CMV as a pathogen of this crop (Dellavalle et al., 1994).

Lettuce mosaic potyvirus (LMV)

Found in Japan (T. Natsuaki, pers. comm.) and in Italy (Dellavalle et al., 1994), it causes slight reduction in growth and green or green-yellow mosaic on leaves (Fig. 45.2; see color plate). LMV infections appear to be frequent in *D. sinuata*. The isolate found in Italy was seed-transmitted in lettuce, but there is no information of seed transmission in *D. sinuata*.

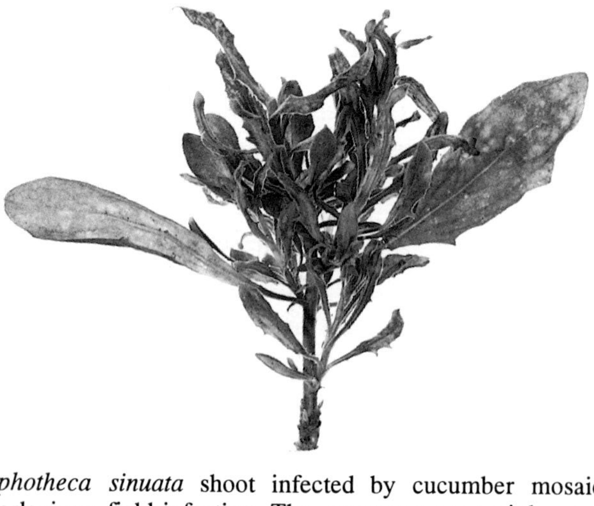

Fig. 45.1. *Dimorphotheca sinuata* shoot infected by cucumber mosaic virus and the *Dimorphotheca* carlavirus; field infection. The symptoms are mainly caused by cucumber mosaic virus. Photo Istituto di Fitovirologia applicata CNR, Torino, Italy.

Tomato spotted wilt tospovirus (TSWV)

Outbreaks of TSWV on *D. sinuata* have occurred in southern France (K. Gebre-Selassie and G. Marchoux, pers. commun.) and in northern Italy (Ramasso et al., 1994). The disease caused by TSWV is severe and widespread, and when it occurs in nurseries, losses are significant. Infected plants show generalized yellowing, necrosis and mosaic on leaves, cessation of growth, and sometimes top necrosis (Fig. 45.3; see color plate). Flowers are generally not affected. If plants are infected at an early stage, the disease is often lethal, while plants infected later can survive but are unmarketable.

Dimorphotheca *carlavirus*

This virus, not yet fully identified, has been repeatedly isolated in Italy. It is serologically related to some carlaviruses (carnation latent, cactus virus 2, helenium virus S, and red clover vein mosaic) but is distinct from them. It also differs from these viruses in experimental host range. No information is available on the field symptoms since the carlavirus has always been found in association with other viruses. *D. sinuata* seedlings experimentally infected were less vigorous than healthy ones but showed no specific symptoms. Infection by this virus might occur unnoticed in crops and possibly increase the severity of symptoms in case of mixed infections (Dellavalle et al., 1994).

To complete the data on virus diseases of *D. sinuata*, it may be worth noting that a different *Dimorphotheca* species, *D. pluvialis*, has been experimentally infected by *bidens mottle potyvirus* (BiMV) in the US (Logan et al., 1984).

MLO Diseases

A disease of *D. sinuata* associated with MLOs has been reported in Sicily. Infected plants showed virescence and phyllody, shortened internodes, diffuse chlorosis, and reduced size of basal leaves. Symptoms were not uniformly distributed in the plants since a stalk could bear virescent and normal flowers; similarly, MLOs were not uniformly distributed in phloem tissues of infected plants (Grimaldi and Grasso, 1988).

Identification and Assay Methods

All the viruses described are sap transmissible and infect the commonly used indicator plants. *Chenopodium amaranticolor* and *C. quinoa* are good assay hosts and can provide some diagnostic information. Both hosts react similarly to all viruses, with the following symptoms:

- AMV — chlorotic local lesions and systemic flecking, green mosaic and stunting
- BiMV — chlorotic local lesions and systemic chlorotic spots and flecks
- CMV — necrotic local lesions, generally no systemic infection
- LMV — chlorotic local lesions, yellowish mosaic and twisting of young leaves
- TSWV — necrotic local lesions, no systemic infection
- *Dimorphotheca* carlavirus — chlorotic local lesions, systemic vein banding and leaf roll

Specific antisera are available for all the viruses except the carlavirus, and serological diagnosis can be applied. ELISA can be used for large-scale routine diagnosis. The carlavirus can be detected, at least by immunoelectron microscopy or by the slide precipitin test, using a serum to carnation latent virus.

The MLO-associated disease can be identified by the techniques indicated in Chapter 7.

Transmission and Control

All the viruses described in *D. sinuata* are transmitted in the field by winged vectors; aphids for AMV, CMV, LMV, the carlavirus, and BiMV; and thrips for TSWV. The MLO is transmitted by leafhoppers. Control of aphids and thrips in the field is difficult, but in nurseries they can be controlled by careful phytosanitization. Similar measures can be applied to control MLO transmission. No information is available on seed transmission which may possibly occur with LMV, which is transmitted at high rates through the lettuce seeds. Propagation of cuttings taken from infected mother plants may be responsible for the spread of LMV and the carlavirus, as these generally do not produce severe symptoms on older plants; mother plants should therefore be chosen with care and indexed for viruses. No information is available on production of virus-free clones through conventional techniques such as thermotherapy and meristem-tip culture.

Selected Literature

Dellavalle, G., Masenga, V. and Lisa, V. 1994. Cucumber mosaic, a potyvirus and a carlavirus from *Dimorphotheca sinuata*. Acta Hortic. 377:91–98.

Grimaldi, V. and Grasso. S. 1988. Mycoplasma diseases of *Dimorphotheca sinuata* and *Catharanthus roseus*. Acta Hortic. 234:137–141.

Kleinhempel, K. 1991. Virus infection of outdoor ornamental plants. In: Rep. for 1990, Inst. für Phytopathologie, Aschersleben, p. 36.

Logan, A.E., Zettler, F.W. and Christie, S.R. 1984. Susceptibility of *Rudbeckia*, *Zinnia*, *Ageratum* and other bedding plants to bidens mottle virus. Plant Dis. 68:260–262.

Ramasso, E., Dellavalle, G., Roggero, P. and Lisa, V. 1994. Nuovi ospiti di Tospovirus in piante ornamentali ed ortensi in Liguria. Inf. Fitopatol. 44(1):44–48.

46

Hibiscus

Roger H. Lawson
US Department of Agriculture, Agricultural Research Service, US National Arboretum, Floral and Nursery Plants Research Unit, Beltsville, MD 20705, USA

Introduction

Hibiscus rosa-sinensis is a shrub that grows in many semi-tropical and tropical areas of the world. In the US hibiscus grows outdoors in southern climates and has been popularized as an indoor plant for pot culture. Typically, cuttings are rooted from stock plants grown in the southern US. The rooted cuttings are planted in a six-inch pot and treated with growth retardant. The resulting product has been widely accepted by consumers. The plant can be grown indoors and will continue flowering, even in winter, when adequate light is provided in a sunny location. In late spring, following danger of frost in northern climates, the plants can be grown as patio specimens outdoors during the summer. The wholesale value of the crop for florist and houseplant sales is estimated to be $5–6 million per year. Garden center and landscape uses have an estimated value of $6–7 mil-lion with a total commercial value of $11–13 million for the US crop.

The Viruses

Hibiscus is a large genus that includes not only *H. rosa-sinensis* but the edible crop okra (*H. esculentus* or *Abelmoschus esculentus*) and the fiber plant kenaf (*H. cannabinus*). A whitefly transmitted yellow vein mosaic has been described infecting *H. rosa-sinensis* in India (Capoor et al., 1950). In addition, a disease described as hibiscus leaf curl was reported on *A. esculentus* (Sastry and Singh, 1973). A virus disease causing line pattern in *H. rosa-sinensis* occurs in South Africa (Wolfswinkel, 1966). *H. rosa-sinensis* showing ringspot symptoms has also been reported from Nigeria (Lana, 1974). This ringspot virus was also transmitted to cotton (*Gossypium hirsutum*) and kenaf.

These viruses were only partially characterized, and their relationships to other known viruses was not reported. In this chapter three characterized viruses infecting *H. rosa-sinensis* are described. They include hibiscus chlorotic ringspot, hibiscus latent ringspot, and tomato vein-yellowing virus.

Hibiscus chlorotic ringspot carmovirus (HCRSV)

HCRSV is found naturally infecting *Hibiscus* sp. and many cultivars of *H. rosa-sinensis*. The virus produces a generalized mottle or many small chlorotic spots and rings that may persist throughout the year on some cultivars (Figs. 46.1 and 46.2) (Waterworth, 1980). Vein-banding is present on most leaves of infected plants. Plants are apparently not stunted and flowering is not affected. In contrast, virus inoculated kenaf and hibiscus seedling showed significant growth reduction, but an effect on flowering was not reported (Raju, 1985).

In Florida most infected hibiscus plants showed characteristic symptoms of the virus in spring (Fig. 46.3; see color plate). During the summer, however, symptoms were less prominent and sometimes totally masked (Raju, 1985).

Transmission

HCRSV has a wide host range in the Malvaceae but infected only 8 of 54 species representing 21 other families. In addition to experimental transmission in extracted sap, HCRSV has been transmitted using pruning shears (Raju, 1985). Since no insect vector is known, transmission by tools in the field may be the major cause of virus spread.

Fig. 46.1. ***Hibiscus*** **species from Thailand infected with hibiscus chlorotic ringspot virus** showing ringspot symptoms. (Photo H.E. Waterworth.)

Identification and assay methods

The virus infected 12 of 13 species in the Malvaceae (Waterworth, Lawson and Monroe, 1976). Several genera in the Amaranthaceae, Chenopodiaceae, Leguminosae and Scrophulariaceae formed local lesions. Diagnostic species include *Chenopodium quinoa* that produces chlorotic local lesions after 10 days without systemic infection. *Cyamopsis biflorus* forms necrotic local lesions with systemic invasion. *Hibiscus cannabinus*, *Gossypium hirsutum* and *Phaseolus vulgaris* all produce local lesions, but only *H. cannabinus* is systemically infected. *H. cannabinus* is not a reliable indicator during the summer in Florida (Raju, 1985).

Cyamopsis tetragonotoba and *Dilichos biflorus* are also local lesion assay hosts. *H. cannabinus* is considered the most sensitive host, however, since this plant may show symptoms after 2 to 3 weeks when some inocula fail to produce symptoms on the other two hosts.

HCRSV has been purified from infected kenaf leaves (Waterworth et al., 1976). The virus is composed of a single component with isometric particles

Fig. 46.2. *Hibiscus rosa-sinensis* infected with hibiscus chlorotic ringspot virus showing small chlorotic spots (right) and a healthy plant (left). (Photo H.E. Waterworth.)

about 28 nm in diameter with an RNA molecular weight of 1.55×10^6 daltons. The virus is a good immunogen, and a polyclonal antiserum with a titer of 1:512 was produced (Waterworth et al. 1976). These authors reported that the virus could be reliably detected in crude sap of most experimental species in a gel diffusion test. The virus is also detected in ELISA tests from infected hibiscus, kenaf, and okra (Raju, 1985). Comparing the sensitivity of detection of ELISA with kenaf inoculations, it was shown that ELISA was superior in tests with infected leaves. In contrast, when petals from infected plants were tested, kenaf and ELISA were equally sensitive (Raju, 1985).

Hibiscus latent ringspot nepovirus (HLRSV)

HLRSV has been reported from Nigeria (Brunt et al., 1980). The virus is known to occur naturally only in *H. rosa-sinensis* where it produces faint chlorotic foliage, but is usually symptomless. The virus does not cause any apparent flower abnormality.

Transmission

HLRSV is not transmitted by *Myzus persicae*, and there is no evidence it is seedborne in *C. quinoa* or *N. clevelandii*.

Identification and assay methods

HLRSV has been transmitted mechanically in sap to 22 of 74 species in 20 families (Brunt et al., 1980). Diagnostic species include *Chenopodium murale* that produces necrotic local lesions in 5–7 days without systemic infection. The virus causes chlorotic local lesions in 4–6 days on *C. quinoa*. *Gomphrena globosa* develops necrotic local lesions in 5–6 days with red borders expanding to 4–5 mm. Systemic hosts include *H. cannabinus* and *Trifolium incarnatum*.

Nicotiana clevelandii is a good propagation host that has been used for purification. Virus yields were up to 65 mg/kg of leaf tissue (Brunt et al., 1980). The virus particles are isometric with a diameter of about 28 nm. Three components are formed. This differs from the two components formed by okra mosaic and the single component produced by HCRSV.

HLRSV is a good immunogen, and antiserum with an endpoint titer of 1/8,192 in gel diffusion tests was produced (Brunt et al., 1980). Although some properties of the virus indicate it may be a nepovirus, it was serologically unrelated to 15 known or possible members of this group (Brunt et al., 1980).

Tomato vein-yellowing rhabdovirus (TVYV)

TVYV was reported infecting tomato in Morocco (El Maataoui et al., 1985). The virus was found to naturally occur in *Lycopersicon esculentum*, *Solanum sodomaeum*, *S. melongena*, and *S. nigrum*. TVYV is a rhabdovirus and is morphologically similar to a virus first found in *H. rosa-sinensis* showing vein-yellowing symptoms in the Canary Islands and Greece (Peters, 1981).

Transmission

Vein-yellowing symptoms in hibiscus could not be unequivocally attributed to the associated rhabdovirus because the virus occurred naturally in mixed infections with HCRSV and HLRSV in hibiscus (Lockhart, 1987). HCRSV and HLRSV produce little effect on hibiscus flower quality (Plavsic et al., 1984). The vein yellowing rhabdovirus, however, was associated with yellowing of the calyx, failure of the buds to open fully, and premature flower drop (Lockhart, 1987). It remains to be determined if TVYV alone produces vein-yellowing in hibiscus. Vein yellowing was present, however, in chlorotic leaves where rhabdovirus-like particles were observed in ultrathin sections of parenchyma cells (Plasvic et al., 1985). Spread of the virus probably occurs through infected planting material. It should also be noted, however, that although no vector has been identified, spread of TVYV from weed hosts to cultivated tomato and eggplant indicates that a natural vector may occur.

Identification and assay methods

TVYV can be mechanically transferred from hibiscus to *N. benthamiana*, *N. rustica*, and *N. debneyi*. Symptoms in these hosts were local and systemic chlorotic spotting and vein yellowing. Antiserum prepared to TVYV (El Maataoui et al., 1985) reacted with crude sap from hibiscus virus infected tobacco in gel diffusion tests (Lockhart, 1987).

Selected Literature

Brunt, A.A., Barton, R.J. and Phillips, S. 1980. Hibiscus latent ringspot virus, a newly recognized virus from *Hibiscus rosa-sinensis* (Malvaceae) in Western Nigeria. Ann. Appl. Biol. 96:37–43.

Brunt, A.A., Barton, R.J. and Phillips, S. 1981. Hibiscus latent ringspot virus. CMI/AAB Descriptions of Plant Viruses, No. 233.

Capoor, S.P. and Varma, P.M. 1950. Yellow-vein mosaic of *Hibiscus esculentus* L. Ind. J. Agric. Sci. 20:217-230.

El Maataoui, M., Lockhart, B.E.L. and Lesemann, D.-E. 1985. Biological, serological and cytopathological properties of tomato vein-yellowing virus, a rhabdovirus occurring in tomato in Morocco. Phytopathology 75:109–115.

Lana, A.O. 1974. A new ringspot virus disease in *Hibiscus rosa-sinensis*. Plant Dis. Rep. 58:1040–1042.

Lockhart, B.E.L. 1987. Evidence for identity of plant rhabdoviruses causing vein-yellowing diseases of tomato and *Hibiscus rosa-sinensis*. Plant Dis. 71:731–733.

Peters, D. 1981. Plant rhabdovirus group. CMI/AAB Descriptions of Plant Viruses, No. 244.

Plavsic, B., Eric, Z. and Milicic, D. 1984. Rhabdovirus-like particles associated with vein yellowing on *Hibiscus rosa-sinensis* L. Phytopathol. Mediterr. 23:49–51.

Plavsic, B., Milicic, D. and Eric, Z. 1985. Rhabdovirus-like particles in *Hibiscus rosa-sinensis* L. Acta Hortic. 164:41–43.

Raju, B.C. 1985. Occurrence of chlorotic ringspot virus in commercial *Hibiscus rosa-sinensis* cultivars. Acta Hortic. 164:273–280.

Sastry, K.S.M. and Singh, S.J. 1973. Field evaluation of insecticides for the control of whitefly (*Bemesia tabaci*) in relation to the incidence of yellow vein mosaic of okra (*Abelmoschus esculentus*). Ind. Phytopathol. 26:129–138.

Waterworth, H.E., Lawson, R.H. and Monroe, R.L. 1976. Purification and properties of hibiscus chlorotic ringspot virus. Phytopathology 66:570–575.

Waterworth, H. 1980. Hibiscus chlorotic ringspot virus. CMI/AAB Descriptions of Plant Viruses, No. 227.

Wolfswinkel, L.D. 1966. A virus disease in *Hibiscus rosa-sinensis* L. S. Afr. J. Agric. Sci. 9:483–486.

47

Hydrangea

Roger H. Lawson
US Department of Agriculture, Agricultural Research Service, US National Arboretum, Floral and Nursery Plants Research Unit, Beltsville, MD 20705, USA

Introduction

Hydrangea macrophylla (Thunb.), the florists' hydrangea, is a native plant in Japan and a member of the family Saxifragaceae. The greenhouse forms of hydrangea include the hortensia types with a globular cyme and perfect flowers that form seed only in the inner part of the cyme. The outer portion of the inflorescence contains staminate florets. The flowers have a reduced corolla, but the staminate flowers have large petaloid sepals which compose the showy portion of the inflorescence. The structure of the flower is responsible for the long-lasting flowering characteristic since the petaloid sepals do not rapidly fade.

The florists' hydrangea is considered a specialty crop in the United States since it is produced primarily as a seasonal plant sold for Easter and Mother's Day. Although it is not considered a major crop, it is economically important. In Europe, hydrangea is becoming increasingly important. A total of 3.17 million pot plants were sold in the Dutch auction in 1991. This number increased to 5.46 million in 1993. The crop is grown year-round and is frequently propagated *in vitro*.

The Viruses

Most of the viruses described in hydrangea were identified many years ago (Koenig, 1973). They include hydrangea ringspot, tomato ringspot, tobacco ringspot, tobacco necrosis, cucumber mosaic, and alfalfa mosaic virus. Three additional viruses are hydrangea mosaic, tomato spotted wilt virus, and an undescribed carlavirus. Mycoplasma-like organisms (MLOs) have been associated with different forms of a virescence disease in hydrangea. This chapter presents information about the most important virus and MLO diseases and methods of detection and control.

Hydrangea ringspot potexvirus (HRSV)

Hydrangea ringspot virus (HRSV) has been reported from the US and Europe (Koenig, 1973) but probably occurs wherever hydrangeas are cultivated.

"Running out" of some cultivars has been attributed to the gradual contamination of intolerant selections during propagation (Brierley and Lorentz, 1957). Although many *H. macropylla* plants may be symptomlessly infected with HRSV, some varieties show crinkling, rolling, and asymmetry of leaves and a decreased number of florets in the inflorescence (Koenig, 1973). More tolerant cultivars reportedly included Stafford, Engel's White, Todi, and Monte Forte Perle that usually were asymptomatic during the forcing period. The cv. Strafford frequently showed conspicuous yellow blotches and chlorotic rings in green leaves and occasional reddish-brown rings (Fig. 47.1). Shaded leaves of cv. Merveille often showed a reversal pattern with green rings on a yellow background (Brierley, 1957). The only flower symptom associated with the virus has been a suppression in floret numbers.

Expression of HRSV symptoms may be modified by environment, age of the leaves, and nutrition. Symptoms are often more prominent on plants in the early spring and late summer. A low level of nitrogen favors the development of chlorotic ringspots. High nitrogen fertilizer suppresses ringspot symptoms in varieties that normally show symptoms with reduced levels of fertilizer.

Fig. 47.1. Hydrangea ringspot virus symptoms on leaves of naturally infected hydrangea cv. Strafford. (Photo P. Brierley.)

Transmission

Transmission of HRSV is by cutting knife. Passage of a cutting knife through the stem of a diseased plant, followed by a cut through the base of a healthy cutting, resulted in transmission to 12% of the cuttings (Brierley and Lorentz, 1957).

Identification and assay methods

Hydrangea is apparently the only natural host of HRSV. More than 20 different species in 12 families have been infected by experimental inoculations (Koenig, 1973). The bioassay host *Gomphrena globosa* shows necrotic spot local lesions 3–5 days after inoculation with HRSV (Fig. 47.2). Local lesions rapidly expand and become confluent and collapse on leaves with a high lesion density. There is no systemic reaction. Lesions develop over a range of 15–30°C in a greenhouse, but few or no lesions are formed at 35–40°C (Brierley and Lorentz, 1957). *Chenopodium amaranticolor* and *C. quinoa* are also local lesion hosts for HRSV. No insect vectors are known to transmit HRSV and the virus is not transmitted through seed.

Antiserum has been produced to HRSV of sufficiently high quality that it could be used in a preindexing screening procedure (Hartleb, 1974).

Fig. 47.2. Necrotic local lesions of hydrangea ringspot virus on *Gomphrena globosa* (right); healthy uninoculated leaf (left).

Control

Tip cuttings of the cv. "Charm", taken from HRSV infected plants, have been produced virus-free after plants were grown in a greenhouse compartment at a constant 35°C day and night for 12–19 weeks (Brierley, 1957). None of the tip cuttings were virus-free after four weeks. Subsequent tests with cultivars treated 14 weeks at 35°C showed a variation in the number of virus-free plants from 20–50% with the cv. Strafford not producing any virus-free cuttings. In Denmark hydrangeas were successfully heat-treated at 30°C and 34°C for 240 days and 120 days, respectively (Paludan and Christensen, 1973). Plants were killed at 36°C after heat treatment for 30 days. Brierley (1957) found that whenever heat-treated stock plants were tested, HRSV was detected. The difference in these results and those reported in Denmark may be a result of differences in hydrangea cultivars or the size of the plants treated.

HRSV has been eliminated from commercial hydrangeas by meristem tip culture (Preil et al., 1978). Shoot tips were surface sterilized with sodium hypochlorite and meristem tips 0.3–0.4 mm in size were excised and placed on an agar medium. About 30% of the resulting plants were virus free.

Tomato ringspot nepovirus (TomRSV)

Symptoms of tomato ringspot virus (TomRSV) appear as a dull yellow chlorosis in the leaves with diffuse-bordered dark-green blotches superimposed (Fig. 47.3). The virus is apparently not responsible for all mosaic patterns in hydrangea (Brierley and Smith, 1952). Plants may be stunted and leaves are smaller than normal often appearing narrow and irregular in shape (Brierley, 1954). Leaves may sometimes be stiff and brittle. Flowers are smaller than normal and open irregularly with green and colored florets in a mixed cyme.

TomRSV symptoms in the cv. Strafford appeared as yellowed leaves with green blotches eight months after inarch grafting (Brierley, 1954). Symptoms were apparent when plants were forced and new growth was developing rapidly. The virus was detected in both inoculated and source plants after 7 months and 21 months following inoculation. TomRSV was transferred by sap inoculation

Fig. 47.3. Normal leaf of hydrangea seedling (left). Leaf of hydrangea showing typical symptoms of tomato ringspot virus (right). (Photo P. Brierley.)

from hydrangea to hydrangea with chlorotic ring and yellow vein patterns developing the year after inoculation. The virus could not be isolated from these plants but was recovered from four of ten asymptomatic plants (Brierley, 1954). TomRSV was not transmitted to cuttings made with a knife that was first used on infected plants and then used to make cuttings from healthy plants (Brierley and Lorentz, 1957).

Transmission

Although there are no reports of vector transmission of TomRSV to hydrangea, the nematode *Xiphenema americanum* could possibly transmit the virus in soil. One method of determining the presence of the virus in the vector is to plant seed of a susceptible host such as cucumber in soil that contains the nematode. If the *X. americanum* in the soil contains the virus, they will transmit it to the cucumber where symptoms are expressed.

The principal method of transmission of TomRSV from hydrangea to hydrangea is probably through vegetative propagation.

Identification and assay methods

TomRSV was detected during all seasons of the year in inoculations from hydrangea to *Nicotiana tabacum* when the hydrangeas were in active growth (Brierley, 1954). In addition to tobacco, the virus was transmitted to the following species that all developed systemic symptoms: *Antirrhinum majus*, *Callistephus chinensis*, *Cucumis sativus*, *Helichrysum bracteatum*, *Hydrangea macrophylla*, *Lycopersicon esculentum*, *Petunia* hybrida, *Phlox drummondii*, *Platycodon grandiflorum*, *Senecio cruentus*, *Verbesina encelioides*, and *Zinnia elegans*.

Control

Because there are no above-ground vectors, it should be possible to control the disease by utilizing virus-free stock and planting in soil free of the viruliferous nematode vector.

Tobacco ringspot nepovirus

Tobacco ringspot virus (TobRSV) was isolated from hydrangea macrophylla showing uneven chlorotic crinkled leaves and stunting (Anderson, 1958). There was no evidence that the symptoms described were caused by this virus since Koch's postulates were not fulfilled. TobRSV was also isolated from *H. paniculata* Sieb. "Grandiflora" (Chiko and Godkin, 1986). The foliage was faintly mottled and leaves on some plants showed faint chlorotic ringspots.

Transmission

The information presented for TomRSV is probably also applicable to TobRSV.

Identification and assay methods

TobRSV was transmitted from hydrangea to *Vigna unguiculata*, *Nicotiana glutinosa*, *N. tabacum*, *Datura stramonium*, *Zinnia elegans*, and *Cucumis sativus* (Anderson, 1958).

Control

TobRSV can be controlled by growing plants in soil that does not contain *Xiphenema americanum*, the vector of tobacco ringspot virus. Care should also be taken to avoid mechanical transmission of the virus on cutting tools.

Tomato spotted wilt tospovirus (TSWV)

Transmission

Tomato spotted wilt virus (TSWV) was isolated from field grown *H. macrophylla* cv. Imaculata showing ringspot symptoms (Allen et al., 1983). Symptoms could not be reproduced in *H. macrophylla* cvs. Merritt Supreme or Rose Supreme inoculated with a hydrangea isolate of the virus and the virus could not be recovered (Allen et al., 1985).

Identification and assay methods

TSWV produced local and systemic symptoms on *Datura stramonium*, *Gomphrena globosa*, *Lycopersicon esculentum*, *Nicotiana glutinosa*, *N. tabacum*, *N. hybrida*, and *Vinca rosea*. Only local lesions were expressed on *Cucumis sativus* and *Lathyrus odoratus*.

Control

TSWV is transmitted by the thrips vector *Frankliniella occidentalis* (Pergande), the western flower thrips. This insect commonly transmits TSWV and is also the vector for impatiens necrotic spot virus (INSV).

Control of thrips-transmitted TSWV depends on maintaining clean stock free of this insect. Since the thrips prefer feeding in the flower rather than foliage, efforts should be made to apply control measures to exclude thrips during the flowering period.

TSWV can be detected with specific antisera that is commercially available. The antiserum should be employed in a certification program designed to produce virus-free stock.

Hydrangea mosaic ilarvirus (HdMV)

Hydrangea mosaic virus (HdMV) was first reported from the United Kingdom (Thomas et al., 1983). Diseased hydrangeas showed obvious leaf symptoms of a mottled mosaic of light- and dark-green segments. Flowers were unaffected. Symptoms were reproduced by graft inoculation of the cv Hamburg. The authors speculated that propagation of HdMV-infected plants probably does not occur because of the obvious symptoms. The symptoms on leaves are distinct from the distortions, ringspots, or chloroses induced by HRSV, CMV, TRSV, ToRSV, tomato black ring, arabic mosaic, and tobacco rattle viruses.

Transmission

HdMV was transmitted with difficulty from *H. macrophylla* to *Chenopodium quinoa*. *C. quinoa* exhibited diffuse local circular chlorotic lesions in the inoculated leaves. Seven to 14 days after inoculation the apical leaves showed a flush of merging chlorotic lesions (Thomas, et al. 1983). The chlorotic symptom appeared as a slight chlorosis but the symptoms faded and the plants were symptomless. The virus was transmitted from *C. quinoa* to four other herbaceous species in three families. Symptomless systemic infection was present in *Gomphrena globosa*, *N. benthamiana*, and *N. clevelandii*. The virus was seed transmitted in *C. quinoa* but was not transmitted by *Myzus persicae*.

Identification and assay methods

HdMV is a poor immunogen, but an antiserum was prepared that reacted in gel diffusion tests. The virus was serologically distinct from six ilarviruses and alfalfa mosaic virus (Thomas et al., 1983). The virus does, however, resemble ilarviruses since it has a tripartite genome with physical and biochemical properties similar to ilarviruses. No relationship was found to 41 other representative monopartite, bipartite, or tripartite viruses.

There is no evidence for spread of the virus from hydrangea to hydrangea. It should, therefore, be possible to prevent the propagation of HdMV-infected *H. macrophylla* by maintaining vigorously growing, asymptomatic plants.

Because HdMV has affinities with the ilarvirus group but was serologically unrelated to any of the ilarviruses tested, it is regarded as a distinct virus (Thomas et al., 1983).

Alfalfa mosaic alfamovirus (AMV)

Alfalfa mosaic virus (AMV) was reported as the probable cause of vein clearing of hydrangea in Italy (Belli, 1968). AMV also produces mottling, vein yellowing, and mosaic on apical foliage (Chiko and Godkin, 1986). These authors conclude that additional inoculation tests will be required to determine whether AMV causes mosaic in hydrangea or infection by both AMV and HRSV induces more severe effects than infection by either virus alone.

AMV was detected serologically in extracts from inoculated or systemically infected leaves of tobacco with AMV antiserum.

Other viruses

Although cucumber mosaic virus (Tamura and Komuro, 1967; Welvaert and Samyn, 1973) and tobacco necrosis virus (Welvaert and Samyn, 1973) have been reported in hydrangea, these viruses are not considered sufficiently important to include a more detailed description of the diseases they cause in hydrangea.

Mycoplasma-associated Diseases

Hydrangea virescence

Hydrangea virescence has been reported from Europe (Welvaert et al., 1975) and the US (Hearon et al., 1976). Virescence is characterized by a green flower abnormality that is also referred to as phyllody or proliferation. This disease was reportedly caused by hydrangea ringspot virus (Zeyen and Stienstra, 1973). The disease agent is transmitted by grafting but is not mechanically transmitted (Lawson et al., 1976).

Symptoms of virescence on naturally infected plants differ among cymes on the same plant and between different plants. On a single plant, some cymes may produce florets with a normal pink color while others contain a mixture of pink and green or all green florets. Leaves may appear normal. In contrast, other hydrangeas may show severe symptoms with green cymes and dwarfed florets with leafy shoot-like structures proliferating from the pistils. Plants with severe floral symptoms also show vein yellowing and dwarfed leaves.

Three stable forms of hydrangea virescence were reported (Lawson and Smith, 1980). Plants with severe symptoms are stunted and produce small leaves with vein yellowing and flowers are abnormal with small green cymes (Figs. 47.4 and 47.5; see color plate). An intermediate form of the disease produces cymes that vary from completely green with dwarfed florets to mixed cymes with both pink and green florets. A third mild form of the disease produces leaves that expand normally with large green florets in the cymes. Vein yellowing symptoms never appear on plants with the mild form of the disease.

Identification and assay methods

Virescence is associated with the presence of mycoplasma-like organisms in the phloem (Hearon et al., 1976). The MLO is typical of those structures associated with aster yellows and other MLOs. We failed to transmit the virescence agent to healthy hydrangea or China aster with the aster leafhopper (Lawson, unpublished).

Detection of virescence agent using nucleic acid probes

Methods for detecting virescence have been developed using nucleic acid probes (Bertaccini et al., 1992). Although these procedures have been used primarily to distinguish different forms of MLO, they may be important in the future for detection of hydrangea MLOs in a routine diagnostic procedure.

Control

Effective control of the agents causing the virescence disease should include the planting of new hybrids in areas separated from established plantings where older cultivars are grown. In the absence of identifying a specific insect vector, isolation of the plant material may provide the best protection. This recommendation is based on an unpublished observation that the virescence agent may be transmitted by root grafting. Root grafting was apparently the cause of transmission when a diseased hydrangea with the intermediate form of virescence was grown in a pot adjacent to a healthy seedling. The healthy plant became diseased 18 months after the plants were potted side by side.

Control of hydrangea virescence may also be possible by establishing nuclear stocks through a combination of heat therapy and tissue culture since this procedure has been shown to be therapeutic for other MLO-induced diseases.

Selected Literature

Allen, T.C., McMorran, J.P. and Locatelli, E.A. 1983. Isolation of tomato spotted wilt virus from hydrangea and four weed species. Plant Dis. 67:429–431.

Allen, T.C., McMorran, J.P. and Lawson, R.H. 1985. Detection and identification of viruses in hydrangea. Acta Hortic. 164:85–88.

Anderson, C.W. 1958. Tobacco ringspot virus on florist' hdrangea. Plant Dis. Rep. 42:932–933.

Belli, G. 1968. Researches on a vein clearing of Hydrangea. Phytopathol. Mediterr. 7:70–71.

Bertaccini, A., Davis, R.E., Hammond, R.W., Vibio, M., Bellardi, M.G. and Lee, I.M. 1992. Sensitive detection of mycoplasmalike organisms in field-collected and *in vitro* propagated plants of *Brassica, Hydrangea* and *Chrysanthemum* by polymerase chain reaction. Ann. Appl. Biol. 121:593–599.

Brierley, P. 1954. Symptoms in the florists' hydrangea caused by tomato ringspot virus and an unidentified sap-transmissible virus. Phytopathology 44:696–699.

Brierley, P. 1957. Virus-free hydrangeas from tip cuttings of heat-treated ringspot-affected stock plants. Plant Dis. Rep. 41:1005.

Brierley, P. and Lorentz, P. 1957. Hydrangea ringspot virus, the probable cause of "running-out" of the florists' hydrangea. Phytopathology 47:39–43.

Chiko, A.W. and Godkin, S.E. 1986. Occurrence of alfalfa mosaic, hydrangea ringspot, and tobacco ringspot viruses in *Hydrangea* spp. in British Columbia. Plant Dis. 70:541–544.

Hartleb, H. 1974. Ein kombinierter routinetest bei der grossproduktion von hortensien zym nachiveis des hydrangearingfleckenvirus. Arch. Gartenbau 22(7):411–417.

Hearon, S.S., Lawson, R.H., Smith, F.F., McKenzie, J.T. and Rosen, J. 1976. Morphology of filamentous forms of a mycoplasmalike organism associated with hydrangea virescence. Phytopathology 66:608–616.

Koenig, R. 1973. Hydrangea ringspot virus. CMI/AAB Descriptions of Plant Viruses, No. 114.

Lawson, R.H., Hearon, S.S. and Smith, F.F. 1976. Virescence disease in Hydrangea macrophylla (Florists' hydrangea): Graft transmission of the causal agent and symptoms. Acta Hortic. 59:153–159.

Lawson, R.H. and Smith, F.F. 1980. Three stable types of hydrangea virescence that differ in severity in foliage and flowers. Plant Dis. Rep. 64:659–661.

Preil, Vol W., Kuhne, H. and Hoffman, M. 1978. Hydrangea ringspot virus-freie Hortensien durch "Meristemkultur"-Kritische Betrachtung der Erfolgsaussichten-Nach-richtenblatt des Deutschen Pflanzenschtzd. (Braunschweig) 30:88–90.

Tamura, M. and Komuro, Y. 1967. The mosaic disease of Hydrangea caused by cucumber mosaic virus. Ann. Phytopathol. Soc. Jpn. 33:27–31.

Welvaert, W. and Samyn, G. 1973. De verspreiding van het hydrangea ringspot virus by de *Hydrangea macrophylla* (Thunb.) DC. Mededelingen Faikulteir Landbouwwetenschappen Gent 38:1647–1654.

Welvaert, W., Samyn, G. and Lagasse, A. 1975. Recherches sur les symptomes de la virescence chez l'*Hydrangea macrophylla* Thunb. Phytopathol. Z. 83:152–158.

Thomas, B.J., Barton, R.J. and Tuszynski, A. 1983. Hydrangea mosaic virus, a new ilarvirus from *Hydrangea macrophylla* (Saxifragaceae). Ann. Appl. Biol. 103:261–270.

Zeyen, R.J. and Stienstra, W.C. 1973. Phyllody of florists' hydrangea caused by hydrangea ringspot virus. Plant Dis. Rep. 57:300–304.

48

Impatiens

Roger H. Lawson and Hei-ti Hsu
US Department of Agriculture, Agricultural Research Service, US National Arboretum, Floral and Nursery Plants Research Unit, Beltsville, Maryland 20705, USA

Introduction

Impatiens are grown as both seed produced and vegetatively propagated ornamentals. The seed produced form, *Impatiens wallerana*, is the number one selling bedding plant in the United States and in Europe. New Guinea impatiens are vegetatively propagated and are prized for their flowers and decorative foliage. Seed-produced New Guinea types were recently introduced to the market.

Genetic improvement of impatiens is a remarkable story of plant exploration and breeding. In the early 1970s, collections were made in New Guinea, and by the late 1970s many hybrids of both the seed-produced and New Guinea types were commercialized. The popularity of these plants continued to increase through the late 1980s and 1990s. Although impatiens were first introduced 20 years ago, sales in the US of seed-produced types have increased more than 15% annually.

The popularity of impatiens has been based primarily on the rapid introduction of a wide range of new colors and plant forms. Intensive breeding has resulted in improved self-branching characteristics combined with solid red, orange, pink, rose, and lavender colors and bicolors. Variegated dark-green to bronze-colored leaves including some with red and yellow patterns in the foliage make the New Guineas outstanding bedding plants as well as excellent hanging baskets.

The Viruses

Of the different viruses that can infect impatiens, the most severe are those transmitted by thrips, including tomato spotted wilt and impatiens necrotic spot viruses. Two aphid transmitted viruses — cucumber mosaic and turnip mosaic potyvirus — may also infect impatiens. In addition, tobacco mosaic tobamovirus may occur and can be mechanically transmitted from plant to plant. A virus originally described as a potexvirus that was symptomless in *I. holstii* is now recognized as helenium virus S. Helenium virus S was also associated with nearly symptomless *I. holstii*. Tobacco streak ilarvirus and tobacco ringspot nepovirus have both been reported, but the extent of their distribution and relative importance is unknown.

Importance of Tospoviruses

Tomato spotted wilt virus (TSWV) has a wide host range with more than 500 species infected in more than 50 families (German et al., 1992). TSWV became a major problem in commercial vegetable and flower production in the US and Europe in the late 1980s. Although the virus was known to occur in many different plants, it was not widely distributed. In 1986–87 major outbreaks of Western flower thrips occurred in the fields across the US from the western to the eastern states. The thrips pest also became widespread in greenhouses, and the insect was identified in areas of eastern Canada where it was previously unknown. At the same time the insect was first observed in Europe where it quickly became established. Transport of the virus and vector has apparently occurred on propagation stock as well as finished products. One reason for this rapid spread probably relates to the difficulty of controlling thrips that are often concealed in transported flowers. Although several different tospoviruses have been reported, only tomato spotted wilt and impatiens necrotic spot viruses have been identified in impatiens.

Impatiens necrotic spot virus

Impatiens necrotic spot virus (INSV) is a distinct member of the Tospovirus genus (Law and Moyer, 1990). Although this virus was at one time classified as a strain of TSWV, we now know it has a serologically distinct nucleocapsid protein. Classification of INSV is further complicated by the occurrence of variants derived from cultures of the virus that are no longer serologically related to the original virus from which they were derived (Lawson et al., 1993). The apparent plasticity of INSV may be reflected in the highly diverse host range of the virus that reportedly infects many different ornamental plants (Hausbeck et al., 1992).

Symptoms of INSV on naturally infected seed type impatiens include small black to brown necrotic flecks and ringspots (Fig. 48.1; see color plate). Small dark purple ringspots occur on leaves, and white rings may be present on flower petals. Necrosis may also appear along the midribs of diseased leaves that often show yellowing, malformation, and stunting. Symptoms of necrosis may also appear in the stems of affected plants. Infected leaves often drop prematurely.

Symptoms on INSV infected New Guinea impatiens may be similar but are more difficult to diagnose because of the abundance of anthocyanin pigments in the leaves. Ringspots and chlorosis and necrosis appear on infected plants, but leaves on some branches of a diseased plant are frequently asymptomatic.

Temperature has a significant effect on symptom expression of INSV in naturally infected seed-produced and New Guinea impatiens. In general, high light above 1500 foot candles and high temperatures above 27°C will suppress symptom development. In a greenhouse symptoms are often more easily observed in plants growing during the winter months.

Tomato spotted wilt virus

Tomato spotted wilt virus (TSWV) was for many years recognized as the only member of the genus Tospovirus. TSWV is the type member of this genus and possesses a serologically distinct nucleoprotein (German et al., 1992). Symptoms produced by TSWV infection on impatiens are similar to those induced by INSV. TSWV has been described from impatiens in the Netherlands (de Avila et al., 1992).

Thrips transmission

Thrips are vectors of tospoviruses. INSV and TSWV are transmitted by the Western flower thrips, *Frankliniella occidentalis* (Pergande). Acquisition of the viruses by thrips occurs when the insects feed on infected plants in the larval stage (German et al., 1992). The infection cycle is initiated only when the female adult thrips lays eggs on tospovirus-infected leaves that are suitable hosts for egg and larval development. TSWV may be acquired after 30 minutes of larval feeding, but acquisition may occur throughout larval development. Tospoviruses are retained by thrips through molting and pupation and are transferred to the adult insect. Adult thrips may inoculate plants with TSWV soon after emergence or several days may be required for the insects to initiate transmission.

In addition to *F. occidentalis*, TSWV is transmitted by *Frankliniella schultzei* (Trybom), blossom thrips, *Thrips tabaci* (Lindeman), onion thrips, and *Frankliniella fusca* (Hinds), the tobacco thrips. There is no evidence of transovarian transmission by any of these vectors. *Thrips palmi* Karny and *Frankliniella intonsa* (Trybom) have been reported as vectors of tospoviruses in Taiwan and Japan, respectively.

Mechanical transmission

TSWV and INSV can be transmitted to a wide range of indicator plants. *Nicotiana* spp. have been commonly used as test plants and as a source for maintaining and purifying both viruses. Differences in symptom severity have been associated with different isolates of TSWV, and similar differences occur with INSV. For example, symptoms on leaves of impatiens "Accent Salmon" inoculated with a severe isolate of INSV derived from a defective isolate repeatedly passaged at high temperature appeared earlier and were more severe than those produced on the same variety inoculated with a more mild non-defective isolate. The severe isolate produced large expanding necrotic spots and rings (Fig. 48.2). The mild isolate produced small chlorotic spots without severe necrosis (Fig. 48.3). Both sets of plants were grown at 27/24°C day/night temperature with 1500 foot candles of fluorescent illumination. The severe isolate that produced large spreading necrotic local lesion symptoms also produced necrotic midribs and stem necrosis on some inoculated plants (Fig. 48.4; see color plate). Stem necrosis was rare on plants infected with the mild form of INSV.

Identification and assay methods

Bioassay — INSV and TSWV can be detected by mechanical inoculation using inoculum prepared by grinding infected leaf or stem material in 0.05 M phosphate buffer at pH 7.0–7.2. Reducing agents such as sodium sulfite or mercaptoethanol

Fig. 48.2. Large spreading necrotic local lesions on leaves of *Impatiens* "Accent Salmon" 10 days after inoculation with an isolate of impatiens necrotic spot virus. The isolate was derived after repeated passage of a defective isolate at high temperature. Note the severity of the reaction. The plant was incubated at 27/24°C day/night temperature with 1500 foot candles of fluorescent illumination following inoculation.

Fig. 48.3. Small restricted necrotic and chlorotic local lesions on leaves of *Impatiens* "Accent Salmon" inoculated with a second non-defective isolate of impatiens necrotic spot virus. Note the mild reaction compared to the isolate illustrated in Fig. 48.2. The plant was incubated at 27/24°C day/night temperature with 1500 foot candles of fluorescent illumination following inoculation.

are usually added. *Nicotiana benthamiana* and *Petunia hybrida* can be used to detect the virus. Necrotic spot local lesions are formed on the leaves of *N. benthamiana* followed by systemic vein yellowing and necrosis. Petunia leaves show necrotic lesions on inoculated leaves without systemic symptoms. TSWV and INSV can usually be isolated from symptomatic tissue, but isolation of these viruses from asymptomatic tissue is less reliable.

Petunia indicator

In spite of the absence of systemic invasion of petunias, some varieties can be used to detect thrips transmitted virus (Allen and Matteoni, 1991). For example, young Calypso petunia plants produce necrotic spot local lesions when infected with TSWV or INSV. These plants are attractive to thrips, and white feeding scars are produced by the insect. When the thrips transmits a tospovirus, the tissue surrounding the white scar shows a necrotic ring symptom. This reaction usually occurs 6–7 days after infection. Some growers place the petunia indicator plants among those plants they would like to survey for the presence of INSV or TSWV. This is an inexpensive way of monitoring for the presence of these viruses and of determining the relative distribution of each type of virus. Before establishing this type of bioassay, an effort should be made to identify the variety of petunia that gives the best local lesion response and also maximizes the opportunity for thrips feeding.

Serology

Polyclonal antisera and mouse monoclonal antibodies have been made to INSV and TSWV. These antisera have been used in the enzyme-linked immunosorbent assay (ELISA) to detect the viruses. The ELISA test may not, however, be the most sensitive procedure. A comparison of dot-blot immunoassay (DBIA) and ELISA showed that DBIA was nearly sixteen times more sensitive than ELISA in detecting TSWV (Fig. 48.5) (Hsu and Lawson, 1991). In addition, a method of direct tissue blotting was found useful for TSWV detection. Using this procedure, TSWV can be detected in asymptomatic tissue from infected impatiens by closely examining the blot image at 10–20 magnification with a dissecting microscope. Tissue extracts from the same asymptomatic samples failed to react positively in ELISA (Hsu and Lawson, 1991).

Because serological diagnosis of impatiens infected with TSWV or INSV is most accurate if samples are taken from symptomatic tissue, leaves from asymptomatic plants should be sampled from more than one location on the plant. In addition, plants should be sampled more than one time.

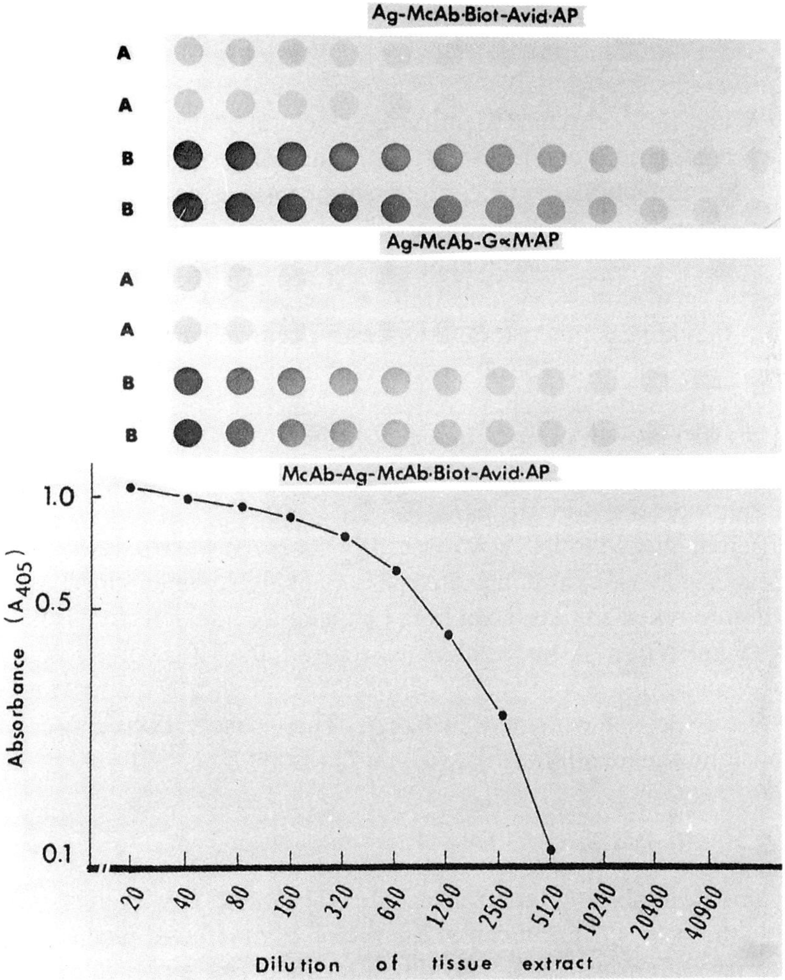

Fig. 48.5. DBIA and ELISA titrations of tomato spotted wilt virus antigens in infected leaves of *Nicotiana benthamiana*. Two-fold dilutions, from left to right, of healthy plant extracts (A) or TSWV-antigens (B) were applied to a nitrocellulose membrane or to a mouse antibody coated polyvinyl chloride plate. Biotinylated monoclonal antibodies and avidin-enzyme conjugate (Ag-McAb·Biot-Avid·AP) or monoclonal antibodies and goat anti-mouse immunoglobulin enzyme conjugate (Ag-McAb-GαM·AP) were used to detect viral antigens in DBIA. Biotinylated monoclonal antibodies followed by avidin-enzyme conjugate (McAb-Ag-McAb·Biot-Avid·AP) were used in ELISA.

Electron microscopy — TSWV and INSV

Electron microscopy

TSWV and INSV produce identifying structural characteristics. Both viruses produce accumulations of enveloped virions that range in size from 80–110 nm in diameter. Accumulations of nucleoprotein in the form of electron-dense viroplasms that appear identical in thin section are also formed. A distinguishing structural feature is the presence of filamentous inclusions in the form of paracrystalline arrays associated with INSV infection. Although TSWV produces filamentous inclusions that appear in clusters, they never occur in paracrystalline form.

Virions may not always be observed in some isolates of TSWV and INSV. The absence of virus particles is characteristic of defective isolates of the virus (Urban et al., 1991). Defective forms of the virus usually develop as a result of repeated mechanical passage of an isolate. Although viroplasm and nucleoprotein inclusions are present, there are no virus particles.

A different change in INSV was recently observed when a defective isolate of INSV developed virions after repeated passage at high temperature (Lawson et al., 1993). Before repeated passage at high temperature, the defective isolate reacted with INSV antiserum. After the change occurred, the virus culture failed to react with INSV antiserum (Lawson et al., 1993). A serologically distinct nucleocapsid protein was associated with this new isolate. These observations may be important in establishing the identity of new tospovirus isolates.

Control of tospoviruses

Three primary methods of control include: (1) sanitation with appropriate weed control in the field, (2) exclusion of the vector from infected plants, and (3) control of thrips with insecticides. Roguing is an important approach to sanitation. It is essential to remove all suspect plants from a greenhouse after the cropping season is com-pleted. Removal of weeds that harbor thrips is also important to control greenhouse spread.

All infected plants must be destroyed. Any new propagation material must be derived only from plants that are certified virus-free. New plants brought into the greenhouse or other growing areas must also be inspected and care taken to eliminate thrips.

Greenhouse structures can be covered with 400 mesh screens that will prevent entry of thrips. Thrips can be monitored in the greenhouse with the use of yellow sticky cards. Cards should be checked once a week and insecticides applied as needed. Populations of western flower thrips have developed resistance to insecticides in many greenhouses. It is, therefore, important to rotate the class of insecticide applied. Rotation of pyrethroids, carbamates, chlorinated hydrocarbons, organophosphates, and soaps will help reduce the build-up of resistant thrips populations.

Tobacco streak ilarvirus

Tobacco streak virus (TSV) was isolated from naturally infected impatiens (*I. holstii*) showing leaf twisting, deformation, and stunting (Lockhart and Betzold, 1980). Identical symptoms were reproduced on virus-free impatiens with a single lesion isolate of the virus. Infected plants produced fewer and smaller flowers than healthy plants. No deformation was observed in the flowers.

Identification and assay methods

The host range of the impatiens isolate was similar to that reported for other strains of the virus. The virus produced necrotic or chlorotic local lesions without systemic infection on *Phaseolus vulgaris* "Bountiful" and "Red Kidney", *Cucumis sativus* "National Pickling", *Cucurbita pepo* "Fordhook Zucchini", and *Beta patelaris*. Local and systemic symptoms were produced on *Chenopodium quinoa*, *Petunia hybrida*, *Gomphrena globosa*, *Glycine max*, *Datura stramonium*, *Ocimum basilicum*, *Lactuca sativa*, *Lycopersicum esculentum*, *Vigna ungiuculata*, *Zinni a elegans*, and *Nicotiana* species.

The virus was identified serologically in gel immunodiffusion tests using crude sap from infected *N. clevelandii*. No data have been presented on serological detection of the virus directly from impatiens.

Tobacco ringspot nepovirus

Tobacco ringspot virus (TobRSV) was isolated from *Impatiens holstii* showing stunting, leaf mottling, and leaf abscission (Lockhart and Pfleger, 1979). An isolate of the virus observed in the natural infection produced chlorotic or necrotic ringspots on inoculated leaves of impatiens. Systemic symptoms consisted of mosaic and ringspots. Leaf abscission observed in the original plants was not reproduced.

Identification and assay methods

The host range of the impatiens isolate was similar to that reported for other strains of the virus. The plants included *Antirrhinum majus*, *Nicotiana tabacum*, *N. clevelandii*, *Cucumis sativus*, *Cucurbita pepo*, *Physalis flordana*, *Petunia hybrida*, *Glycine max*, *Pisum sativum*, *Vigna unguiculata*, and *Phaseolus vulgaris*. *Chenopodium amaranticolor* and *C. quinoa* could be systemically infected when plants were grown on long days at high temperature.

TRSV was serologically identified in gel double-diffusion tests in purified form. There is no report of serological detection of the virus from crude sap of naturally infected plants.

Symptoms of TRSV were more prominent in impatiens plants grown at 28–30°C than in plants grown at 20–24°C. It may, therefore, be important to grow stock plants at elevated temperatures and conduct visual inspection for the presence of symptoms at the same time that bioassays are performed to indicator hosts.

Impatiens latent virus = Helenium S carlavirus

A rod-shaped virus 520 μm long was isolated from impatiens and transmitted from *I. holstii* to *Chenopodium quinoa* producing local lesions and systemic mottle (Lockhart and Betzold, 1980). Similar symptoms were produced on *C. amaranticolor* and *C. album*. Local lesions without systemic symptoms were formed on *Gomphrena globosa* and *Tetragonia expansa*. In particle morphology, physical properties, intracellular appearance, and absence of aphid transmissibility, the virus exhibited characteristics typical of the potexvirus group. Transmission of the virus to impatiens failed to produce symptoms. No serological relationship was found when the virus was compared with six other potexviruses (Lockhart and Betzold, 1980). More recently, apparently the same virus was isolated from the same species of impatiens and was identified as Helenium virus S carlavirus (Koenig and Lesemann, 1983; Plese et al., 1988).

Cucumber mosaic cucumovirus

Impatiens sultani was found naturally infected with a virus that produced stunting, mosaic, and deformation of leaves (Herold, 1964). The virus was identified as cucumber mosaic virus based on test plant symptoms and serology. Infected test plants included *Nicotiana glutinosa* and *Petunia hybrida* inoculated directly from diseased impatiens. CMV was transmitted mechanically and by *Aphis gossypii* from young leaves of young cuttings but was difficult to transmit from older plants. Symptoms appeared as faint mosaic and slight leaf deformation of the youngest leaves about 16 days after mechanical inoculation and six weeks after aphid transmission.

Selected Literature

Allen, W.R. and Matteoni, J.A. 1991. Petunia as an indicator plant for use by growers to monitor for thrips carrying the tomato spotted wilt virus in greenhouses. Plant Dis. 75: 78–82.

de Avila, A.C., deHaan, P., Kitajima, E.W., Kormelink, R., Resende, R. de O., Goldbach, R. and Peters, D. 1992. Characterization of a distinct isolate of tomato spotted wilt virus (TSWV) from *Impatiens* sp. in the Netherlands. J. Phytopathol. 134:133–151.

German, T.L., Ullman, D.E. and Moyer, J.W. 1992. Tospoviruses: diagnosis, molecular biology, phylogeny, and vector relationships. Annu. Rev. Phytopathol. 30:315–348.

Hausbeck, M.K., Welliver, R.A., Derr, M.A. and Gildow, F. 1992. Tomato spotted wilt virus in greenhouse ornamentals in Pennsylvania. Plant Dis. 76:795–800.

Herold, F. 1964. Natural infection of Impatiens sultani with cucumber mosaic virus. Plant Dis. Rep. 48:603–605.

Hsu, H.T. and Lawson, R.H. 1991. Direct tissue blotting for detection of tomato spotted wilt virus in *Impatiens*. Plant Dis. 75:292–295.

Koenig, R. and Lesemann, D.-E. 1983. Helenium virus S. CMI/AAB. Descriptions of Plant Viruses, No. 265.

Law, M.D. and Moyer, J.W. 1990. A tomato spotted wilt-like virus with a serologically distinct N protein. J. Gen. Virol. 71:933–938.

Lawson, R.H., Dienelt, M.M. and Hsu, H.T. 1993. Effects of passaging a defective isolate of impatiens necrotic spot virus at different temperatures. Phytopathology 83:662–670.

Lockhart, B.E.L. and Betzold, J.A. 1980. Latent infection of impatiens by a potexvirus. Acta Hortic. 110:81–83.

Lockhart, B.E.L. and Betzold, J.A. 1980. Leaf-curl of impatiens caused by tobacco streak virus infection. Plant Dis. 64:289–290.

Lockhart, B.E.L. and Pfleger, F.L. 1979. Identification of a strain of tobacco ringspot virus causing a disease of impatiens in commercial greenhouses. Plant Dis. Rep. 63:258–261.

Pleše, N., Erić, Ž. and Krajačić, M. 1988. Further information of infection of *Impatiens holstii* with Helenium virus S. Acta Hortic. 234:477–484.

Urban, L.A., Huang, P.-Y. and Moyer, J.W. 1991. Cytoplasmic inclusions in cells infected with isolates of L and I serogroups of tomato spotted wilt virus. Phytopathology 81: 525–529.

Pot and Bedding Plants

49

Kalanchoë blossfeldiana

Karen Bech and Karen Husted
Danish Institute of Plant and Soil Science,
Department of Plant Pathology and Pest Management, Lyngby, Denmark

Introduction

The *Kalanchoë* production in 1993 has been estimated to amount to approximately 120 million potted plants. The cultivation takes place mainly in Europe and the US. The Netherlands and Denmark contribute the largest number of plants, 50 and 35 million, respectively. *Kalanchoë* is the most important potted crop in Denmark in terms of the number being produced, and the second most important when market value is considered. The great interest in the production of *Kalanchoë* started in Denmark in the late 1960s.

K. blossfeldiana cultivars constitute approximately 98% of the *Kalanchoë* production worldwide, and the remaining 2% consists of other *Kalanchoë* species and *Kalanchoë* hybrids, e.g. "Tessa" (*K. gracilipes* × *K. manginii*).

Kalanchoë is a succulent member of the family *Crassulaceae*. The genus originates from tropical Africa, mainly from Madagascar where several species grow natively. The present form of *K. blossfeldiana* is a result of mutations, selection, and cross-fertilization among cultivars of the original species introduced into European botanical gardens and brought to flower in 1927 when Robert Blossfeld, a dealer in seeds, became interested in the horticultural value of this species and improvement by breeding began (Anon., 1958). *K. blossfeldiana* is vegetatively propagated by cuttings. Reproduction by seeds is now only used in breeding.

Viruses are often found in a relatively low concentration and are generally unevenly distributed in *Kalanchoë* which makes the testing more difficult.

The Viruses

Tomato spotted wilt (TSWV)/Impatiens necrotic spot (INSV) tospoviruses

Symptoms

Kalanchoë blossfeldiana can be systemically infected with TSWV and INSV. TSWV infected plants show diffuse chlorotic spots on young leaves (Fig. 49.1a; see color plate). Besides chlorosis, necrosis is sometimes seen on leaves, petioles, or stems in many cultivars (Verhoeven, 1991). Infections may also result in necrosis alone (Inge Bouwen, pers. comm.).

INSV causes yellowish chlorotic rings (Fig. 49.1b), spots, or patterns on leaves; some leaves may wilt (B. Lockhart, pers. comm.), and necrosis has been registered in a few cases (Hausbeck et al., 1992). Symptoms are often transient.

TSWV is not evenly distributed in the inoculated *Kalanchoë* plants (Paludan, 1991) and environmental factors, time of the year, etc. may influence the expression of tospovirus symptoms (German et al., 1992).

For a description of the virus, see tospovirus (Chapter 5).

Geographical distribution and damage

Tomato spotted wilt virus is widespread in tropical, temperate, and especially subtropical regions of the world where it causes comprehensive damage in an extremely wide host range including vegetables as well as ornamental plants and weeds (Ie, 1970; German et al., 1992).

A fair frequency of infection is seen in *Kalanchoë* in countries where TSWV and INSV are found. Infected *Kalanchoë* plants should be destroyed.

Methods of assay

The tospoviruses are mechanically sap-transmitted to a number of indicator plants: *Nicotiana* species, *Lycopersicon esculentum*, *Petunia* hybrida, *Datura stramonium*, and *Cucumis sativus* (Ie, 1970; German et al., 1992; Vaira et al.,

Fig. 49.1b. *Kalanchoë blossfeldiana* "Attraction" inoculated with impatiens necrotic spot tospovirus (INSV). (Compare to TSWV infection, Fig. 49.1a; see color plate.)

1993). TSWV is systemic in the genus *Nicotiana* where *N. occidentalis* "P1" or *N. rustica* is a suitable host. INSV gives systemic symptoms in *N. benthamiana* (German et al., 1992; Vaira et al., 1993).

ELISA is currently the method of choice for routine detection of tospoviruses, using either monoclonal antibodies specific to capsid proteins and glycoproteins or polyclonal antibodies (German et al., 1992). TSWV and INSV are serologically distinct.

The ELISA method is very sensitive, but other testing methods (PCR, indicator plants) should also be used, e.g. to detect new strains.

Natural and experimental transmission

The spread of the virus and the thrips vectors on latently infected bedding plants or nursery stock presents a significant threat. Seven thrips species are able to acquire and transmit TSWV, whereas *F. occidentalis* is the only thrips species that has been tested as a vector for INSV in North America, probably because of dominance of this species in greenhouse crops where INSV is important (German et al., 1992). *F. occidentalis* is extremely polyphagous, and in addition to this, resistance to insecticides makes control difficult (Brødsgaard, 1989a; Marchoux et al., 1993).

TSWV and INSV can be mechanically transmitted by sap from infected plants.

Control methods

Precautions against the virus as well as its vector include quarantine of imported plant material, blue sticky traps (Brødsgaard, 1989b), and fine-meshed netting in embrasure in greenhouses combined with inspection for visual symptoms and random testing. Testing is necessary because of the transient symptom expression (Hausbeck et al., 1992), and infected *Kalanchoë* plants should be destroyed.

A breeding system where the best *Kalanchoë* clone is selected and repeatedly tested for latent infections before having the status of a nuclear stock plant is strongly recommended. In the Danish certification system nuclear stock plants are grown in a separate house under strict sanitary conditions. The plants are renewed every year and simultaneously retested for pathogens. The nuclear stock plants and the plants taken from these (elite plants) are regularly inspected for symptoms and randomly tested. In this way it is possible to secure delivery of healthy certified cuttings from specialized nurseries.

Kalanchoë top-spotting badnavirus (KTSV)

Symptoms

During periods of vigorous vegetative growth, infected *Kalanchoë* plants show coalescing chlorotic spots and ring spots on emerging leaves (Fig. 49.2). Spots usually fade as the leaves expand, but circular depressions often develop on the upper leaf surface at sites that formerly showed chlorotic spots. The intensity and duration of the spotting or mottling vary among cultivars (Hearon and Locke, 1982, 1984; Lockhart and Ferji, 1988). The symptoms are especially distinct in the spring and autumn (Hearon and Locke, 1982).

For a description of badnavirus, see Chapter 5.

Top-spotting is caused by a nonenveloped bacilliform virus 20–25 × 50–100 nm in size (Hearon and Locke, 1984) having a double-stranded DNA genome of approximately 7.3 kbp (Olszewski et al., 1989; Lockhart, 1990). A second badnavirus quite unrelated to KTSV serologically and genomically has been isolated from *K. fedtschenkoi* (Lockhart and Olszewski, 1993). When transferred to *K. blossfeldiana*, it gave symptoms similar to those described for KTSV.

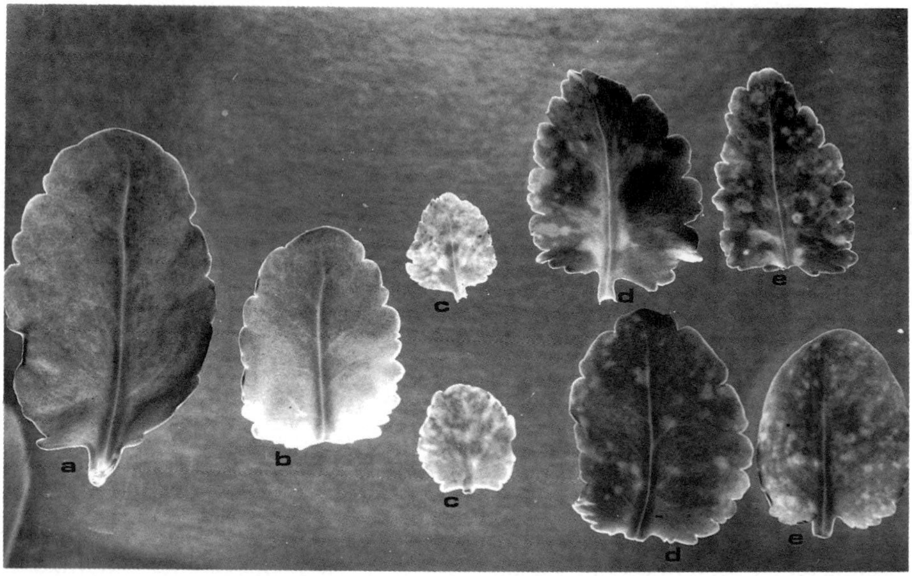

Fig. 49.2. Comparison of leaf symptoms in *Kalanchoë blossfeldiana* Poelln. (a) Kalanchoë latent carlavirus (KV-1) infected leaf, (b) healthy leaf, (c) and (d) Kalanchoë top-spotting badnavirus (KTSV) infected leaves, (e) leaves doubly infected with KTSV and KV-1 (Photo: Ferji, 1988).

Geographical distribution and damage

Symptoms of top-spotting occur regularly on *Kalanchoë* in commercial greenhouses in the US (Hearon and Locke, 1982, 1984; Lockhart and Ferji, 1988). No intensive geographical studies have been carried out, but KTSV is probably found in all countries where *Kalanchoë* is grown. In countries with systematic selection of symptomless cultivars, KTSV is seldom seen.

Methods of assay

Grafting of infected scions on non-infected *K. blossfeldiana* gave rise to symptoms within 3 months (Hearon and Locke, 1984). A purified preparation of bacilliform virus particles was mechanically transmitted to *Kalanchoë* where typical top-spotting appeared on emerging leaves 12–14 days after inoculation, and particles were detected in negatively stained leaf dip preparations (Lockhart and Ferji, 1988). Hearon and Locke (1984) detected bacilliform particles in ultrathin sections. Bacilliform particles were most prevalent in vascular tissue.

The bacilliform virus did not produce symptoms in traditional diagnostic species such as *Chenopodium* and *Nicotiana* (Hearon and Locke, 1982; Lockhart and Ferji, 1988). Antiserum is not commercially available, and the preparation of a good antiserum for routine detection is complicated due to the serological variability of KTSV isolates as seen for other badnaviruses (Lockhart and Olszewski, 1994). This suggests that the PCR technique should be further investigated (Olszewski and Lockhart, 1992).

In plants infected with a badnavirus flexuous virus-like particles which are in fact proteinaceous, structures are seen. This occurs especially if plants show pronounced symptoms but contain low levels of normal bacilliform particles. These structures with rounded ends are 13–15 nm in width, vary greatly in length, and are serologically unrelated to the capsid protein of the badnavirus (Lockhart and Olszewski, 1994).

Natural and experimental transmission

Top-spotting is transmitted by infected cuttings, pollen and seed, and by grafting (Hearon and Locke, 1982, 1984). Transmission experiments with mealybugs, particularly *Planococcus citri*, were successful in two *Kalanchoë* species: *K. daigremontiana* and *K. fedtschenkoi* (Lockhart and Olszewski, 1994).

Control methods

Methods for routine testing are not available, and therefore selection of symptomless plants is recommended.

Kalanchoë mosaic potyvirus (KMV)

Symptoms

In *K. blossfeldiana* a systemic mosaic is registered in many cultivars (Fig. 49.3). Severity varies among cultivars from a mild mosaic in young leaves to a pronounced light green mosaic with characteristic "green islands" in other cultivars.

Symptom expression varies with season being most distinct in periods of vigorous growth and in young plants. Cultivars with dark green thick-skinned leaves are more likely than others to be symptomless (Husted et al., 1994). Infected plants often show reduced growth compared to non-infected ones.

For a description of potyvirus, see Chapter 5.

KMV particles have a mean length of 724 nm, and the capsid protein has a M_r of 33 kDa (Husted et al., 1994).

Fig. 49.3. *Kalanchoë blossfeldiana* "Attraction". Left: healthy leaf. Right: leaf systemically infected with kalanchoë mosaic potyvirus (KMV).

Geographical distribution and damage

KMV is registered in European countries growing *K. blossfeldiana* (Husted et al., 1994), and the potyvirus seen in the US (Hearon, 1981) is probably identical to KMV. Infection by this virus indirectly reduces the quality of the plants.

Methods of assay

Grafting of healthy *K. blossfeldiana* cuttings to symptomless infected plants was previously used as an assay for KMV (Paludan, 1991). After purification of the virus, antiserum was raised against KMV, and now ELISA is recommended for routine testing (Husted et al., 1994). Another assay is sap-transmission to virus free *K. blossfeldiana* "Attraction" where symptoms appear after about 3 weeks. KMV also infects *C. quinoa* and *C. amaranticolor* locally, but transmission from *K. blossfeldiana* often results in only a few chlorotic lesions due to, among other things, a low concentration and uneven distribution of the virus in *Kalanchoë*.

Natural and experimental transmission

The KMV virus is mainly transmitted through infected cuttings and is mechanically transmitted by sap. Aphids transfer the virus in a non-persistent manner.

Several *Kalanchoë* spp. inoculated with KMV are latently infected, e.g. *K. tubiflora, K. daigremontiana*, and *K. fedtschenkoi* (Husted and Bech, unpublished data) and therefore could be a source of infection.

Control methods

By selection of symptomless plants and repeated testing of these mother plants by ELISA or inoculation to *K. blossfeldiana* "Attraction", damage by KMV may be prevented.

Viruses of minor importance

Kalanchoë *latent carlavirus (KV-1 and KV-2)*

Kalanchoë latent virus is known to cause symptomless infection in several cultivars of *K. blossfeldiana* worldwide (Hearon, 1982, 1984; Lockhart and Ferji, 1988). Two strains have been identified: KV-1 and KV-2 (Hearon, 1984). KV-1 is often found in combination with other viruses, and its concentration varies between cultivars (Hearon, 1981; Lockhart and Ferji, 1988).

Double infection by KV and another virus might intensify the symptoms of the second virus (Inge Bouwen, pers. comm.), which is demonstrated by the leaves doubly infected with KV-1 and KTSV in Fig. 49.2.

For a description of carlavirus, see Chapter 5.

KV-1 and KV-2 particles have a mean length of 620 and 650 nm, respectively (Hearon, 1984).

Methods of assay

The two strains of Kalanchoë latent virus are biologically and serologically distinguishable (Hearon, 1985). KV-1 gives rise only to local symptoms in *Chenopodium* spp. while KV-2 causes local as well as systemic infections in *Chenopodium* spp. (Hearon, 1984). *C. quinoa* is the preferred host (Hearon, 1985); however, the first transmission from *K. blossfeldiana* often results in only a few lesions in *C. quinoa*.

ELISA and EM tests are reliable detection methods, although the KV virus concentration in *Kalanchoë* leaf tissue is low (Hearon, 1985).

Natural and experimental transmission

The virus is spread by naturally infected cuttings and is mechanically transmitted through infected sap, e.g. on contaminated tools. Carlaviruses are normally aphid transmitted, but information on KV aphid transmission is lacking.

Control methods

A testing and certification system like the one mentioned under TSWV is recommended. It is possible to inactivate KV by heat treatment and meristem tip culture (Paludan, 1985).

Other viruses of minor importance

Kalanchoë spp. are naturally infected or susceptible to infection by other viruses, including alfalfa mosaic virus (unpublished, in Hearon, 1984), a strain of tobacco streak ilarvirus (R.W. Fulton, pers. comm., in Hearon 1984; B. Lockhart, pers. comm.), potato yellow dwarf rhabdovirus (Francki, 1973), and another yet uncharacterized rhabdovirus serologically unrelated to other rhabdoviruses (Inge Bouwen, pers. comm.; B. Lockhart, pers. comm.). Finally isometric virus-like particles have been isolated from three different *Kalanchoë* species (Izaguirre-Mayoral et al., 1990).

Selected Literature

Anon. 1958. *Kalanchoë* Adans. In: F. Encke (ed.), Pareys Blumengärtnerei. Beschreibung Kultur und Verwendung der gesamten gärtnerischen Schmuckpflanzen. Paul Parey, Berlin, Hamburg, pp. 721–724.

Brødsgaard, H.F. 1989a. *Frankliniella occidentalis* (*Thysanoptera*; *Thripidae*) — a new pest in Danish glasshouses. A review. Danish J. Plant Soil Sci. 93:83–91.

Brødsgaard, H.F. 1989b. Coloured sticky traps for *Frankliniella occidentalis (Thysanoptera: Thripidae)* in glasshouses. J. Appl. Entomol. 107:136–140.

Ferji, Z. 1988. Role of a small bacilliform virus in the etiology of top-spotting disease of *Kalanchoë blossfeldiana*, and some properties of the virus. MS Thesis, University of Minnesota.

Francki, R.I.B. 1973. Plant rhabdoviruses. Adv. Virus Res. 18:257–345.

German, T.L., Ullman, D.E. and Moyer, J.W. 1992. Tospoviruses: diagnosis, molecular biology, phylogeny, and vector relationships. Annu. Rev. Phytopathol. 30:315–348.

Hausbeck, M.K., Welliver, R.A., Derr, M.A. and Gildow, F.E. 1992. Tomato spotted wilt virus survey among greenhouse ornamentals in Pennsylvania. Plant Dis. 76:795–800.

Hearon, S.S. 1981. Detection of viruses in ultrastructural studies of naturally infected *Kalanchoë blossfeldiana*. Phytopathology 71:767. (Abstr.)

Hearon, S.S. 1982. A carlavirus from *Kalanchoë blossfeldiana*. Phytopathology 72:838–844.

Hearon, S.S. 1984. Comparison of two strains of *Kalanchoë* latent virus, carlavirus group. Phytopathology 74:670–676.

Hearon, S.S. 1985. Detection of *Kalanchoë* latent virus in *Kalanchoë blossfeldiana* by enzyme-linked immunosorbent assay (ELISA). Acta Hortic. 164:77–84.

Hearon, S.S. and Locke, J.C. 1982. Graft and seed transmission of the causal agent of top spotting in *Kalanchoë blossfeldiana*. Phytopathology 72:990. (Abstr.)

Hearon, S.S. and Locke, J.C. 1984. Graft, pollen, and seed transmission of an agent associated with top spotting in *Kalanchoë blossfeldiana*. Plant Dis. 68:347–350.

Husted, K., Bech, K., Albrechtsen, M. and Borkhardt, B. 1994. Identification, partial sequencing, and detection of a potyvirus from *Kalanchoë blossfeldiana*. Phytopathology 84:161–164.

Ie, T.S. 1970. Tomato spotted wilt virus. CMI/AAB Descriptions of Plant Viruses, No. 39.

Izaguirre-Mayoral, M.L., Carballo, O. and Gil, F. 1990. Purification and partial characterization of isometric virus-like particles in *Kalanchoë* species. J. Phytopathol. 130: 303–311.

Lockhart, B.E. 1990. Evidence for a double-stranded circular DNA genome in a second group of plant viruses. Phytopathology 80:127–131.

Lockhart, B.E. and Ferji, Z. 1988. Purification and mechanical transmission of *Kalanchoë* top-spotting-associated virus. Acta Hortic. 234:72–77.

Lockhart, B.E. and Olszewski, N.E. 1993. Serological and genomic heterogenity of banana streak badnavirus: implications for virus detection in *Musa* germplasm. In: J. Ganry (ed.), Breeding Banana and Plantain for Resistance to Diseases and Pests, CIRAD-INRA/INIRAP, Montpellier, France.

Lockhart, B.E. and Olszewski, N.E. 1994. Plant badnavirus group. In: A. Granoff and R.G. Webster (eds.), Encyclopedia of Virology, Vol. 1, Academic Press, New York, pp. 139–143.

Marchoux, G., Gébré-Sélassié, K. and Berling, A. 1993. Le tomato spotted wilt virus. Evolution du problème et des acquis de la recherche. Phytoma 449:35–40.

Olszewski, N.E. and Lockhart, B.E. 1992. Use of PCR technology for the detection and analysis of badnaviruses. Phytopathology 82:1087 (Abstr.).

Olszewski, N.E., Medberry, S.L. and Lockhart, B.E. 1989. Molecular characterization of non-enveloped bacilliform viruses. Phytopathology 79:1158–1159 (Abstr.).

Paludan, N. 1985. Inactivation of *Kalanchoë* latent virus strain 1 in *Kalanchoë* using heat treatment and meristem tip culture. Danish J. Plant Soil Sci. 89:191–195.

Paludan, N. 1991. Virus diseases of ornamental plants, In: Plant Diseases, Pests and Weeds in Denmark 1991, 108th Annu. Rep., Research Centre for Plant Protection, pp. 18–19.

Vaira, A.M., Roggero, P., Luisoni, E., Masenga, V., Milne, R.G. and Lisa, V. 1993. Characterization of two tospoviruses in Italy: tomato spotted wilt and impatiens necrotic spot. Plant Pathol. 42:530–542.

Verhoeven, J.Th.J. 1991. Tomatebronsvlekkenvirus breidt uit. Vakblad Bloemisterij 20: 39–42.

50

Pelargonium

A. Franck and G. Loebenstein
*Department of Virology, Agricultural Research Organization, The Volcani Center
Bet Dagan 50250, Israel*

Introduction

The commonly named geranium, cultivated by horticulturists, belongs to the genus *Pelargonium*. There are more than 200 species, 90% of them originating in the Cape region of South Africa. The cultivated pelargoniums are all hybrids created during a period of almost 300 years. The cultivated pelargoniums belong to three subgenera: *Pelargonium* × *hortorum*, commonly named geranium, originating from crosses between *Pelargonium zonale* and *P. inquinans*; *Pelargonium* × *Hederaefolium*, commonly named *P. peltatum*, from crosses between *P. peltatum* and *P. lateripes*; and *Pelargonium* × *domesticum*=*P.* × *grandiflorum* (Regal

Contribution No. 1365-E, 1994 series, from the Agricultural Research Organization, The Volcani Center, Bet Dagan, Israel.

Geraniums). In 1870, Paul Bruant selected from *P. hortorum* a tetraploid mutant (Gros Bois) from which most cultivated *P.* × *hortorum* originated.

Pelargonium is probably the largest selling garden ornamental crop in the world. The wholesale value of vegetatively and seed-propagated pelargonium in the US reached approximately $160 million at the beginning of this decade. According to the Agricultural Statistics Board of the USDA (1991), some 111.6 million potted geraniums, valued at approximately $130 million, were sold in 1990, in addition to 3.5 million flats of bedding geraniums, valued at $30 million. In 1992 more than 300 million plants were sold. The total European production in 1993 is estimated at approximately 400 million finished plants, with a wholesale value of $300–350 million (A. Malter, personal communication).

Pelargonium (geranium) is propagated mainly vegetatively by cuttings, although in recent years some geraniums, especially for small pots and market packs, have been grown from seed. Consequently, viral diseases have been and remain a threat to the production and quality of the crop. Economic losses resulting from geranium viruses are often difficult to assess, and symptoms depend on environmental conditions. However, with the foreseen changes in the regulations of the European Union (EU), regarding movement of plant material to and within the EU, improved programs for the production of certified pathogen-tested plants, including viruses, will be required. The European and Mediterranean Plant Protection Organization (EPPO) has proposed guidelines which will serve as the basis for the EU regulations (EPPO, 1992). For these guidelines the following eight viruses should be considered during the certification program: pelargonium flower break virus (PFBV), pelargonium line pattern virus (PLPV), pelargonium leaf curl virus (PLCV), pelargonium ringspot virus (PRSV), cucumber mosaic virus (CMV), tobacco ringspot virus (TobRSV), tomato blackring virus (TBRV), and tomato spotted wilt virus (TSWV). The first three are specific to pelargonium, while the last four are not limited to pelargonium and have a wide host range. PRSV is a catch-all name for a group of diseases caused by different viruses (see below). In addition to these, other virus and virus-like diseases have been reported to occur in pelargonium (Stone, 1980; Nameth and Adkins, 1993).

Surveys in Danish and Belgian geranium nurseries showed the presence of CMV, PFBV, tobacco mosaic virus (TMV) and TobRSV (Paludan, 1976; Welvaert et al., 1982). Approximately 70% of *Pelargonium zonale* plants tested in Belgian nurseries were found to be virus-infected. Surveys of geranium nurseries in Israel during 1991/92 showed infection rates of 80% by PFBV in several of the nurseries which had received planting material from Europe (Franck and Loebenstein, 1994).

The Viruses

Pelargonium flower break carmovirus (PFBV)

This virus is widely distributed in the world and has been reported to occur in European countries, the US, and Israel, but is probably present wherever geraniums are grown. It is found mainly in *Pelargonium zonale*, but we found PFBV also in some *P. peltatum* plants. In the Netherlands, also, PFBV was found in *P. peltatum* (Bouwen and Maat, 1992).

Symptoms

Symptoms include mild foliar mottling, chlorotic spots (Fig. 50.1) and flower breaking on both sides of the petals in pink varieties (Fig. 50.2; see color plate). On red varieties flower break symptoms can sometimes be seen on the lower side of the petals.

Causal agent

Particles of PFBV are isometric, 28–30 nm in diameter (Hollings and Stone, 1974; Franck and Loebenstein, 1994). The viral coat contains one protein of approximately 33 kD, and from dsRNA preparations of PFBV- infected plants three bands with M_r of 3.6, 2 and 1.7 kbp were obtained (Franck and Loebenstein, 1994). Purified virus preparations give an $OD_{260}/_{280}$ ratio of 1.75, corrected for light scattering. The extinction coefficient at 260 nm is 4.5–5.0 $(mg/ml)^{-1}$ cm^{-1}.

Fig. 50.1. Chlorotic spots and mottling associated with pelargonium flower break virus (PFBV) on cv. Bruni.

Transmission

The virus can be transmitted by the thrips *Frankliniella occidentalis* (Krczal et al., 1995) by grafting to pelargonium plants and mechanically to various herbaceous plants. For mechanical transmission, leaf and petioles are ground 1:10 (w/v) in 0.15 M phosphate buffer, pH 7, containing 4% polyethylene glycol (PEG 6000). Alternatively, infected leaf material is triturated in 0.05 M Na_2HPO_4-KH_2PO_4 buffer (3 ml/g leaves) containing 4% PEG 6000, pH 7.6, and 100 mg ml^{-1} of activated charcoal. The crude extract can be clarified by centrifugation. After inoculating the plants it is advisable to rinse them with tap water.

PFBV can be transmitted mechanically to *Chenopodium quinoa*, resulting in chlorotic lesions 5–6 days after inoculation, some of which later become necrotic. PFBV can also be transmitted to *Nicotiana clevelandii*, causing a yellowing of the inoculated leaves. The virus is able to infect plants through the roots via recirculating irrigation systems (Krczal et al., 1995).

Identification and assay methods

PFBV can be detected by enzyme-linked immunosorbent assay (Sandwich ELISA), using 0.1 M Tris (hydroxymethyl) aminomethane-HCl buffer, pH 8.6, containing 0.2% Tween 20, 0.2% PVP 40 T and 0.4% BSA (Franck and Loebenstein, 1994). As stated above, *C. quinoa* reacts to infection by PFBV with chlorotic local lesions. These, however, do not enable identification of PFBV, as *C. quinoa* responds similarly to many viruses. Inoculation of *C. quinoa* is nevertheless used for testing pelargonium nuclear stocks for the presence of viruses.

Purification

PFBV can be purified from inoculated leaves of *N. clevelandii* as described by Franck and Loebenstein (1994).

Pelargonium line pattern virus (PLPV)

Pelargonium line pattern virus (PLPV) has been reported to occur in the United Kingdom, Germany, Yugoslavia and Israel.

Symptoms

Symptoms include line patterns along the veins and veinal flecks. However, symptoms often are not apparent and inoculated pelargonium plants may remain symptomless for long periods. In Israel, no symptoms were observed in PLPV-infected plants. PLPV induces intracellular changes similar to those caused by some dianthoviruses (Krajacic and Plese, 1991).

Causal agent

In extracts from lesions on *C. quinoa*, inoculated with PLPV, isometric particles with diameters of approximately 27 and 30 nm were observed. The viral coat protein contains one protein of approximately 34.5 kD, and dsRNA preparations from PLPV-infected *C. quinoa* plants gave two bands with M_r of 3.6 and 2 kbp. Purified virus preparations gave an $OD_{260}/_{280}$ ratio of 1.75, corrected for light scattering. The extinction coefficient at 260 nm is 5–6.5 $(mg/ml)^{-1}$ cm^{-1} (Bouwen and Maat, 1992; Franck and Loebenstein, 1994). PLPV seems to be closely related to pelargonium ring pattern virus (Bouwen and Maat, 1992).

Transmission

There is no evidence of insect transmission. The virus can be transmitted by grafting to pelargonium and mechanically to various herbaceous plants. For mechanical transmission the same procedures as for PFBV (see above) can be used. PLPV can be transmitted mechanically to *C. quinoa*, causing relatively fast growing lesions which became necrotic 6 days after inoculation; and to cowpeas, resulting in chlorotic lesions. PLPV also infects *Nicotiana tabacum cvs.* "White Burley" and "Samsun", *N. sylvestris, N. clevelandii, N. rustica, Datura stramonium, Physalis floridana, Phaseolus vulgaris* and *Pisum sativum* (Bouwen and Maat, 1992; Franck and Loebenstein, 1994).

Identification and assay

PLPV can be detected by ELISA using the same buffer as for PFBV (see above). Mechanical inoculation to *C. quinoa* results in relatively fast growing lesions that become necrotic.

Purification

PLPV can be purified from inoculated leaves of *N. clevelandii* or *C. quinoa* as described by Franck and Loebenstein (1994).

Pelargonium ringspot disease (PRSV)

Pelargonium ringspot disease (PRSV) is a "catch-all" name for a group of diseases which have similar symptoms but which are caused by different viruses (Romaine and Lawson, 1982). PRSV has been reported to occur in Canada, the US, England, and Sweden, but probably exists also in many other locations where pelargonium is grown.

Symptoms

Symptoms include chlorotic rings (Fig. 50.3; see color plate), line patterns, and yellowish spots, which may appear on only a few leaves of the plant. Symptoms disappear when plants age.

Causal agent

PRSV has been associated with infection by either tomato ringspot virus (TomRSV) (Reinert et al., 1963; Paludan, 1976) or TobRSV (Kemp, 1967). Both TomRSV and TobRSV are nepoviruses, with isometric particles 25 and 25–29 nm in diameter, respectively. Particles sediment as three components (see Chapter 15). It is possible, however, that other viruses in addition to TomRSV and TobRSV are involved in the pelargonium ringspot disease. Thus, Stone (1980) purified a spherical virus from pelargoniums with ringspot symptoms.

Transmission

Both TobRSV and TomRSV can be transmitted to pelargonium by grafting. Both viruses are transmitted by nematodes, but no reports on nematode transmission in pelargonium have come to our attention. The two viruses can be transmitted mechanically to herbaceous plants, but only with difficulty, due to the low concentration of the viruses in pelargonium. For mechanical transmission a phosphate buffer containing PEG and charcoal, as outlined above, can be used. Both viruses can be transmitted to *C. quinoa*, causing chlorotic or necrotic local lesions. On this host TomRSV gives a systemic apical necrosis while TobRSV usually gives no systemic infection. From *C. quinoa* the two viruses can be transmitted to other hosts. TomRSV causes local necrotic flecks and later systemic mottle and necrosis on tomato, while TobRSV infects tomato only with difficulty, causing small necrotic spots on inoculated leaves and later systemic vein necrosis and epinasty. *N. clevelandii* and *N. tabacum* are good sources for purification of both viruses.

Transmission of these two viruses through pelargonium seeds and of TomRSV by pollen has been reported (Scarborough and Smith, 1977).

Identification and assay

Identification of PRSV in pelargonium is still difficult, probably due to the low concentration of the virus in the plant. Detection by ELISA is therefore often erratic. Furthermore, with TomRSV there are several serogroups and the serological reagents often do not detect a broad spectrum of isolates. Some of the antisera are also of low potency. TomRSV in pelargonium can also be detected with monoclonal antibodies, although polyclonal antibodies generally reacted better than the monoclonals (Powell, 1990). Dot immunobinding assay (DIA) with polyclonal antisera was more sensitive than ELISA in discriminating between healthy and infected pelargonium (Powell, 1987). Improved assay methods are therefore required and can be achieved perhaps by developing sensitive nucleic acid probes or immunocapture polymerase chain reaction (PCR) technologies.

C. quinoa reacts to infection by TomRSV and TobRSV with chlorotic or necrotic local lesions. However, as this reaction is not specific to these viruses, it is used only for testing nuclear stocks for the presence of viruses.

Purification

For purification of TobRSV, see the method of Rezaian and Francki (1973); and for TomRSV, see the method of Stace-Smith (1984).

Pelargonium leaf curl tombusvirus (PLCV)

The leaf curl disease, also called crinkle, is widespread in Europe and the US. In Israel it was identified in only one case, perhaps because of the sensitivity of the virus to high temperatures, which are prevalent there in several months of the year. The disease occurs in *P. zonale, P. peltatum* and *P. domesticum*.

Symptoms

Symptoms are expressed only at the end of winter, after a cold season. Small stellate yellow spots which later become necrotic appear on the young leaves. The necrotic zone later falls out, leaving “shot-holes”. The leaves become wrinkled and curled. Leaves developing in summer and autumn do not show symptoms and pelargonium plants grown at high temperatures do not express symptoms in the following winter (Hollings, 1962).

Causal agent

Particles of PLCV are isometric, 28 nm in diameter. The viral coat contains one protein of approximately 37.5 kD (unpublished) and single-stranded RNA. The molecular weight of the virus particle is $8.5–9.6 \times 10^6$ kD and its sedimentation coefficient is 125–129 $S_{20,w}$ (Molina-Garcia et al., 1990). Purified preparations give an $OD_{260}/_{280}$ ratio of 1.55, corrected for light scattering.

The virus is related to the Tombus group, giving a serological cross-reaction with TBSV (Hollings and Stone, 1965). Similarly, PLCV resembles TBSV in its physicochemical (Molina-Garcia et al., 1990) and intracellular properties (Martelli and Russo, 1972).

Transmission

The virus can be mechanically transmitted from pelargonium leaves showing symptoms to *Vigna unguiculata* (cowpea) (brown lesions), *C. quinoa*, and *C. amaranticolor* (necrotic and faint necrotic lesions, respectively) by procedures described for PFBV. The virus can easily be subtransmitted to *Gomphrena globosa* (brown-ringed necrotic lesions), *P. vulgaris* (brown lesions), and *D. stramonium* (chlorotic lesions). *Nicotiana clevelandii* is a good propagation species but with some strains of the virus the leaves collapse three days after inoculation.

The virus is graft-transmissible from pelargonium to pelargonium if grafting is done in winter. There is no indication of transmission by aphids, whiteflies, or seeds.

Identification and assay methods

PLCV can be detected in diseased pelargonium only if the plants are tested in winter when the virus level is apparently high. Testing can be done by ELISA or by test plants.

Purification

Inoculated leaves of *N. clevelandii* 4 days after inoculation are homogenized as described for PFBV. After filtration through Kimwipes, the juice is emulsified with an equal volume of 1:1 butanol-chloroform. After centrifugation the aqueous phase is further treated as previously described for PFBV.

Other viruses

Tomato spotted wilt tospovirus

This virus (Tospo group, see Chapter 5) has been reported to occur in pelargonium in Europe and the US in both *P. peltatum* and *P. zonale*. Due to the spread of TSWV by the thrips *Frankliniella occidentalis* in many countries, it is expected that TSWV incidence in pelargonium will increase. TSWV is on the list of the EPPO (1992) certification scheme.

On infected pelargonium leaves, chlorotic and brown rings are observed (Fig. 50.4; see color plate). The virus is transmitted by grafting and in a persistent manner by several thrips species. It can be transmitted mechanically but not by contact between plants.

Testing methods for TSWV by ELISA are not fully reliable and generally give positive reactions only with leaves showing symptoms. For ELISA, antisera to several TSWV strains and impatiens necrotic spot virus should be used. TSWV can be mechanically transmitted to *Petunia hybrida*, giving necrotic local lesions without systemic infection. On *Nicotiana glutinosa* or *N. clevelandii*, necrotic lesions are observed on the inoculated leaves, with subsequent systemic necrotic patterns and leaf deformation on the upper leaves.

Pelargonium tissue should be ground in inoculation buffer (1 g leaves in 5 ml buffer). The inoculation buffer consists of 0.02 M phosphate buffer, pH 7.4, containing 2% (w/v) polyvinylpyrrolidone (PVP) and 1% (w/v) Na_2SO_3. Mechanical transmission, however, is also problematic, due to the uneven distribution of the virus in the plant. Visual inspections therefore have to be combined with the test methods.

Cucumber mosaic cucumovirus

This virus (Cucumo group, see Chapter 5), thought to be the cause of pelargonium leaf breaking (Reinert et al., 1963), has been reported from Denmark, Spain, and the US and is on the list of the EPPO (1992) certification scheme. Symptoms include a reduction or change in the anthocyanin pigments of the leaves, the normal zonal pattern disappears, and purple spotting appears along the veins. CMV has also been associated with stunting, mosaic symptoms, chlorotic mottling and vein clearing (Paludan, 1976) (Fig. 50.5), symptoms differing from leaf breaking.

Testing for CMV is done by ELISA and by inoculation to *C. quinoa* and *V. unguiculata* causing chlorotic or reddish necrotic lesions, respectively; or to *Cucumis sativus*, causing systemic mosaic.

CMV is transmitted non-persistently by aphids as well as mechanically.

Pelargonium curly top

This disease, which causes leaf cupping, is rarely seen. It has been reported from Ontario, Canada, and California. The disease is transmitted by the leafhopper *Circulifer tenellus*.

Fig. 50.5. Chlorotic mottling associated with cucumber mosaic virus (courtesy of Drs. N. Paludan and J. Begtrup, Denmark).

Pelargonium ring pattern

Pelargonium ring pattern virus (PelRPV) has been reported from England, especially in older pelargonium cultivars. On *P. zonale* yellow ringspots are observed on older leaves. No vector has been discovered (Stone, 1980). Recently, Bouwen and Maat (1992) found a close serological relationship between PelRPV and pelargonium line pattern virus.

Yellow net vein

Yellow net vein has been reported from Canada, Denmark, Israel, and the US. As its name implies, yellowing appears along the veins (Fig. 50.6). The causal agent is not known, but the disease can be transmitted by grafting.

Pelargonium zonate spot

This disease causes concentric chrome yellow spots on *P. zonale*. It has been reported from southern Italy. The virus can be transmitted to a wide range of test plants and also infects tomato in the field (Gallitelli, 1982). The virus particles are isometric, ranging from 25–35 nm in diameter. By some of its biological and physicochemical properties the virus could be related to the Ilar group.

Pelargonium vein clearing

This disease has been reported from southern Italy (Russo et al., 1979). The virus belongs to the Rhabdo group.

Fig. 50.6. Vein-netting symptoms in pelargonium cv. Palais.

Control

Heat treatment and meristemming

Virus-tested plants can be obtained by seedling propagation, but more often from meristem tips after heat treatment of the plant. Plants are exposed to alternating day (16 h) and night (8 h) temperatures beginning at 24 °C and 21 °C, respectively. Every 24 h the temperatures are raised by 3 degrees until 38°C and 35°C, respectively, are reached. Plants are kept under these conditions for 4 weeks, after which meristem tips are removed for tissue culture. Plants are raised from these cultures and are then tested carefully for viruses as nuclear stock candidate plants.

Certification scheme

The main, if not only, way to control viruses in pelargonium is to use virus-tested propagation material. According to the EPPO (1992) scheme, propagation material is produced in three stages: nuclear, propagation, and certified stocks.

Production of nuclear stocks

The candidate material for nuclear stock status should be kept under quarantine in an isolated, aphid-proof house, separately from the nuclear stock. All plants are individually tested for CMV, PFBV, PLCV, PLPV, PRSV, TobRSV, TomRSV, TSWV and TBRV, in addition to tests for *Xanthomonas campestris pelargonii* and *Verticillium* spp., as well as being subjected to visual inspections for trueness-to-type. It is advisable to conduct at least two to three tests during each of two growing seasons. All plants should be grown in individual pots. Sometimes meristemming is necessary to obtain virus-free candidate plants.

Cuttings from plants passing this stage are planted in the nuclear stock aphid-proof house. They should be maintained under the same conditions and with the same checks on freedom from pathogens as candidate nuclear stock plants. The useful life of a nuclear stock plant of pelargonium generally does not exceed 18 months. Cuttings taken from nuclear stock plants can also be considered as nuclear stock, provided they are grown under the same conditions and are individually tested at least for PFBV and PLPV and also for *X. campestris*. If nuclear stock plants are kept for over 12 months, they should be individually retested once a year for the pathogens just mentioned. Any plant giving a positive result or showing symptoms of any disease (fungal, bacterial, viral) should be rejected.

Propagation stock

Propagation stock nurseries are planted from nuclear stock cuttings. The plants should be kept in isolated houses and grown either in individual containers or in a system of small growing units ensuring adequate isolation. The filiation of the plants should be recorded so that each lot is known to be derived from nuclear stock.

General precautions against diseases should be maintained. Plants should be randomly tested for PFBV and *X.c. pelargonii* and examined visually for the other pathogens. Any plant testing positive or showing symptoms should be eliminated.

The number of generations of propagation stock should not exceed two and the useful life of a propagation plant generally does not exceed 18 months.

Certified stock

The cuttings taken from the propagation stock become the mother plants for the certified stock (cuttings) for sale. General precautions against diseases should be maintained. The plants should be checked regularly by visual inspection. Any plant showing symptoms of disease (fungal, bacterial, viral) should be eliminated. The useful life of these mother plants generally does not exceed one year.

Such schemes are presently in operation in several countries. In Israel a propagation facility working by this scheme was able to produce certified plant material with very low virus incidence (unpublished).

Remarks

Pelargonium ringspot is probably one of the more difficult diseases to control. The causal agents have not been characterized completely, serological methods are only partially reliable, and the concentration of the causal agents is low and fluctuates. It is advisable to remove symptom-expressing plants and to test nuclear stocks carefully both by ELISA and by inoculation to *C. quinoa*.

PLCV is not a threat to commercial cultivation of pelargonium as heat therapy and the use of virus-indexed propagation material eliminate infection.

Acknowledgement

This research was supported in part by a grant from the Chief Scientist, Israeli Ministry of Agriculture, and by the United States-Israel (Binational) Agricultural Research and Development Fund (BARD).

Selected Literature

Bouwen, I. and Maat, D.Z. 1992. Pelargonium flower-break and pelargonium line pattern viruses in the Netherlands; purification, antiserum preparation, serological identification, and detection in pelargonium by ELISA. Neth. J. Plant Pathol. 98:141–156.

European and Mediterranean Plant Protection Organization (EPPO). 1992. Certification scheme. Pathogen tested material of pelargonium. EPPO Bull. 22:285–296.

Franck, A. and Loebenstein, G. 1994. Virus and virus-like diseases of pelargonium in Israel. Acta Hortic. 377:31–39.

Gallitelli, D. 1982. Properties of a tomato isolate of pelargonium zonate spot virus. Ann. Appl. Biol. 100:457–466.

Hollings, M. 1962. Studies of pelargonium leaf curl virus. I. Host range, transmission and properties in vitro. Ann. Appl. Biol. 50:189–202.

Hollings, M. and Stone, O.M. 1965. Studies of pelargonium leaf curl virus. II. Relationships to tomato bushy stunt and other viruses. Ann. Appl. Biol. 56:87–98.

Hollings, M. and Stone, O.M. 1974. Pelargonium flower break virus. CMI/AAB Descriptions of Plant Viruses, No. 130.

Kemp, W.G. 1967. Natural occurrence of tobacco ringspot virus in pelargonium in Ontario. Can. J. Plant Sci. 47:295–300.

Krajacic, M. and Plese, N. 1991. Some molecular characteristics of pelargonium line pattern virus. Acta Bot. Croat. 50:1–5.

Krczal, G., Albouy, J., Damy, I., Kusiak, C., Moreau, J.P., Berkelman, B. and Wohanka, W. 1995. Transmission of pelargonium flower break virus (PFBV) in irrigation systems and by thrips. Plant Dis. 79:163–166.

Martelli, G.P. and Russo, M. 1972. Pelargonium leaf curl virus in host leaf tissues. J. Gen. Virol. 15:193–203.

Molina-Garcia, A.D., Harding, S.E. and Fraser, R.R.S. 1990. Physicochemical studies on tomato bushy stunt virus variants. Biopolymers 29:1443–1452.

Nameth, S.T. and Adkins, S.T. 1993. Viral diseases. in: Geraniums IV. J.W. White (ed.), Ball, Geneva, IL, pp. 267–275.

Paludan, N. 1976. Virus diseases in *Pelargonium hortorum* specially concerning tomato ringspot virus. Acta Hortic. 59:119–130.

Powell, C.A. 1987. Detection of three plant viruses by dot immunobinding assay. Phytopathology 77:306–309.

Powell, C.A. 1990. Detection of tomato ringspot virus with monoclonal antibodies. Plant Dis. 74:904–907.

Reinert, R.A., Hildebrandt, A.C. and Beck, G.E. 1963. Differentiation of viruses transmitted from *Pelargonium hortorum*. Phytopathology 53:1292–1298.

Rezaian, M.A. and Francki, R.I.B. 1973. Replication of tobacco ringspot virus. I. Detection of a low molecular weight double-stranded RNA from infected plants. Virology 56:238–249.

Romaine, C.P. and Lawson, R.H. 1982. Viruses. in: Geraniums III, J.W. Mastalerz and E.J. Holcomb (eds.), PA Flower Growers, University Park, PA, pp. 242–252.

Russo, M., Castellano, M.A. and Martelli, G.P. 1979. Rhabdovirus-like particles in English ivy (*Hedera helix*) and ivy-leafed geranium (*Pelargonium peltatum*). Phytopathol. Z. 96:122–131.

Scarborough, B.A. and Smith, S.H. 1977. Effects of tobacco- and tomato ringspot viruses on the reproductive tissues of *Pelargonium* x *hortorum*. Phytopathology 67:292–297.

Stace-Smith, R. 1984. Tomato ringspot virus. CMI/AAB Descriptions of Plant Viruses, No. 290.

Stone, O.M. 1980. Nine viruses isolated from pelargonium in the United Kingdom. Acta Hortic. 110:177–182.

Welvaert, W., Samyn, G. and Simpelaere, P. de. 1982. A virus survey on pelargonium in Belgian nurseries. Meded. Fac. Landbouwwet. Rijksuniv. Gent 47:1033–1038.

51

Alphabetical Listing of Viruses, Viroids, and Mycoplasma-like Organisms Infecting Ornamental Plants

J. Cohen

Department of Virology, Agricultural Research Organization, The Volcani Center, Bet Dagan 50250, Israel

Virus	Hosts (pages)
Alfalfa mosaic alfamovirus (AMV)	*Dimorphotheca sinuata* (472), Hydrangea (490), *Kalanchoë blossfeldiana* (512)
Alstroemeria carlavirus (AlCV)	Alstroemeria (245)
Alstroemeria mosaic potyvirus (AlMV)	Alstroemeria (238)
Alstroemeria streak potyvirus (AlSV)	Alstroemeria (242)
Apple mosaic ilarvirus (ApMV)	Roses (429)
Arabis mosaic nepovirus (ArMV)	Alstroemeria (248), Begonia (450), Carnation (389), Crocus (263), Gladiolus (288), Hyacinth (301), Lily (319), Narcissus (328), Nerine (340), Roses (430), Tulip (372)
Argentine plantago potexvirus (APlaV)	Aster (258)
Asclepias rhabdovirus	Asclepias (440)
Aster chlorotic stunt carlavirus (AsCSV)	Aster (258)

Virus	Hosts (pages)
Bean yellow mosaic potyvirus (BYMV)	Crocus (261), Freesia (275), Gladiolus (286), Iris (bulbous) (307), Lisianthus (445), Orchids (412)
Bidens mottle potyvirus (BiMV)	*Dimorphotheca pluvialis* (473)
Broad bean wilt fabavirus (BBWV)	Begonia (450), Iris (rhizomathus) (309), Lisianthus (445), Narcissus (324)
Carlavirus (not characterized)	Hydrangea (483)
Carnation bacilliform rhabdovirus	Carnation (389)
Carnation cryptic cryptovirus (CCV)	Carnation (389)
Carnation etched ring caulimovirus (CERV)	Carnation (387), *Dianthus barbatus* (461)
Carnation Italian ringspot tombusvirus (CIRSV)	Carnation (389)
Carnation latent carlavirus (CarLV)	Carnation (388), *Dianthus barbatus* (461)
Carnation mottle carmovirus (CarMV)	Begonia (450), Carnation (386), *Dianthus barbatus* (461)
Carnation necrotic fleck closterovirus (CNFV)	Carnation (388
Carnation ringspot dianthovirus (CRSV)	Carnation (388), *Dianthus barbatus* (461)
Carnation vein mottle potyvirus (CVMV)	Carnation (387), *Dianthus barbatus* (460)
Carnation yellow stripe necrovirus (CYSV)	Carnation (389)
Chrysanthemum B carlavirus (CVB)	Chrysanthemum (397)
Citrus tatter leaf virus (CiTLV)	Lily (319)
Clover yellow vein potyvirus (CLYVV)	Orchids (414)
Cucumber mosaic virus (CMV)	Alstroemeria (245), Anemone (252), Asclepias (440), Aster (257), Begonia (452), Carnation (390), Crocus (263), Chrysanthemum (406), Cyclamen (455), Dahlia (267), *Dianthus barbatus* (462), *Dimorphotheca sinuata* (472), Freesia (277), Gladiolus (283), Hippeastrum (Amaryllis) (294), Hyacinth (301), Hydrangea (490), Impatiens (502), Iris (rhizomatous) (309), Lily (316), Lisianthus (445), Narcissus (324), Nerine (338), Orchids (414), Pelargonium (522), Phlox (422), Ranunculus (347), Tulip (368), Zantedeschia (Calla lilies) (465)

Virus	Hosts (pages)
Cymbidium mild mosaic virus	Orchids (414)
Cymbidium mosaic potexvirus (CyMV)	Orchids (410)
Cymbidium ringspot tombusvirus (CyRSV)	Orchids (414)
Cypridium filamentous virus	Orchids (412)
Dahlia mosaic caulimovirus (DaMV)	Dahlia (266)
Dasheen mosaic potyvirus (DsMV)	Caladium (465), Diffenbachia (465), Zantedeschia (Calla lilies) (465)
Dendrobium mosaic potyvirus (DeMV)	Orchids (414)
Dendrobium rhabdovirus	Orchids (413)
Dendrobium vein necrosis virus (DVNV)	Orchids (414)
Dimorphotheca carlavirus	*Dimorphotheca sinuata* (473)
Freesia leaf necrosis virus-like agent (FLN)	Freesia (278)
Freesia mosaic potyvirus (FreMV)	Alstroemeria (248), Freesia (275)
Grammatophyllum bacilliform virus	Orchids (413)
Hibiscus chlorotic ringspot carmovirus (HCRSV)	Hibiscus (477)
Hibiscus latent ringspot nepovirus (HLRSV)	Hibiscus (479)
Hippeastrum mosaic potyvirus (HiMV)	Hippeastrum (Amaryllis) (294)
Hippeastrum streak virus	Hippeastrum (Amaryllis) (296)
Hyacinth mosaic potyvirus (HyMV)	Hyacinth (299)
Hydrangea mosaic ilarvirus (HdMV)	Hydrangea (489)
Hydrangea ringspot potexvirus (HRSV)	Hydrangea (483)
Impatiens latent=Helenium S carlavirus (HeVS)	Impatiens (502)
Impatiens necrotic spot tospovirus (INSV)	Alstroemeria (245), Begonia (450), Cyclamen (455), Dahlia (268), Impatiens (495), Iris (bulbous) (308), *Kalanchoë blossfeldiana* (505), Orchids (413), Ranunculus (346)
Iris fulva mosaic potyvirus (IFMV)	Iris (rhizomatous) (309)
Iris mild mosaic potyvirus (IMMV)	Iris (bulbous) (304)
Iris severe mosaic potyvirus (ISMV)	Crocus (261), Iris (bulbous) (306)
Kalanchoë latent carlavirus (KLV)	*Kalanchoë blossfeldiana* (511)
Kalanchoë mosaic potyvirus (KMV)	*Kalanchoë blossfeldiana* (509)
Kalanchoë top spotting badnavirus (KTSV)	*Kalanchoë blossfeldiana* (507)
Leelia red leafspot virus	Orchids (413)
Lettuce mosaic potyvirus (LMV)	*Dimorphotheca sinuata* (472), Lisianthus (445)
Lilac chlorotic leafspot closterovirus	Iris (rhizomatous) (305)
Lily mottle potyvirus (LMoV)	Lily (315)
Lily symptomless carlavirus (LSV)	Lily (314), Tulip (357)

Virus	Hosts (pages)
Lily X potexvirus (LVX)	Lily (317)
Lisianthus line pattern ilarvirus (LLPV)	Lisianthus (444)
Lisianthus necrosis necrovirus (LNN)	Lisianthus (444)
Long orchid rhabdovirus	Orchids (413)
Masdevallia isometric virus	Orchids (412)
Narcissus degeneration potyvirus (NDV)	Narcissus (325)
Narcissus late season yellows potyvirus (NLSYV)	Narcissus (326)
Narcissus latent carla (maclura) virus (NLV)	Gladiolus (290), Iris (bulbous) (305), Narcissus (325), Nerine (340)
Narcissus mosaic potexvirus (NMV)	Lily (319), Narcissus (329), Nerine (340)
Narcissus Q virus (NVQ)	Narcissus (324)
Narcissus tip necrosis carmovirus (NTNV)	Narcissus (329)
Narcissus white streak virus (NWSV)	Narcissus (324)
Narcissus yellow stripe potyvirus (NYSV)	Narcissus (326), Nerine (340)
Nerine latent carlavirus (NeLV)	Hippeastrum (Amaryllis) (294), Nerine (337)
Nerine X potexvirus (NVX)	Nerine (337)
Nerine yellow stripe potyvirus (NeYSV)	*Amaryllis belladona* (296), Nerine (336)
Onion yellow dwarf potyvirus (OYDV)	Narcissus (324)
Orchid fleck rhabdovirus (OFV)	Orchids (413)
Orchid tobacco mosaic tobamovirus (TMV-O)	Orchids (411)
Ornithogalum mosaic potyvirus (OMV)	Ornithogalum (343)
Pelargonium flower break carmovirus (PFBV)	Pelargonium (516)
Pelargonium leaf curl tombusvirus (PLCV)	Pelargonium (520)
Pelargonium line pattern virus (PLPV)	Pelargonium (517)
Pelargonnium ring pattern virus (PelRPV) = pelargonium line pattern virus	Pelargonium (523)
Pelargonium vein clearing rhabdovirus (PVCV)	Pelargonium (523)
Pelargonium zonate spot virus (PZSV)	Pelargonium (523)
Pelargonium ringspot disease: 1) Tobacco ringspot nepovirus (TobRSV) 2) Tomato ringspot nepovirus (TomRSV)	Pelargonium (518)
Phalaenopsis-a bacilliform virus from (unnamed)	Orchids (413)
Potato X potexvirus (PVX)	Cyclamen (455), Tulip (357)
Potato Y-R potyvirus (PVY-R)	Ranunculus (347)
Potato yellow dwarf rhabdovirus (PVDV)	*Kalanchoë blossfeldiana* (512)
Prunus necrotic ringspot ilarvirus (PNRSV)	Roses (429)
Raspberry ringspot nepovirus (RRSV)	Anemone (254), Narcissus (324), Phlox (422)
Rhabdovirus (not characterized)	*Kalanchoë blossfeldiana* (512)
Rhabdovirus (unnamed)	Orchids (413)

Virus	Hosts (pages)
Short orchid rhabdovirus	Orchids (413)
Soybean mosaic potyvirus (SoyMV)	Chrysanthemum (406), Gladiolus (290)
Strawberry latent ringspot nepovirus (SLRV)	Gladiolus (288), Lily (319), Narcissus (324), Phlox (422), Roses (432)
Tobacco mosaic tobamovirus (TMV)	Cyclamen (455), Gladiolus (290), Hippeastrum (Amaryllis) (296), Impatiens (494), Iris bulbous) (309), Lily (319), Lisianthus (446), Pelargonium (515), Tulip (364)
Tobacco necrosis necrovirus (TNV)	Anemone (253), Begonia (450), Gladiolus (290), Hydrangea (490), Narcissus (324), Ranunculus (347), Tulip (369)
Tobacco rattle tobravirus (TRV)	Alstroemeria (247), Aster (258), Crocus (262), Cyclamen (455), Freesia (277), Gladiolus (288), Hyacinth (300), Iris (bulbous) (309), Lily (319), Narcissus (326), Nerine (337), Orchids (412), Phlox (422), Ranunculus (346), Tulip (365)
Tobacco ringspot nepovirus (TobRSV)	Anemone (253), Begonia (450), Crocus (263), Gladiolus (287), Hyacinth (301), Hydrangea (487), Impatiens (501), Iris bulbous) (309), Iris (rhizomatous) (309), Lily (319), Narcissus (324), Roses (434), Tulip (374)
Tobacco streak ilarvirus (TSV)	Dahlia (269), Gladiolus (290), Impatiens (501), *Kalanchoë blossfeldiana* (512), Roses (434)
Tobamovirus	Roses (434)
Tomato aspermy cucumovirus (TAV)	Chrysanthemum (400), Cyclamen (455)
Tomato black ring nepovirus (TBRV) (TBRSV)	Gladiolus (288), Narcissus (324), Pelargonium (515), Phlox (423), Tulip (372)
Tomato bushy stunt tombusvirus (TBSV)	Tulip (376)
Tomato mosaic tobamovirus (ToMV)	Lisianthus (446)
Tomato ringspot nepovirus (TomRSV)	Gladiolus (287), Hydrangea (486), Narcissus (327), Orchids (413)

Virus	Hosts (pages)
Tomato spotted wilt tospovirus (TSWV)	Alstroemeria (245), *Amaryllis belladona* (296), Anemone (252), Asclepias (440), Aster (275), Begonia (450), Chrysanthemum (402), Cyclamen (455), Dahlia (268), *Dimorphotheca sinuata* (473), Gladiolus (289), Hippeastrum (Amaryllis) (295), Hydrangea (488), Impatiens (495), Iris (bulbous) (308), *Kalanchoe blossfeldiana* (505), Lisianthus (445), Narcissus (324), Nerine (339), Orchids (413), Pelargonium (521), Ranunculus (348), Zantedeschia (Calla lilies) (465)
Tomato vein yellowing rhabdovirus (TVYV)	Hibiscus (480)
Tomato yellow leaf curl geminivirus (TYLCV)	Lisianthus (446)
Trichopillia isometric virus	Orchids (412)
Tulip breaking potyvirus (TBV)	Tulip (359)
Tulip chlorotic blotch potyvirus (TCBV)	Tulip (362)
Tulip halo necrosis ungrouped virus (THNV)	Tulip (378)
Tulip mild mosaic virus	Tulip (357)
Tulip necrotic streak	Tulip (359)
Tulip X potexvirus (TVX)	Tulip (355)
Turnip mosaic potyvirus (TuMV)	Anemone (251), Impatiens (494), Iris (bulbous) (309), Orchids (414)
Vanilla mosaic potyvirus virus	Orchids (414)
MLOs	Cyclamen (457), *Dianthus barbatus* (462), *Dimorphotheca sinuata* (473), Gladiolus (290), Hyacinth (301), Hydrangea (490), Lily (319), Ranunculus (349)
Viroids	
Chrysanthemum chlorotic mottle viroid (CCMVd)	Chrysanthemum (406)
Chrysanthemum stunt viroid (CSVd)	Chrysanthemum (404)

Index